Chemical evolution and the origin of life

Molecular evolution
I

Chemical evolution and the origin of life

Proceedings of the International Conference on the Origin of Life

Edited by

R. Buvet
*Faculté des Sciences de Paris,
Laboratoire d'Energetique Electrochemique,
Paris Ve,
France*

and

C. Ponnamperuma
*NASA Ames Research Center,
Moffett Field,
California, USA*

1971

NORTH-HOLLAND PUBLISHING COMPANY – AMSTERDAM · LONDON
AMERICAN ELSEVIER PUBLISHING COMPANY, INC. – NEW YORK

Library of Congress Catalog Card Number: 75-146189

ISBN North-Holland – Series: 0 7204 4082 3

– volume: 0 7204 4083 1

190 graphs and illustrations, 65 tables

Published by:
NORTH-HOLLAND PUBLISHING COMPANY – AMSTERDAM

Sole distributors for the U.S.A. and Canada:
AMERICAN ELSEVIER PUBLISHING COMPANY, INC.
52 VAN DER BILT AVENUE
NEW YORK, N.Y. 10017

Printed in the Netherlands

EDITORS' PREFACE

Since the publication of the first scientific treatises on the origin of life, by Oparin in 1924 and Haldane in 1928, there has been a growing interest in a problem which touches every aspect of human existence. Many speculations and few experiments characterized the early era of chemical evolution. At the first international conference on the origin of life on earth, held in Moscow in 1957, under the auspices of the International Union of Biochemistry the sanguine hopes and expectations of the enthusiastic students of the subject were revealed.

The growing interest in the exploration of space, and the imminent possibility of the discovery of extraterrestrial life have led investigators around the world to concentrate on experimental studies on abiogenesis. The retracing of the path of chemical evolution on earth would lend great support to the belief in the existence of extraterrestrial life. Indeed, this search, which has been described as the prime goal of exobiology, is only one aspect of the scientifically broader question of the origin of life in the universe. The second international conference held at Wakulla Springs, Florida, under the auspices of the Institute for Space Biosciences of the Florida State University, and the U.S. National Aeronautics and Space Administration, brought together a small number of individuals involved in the study of the problem, for a close scrutiny of the latest developments in the theory and practice of chemical evolution.

Since 1963 numerous meetings have been held on various aspects of the problem of the origin of life. But most of these have been limited in scope and regional in character. The seven years of fruitful labor after Wakulla Springs demanded a review and appraisal at a third international gathering.

Since the range of the subject had expanded vastly, sessions were held on the related fields of organic geochemistry, lunar analysis, and planetary exploration. There was special emphasis on the biochemical aspects of early evolution and on problems related to energy transfer and the origin of metabolism.

All scientists invited to participate in the conference were entitled to submit their contributions for inclusion in this volume. The scientific content was deemed to be the responsibility of the authors and not of the editors. Some of the opinions expressed, therefore, appear at times to be controversial and even widely divergent.

We owe the success of this conference to the scientific contributions made by the participants. A special debt of gratitude is due to Alexander Ivanovich Oparin, the father of the field of chemical evolution, to whom the meeting was dedicated. We are indebted to the Ministry of Education of the Republic of France and to the Faculty of Sciences of the University of Paris for the generous financial support. To Madame Francine Godin, our heartfelt appreciation for the painstaking preparation of the manuscripts for publication.

R. Buvet
C. Ponnamperuma

CONTENTS

Part III. *Syntheses of small molecules*

Part IV. *Oligomers and polymers*

Part VI. *Origin of biological structures*

Part VII. *Primitive biochemistry and biology*

PART I

INTRODUCTORY SESSION

Chemical Evolution and the Origin of Life, eds. R. Buvet and C. Ponnamperuma

PROBLEM OF THE ORIGIN OF LIFE: PRESENT STATE AND PROSPECTS

A.I. OPARIN
A.N. Bakh Institute of Biochemistry,
USSR Academy of Sciences, Moscow, USSR

The problem of the origin of life is one of the fundamental problems, which has attracted the minds of mankind for many centuries.

I would like to give here a picture of an alabaster vase (fig. 1) found during the excavations of the Eonn sanctuary in Uruck, one of the most ancient towns of Sumer, South Mesopotamia.

This vase was made almost three thousand years B.C. At its lower end, near the bottom, sea or river waves are seen. Higher, the plants arise from them. Above these the animals and then the people appear. The whole picture is crowned by an image of Ishtar, the goddess of fertility and the continuous origin of life.

Thus, the picture on the Uruck vase illustrates an ancient Sumer legend on the appearance of life from water, the opinion which we now share, although previously it was founded only on intuition and on superficial and naive observations of the appearance of organisms in the mud and silt of irrigation canals.

The suggestion of the direct origin of the organisms from dead inorganic matter was accepted by Babylonian scientists, ancient Greek philosophers who had lived in Asia Minor, Milesians, and later by Aristotle, whose ideas dominated the human mind for almost two thousand years.

A rational approach to the problem has appeared only in our century when the "generatio aequivoquia" principle was disproved in Pasteur's studies. Further experiments have demonstrated a possibility of the synthesis of organic compound abiogenetically under conditions simulating the earth and other celestial bodies.

It has now become quite clear that the origin of life was not the result of some "happy chance" as was thought till quite recently, but a necessary stage in the evolution of matter. The origin of life is an inalienable part of

Figure 1

the general process of development of the Universe and, in particular, the development of our earth. Hence, the phenomenon is accessible to science.

It is very important that at the same time experimental approaches to the problem have been worked out. In this respect the cosmic and geological researches are very promising. Besides these, we are now able to synthesize abiogenetically more and more complex organic substances of biological importance by reproducing in the laboratory the conditions of the primeval earth.

In addition we can observe a formation of multimolecular complexes, some of them can be studied with a light microscope. They are capable of interaction and exchange with the material present in the water solution which surrounds them, thus modelling the processes taking place in the primordial hydrosphere of the earth. Finally, comparative cytology and biochemistry, a comparative study of protoplasmic structures and initial steps of the contemporary metabolism, are of great importance for our cognition of the pathways of the origin of life, since they reveal an order in the formation and evolution of the structures and metabolism. Recently abundant information has been derived from the study of Precambrian paleontology and paleobiochemistry.

An analysis of all these results allows us to form an idea of the successive stages in the process studied. Experiments in several branches of science, carried out by many scientists in various countries, allow us now to imagine in general outline the following principal stages in the evolution of carbon compounds, underlying the pathway to the origin of living things on our earth.

Its first stage includes an appearance of hydrocarbons, cyanides and their close derivatives in space, during the course of the formation of the earth as a planet and during the formation of its crust and hydrosphere.

In the second stage, abiogenetic synthesis of more and more complicated organic substances has proceeded in interplanetary space, and on the surface of the planets the so-called "primeval broth" has been formed.

Then, in the third stage, individual open systems, the ones capable of interaction with the environment, of growth and reproduction have formed in the broth: the so-called "protobionts" appear.

The fourth stage includes further evolution of "protobionts" by prebiotic selection which has resulted in the improvement of their metabolism, intermolecular and supermolecular organisation. The primeval organisms have appeared.

Such evolution is mostly considered to proceed under a definite plan or along a nearly linear definite pathway, so that a molecule or a supermolecular system A turns to the system B, this one to C, etc.

$$A \rightarrow B \rightarrow C \rightarrow \ldots$$

If, however, our studies do not assume any pre-existing plan, or, in general, any interference from outside, application of this scheme will entail great difficulties. "Such arguments", says Bernal, "for instance, as that some particular stage (of the genesis of life a.o.) happened by chance are always laid open to a straightforward, statistical calculation of what that chance is, and it is usually very low."

The thermodynamic potential of any system, even that of an organic molecule, is very high and the probability that A just turns to B is only a small part among all other ways for realization of that potential. Since probabilities should not be summed, but multiplied, any multistage process, even the natural formation of complicated organic molecules (to say nothing of organisms) seems to be of extremely low probability.

Besides, any transition from one stage of biopoesis to the next usually entails the growth of a complex and organized system. After the second law of thermodynamics a reverse decomposition process is much more probable than the direct synthetic one.

Numerous calculations of a thermodynamic equilibrium, which would occur when the organic substances dissolved in the primordial hydrosphere were irradiated with ultraviolet or were provided with some other energy sources, show that no primeval broth could exist there, since under ultraviolet light, decomposition processes would prevail over synthetic ones.

As we now know interplanetary space is very rich in organic matter. In particular carbonaceous chondrites appear to be more abundant in space than we believed until recently. However, according to the thermodynamic equilibrium calculation of Dayhoff, the concentration of organic substances in the planets' atmosphere can only be insignificant.

This contradiction may be partially eliminated after a more careful analysis of the numerous experiments modelling abiogenetic synthesis of organic substances.

The significant accumulation of organic substances, sometimes rather complicated ones, in such experiments is related to their removal from the sphere of action of the energy source which caused their formation.

For instance, in Miller's experiments, amino acids which have been formed in an electric discharge, have rapidly moved from the site of their formation and accumulated in the adjoining vessel. Fox and coworkers have reproduced volcanic conditions heating an amino acid mixture to 170°C on a piece of lava. Over a long incubation at such temperature proteinoids formed

Figure 2

under these conditions would necessarily undergo decomposition. But a rain included by Fox in the next step of his experiment washes out and cools the products before they are decomposed.

The samples of moon dust brought back by the Apollo 11 astronauts do not contain a significant amount of organic matter. These compounds might be completely decomposed by various energy sources of cosmic space on the absolutely unprotected moon surface. However, if they had been covered with dust, they would have been able to remain and enter the process of evolution. It is of interest that moon dust does contain traces of porphyrins.

These compounds, however, are not the moon's material, but are thought to have been derived from the rocket gases.

Many other examples could be given. They show that rather general calculations of the thermodynamic equilibrium do not take into account the vast opportunities for a transfer of the substance and systems from one environment to the other. Nature, however, provides a great variety of such environmental shifts. For instance, the calculations made for the primordial ocean are correct for a relatively thin surface layer. But the compounds formed there may be submerged in depths as a result of currents or due to formation of heavy complex formation due to absorption on minerals (e.g. clays), etc.

We have to reject the concept of a straightforward type of evolution, and here I would like to propose another pattern illustrated by a diagram (fig. 2) which is in principle adopted from "Chemical evolution" by M. Calvin:

When discussing the first stages of evolution of carbon compounds, we should not think of any common starting material or any common general pathway for further development. The formation and further complication of organic substances could occur in space and on the earth under very different conditions and be fed by different energy sources. Therefore, we have not a sole stem, but a bush of various pathways at the bottom of the diagram. Most of the lines have reached a dead end. They have come to complete decomposition or evaporation into space (that, probably, occurs on the moon surface), or they have been buried as very resistant high molecular compounds, incapable of further changes (like meteorite bitumen).

In either environment some of the starting organic materials should necessarily be transformed to more and more complex, high molecular substances. We can not consider this process as a lucky chance. Certainly, it might or might not happen at a definite site and on a definite moment, just as it may be raining or not tomorrow. But on a world scale and for long time intervals the starting material should necessarily and repeatedly undergo a rise to a new level of organization. The forms of this organization could be very different. They should not be considered as molecules with a perfect internal structure, like proteins or nucleic acids. They could be random polypeptides and polynucleotides, carbohydrates, lipids, and even compounds not present in comtemporary organisms. The initiation of phase relations in the originally uniform primaeval broth was the only capacity they needed to enter the new stage of evolution.

The individual multimolecular systems, separated from the environment by a partition surface but capable of interaction with it as open systems, could arise on this basis only.

Formation of such systems was the bridge between a chemical evolution and a biological one. The first step of biological evolution was to overcome the growth of the increase of entropy in the systems which were appearing.

As you know, in contrast to the processes in dead nature leading in general to the growth of entropy, the living systems are featured by a decrease of entropy and an appearance of order from the chaotic thermal movement of molecules. It is caused by the fact that the living organism is an open system which needs permanent interaction with the environment to maintain its existence. Streams of matter, energy, and information pass through it, and it is fed by negative entropy derived from the environment, as Shroedinger said.

A great variety of such systems, differing from each other by decomposition and nature of intermolecular bonds, could and must have appeared on the earth.

In both these respects most of them were principally different from the contemporary organisms. For instance, the left part of the diagram illustrates the development of the systems formed mainly of one sort of polymer (i.c. protein or proteinoid).

Our modelling experiments with the coacervate drops showed that the further evolution of such individual open systems ("Protobionts") gradually leading to the improvement of their interaction with the environment and to growth and multiplication could be based on prebiotic natural selection.

Though the organization level of certain protobionts allowed them to exist in the given environment, they were unable to compete with the more perfect ones and disappeared for ever. For example, an appearance of code relations between polynucleotides and polypeptides in some protobionts, containing these compounds, favored development of the mechanisms for obtaining and keeping information. It enhanced their development and at the same time suppressed all other systems, including homopolymeric ones drawn in the left part of the diagram.

On this basis, the adaptability to the special functions of the integral parts of the organisms (its molecules, cell organoids, and organs) and the adaptability of the whole organism to permanent self-preservation and self-reproduction in a given environment, the features so specific for all living beings, appeared.

Certainly, all these suggestions are of a very hypothetic nature, but they show that the origin of life was not the kind of improbable event as some sceptics try to prove.

Chemical Evolution and the Origin of Life, eds. R. Buvet and C. Ponnamperuma

THE PRESENT STATE OF MOLECULAR PALEONTOLOGY

Marcel FLORKIN
Department of Biochemistry, University of Liège, Belgium

By molecular paleontology, Calvin [14] proposes to define the study of the organic molecules which are left in rocks and which may have been parts of molecular aggregates before biological evolution. This is, of course, a most interesting aspect and we shall return to it, but while I believe that such prebiological aspects may, of course become part of molecular paleontology, the whole domain of this science concerns preserved molecules of the past found in rocks and sediments. Let us hope that, in order to avoid utter confusion, the study of fossil organic molecules, i.e. taken out of the earth, either free or in organic remains, will still constitute the subject of the molecular aspects of paleontology, which means the study of fossils (unearthed remains). It has been a rather unexpected discovery that proteins may be preserved in fossils for periods of millions of years [27]. This finding came from a study of Molluscan shells inaugurated in my laboratory in the fifties, and I shall start with this topic.

When I became interested in the comparative biochemistry of the organic matrix of Molluscan shells, I was fortunate enough to obtain the collaboration of my colleagues, Mme Duchâteau-Bosson for the chemical studies, and Prof. Ch. Grégoire, who is the head of the section of electron microscopy in my department. In the extensive paper we have published on shells [40], we described the dissociation of the soft iridescent membranes resulting from the demineralization of mother of pearl (nacre conchiolin) into thin membranes permeable to electrons, of reticulated lace-like structures, the pattern of which is characteristic of the class of Mollusks studied (figs. 1, 2, 3). In the chemical section of the paper, we stated that nacre conchiolin appears to consist of a core of fibrils (nacroine) surrounded by a scleroproteinic layer (nacrosclerotine), these constituents forming the lace-like structure itself coated by a water-soluble protein (nacrine). I may note here that, from subsequent studies in my laboratory, it has been shown that if nacrine is a distinct constituent, the separation between nacroine and nacrosclerotine is an

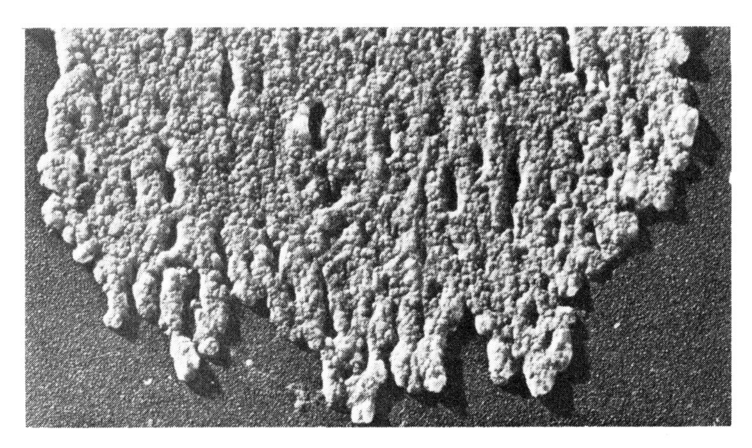

Fig. 1. *Nautilus pompilius* Lamarck. Conchiolin from mother of pearl. Nautiloid pattern of structure. × 37,500 [39].

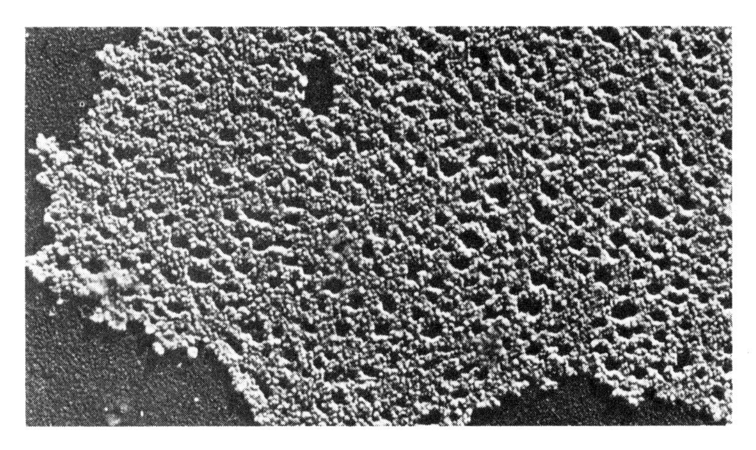

Fig. 2. *Angaria delphinus* Linne. Conchiolin from mother of pearl. Gastropod pattern of structure. × 37,500 [35].

artefact [69] and I therefore propose dividing the constituent of nacre conchiolin into the water-soluble nacrine and the insoluble nacroine, a scleroprotein. As shown in my laboratory by Goffinet and Jeuniaux [31], nacroine is a chitinoproteic complex.

Grégoire independently extended his studies on the comparative submicro-

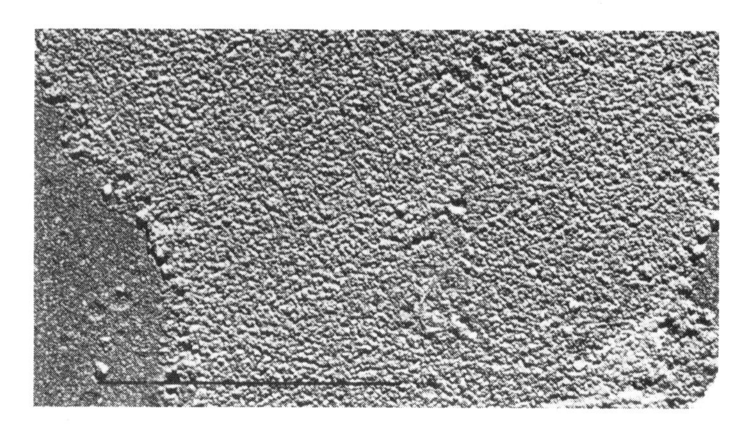

Fig. 3. *Mytilus crenatus* Linne. Conchiolin from mother of pearl. Pelecypod pattern of structure. × 37,500 [35].

scopic morphology of shells by electron microscopy to a broad spectrum of species and in the course of these researches, [33,34] he recognized in fossil shells lace-like structures showing, more or less altered, the same morphology as those we had described in Recent shells [37]. Grégoire did not commit himself to whether or not these structures were turned into stone.

When we were able to obtain enough material of such nacreous remains, a chemical study was initiated with the collaboration of my colleagues S. Bricteux-Grégoire and E. Schoffeniels, while Grégoire performed a study of samples of the same material by electron microscopy [27] (fig. 4). After isolating particles of mother of pearl from the fossil shells and carefully removing the remains of the prismatic structure, these particles were powdered in a mortar. The resulting powder was washed with boiling water until free amino acids had been completely removed and then treated with successive portions of HCl 6 N until complete decomposition of the mineral constituents (end of effervescence). The residue was dialyzed against running water and for 4 hr against distilled water. It was evaporated to dryness and reflux condensed in boiling HCl 6 N for 24 hr, evaporated under reduced pressure, treated by active charcoal and brought to a known volume of which aliquots were used for amino acid determinations by chromatography on Dowex 50 according to the method of Moore and Stein. We obtained a mixture of amino acids, the pattern of which corresponded to the pattern of nacre conchiolin (table 1, fig. 5). The older the fossil, the larger the amount

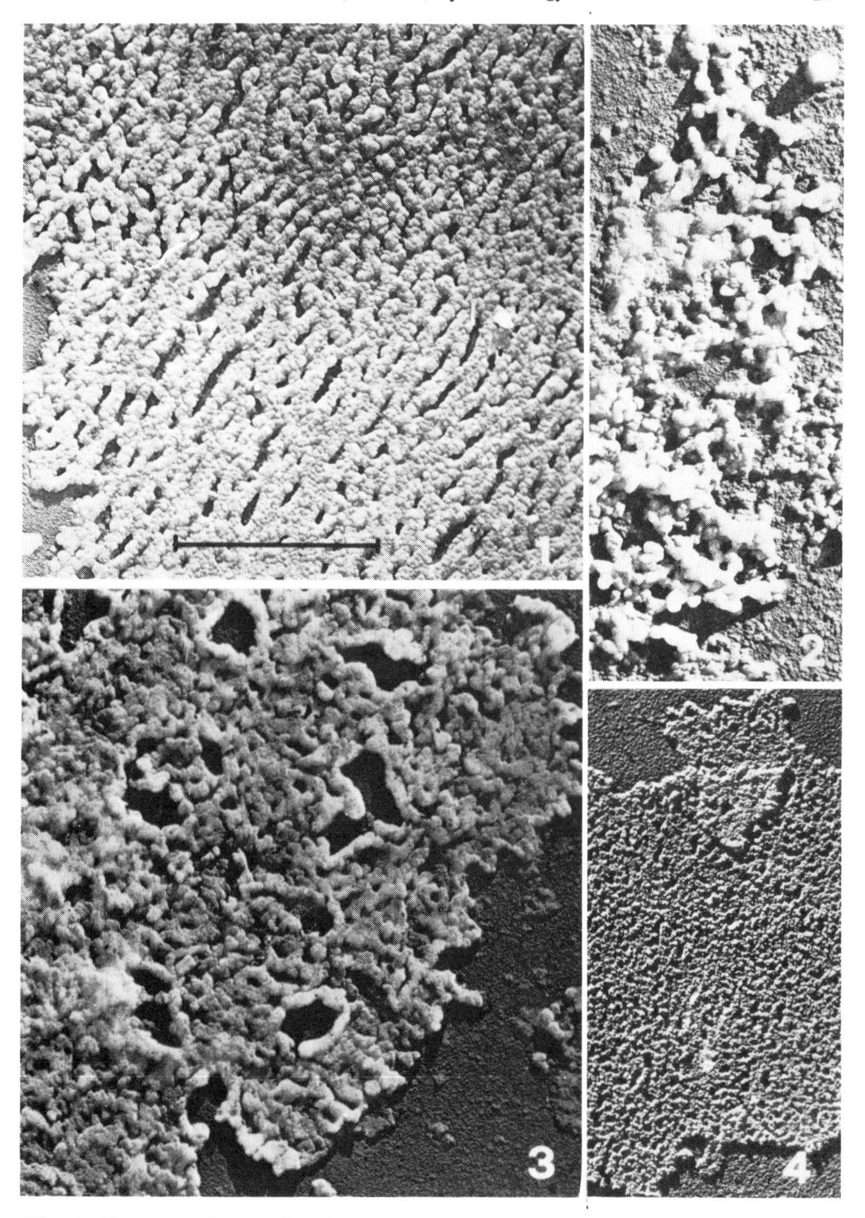

Fig. 4. Electron micrographs of nacreous organic remnants after decalcification with chelating agents. Fragments of lace-like reticulated sheets of conchiolin. (1) *Nautilus macromphalus* Sow. (Cephalopoda, Nautiloida). Recent (Nautiloid pattern). X 27,000. (2) *Nautilus* sp. (Cephalopoda, Nautiloida). Eocene (60 million years). Nautiloid pattern still recognizable in some regions. X 27,000. (3) *Aturia* sp. (Cephalopoda, Nautiloida). Oligocene (40 million years). Nautiloid pattern still recognizable. X 27,000. (4) *Iridina spekii* Woodward. (Pelecypoda, Mutelida). Holocene (10,000 years). Pelecypod pattern X 27,000 [34].

Table 1
"Conchiolin" of mother of pearl. Amino acid composition: molecular fraction per 100 molecules of amino acids [27].

	Nautilus	*Aturia*	*Iridina*	*Nautilus*
	Eocene (60 million years)	Oligocene (40 million years)	Holocene (10,000 years)	Modern
Aspartic acid	8.7	9.0	10.1	9.0
Threonine	5.6	3.8	1.6	1.5
Serine	24.0	16.7	7.8	10.9
Glutamic acid	15.6	11.9	5.0	5.5
Proline	3.7	4.8	2.0	1.8
Glycine	20.8	23.3	29.8	35.7
Alanine	9.7	15.2	28.2	27.2
Valine	3.1	5.7	4.0	2.2
Isoleucine	3.1	3.3	2.4	1.8
Leucine	5.6	6.2	4.0	1.9
Tyrosine	0	0	0	2.4
Phenylalanine	0	0	0	2.4
Histidine	0	–	0	0
Lysine	0	–	2.3	0
Arginine	0	–	1.9	0

of fossil material necessary to obtain a quantity of nitrogen between 132 and 155 μg N. In spite of the constancy of the amounts of nitrogen, the amounts of amino acids obtained were smaller in the older fossils, the difference being accounted for by ammonia liberated in the course of hydrolysis. The older the fossil, the lower the proportions of glycine and alanine. We also found that particles of the decalcified material showed positive biuret and amido-schwarz reactions. A year before, Jones and Vallentyne [34] had shown the presence of conchiolin remains in a *Mercenaria* shell a 100,000 years old. At that time, the current view was that we could not expect to find preserved proteins in a fossil shell older than 100,000 years. The paper by Florkin et al. [27] demonstrated the preservation of shell proteins in fossils for much longer periods, and in one example, in a fossil of the Eocene (60 million years).

This evidence was confirmed and broadened by other observations. Grégoire [32,38], applying the biuret reaction, as introduced by Florkin et al. [27], to the remains of nacre conchiolin of more than 250 shells of Cenozoic, Mesozoic and Paleozoic Mollusks, observed a positive reaction in all cases. Akiyama [7], Degens and Love [18], Mitterer [58], Bricteux-Grégoire et al. [13] detected proteins in a number of Molluscan fossil shells.

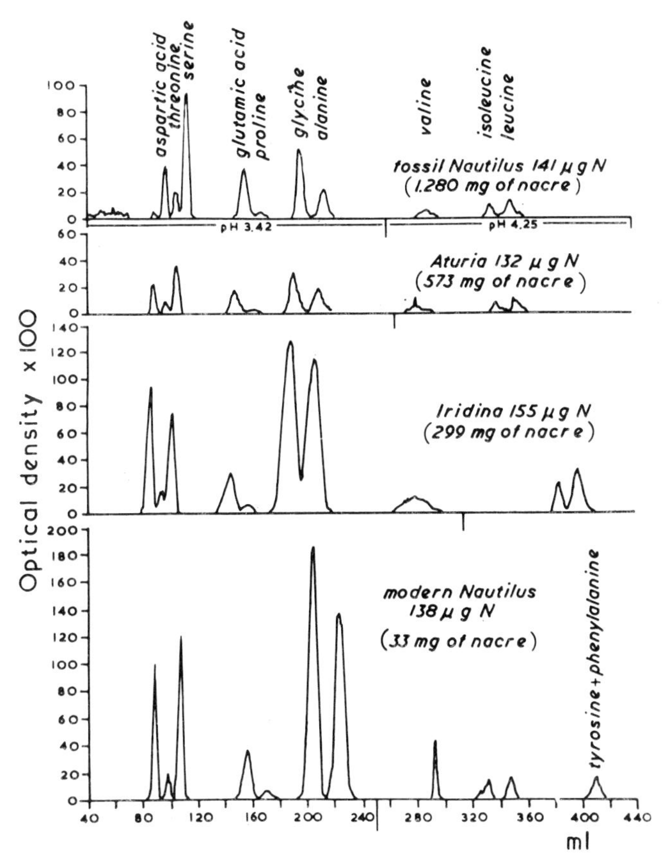

Fig. 5. Separation by chromatography on Dowex 50, using the Moore and Stein technique, of amino acids resulting from the hydrolysis of the residue of nacre after washing out the free amino acids, decalcification and dialysis, of modern *Nautilus*, fossil (Eocene) *Nautilus*, Recent (Holocene) *Iridina* and fossil (Oligocene) *Aturia* [27]. The ultrastructure of the samples used are shown in fig. 4.

As already pointed out by Florkin et al. [27], these authors observed a reduction in the amount of protein preserved, proportional to the age of the fossil studied.

In very extensive research on 22 Ammonoids (Jurassic, Cretaceous) and Nautiloides (Devonian → Miocene) carried out in my laboratory, Voss-Foucart (unpublished) consistently found protein remains that she identified by amino acids liberated by the hydrolysis of the samples washed and dialyzed.

The assertions of Abelson [4] according to which only individual amino acids can persist in a shell 25 million-years old, are contradicted by the persistence of the polypeptide structures during longer periods of up to 500 million years. This was first shown by Florkin et al. [27] and later by Degens and Love [18] and by Ho [44] on tertiary Molluscan shells, by Bricteux-Grégoire et al. [13] on the shell of *Pinna,* Eocene and *Inoceramus* Cretaceous, and by Voss-Foucart (unpublished) on Ammonoids of the Jurassic and the Cretaceous and on Nautiloids from the Devonian to the Miocene. The former assertions have also been contradicted by a number of studies on other fossils.

Foucart, Bricteux-Grégoire, Jeuniaux and myself [28] have carried out analyses on three different samples of fossils belonging to the order of graptoloidea (Silurian *Pristiograptus gotlandicus, Pristiograptus dubius*; Silur-

Table 2
Amino acids in hydrolysates of graptolites, after decalcification and washings [28].

	Pristiograptus gotlandicus and *P. dubius* (Silurian)		*Monograptidae* gn. sp. (Silurian)		*Climacograptus typicalis* (Ordovician)	
	mg/g	mole-fraction per 100	mg/g	mole-fraction per 100	mg/g	mole-fraction per 100
Aspartic acid	218	9	560	8.6	68	10
Threonine	108	4.9	290	4.9	26	4.3
Serine	214	11	550	10.6	122	22.8
Glutamic acid	380	13.9	1,100	15.3	96	12.8
Proline	(83)	(3.9)	(340)	(6)	trace	
Glycine	280	20.1	760	20.8	89	23.4
Alanine	103	6.3	410	9.5	40	8.9
Valine	116	5.3	(240)	(4.1)	trace	
Isoleucine	97	4	230	3.6	(15)	2.3
Leucine	190	7.8	390	6.1	(13)	(2)
Tyrosine	(42)	(1.2)	(90)	(1.1)	–	
Phenylalanine	(87)	(2.9)	(110)	(1.3)	–	
Lysine	96	3.5	(310)	(4.4)	70	9.4
Histidine	60	2.2	(84)	(1.1)	29	3.7
Arginine	128	4	(140)	(2.2)	trace	
Ammonia	223					

[a] The amounts indicated in brackets could only be calculated approximately on account of their low values.

ian *Monograptidae, gn.sp.*, contaminated with a few *Retiolitidae*; Ordovician *Climacograptus typicalis*, provisional identification).

The calcareous rocks containing the fossils were dissolved with cold HCl (0.5 N). The graptolites were separated with micropipettes and washed repeatedly with cold HCl and distilled water in order to remove the soluble components such as free amino acids. As shown in table 2, in addition to high amounts of ammonia, amino acids are present in the hydrolysates of the three samples studied (*Pristiograptus* 2,202 μg/g; *Monograptidae* 5,504 μg/g; *Climacograptus* 568 μg/g). As soluble material had been previously removed by repeated treatments, and as possible sources of contamination were avoided, we consider that the amino acids contained in the hydrolysates are of protein origin. An identification of peptide linkages by means of biuret reaction, as used by Florkin et al. [27] in the case of "conchiolin" remnants, was not possible because of the dark colour of the graptolite fragments.

Table 3
Comparison between the amounts of amino acids of protein origin in graptolites removed from a rock with those found in the rock itself [28].

	mg/g		Mole-fraction	
	graptolites [a]	rock [b]	graptolites	rock
Aspartic acid	68	12.8	10	7.4
Threonine	26	8	4.3	5.1
Serine	122	45.4	22.8	33.4
Glutamic acid	96	9.6	12.8	5.1
Proline	tr	–	tr	–
Glycine	89	30	23.4	30.7
Alanine	40	12	8.9	10.4
Valine	trace	–	trace	–
Isoleucine	(15)	5.6	(2.3)	3.3
Leucine	(13)	7.4	(2)	4.3
Tyrosine	–	–	–	–
Phenylalanine	–	–	–	–
Lysine	70	–	9.4	–
Histidine	29	–	3.7	–
Arginine	trace	–	trace	–
Total	568	130.8	99.6	99.7

[a] *Climacograptus typicalis.*
[b] Ordovician limestone containing *Climacograptus typicalis* (from Ohio) after decalcification and removal of the graptolite fragments.

The question arose whether these proteins were contained in the graptolites themselves or whether they were present in the sediment from which the graptolites had been removed. The analysis of one graptolitic rock, when compared with that of the graptolites extracted from the same rock, allows us to eliminate the latter hypothesis. The results (table 3) show that the amino acid concentrations are much lower than those of the graptolites. On the other hand, some amino acids present in the graptolites have not been detected in the rock. The organic composition of the graptolite remnants is thus quantitatively and qualitatively distinct from that of the kerogen of the parent sedimentary rock. Contamination of the graptolites themselves is also eliminated. Observation of the fragments with the electron microscope did not reveal appreciable proportions of typical contaminants such as bacteria, algal filaments, "conchiolin" remnants. On the other hand, parallelism between the results obtained for the three samples of different origin and geological age could hardly be consistent with the possibility of a contamination by exogenous organic material. The amino acids observed in the hydrolysates of washed and decalcified fragments of graptolites can certainly be considered as originating in remnant fossil proteins belonging to the test of these animals. The three species of graptolites so far examined show a similar amino acid pattern, characterized by high amounts of serine (mole-fraction: 10.6 to 22.8), alanine (6.3 to 9.5), glycine (20.1 to 23.4), aspartic acid (8.6 to 10) and glutamic acid (12.8 to 15.3). Such a composition, particularly the high total amount of glycine, serine and alanine, suggests that the graptolitic proteins are of a scleroprotein nature. The nature of this material was not revealed as chitinic by the specific enzymatic method of Jeuniaux [51] and the chromatograms of the hydrolysates did not show the presence of glucosamine. Furthermore, the residues of the graptolite samples did not contain traces of cellulose. The lack of chitin in graptolites contra-indicates a systematic affinity between graptolites and the Bryozoa, Hydrozoa or other Cnidaria, all of which are provided with a Chitinous test.

The ultrastructure of the protein of Graptolites (see electron micrographs by Grégoire [26]) may be described by meshes of bandlike material, the network being more loose or more tight according to the parts considered. The type of ultrastructure is different from that of *Cephalodiscus inaequatus*, a species of Pterobranchia which is considered to show systematic affinities with graptolites. As well as graptolites, the organic exoskeleton (or coenecium) of *Cephalodiscus* (a living Pterobranch), also lacks chitin [29]. Its proteins show a high percentage of glycine as is the case in proteins of graptolites. On the other hand, the coenecium of *Cephalodiscus* and the tests of graptolites differ in their ultrastructure.

Collagen has been extensively studied in fossil bones. It has been detected by X-ray diffraction by Isaacs et al. [50], in bones, or dentine of fossils, certain of which dated from the Devonian, by electron microscopy, by Little et al. [55] in buffalo bones, by Wyckoff et al. [74] in the calcaeneum of a Pleistocene *Equus*, by Shackleford and Wyckoff [62] and by Wyckoff and Doberenz [72,73] in Miocene and Pleistocene bones and teeth, by Pawlicki et al. [61] in the bones of Dinosaurs of the Upper Cretaceous, by Doberenz and Lund [19] in a Lower Crustaceous fossil and by Doberenz and Wyckoff [20] in Pleistocene bones. Fossil collagen composition was first studied by Wyckoff et al. [74] who obtained a series of amino acids, including hydroxyproline, in the hydrolysis product of decalcification residues of Pleistocene mammalian bones. Ho [43,44,45] compared the composition of fossil and Recent collagens and observed that fossil collagen generally contains less leucine, phenylalanine and tyrosine and more glycine.

Miller and Wyckoff [57] and Wyckoff [71] studied the proteins preserved in more ancient bones (bones and teeth of Jurassic and Cretaceous Dinosaurs). As shown in my laboratory by Voss-Foucart [70], the slow decalcification of dinosaur egg shells (very likely Theropods belonging to the genus *Megalosaurus*) of the Upper Cretaceous of Provence, liberates two different structures: one composed of white ribbons forming an external and an internal net joined by thin hollow tubes which correspond to the air channels; the other consisting of thin brown sheets superposed in the mass of the shell. These two structures still contain protein, the ultrastructure of which has been determined by electron microscopy (figs. 6 and 7). The determination of their amino acid composition, after washing, dessiccation, and hydrolysis, revealed proteins in the two structures, and differences in their composition.

Preserved proteins have also been detected in fossil shells of Brachiopods from the Ordovician and Carboniferous [53,54]. Carlisle [15] found protein remains in a Cambrian fossil of a Pogonophore, *Hyolithellus*. Evidence for the persistence of proteins in fossils for several million years, since it was stated by Florkin et al. [27], has repeatedly been confirmed on fossils of various kinds.

Since 1961, this notion has, of course, been critically discussed. It was at first suggested that free amino acids, liberated from proteins in the fossil, may associate with mineral constituents in the form of carbamates and be accumulated locally. But it is clear that the demineralization of the fossil (decalcification by hydrochloric acid) would eliminate such compounds. Small peptides are also eliminated by the dialysis of the samples before hydrolysis.

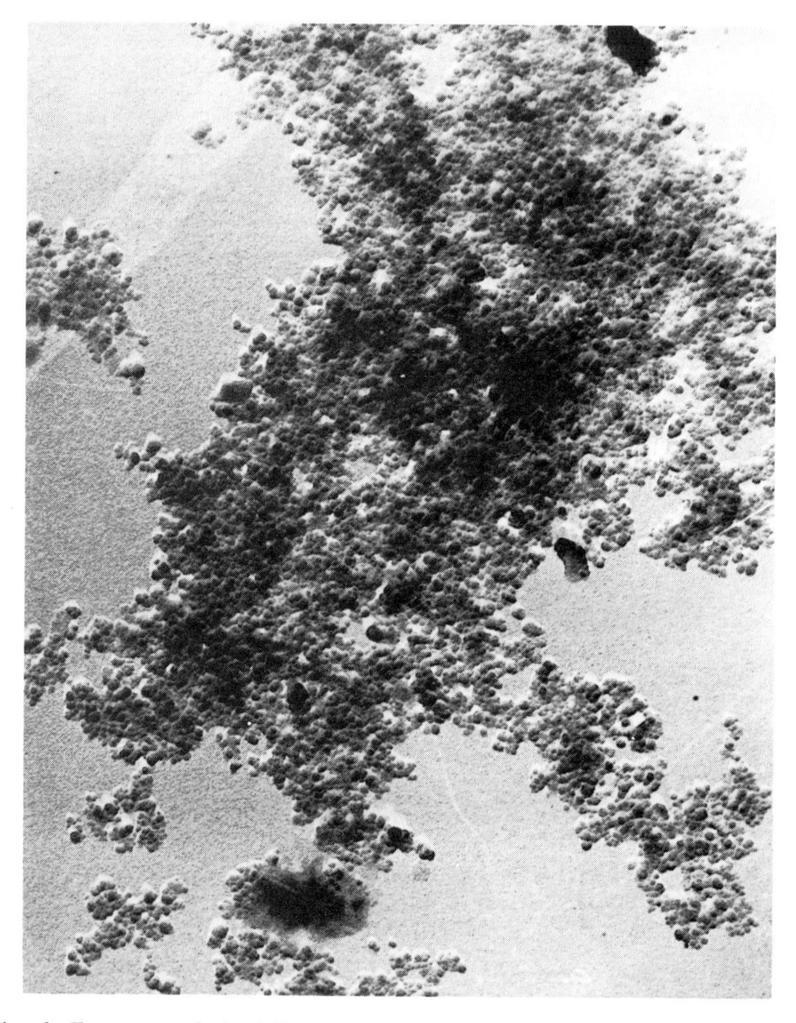

Fig. 6. Fragments of decalcified Dinosaur egg shells (very likely *Megalosaurus*). Corpuscles of the thin stratified sheets. × 42,000 [70].

Contamination may also introduce three different kinds of artefacts:
(1) Organisms which are contemporaries of the fossilified organisms;
(2) Contamination by substances surrounding the fossil;
(3) Contamination during the manipulation and the analyses.
In order to eliminate the possibility of a contamination by parasites,

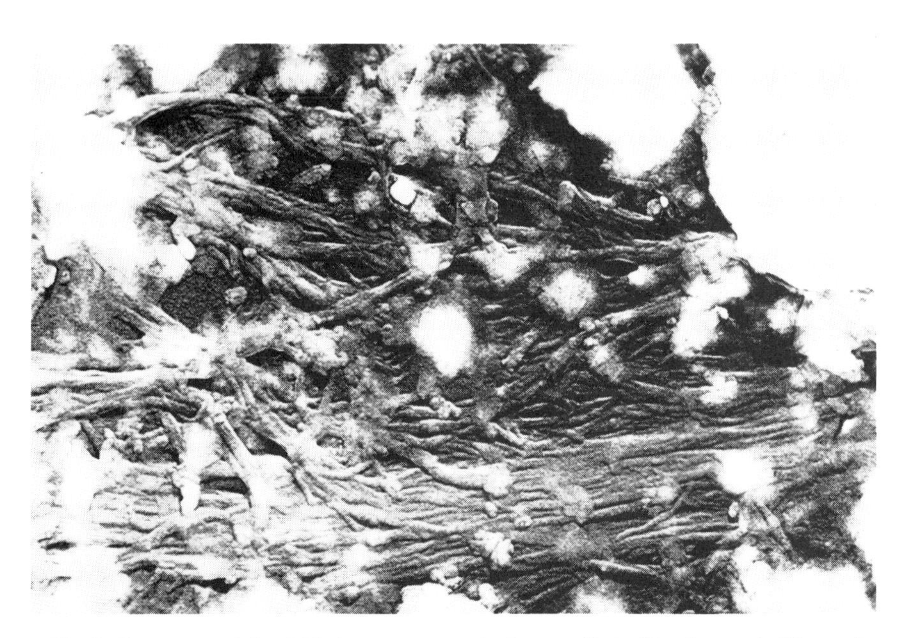

Fig. 7. Fragments of decalcified Dinosaur egg shells (very likely *Megalosaurus*). Fibres of the white ribbons forming the superficial networks and the walls of the air channels. Negative X 48,000 [70].

epibionts, etc., control by electron microscope is of great interest. It is also necessary to handle the material with caution and to avoid any bacterial or material contamination.

Another objection raised by Abelson [5] and by Mitterer [58] was that recrystallization of shells would make preservation of proteins inconceivable.

Grandjean et al. [32] and Hall and Kennedy [41] have shown by X-ray diffraction that in the shells of the Tertiary, recrystallization of the aragonite into calcite is rare. It is more common in secondary specimens (particularly of the Jurassic) and it is generalized in Primary shells. A large amount of data shows that recrystallization replacing aragonite by calcite in shells has taken place in situ [9,38,47]. Voss-Foucart (unpublished data) in my laboratory compared the shells of Ammonoids and Nautiloids whose original aragonite had been preserved with those in which aragonite had been replaced by calcite. She showed that not only protein remains persist in recrystallized shells, appearing in electron microscopy as lace-like structures, but that the patterns of amino acid compositions are similar in aragonitic and calcitic shells of these groups. As Hudson [48] says, "the conditions which favour the preservation of organic matrix are not the same as those which favour the preservation of original carbonate mineralogy, especially in aragonitic forms."

One of the results of extensive electron microscopy of shells pursued by Gregoire in my laboratory, has been the recognition of alterations of structure in the lace-like proteins of fossil shells [38] which can be compared but not on the same scale, to chemical differences observed in the loss of alanine and glycine with the increase in age of fossil shells, shown by Florkin et al. [27]. Attacking the problem experimentally, Grégoire [37] exposed fragments of modern *Nautilus* shells to a few of the factors involved in paleization processes such as heat and pressure. He concluded that "heat alone in open air, or associated with pressure produced various stages of ultrastructure degradation of chonchiolin, similar or identical on the electron microscope scale to those observed consistently in remnants of decalcification of Paleozoic or Mesozoic Nautiloid and Ammonoid shells". In all the specimens hydrolyzed, particles persist which show the biuret reaction, even if heated to 900 °C [37–39].

From an extensive study of Ammonoid and Nautiloid fossils in my laboratory, Voss-Foucart (unpublished) has concluded, from composition determinations, that the conchiolins of the Ammonoids (Jurassic and Cretaceous) and of Nautiloids (Miocene and Devonian) clearly differ from Recent *Nautilus*. On the other hand, she found no important difference between the Ammonoids of the Jurassic and of the Cretaceous. Furthermore, she concludes that all fossil proteins have a similar percentage composition, whether they be in Molluscan shells, graptolites, brachiopods, etc. Moreover, this uniform composition corresponds to the composition of the pyrolyzed modern corresponding proteins. Voss-Foucart attributes these changes to an effect of temperature during long periods of time.

Considering Voss-Foucart's results, we may suggest that, all the proteins concerned being scleroproteins, they probably have a similar background which is responsible for their high insolubility. Paleization and pyrolysis may separate lateral chains or act on weaker bonds and reduce all these scleroproteins to a similar hard core.

When Abelson discovered free amino acids in fossils, he concluded from the distribution of these free amino acids in fossils of different ages [2,5] that free amino acids must be distributed in three groups by order of decreasing stability:

(1) alanine, glycine, glutamic acid, leucine, isoleucine, proline, valine;
(2) aspartic acid, lysine, phenylalanine;
(3) serine, threonine, arginine, tyrosine.

According to Abelson, fossils of the end of the Tertiary are deprived of unstable free amino acids such as serine or threonine. The whole of our knowledge of the preserved proteins of fossils shows that these proteins contain the "unstable" amino acids and this is confirmed by all the data obtained. On the other hand, they show a decrease of "stable" amino acids such as alanine and glycine [27]. It must be stressed here that there is a difference between the resistance of a free amino acid to physical factors and the separation of the same amino acid from a polypeptide chain. This has been confirmed by Vallentyne [65–68]. The process of paleization, leading from native protein to paleoprotein, must still be studied experimentally.

Generally, it must be kept in mind that the tendencies of the studies of Abelson, of Degens and his collaborators, of Vallentyne, Jones, Mitterer, Hare and others, are of a geochemical nature, leading to the understanding of the history of sediments. Protein may form a small portion of the matter of sediments and therefore present little interest for the geochemist. The study of shells, recent and fossil, with the electron microscope, as inaugurated by Grégoire, is another branch of research of great value in comparative submicroscopical morphology. On the other hand, for us, and for those who have pursued the chemical study of paleoproteins since we discovered them in 1961, it is the biochemical aspect which comes to the fore with the purpose of obtaining new data of importance in the phylogeny of organisms. Fractures or loss of terminal or lateral parts do not jeopardize the interest of protein remains, as the primary structure of paleoproteins (amino acid sequence), which is the only reliable test of homology, is unlikely to be modified in paleization except with respect to the length of chains. Studies on the amino acid sequences in fossil proteins should be very rewarding when we succeed in decomposing them into smaller peptides. The possibility of reproducing by pyrolysis results similar to those of paleization offers opportunities to the protein chemist for the study of paleization. In this field, the interest of the electron microscope studies is to demonstrate lack of contamination and they show morphological alterations without of course revealing the molecular nature of the alterations, which can only be observed by chemical studies.

Proteins are not the only organic constituents found in fossils. Free amino acids were first detected by the pioneer studies of Abelson. There is abundant literature on the subject: Abelson [2–5], Ezra and Cook [24], Ijiri and Fujiwara [49], Drozdova and Kochenov [22], Drozdova and Blokh [21], Armstrong and Tarlo [8], Oekonomidis [60] have detected free amino acids in fossil bones or teeth of various ages; Abelson [4], Fujiwara [30], Akiyama and Fujiwara [7], Hotta [46], Oekonomidis [60], Hare and Mitterer [42], Szoor [63] have detected free amino acids in fossil shells;

Manskaya and Drozdova [56] in Ordovician graptolites; Armstrong and Tarlo [8] in Devonian Conodonts.

Chitin has been detected in a Cambrian fossil by Carlisle [15]. Porphyrins have been identified in Crocodile coprolites [25]. Polyhydroxyquinonoic pigments have been detected in Jurassic Crinoids [10–12]. Olefins have been found in fossil shells of Mollusks or Brachiopods [64] and cellulose in fossil wood. Chitin is found in insect fossils of the Middle Eocene [1]. In Pleistocene bones, cholesterol esters, fatty acid esters and phospholipids are found, but no triglyceride [16], while in older fossils only fatty acids are found. In fossils, Everts et al. [23] have found that the fatty acids are essentially non-saturated, contrary to those found in modern bones.

Macromolecules preserved in situ in the anatomical situation present the greatest guarantee of being indigenous to those parts. A critical discussion must be applied in the case of smaller molecules which may have been introduced from outside the fossils.

When considering isolated molecules in sediments, the possibility of introduction from other sources is also great. Large series of organic constituents have been isolated from sediments (see [17]). We cannot exclude that, in very old sediments, products of the degradation of primitive prebiotic aggregates may be found. Precambrian rocks dating from over 3000 million years have been found to contain hydrocarbons such as pristane and phytane, which are considered to be of biological origin.

In a lecture delivered in Leicester in 1968, Calvin mentioned how these polyisoprenoids could be obtained by abiogenesis [59]. He therefore raised the methodological problem of finding internal criteria characteristic of abiogenesis or biogenesis in the case of the mixtures of polyisoprenoids found in such old sediments. An answer to this problem could transfer chemical evolution from the experimental field to the domain of natural observations, and it would also be extremely useful when, in the near future, a study of chemical evolution is performed on material obtained from planets, and from Mars in particular.

References

[1] Abderhalden E. and K. Heyns, Biochem. Z. 259 (1933) 320.
[2] Abelson P.H., Carnegie Inst. Wash. Yearbook 53 (1954) 97.
[3] Abelson P.H., Carnegie Inst. Wash. Yearbook 54 (1955) 107.
[4] Abelson P.H., Sci. Ann. 195 (1956) 83.
[5] Abelson P.H., Ann. N.Y. Acad. Sci. 69 (1957) 276.
[6] Akiyama M., J. Geol. Soc. Japan 70 (1964) 508.

[7] Akiyama M. and T. Fujiwara, Misc. Rep. Res. Inst. Natl. Resources 67 (1966) 67.
[8] Armstrong W.G. and L.B.H. Tarlo, Nature 210 (1966) 481.
[9] Bathurst R.G.C., in: Approaches to Paleoecology, eds. J. Imbrie and N. Newell (Wiley, New York, 1964).
[10] Blumer M., Mikrochemie 36 (1951) 1048.
[11] Blumer M., Nature 188 (1960) 1100.
[12] Blumer M., Geochim. Cosmochim. Acta 26 (1962) 225.
[13] Bricteux-Grégoire S., M. Florkin and C. Grégoire, Comp. Biochem. Physiol 24 (1968) 567.
[14] Calvin M., Trans. Leicester Lit. Philos. Soc., 62 (1968) 45.
[15] Carlisle D.B., Biochem. J. 90 (1964) 1c–2c.
[16] Das S.K., A.R. Doberenz and R.W.G. Wyckoff, Comp. Biochem. Physiol. 23 (1967) 519.
[17] Degens E.T., in: Diagenesis in Sediments. Developments in Sedimentology, Vol 8., eds. G. Larsen and G.V. Chilingar (Elsevier, Amsterdam, 1967).
[18] Degens E.T. and S. Love, Nature 205 (1965) 876.
[19] Doberenz A.R. and R. Lund, Nature 212 (1966) 1502.
[20] Doberenz A.R. and R.W.G. Wyckoff, Proc. Natl. Acad. Sci. U.S. 57 (1967) 539.
[21] Drozdova T.V. and A.M. Blokh, Geochemistry 3 (1966) 530.
[22] Drozdova T.V. and A.V. Kochenov, Geokhimiya (1960) 748.
[23] Everts J.M., A.R. Doberenz and R.W.G. Wyckoff, Comp. Biochem. Physiol. 26 (1968) 955.
[24] Ezra H.C. and S.F. Cook, Science 126 (1957) 80.
[25] Fikentscher R., Zool. Anz., 103 (1933) 289.
[26] Florkin M., in: Organic Chemistry, eds. G. Eglington and M.T.J. Murphy (Springer, Berlin, 1969) p. 506.
[27] Florkin M., C. Grégoire, S. Bricteux-Grégoire and E. Schoffeniels, Compt. Rend. 252 (1961) 440.
[28] Foucart M.F., S. Bricteux-Grégoire, C. Jeuniaux and M. Florkin, Life Sci. 4 (1965) 467.
[29] Foucart M.F. and C. Jeuniaux, Ann. Soc. Roy. Zool. Belg. 95 (1965) 39.
[30] Fujiwara T., J. Geol. Soc. Jap. 67 (1961) 97.
[31] Goffinet M., and Ch. Jeuniaux, Comp. Biochem. Physiol. 29 (1969) 277.
[32] Grandjean J., C. Grégoire and A. Lutts, Bull. Classe Sci. Acad. Roy. Belg. 50 (1964) 562.
[33] Grégoire C., Arch. Intern. Physiol. Biochim. 66 (1958) 674.
[34] Grégoire C., Bull. Inst. Roy. Sci. Natl. Belg. 35 (1959) 14 pp.
[35] Grégoire C., Bull. Inst. Roy. Sci. Natl. Belg. 36 (1960) 22 pp.
[36] Grégoire C., Bull. Inst. Roy. Sci. Natl. Belg. 38 (1962) 71 pp.
[37] Grégoire C., Nature 203 (1964) 868.
[38] Grégoire C., Bull. Inst. Roy. Sci. Natl. Belg. 42 (1966) (39) 36 pp.
[39] Grégoire C., Bull. Inst. Roy. Sci. Natl. Belg. 44 (1968) (25) 69 pp.
[40] Grégoire C., G. Duchâteau and M. Florkin, Ann. Oceanog. (Paris) 31 (1955) 1.
[41] Hall A . and W.J. Kennedy, Proc. Roy. Soc. London Ser. B 168 (1967) 377.
[42] Hare P.E. and R.M. Mitterer, Carnegie Inst. Wash. Yearbook 65 (1966) 362.
[43] Ho T.Y., Proc. Natl. Acad. Sci. U.S. 54 (1965) 26.
[44] Ho T.Y., Comp. Biochem. Physiol. 18 (1966) 353.
[45] Ho T.Y., Biochim. Biophys. Acta 133 (1967) 568.

[46] Hotta S., J. Geol. Soc. Jap. 71 (1965) no 842.
[47] Hudson J.D., Geol. Mag. 99 (1962) 6, 492.
[48] Hudson J.D., Geochim. Cosmochim. Acta 31 (1967) 2361.
[49] Ijiri S. and T. Fujiwara, Proc. Japan Acad. 34 (1958) 280.
[50] Isaacs W.A., K. Little, J.D. Currey and L.B.H. Tarlo, Nature 197 (1963) 192.
[51] Jeuniaux C., Chitine et Chitinolyse, Un chapitre de biologie moléculaire (Masson, Paris, 1963).
[52] Jones J.D. and J.R. Vallentyne, Geochim. Cosmochim. Acta 21 (1960) 1.
[53] Jope M., Comp. Biochem. Physiol. 20 (1967) 601.
[54] Jope M., Comp. Biochem. Physiol. 30 (1969) 225.
[55] Little K., M. Kelly and A. Courts, Bone and Joint Surgery 44B (1962) 503.
[56] Manskaya S.M. and T.V. Drozdova, Geochemistry (1962) 1077.
[57] Miller M.F. II and R.W.G. Wyckoff, Proc. Natl. Acad. Sci. U.S. 60 (1968) 176.
[58] Mitterer R.M. Florida State Univ. Ph. D. Geology (1966).
[59] Munday C., K. Pering and C. Ponnamperuma, Nature 233 (1969) 867.
[60] Oekonomidis A., Folia Bioch. Biol. Graeca 5 (1968) 6.
[61] Pawlicki R., A. Korbel and H. Kubiak, Nature 211 (1966) 655.
[62] Shackleford J.M. and R.W.G. Wyckoff, J. Ultrastruct. Res. 11 (1964) 173.
[63] Szoor G., Acta Biol. Debrecina 5 (1967) 111.
[64] Thompson R.R. and W.B. Creath, Geochim. Cosmochim. Acta 30 (1966) 1137.
[65] Vallentyne J.R., Geochim. Cosmochim. Acta 28 (1964) 157.
[66] Vallentyne J.R., Science 151 (1966) 214.
[67] Vallentyne J.R., Geochim. Cosmochim. Acta 32 (1968) 1353.
[68] Vallentyne J.R., Geochim. Cosmochim. Acta 33 (1969) 1453.
[69] Voss-Foucart M.F., Comp. Biochem. Physiol. 26 (1968) 877.
[70] Voss-Foucart M.F., Comp. Biochem. Physiol. 24 (1968) 31.
[71] Wyckoff R.W.G., Compt. Rend. Acad. Sci. Paris 269 (1969) 1489.
[72] Wyckoff R.W.G. and A.R. Doberenz, Proc. Natl. Acad. Sci. U.S. 53 (1965) 230.
[73] Wyckoff R.W.G. and A.R. Doberenz, J. Microscop. 4 (1965) 271.
[74] Wyckoff R.W.G., E. Wagner, P. Matter III and A.R. Doberenz, Proc. Natl. Acad. Sci. U.S. 50 (1963) 215.

PART II

GENERAL AND THEORETICAL PROBLEMS

Chemical Evolution and the Origin of Life, eds. R. Buvet and C. Ponnamperuma

COHERENT STRUCTURES AND THERMODYNAMIC STABILITY

I. PRIGOGINE and A. BABLOYANTZ *
Université Libre de Bruxelles, Faculté des Sciences, Bruxelles, Belgium

The object of this talk is to present a short review of recent developments in irreversible thermodynamics and its applications to the study of physico-chemical and biological phenomena. These ideas have been developed recently by our group (for a general account see [3,14,15]).

The starting point is the extension of methods of classical thermodynamics to treat situations arbitrarily far from equilibrium. This can be achieved if one assumes that the thermodynamic state functions such as temperature T, pressure p, entropy S, chemical potential μ_i of component i etc. at a given point in space of a globally non-equilibrium system depend on the local values of the thermodynamic variables energy E, volume V and mole fraction n_i of component i in exactly the same way as in equilibrium. In other words, one assumes that a local formulation of irreversible thermodynamics is possible and that in this formulation a state of "local equilibrium" prevails in the neighborhood of a given point in the system.

In particular if ns is the entropy density, ne the energy density and v the specific volume, the well known Gibbs relation will hold locally

$$s = s(ne, v, n_i) \tag{1}$$

$$T\,\mathrm{d}(ns) = \mathrm{d}(ne) + \frac{p}{v}\mathrm{d}v - \Sigma\, \mu_i\, \mathrm{d}n_i \tag{1a}$$

Relation (1) enables us to formulate a local non-equilibrium thermodynamics [3,15].

We notice that the local equilibrium assumption is valid if dissipative processes dominate over purely mechanical ones [3,15].

* Presented by A. Babloyantz.

Non-linear thermodynamics

The local formulation of irreversible thermodynamics has been extended recently to the non-linear region by Glansdorff and Prigogine. We briefly recall here the main results of these investigations. For further details we refer to the original literature [2,3,18].

(i) In the whole domain of phenomena which may be described adequately by a local theory it is possible to construct a differential expression $\mathrm{d}\Phi$ depending on the state variables, such that

$$\frac{\mathrm{d}\Phi}{\mathrm{d}t} \leqslant 0 \tag{2}$$

the equality being applicable at the steady state. $\mathrm{d}\Phi$ comprises both dissipative and convective processes.

This inequality provides an evolution criterion. For systems in mechanical equilibrium this reduces to

$$\int \mathrm{d}V \sum_i J_i \, \mathrm{d}X_i \leqslant 0 \tag{3}$$

where J_i are the flows and X_i the corresponding generalized forces. This expression is closely related to entropy production P. Indeed in domain of validity of local equilibrium it can be shown that [3,15]

$$P = \sum_i J_i X_i \tag{4}$$

In general $\mathrm{d}\Phi$ is a non total differential and does not represent the variation of a thermodynamic state function.

(ii) As the steady states far from equilibrium are not characterized any longer by the minimum of a thermodynamic potential, the stability of these states is not always insured. One has to look therefore for a separate stability criterion.

Recently a complete infinitesimal stability theory of non-equilibrium states has been worked out in Brussels [2,3,18]. The main results are as follows (for simplicity we only consider systems in mechanical equilibrium). Consider the entropy, S, of a non-equilibrium state and expand it around some reference state:

$$S = S_0 + \delta S + \tfrac{1}{2}\delta^2 S + \ldots \tag{5}$$

δ denotes a variation due to fluctuations in the system. Within the domain of validity of local thermodynamics one can show that the second order term

$\delta^2 S$ is a negative definite quadratic form provided that the equilibrium state of the system is stable:

$$\delta^2 S \leqslant 0 \tag{6}$$

In the limit of small fluctuations if one has in addition

$$\frac{\partial}{\partial t} \delta^2 S > 0 \tag{7}$$

the non-equilibrium reference state will be stable.

This formulation is closely related to the ideas underlying Lyapounov's stability theory. Relations (6) and (7) constitute a thermodynamic stability criterion.

Chemical instabilities

A specially interesting case is that of chemical reactions, where the laws between reaction rates and chemical affinities are generally non linear. The stability condition becomes [3,16]

$$\sum_{\rho} \delta W_{\rho} \, \delta(A_{\rho} T^{-1}) \geqslant 0 \tag{8}$$

where W_{ρ} are the chemical reaction rates, A_{ρ} the corresponding affinities. We see that the quantity which expresses the stability properties here is the excess entropy production of the system around the reference state. It can be shown that in order to have instabilities we need autocatalytic or cross-catalytic steps in the reaction mechanism. These steps contribute negative terms to the excess entropy production [3,18].

Let us study some specific, chemical examples. Their behavior might give some indications concerning the actual biological systems. We first observe that in the neighborhood of instabilities we may distinguish between three possible situations [3,14].

1) oscillations around steady states

2) symmetry breaking transitions

3) multiple steady state transitions.

We shall discuss here a few representative examples corresponding to cases (1) and (2).

Oscillations

The occurrence of non-equilibrium instabilities in chemical reactions has been verified in detail in model systems. It has been shown that when the system is maintained spatially uniform it may evolve beyond instability, to a new stable regime which is time-dependent [3,11,14]. This regime is characterized by well-defined values of period and amplitudes of oscillations, independent of the initial conditions. It corresponds to what is known in the theory of non-linear oscillations as a *limit cycle*.

This behavior has been shown recently to explain certain features of biochemical systems. It has been known for several years that the glycolytic cycle [6] and the peroxidase reaction [13] present sustained oscillations in the concentrations of certain intermediate products. Higgins [8] and Sel'kov [21] have proposed models which predict such a behavior. Sel'kov's scheme reads

$$\xrightarrow{\nu_1} X$$

$$X + E_1 \rightleftharpoons V_1$$

$$V_1 \rightarrow Y + E_1$$

$$\gamma Y + E_2 \rightleftharpoons E_1$$

$$Y \xrightarrow{k}$$

where X = ATP Y = ADP, E_1 and E_2 are respectively the active and inactive forms of the enzyme phosphofructokinase. ν_1 the rate of entry of ATP and γ is a constant which expresses the degree of activation of the enzyme by the product (it is assumed that $\gamma > 1$).

It has been shown that for $\nu_1 < \nu_{1c}$ the steady state of the system becomes unstable and subsequently the system evolves to a limit cycle. Using realistic values for the kinetic constants Sel'kov has been able to show that the periods and other properties predicted by the model are close to the observed values [6,7]. Sel'kov's model is phenomenological, in the sense that it introduces a parameter γ which expresses the degree of activation. We would like to report here a more realistic model for glycolysis which has been worked out recently in Brussels by Goldbeter and Lefever [4]. The main feature of the model, is to take explicitly into account the allosteric character of the enzyme phosphofructokinase [12]. The activation of the enzyme by the product, the role of the substrate and the cooperativity on the macromolecular level are described in terms of three independent parameters which are directly related to experiment.

In fact the model is more general and applies to other reactions catalyzed by allosteric enzymes; here we shall mainly concentrate on the application to oscillations in glycolysis.

The model

Fig. 1 gives a schematic representation of the model. The main steps are as follows:

(a) The substrate enters in the system at a constant rate ν_1;

(b) The enzyme is a dimer which can exist at equilibrium in the form R (active) and T (inactive);

(c) The substrate may react with both forms whereas the product (positive effector) can only bind with the species R;

(d) The enzymatic species bound with the substrate decompose irreversibly to give a product;

(e) The product leaves the system at a rate proportional to its concentration.

We observe that steps (a), (d) and (e) are irreversible; this introduces into

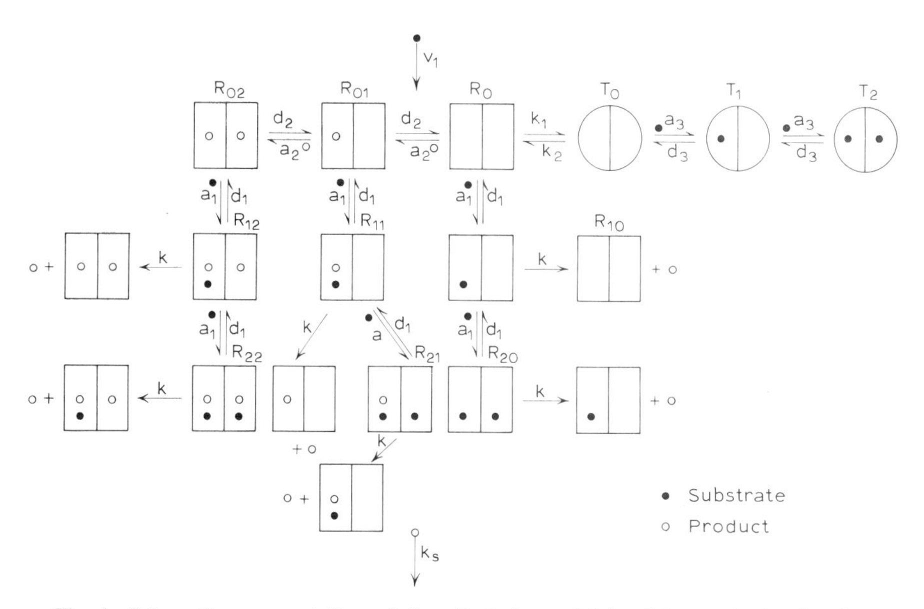

Fig. 1. Schematic representation of the allosteric model involving activation by the product.

ν_1:	rate of entry of the substrate.
R, T:	respectively active and inactive forms of the enzyme.
k_1, k_2:	kinetic constants for the $R_0 \rightleftharpoons T_0$ reaction.
a_i, d_i:	kinetic constants for the reactions involving the binding of the substrate on an enzymatic species.
k:	kinetic constant for the (irreversible) decomposition of an enzymatic intermediate into an enzyme and a product.
k_s:	kinetic constant for the (irreversible) decomposition of the product.

the problem the non-equilibrium properties which will be responsible for the occurrence of oscillations. Starting from these assumptions and introducing additional simplifications based on experimental data one obtains a set of two coupled equations for the evolution of the substrate and product concentrations. These equations admit a single physical uniform steady state solution which becomes unstable provided a set of conditions between the parameters is satisfied. In particular the instability condition is fulfilled for large values of the allosteric constant [12], which agree with the experimentally observed values. It is also observed that the domain of instability is much wider in the

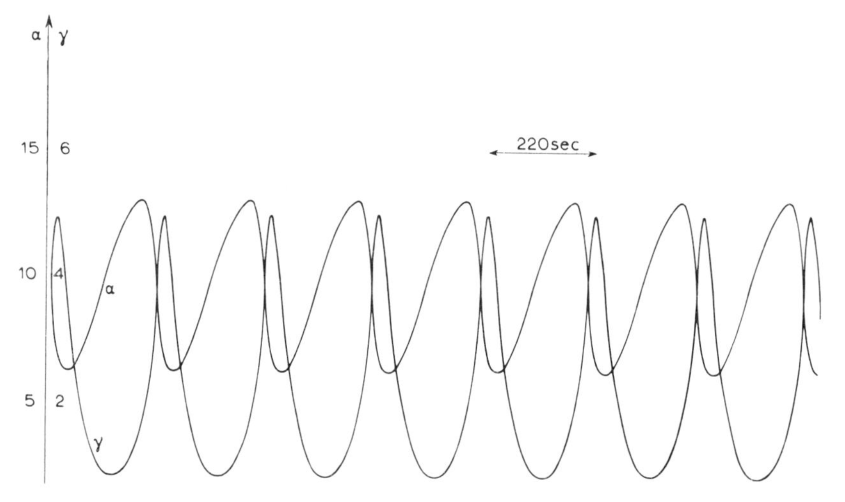

Fig. 2. The time variation of the ATP and ADP concentrations for glycolysis

$$\alpha = [\mathrm{ATP}]\frac{a_1}{d_1} \qquad \gamma = [\mathrm{ADP}]\frac{a_2}{d_2}$$

The following numerical values of the parameters have been chosen:

$$\mathcal{L} = \frac{T_0}{R_0} = 7.5 \times 10^6 \qquad \text{(allosteric constant)}$$

$$a_i = a = 10^7/\text{mM sec}$$
$$d_i = d = 5 \times 10^5/\text{sec}$$
$$\sigma_1 = v_1/d = 10^{-8}\ \text{mM}$$
$$\sigma_2 = k_s/a = 5 \times 10^{-9}\ \text{mM}$$
$$\epsilon = k/d = 10^{-1}$$

Total enzyme concentration

$$D_0 = 5 \times 10^{-4}\ \text{mM}$$

case the substrate activates the enzyme. In the presence of substrate inhibition instabilities are still possible but disappear beyond a threshold value of the degree of inhibition.

The study of the evolution of the system beyond instability has shown that the system tends to a new stable regime in which the concentrations of the chemicals exhibit sustained oscillations [9]. Fig. 2 shows the variation of concentrations in time in the case of glycolysis *. It is seen that when the parameters are replaced by the experimentally known values the period of oscillations is of a few minutes and agrees therefore with the experimental data [6,7].

Symmetry-breaking transitions

Model calculations have shown that non-equilibrium instabilities are also possible in non-uniform chemical systems [9,10,16–19]. In this case the system may evolve, beyond an instability point of the uniform steady state, to a new time-independent stable regime corresponding to a spatial distribution of the chemicals. The system changes its spatial symmetry as a result of an instability and is able to transform the energy received from the outside world into order. However, this transformation is only possible under non-equilibrium conditions. For this reason the structures which emerge in systems far from equilibrium will be termed *dissipative structures*.

Recently these concepts have been substantiated by an experiment involving the oxydation of malonic acid [22]. Dissipative structures have been observed in this system. For more details see references [1] and [5].

In the domain of biochemical reactions it has been shown theoretically [20] that glycolysis and other multienzyme schemes exhibit unstabilities leading to a space dependent state, such that the substrates and products show some spatial organization around the enzymes.

Conclusions

We have seen that the emergence of temporal or spatial order in a previously uniform system at a steady state is possible, provided the system is non-linear and operates beyond an instability.

This behavior seems to provide a satisfacory interpretation of the experi-

* Experimental studies seem to indicate that phosphofructokinase is a tetramer. We expect however that, qualitatively, the behavior is well approximated by the dimer model considered in this section.

mental data on certain types of biochemical oscillatory reactions. On the other hand until now, the occurrence of spatial dissipative structures has been verified experimentally only for the Zhabotinski reaction [1,5,22]. It would be very interesting to find a larger class of organic and possibly biochemical systems showing a similar behavior and relate these effects to observable biological phenomena.

References

[1] Büsse H., J. Physiol. Chem. 73 (1969) 750.
[2] Glansdorff P. and I. Prigogine, Physica 46 (1970) 344.
[3] Glansdorff P. and I. Prigogine, Thermodynamics of Structure, Stability and Fluctuations (John Wiley Interscience, 1970) in press.
[4] Goldbeter A. and R. Lefever, to be published.
[5] Herschkowitz M., Compt. Rend. Acad. Sci. (Paris) 270C (1970) 1049.
[6] Hess B. and A. Boiteux, in: Regulatory Functions of Biological Membranes, ed. J. Järnefelt (Elsevier, Asmterdam, 1968).
[7] Hess B., A. Boiteux and J. Krüger, Advan. Enzyme Regulation 7 (1969) 149.
[8] Higgins J., Proc. Natl. Acad. Sci. U.S. 51 (1964) 989.
[9] Lefever R., J. Chem. Phys. 49 (1968) 4977.
[10] Lefever R., Bull. Cl. Sci., Acad. Roy. Belg. 54 (1968) 712.
[11] Lefever R. and G. Nicolis, J. Theor. Biol., submitted.
[12] Monod J., J. Wyman and J.P. Changeux, J. Mol. Biol. 12 (1965) 88.
[13] Nakamura S., K. Yokota and I. Yomazaki, Nature 222 (1969) 794.
[14] Nicolis G., Advances in Chemical Physics Interscience, (1970) in press.
[15] Prigogine I., Introduction to Thermodynamics of Irreversible Processes (John Wiley, New York, 1967).
[16] Prigogine I. and G. Nicolis, J. Chem. Phys. 46 (1967) 3542.
[17] Prigogine I. and R. Lefever, J. Chem. Phys. 48 (1968) 1695.
[18] Prigogine I., Structure dissipation and life, in: Theoretical Physics and Biology, ed. A. Marois (North-Holland, Amsterdam, 1961).
[19] Prigogine I., Dissipative structures in biological systems, in: Theoretical Physics and Biology (Versailles, 1969).
[20] Prigogine I., R. Lefever, A. Goldbeter and M. Herschkowitz, Nature 223 (1969) 913.
[21] Sel'kov E.E., European J. Biochem. 4 (1968) 79.
[22] Zhabotinsky A.M., Biofizika 2 (1964) 306.

Chemical Evolution and the Origin of Life, eds. R. Buvet and C. Ponnamperuma

AN ENERGETIC APPROACH TO PREBIOLOGICAL CHEMISTRY

Harold J. MOROWITZ
Department of Molecular Biophysics and Biochemistry, Yale University, New Haven, Connecticut, USA

When inquiring about the origin of life from a reductionist point of view one seeks those principles of physics and chemistry which act so as to order a system and free it from the equilibrium constraints of maximum disorder which are clearly inconsistent with life processes. The researches of the last twenty years on energy driven organic synthesis (many of which are reported on or reviewed in this volume) make it clear that spontaneous nonequilibrium processes can lead to a rich variety of biochemicals including macromolecules. The problem with the organic synthesis approach, at the moment, is an embarrassment of riches; too many possibilities present themselves. At the present time a number of diverse geochemical and planetological hypotheses lead to rather similar distributions of prebiotic organic intermediates. It is becoming apparent that no experiment or series of experiments about the origin of life will be entirely convincing without establishing a theoretical framework that provides guidance in choosing between the many possibilities.

One of the roles of physical science at this stage in the inquiry is to examine how order arises in model physical systems and try to relate this to the ordering which must have occurred in the origin of life. Living systems are obviously far from equilibrium which immediately supplies some clues. In general any non-equilibrium system (or structure) if left by itself will begin to decay to equilibrium. This follows from the second law of thermodynamics when viewed macroscopically and follows from random thermal motion when viewed from the microscopic or molecular point of view. Any non-equilibrium structure is thus a dissipative structure in the sense that structural energy is constantly being converted into thermal energy by irreversible processes and energy must be supplied in other forms to maintain the integrity of the structures.

Thus in assessing schemes for the origin of life it is necessary to consider not only that energy must be supplied to initiate the process (to synthesize the organic structures) but must be continually supplied to maintain the structures once they have been synthesized. Very frequently the energy sources proposed for synthesis are different than the maintenance source. This would have necessitated at some time in the past a shift of biospheric energy sources. At the present the energy source is sunlight and this has certainly been the case for most of the history of the biosphere. If we assume that the early structures were built up and maintained by other than solar energy (lightning, volcanoes, shockwaves, radioactivity etc.), then we must assume another stage in evolution which shifted the energy source from the primordial one to the sun. If all biological energy input had been mediated by some ubiquitous chemical species (perhaps polyphosphates and eventually specific polyphosphates such as ATP) then multiple energy sources could have contributed to the synthesis of the intermediate and the shift from one source to another can be more easily understood. In any case it is now clear that proposed schemes for the origin of life must be capable of accounting for both the synthesis and maintenance of structure.

Continuing non-equilibrium systems must thus be maintained by a flow of energy through the systems which counters the disordering due to dissipative processes. This is a statement of a necessary condition. There appears to be associated with this a statement of sufficiency that the flow of energy from a source to a sink through an intermediate system always leads to an ordering of the intermediate system. From a physical point of view organized systems are those systems which store energy in modes other than thermal. Hydrodynamic systems provide an example where macroscopic structures store energy both as potential and kinetic mechanical energy. Viscous forces constantly degrade the energy to heat and mechanical energy must constantly be supplied. Hydrodynamic structures may arise spontaneously, may be arbitrarily elaborate and are maintained by an energy flux. The array of hexagonal flow cells in the Benard problem is an example of hydrodynamic structure. A tornado is another example of such a structure which is maintained by the flux of solar energy.

In biological systems the primary mode of energy storage is in chemical bond energy which is effectively the potential energy of electrons. The photosynthetic energy input serves to build up these high potential energy structures while catabolic processes act so as to degrade this energy to thermal energy.

We have examined some of the consequences of energy storage in covalent bonds in an effort to explore some of the chemical consequences of this process. The model system that has been employed consists of a box containing carbon, hydrogen, nitrogen, oxygen, phosphorus and sulfur. The box is in contact with an isothermal reservoir at temperature T. If the contents of the box are allowed to equilibrate it is possible to calculate the distribution of chemical species within the box and Dayhoff and coworkers have designed a very elegant computer program to use thermochemical data to calculate this distribution [1].

Next place a high potential energy source next to the box. By a high potential source we mean one that transfers energy to the box in quanta which have an average energy large in comparison with kT. The distribution of chemical species will change as the absorption of energy leads to intermediate chemistry. The high energy compounds formed are constantly being degraded leading to a net flow of heat to the reservoir. If energy flows in from the source at a fixed rate, a steady state is ultimately reached where the influx of high frequency energy is just balanced by the flow of heat to the reservoir.

The theoretical solution of the chemical distribution in the steady state is a very complex problem in kinetics which, for any real system, cannot be solved. The solution depends on atomic composition, temperature and source spectral distribution. There is, however, one limiting case for which a theoretical analysis is possible. If the spectral distribution is sufficiently on the high frequency side the absorption of energy by molecules is independent of the molecular structure and we may assume that the steady state corresponds to the most random possible distribution of energy among covalently bonded species of molecules [3]. The problem of how much energy is stored can be approached by noting that in the steady state there will be a range in which the stored energy in covalent bonds will be a monotonic increasing function of the energy flux. This is a phenomenological relation for each energy source. Its existence makes it possible to set an energy level ΔE above the equilibrium state. The program of Dayhoff et al. [1] can now be reworked in the following way. An additional constraint is introduced, which is the increase of internal energy of the non-equilibrium system above the energy of the corresponding equilibrium system. The entropy of the system is now maximized subject to constant pressure and atomic composition. This maximization is made with respect to concentration of chemical species so that it is possible to predict the chemical composition for the states under consideration [2].

Tables 1 and 2 show the results of such a computation for a system at 300 °K. The atomic composition of the system is carbon 4828, hydrogen 9489, nitrogen 537, oxygen 14,347, phosphorus 78 and sulfur 35. This is a highly oxidizing chemical composition and may be considered the worst case

Table 1
Mole fraction composition (normalized). Decreasing composition.

CARBON DIOXIDE	0.4728E 00	NIO2	0.9914E−27
WATER	0.4622E 00	H2O2	0.6766E−27
N2	0.5373E−01	NITROUS OXIDE	0.7735E−28
H3PO4	0.7834E−02	P4O10	0.1838E−28
SIO3	0.3456E−02	NITRIC ACID	0.2764E−29
SULFUR DIOXIDE	0.9942E−06	H2	0.3463E−31
SULFURIC ACID	0.8263E−11	H2SIO3	0.1577E−35
O2	0.4502E−17	CARBON MONOXIDE	0.3923E−36
NITRIC OXIDE	0.4105E−24	HYDROXYLAMINE	0.1024E−36
NITROUS ACID	0.8152E−26		

Table 2
Energetically enriched state.

WATER	0.3411E 00	S2	0.1485E−07
CARBON MONOXIDE	0.3137E 00	FORMIC ACID	0.1151E−07
O2	0.1735E 00	H1C1N1	0.7544E−08
CARBON DIOXIDE	0.6916E−01	CYANIC ACID	0.5554E−08
H2	0.4004E−01	C1S1	0.4896E−08
N2	0.3627E−01	P2	0.3837E−08
NITRIC OXIDE	0.1446E−01	HYDROXYLAMINE	0.7315E−09
P1O2	0.3942E−02	PHOSPHINE	0.3163E−10
H2O2	0.2591E−02	N1O3	0.1135E−10
P1O1	0.2330E−02	P1C1	0.5545E−11
S1O1	0.1602E−02	NITRIC ACID	0.5522E−11
SULFUR DIOXIDE	0.1129E−02	H2S1O3	0.4853E−12
S1	0.5592E−04	METHANE	0.2505E−12
P1H2	0.5238E−04	H1N1C1S1	0.1159E−13
S1H1	0.1186E−04	C1S2	0.6326E−14
P1N1	0.1158E−04	METHANOL	0.5765E−14
P1	0.7106E−05	SULFURIC ACID	0.5214E−14
N1O2	0.5692E−05	C1H2S1O1	0.1766E−14
H2S1	0.5911E−06	ACETYLENE	0.1406E−14
P1H1	0.5262E−06	KETENE	0.1271E−14
C	0.3106E−06	GLYOXAL	0.6023E−15
S1O3	0.1615E−06	CYANAMIDE	0.2921E−15
OZONE	0.1106E−06	GLYOXALIC ACID	0.8691E−16
P1S1	0.1069E−06	H3PO4	0.3372E−19
NITROUS OXIDE	0.1018E−06	DIMETHYL ETHER	0.2207E−23
S1N1	0.7169E−07	C2H3N1S1	0.5054E−27
NITROUS ACID	0.6170E−07	METHYL ACETATE	0.4872E−30
C1O1S1	0.4209E−07	THIOGLYCOL	0.2423E−35
AMMONIA	0.3439E−07	FORMYLGLYCINE	0.3069E−37
FORMALDEHYDE	0.2007E−07		

for synthesis of organic intermediates. Table 1 gives the equilibrium distribution at 300 °K and is of course dominated by completely oxidized compounds. In table 2 the temperature has been maintained at 300 °K but the total energy of the system has been increased by an amount roughly half of that found if the system were to exist as an organism and free oxygen. Free oxygen appears as well as a number of organic compounds. Randomly pumping energy into the system provides an interesting first step on the way to organic complexity.

These calculations can be extended to more exacting boundary conditions in order to examine various models of prebiological chemistry.

References

[1] Dayhoff M.O., E.R. Lippincott, R.V. Eck and G. Nagarajan, Thermodynamic Equilibrium in Prebiological Atmospheres of C, H, O, N, P, S, and Cl (NASA publication No. SP-3040, 1967).

[2] Huang H.W. and H.J. Morowitz, unpublished results.

[3] Rider K. and J. Morowitz, J. Theoret. Biol. 21 (1968) 278.

Chemical Evolution and the Origin of Life, eds. R. Buvet and C. Ponnamperuma

THE RECOGNITION OF DESCRIPTION AND FUNCTION IN CHEMICAL REACTION NETWORKS

H.H. PATTEE
W.W.Hansen Laboratories of Physics, Stanford University, Stanford, California 94305, USA

The problem of the origin of life must be approached along many paths. The greatest advances so far have been in our understanding of possible prebiological chemical syntheses, and discovery of the earliest geological history of living organisms. My own approach is a search for coherence in chemical phenomena or what the classical biologist might call the origin of functional behavior. As I shall explain, in the context of origins, I regard description as inseparable from function. Description and function are biological terms, and we must rightly be suspicious of introducing biological concepts without explaining their non-biological physical and chemical realizations. The reduction of biological activities to their chemical structure has been the central goal of molecular biology, – a goal which has nearly been reached, at least at the cellular level. In fact, the very idea of an explanation of living matter has recently come to mean little more than a detailed description of each chemical structure which underlies an observable, well-established function. This strategy may be adequate to explain "how things work" but I doubt if it is sufficient to understand "how things originated".

Most origin of life studies have, up to now, followed this structural approach under the assumption that similar chemical structures will inherently execute similar functions. Consequently discussions among origin of life workers have not been over strategies but over matters of the reasonableness of initial conditions, reactants, and energy sources, and the similarity of the synthesized structures to present biological structures.

As a stimulus to origin of life study, these abiogenic synthesis experiments have been extremely rewarding and to some extent reassuring, since the results have suggested that there are many possible initial conditions and energy sources for most of the general chemical structures that are found in living organisms. In spite of this primary success, I would like to give some arguments why I think an exclusive emphasis on chemical synthesis aimed at matching existing biomolecular or cellular structures may evade the central origin problem.

What are the facts of life?

In one way or another, origin of life experiments are searches for some type of life-like characteristic from a reasonably likely non-life-like matrix. What I have been saying is that "life-like" should not be limited to only structural statements such as "made of polynucleotides and polypeptides". While I certainly accept this as *one* fact of life (I am not proposing an alternative chemical basis for life) there are also other types of facts of life which may be relevant to our search. In particular, there is a more classical, evolutionary fact of life which we may also need to bear in mind when we design origin experiments. For this brief discussion let me state it in the following form: no complex chemical structure persists or evolves unless it has a simple function and a simple description. Of course this is not a precise statement, but I believe it conveys a fundamental property of life which origin experimenters must recognize. It implies that the search for more and more complex chemical structures will cease to have any evolutionary, and hence biological, significance unless such structure is part of a simpler description-function process.

Searches for complexity from simplicity

The idea of simplicity and complexity are very fundamental here. Our own molecular description of any biological structure, whether it is an enzyme, a muscle, an eye, a fish or a bird, is always much more complicated than the description of the basic function associated with the structure. Because of this, from the structural point of view, origin problems become a search for growing complexity, whereas from the description-function point of view, the origin problem is a search for coherent simplicity.

Many of us picture the process of chemical evolution as beginning with the simplest molecules like carbon dioxide, ammonia, and water, and progressing continuously through increasingly complex stages of molecular complexity to polynucleotides, polypeptides and finally to intricate structures and networks of reactions which begin to "look like" a cell. This spontaneous development of complexity from very simple beginnings does indeed appear to satisfy one aspect of the origin problem. Any "limited heterogeneity" or natural constraints on the conceivable variety of macromolecular structures can also appear to be a source of satisfaction if they are interpreted as the beginnings of informational constraints which are so essential for living matter.

However, it is my present view that these satisfactions are deceptive because they do not distinguish description, structure and function, which I consider fundamental properties of life. This is not simply a semantic difference in the way we interpret present abiogenic synthesis experiments, but a different physical theory of life. This theory requires an entirely different type of experiment for verification. Such experiments must begin with much more complex and natural environments and look for simple, persistent chemical regularities.

Some physical conditions for function and description

What I mean by the growth of simplicity here is not the degradation of complex molecules into simple molecules. Nor do I mean choosing to look at only a small part of the system. What I mean is finding a new, collective *regularity* in the overall behavior of the complex system – a regularity which is not easily recognizable from only the detailed structure or dynamics. This simple, regular behavior in a complex dynamical system is one necessary condition for what I call *function*. But this condition is not sufficient since a simple regularity may appear only by affecting the mind of an intelligent observer, and such an "outside" observer does not alter the chemical development of the system. In order to make our concept of function autonomous and objective we must add the condition that the simple, regular behavior of the complex dynamical system must affect the behavior of the system itself. In other words, *function must appear as a constraint on the detailed dynamics of the system which displays the simple regularity*.

Now I come to a crucial step in my theory of living matter. I claim that an objective concept of a *constraint* in a physical system can only be made distinctly in terms of some form of *description*. Therefore my concept of function, which is a type of a constraint, can only have objective meaning if there is a *description* of the function. In other words, function can have no objective meaning unless its simple regularity is embodied in a description. Or alternatively, we could say that we recognize descriptive behavior of matter because of its correlation with simple, regular functions.

The necessity of alternative descriptions for representing constraints in physics I have explained in more detail in my discussions of hierarchical control systems [7–9]. In particular it is necessary to distinguish structural constraints. which permanently freeze out degrees of freedom, from control constraints which correlate selected degrees of freedom in a conditional, time-dependent way. The necessity for alternative description comes about because physical laws at the microscopic dynamical level are assumed to be complete. That is, the laws of motion predict the behavior of the system completely, in so far as it is deterministic, so that additional constraints at this detailed level are incompatible with these laws. Equations of constraint are always simplifications of the detailed dynamics which express some additional regularity of the collective system, and therefore these equations cannot be at the same level of description as the laws of motion.

Living matter, I assume, obeys precisely the same laws of motion as non-living matter. Furthermore, as I have said, it is possible for us to recognize all kinds of constraints in non-living matter which we represent by alternative descriptions to our expressions of laws of motion. But these are the physicist's descriptions and they are clearly outside the physical system which the physicist describes. I believe that the only useful objective criterion for distinguishing living from non-living matter is to say that *the constraints of living matter must contain their own descriptions*.

Self-replication requires description and function

This above criterion is as close as I have come to finding a physical basis for the evolutionary requirement of a separate genotype and phenotype – the genotype being essentially the cell's description of its constraints, and the phenotype the cell's construction of functional structures from this description. This separation is of course the fundamental condition for the theory of evolution by natural selection [e.g., 2].

This requirement for the simultaneous occurrence of both description and function, which is necessary to make sense out of either concept in an isolated physical system, also clarifies the concepts of self-organization and information. The mere occurrence of constraints in the ordinary physical or stereochemical sense should not be called by the same names as the constraints of a description. In other words, there is no doubt that limited heterogeneity in copolymers synthesized with no specific condensing agents or selective catalysts is a kind of order. But this is not self-ordering unless one can show that the prefix "self" stands for something – i.e., a description of the order. Similarly, there is no inherent informational process in a closed physical system unless we can specify an objective separation of the physical structures representing messages from the physical structures constrained by these messages.

Finally, I would also apply this same criterion to the concept of self-replication. Without a separate description of what is being replicated I again can see no value in using the prefix "self". Many types of repetitive growth, fissioning, proliferation and template copying may certainly be called replication, but unless there is a corresponding description there can be no function and hence no evolution by natural selection [cf. 4,5].

Searches for simplicity from complexity

How does this evolutionary point of view alter our empirical outlook? We suggested earlier that the aim of most abiogenic synthesis experiments is to find increasingly complex biochemical structures beginning with a simple matrix. In particular, there have been searches for amino acids from the simplest compounds, for copolymers from simple monomers, and for cell-like structures from copolymers. It is generally assumed that by bringing together several of these diverse lines of synthesis there will emerge an even more complex organization which we shall recognize as more life-like than the stage before.

Of course we all recognize the necessity of greater complexity in reaching the genotype-code-phenotype level of evolution; but stated as "bringing together" the necessary *structures* in the form of genetic nucleic acids and enzymatic polypeptides, the problem sounds exceedingly difficult. Certainly if we assume that physical structures such as a nucleic acid molecule or a polypeptide molecule begin with a specific, inherent descriptive or functional property which is solely the result of its own particular chemical structure, then indeed the "bringing together" of a coherent system of descriptions, codes, and functions appears as incredibly unlikely.

But from what I have said, this may be asking the question backwards. I prefer to ask the question beginning with only a very complex but incoherent chemical reaction network with no implication of descriptive or functional behavior. Obviously the question then is not one of "coming together" but how the descriptive and functional aspects of these complex chemical reactions grew to be separated. In this case, the question is equivalent to asking how objective meaning can be attached to *description* and *function* in any complex network of ordinary chemical reactions.

Hierarchical constraints separate description and function

We must bear in mind that the connotations of description and function in ordinary language are exceedingly complex and not entirely clear. A description implies some arrangement of an arbitrary but fixed set of symbols under a fixed set of rules of grammar which can be called a language. At present, no one understands either the origin of language or the minimum number of rules which would suffice for an evolutionary discourse.

The concept of function is also somewhat vague. It implies a coherent, integrated activity of parts acting on some environment, often described as goal-directed or purposive activity. Clearly in the origin of life context we must take only the most primitive and essential characteristics of description and function. In particular we must reduce these concepts to physical language.

The most primitive physical characterization of *function* I can imagine is the idea of a constraint acting on specific degrees of freedom of a relatively complex environment. This function must be realized by a structure which is separate from the environment. Furthermore, according to my definition of constraint, it must have an internal description (assuming no outside description). To keep the picture as primitive as possible I can also characterize *description* by the idea of a constraint; however, this descriptive constraint shall not act directly on the environment but on specific degrees of freedom of the structures which form the functional constraint.

Now of course we have reached a primordial chicken-egg problem, since our descriptive constraint will itself require a description. But this need not be an infinite regress paradox. We have a model of finite self-description in every language. Self-description may be thought of as a small subset of descriptive constraints that are universal. That is, they act on all descriptions in the way that rules of grammar are constraints acting on all strings of words, and generating sentences including the sentences which define the grammar itself. This is similar to a chemical closure property that limits the substrates and their constrained reactions to a fixed and finite number of types even though the possible number of sequences of these reactions is unlimited.

But the essential concept I am trying to illustrate is that of *hierarchical* constraints in a complex network of chemical reactions. Descriptional and functional constraints are on two hierarchical levels with description the higher level in the sense that it can control function. Hierarchical control is the most universal organizational principle of living matter. One must even consider the genotype-phenotype relation as a particular case of hierarchical control. The entire course of evolution can be pictured as a progression of hierarchical levels of control. It is therefore quite reasonable that hierarchical chemical constraints had fundamental prebiological significance.

What experiments need to be done?

The final question, of course, is what kind of experiments do we design to study the origin of hierarchical controls? The basic nature of hierarchical organization is that it imposes simple rules which constrain otherwise complex collections of individual elements. In order to study the spontaneous origin of such rules we must begin with a realistically complex matrix. We must be particularly careful not to impose our own intelligent forms of artificial simplification, since they may very easily override the natural regularities which we would expect to develop gradually. Well-developed hierarchical organizations have clear and definite constraints, but at their origin, like the nucleation process in condensation, the crucial events may be weak and arbitrary.

As one example of a "realistically complex" matrix I have suggested an artificial, sterile seashore experiment as a complement to the many specific abiogenic syntheses which have been demonstrated, [6]. A seashore simulation must have all three phases of matter, the atmosphere, the ocean and the beach, also the primeval solar radiation with normal night and day periods, as well as waves and tides washing the sand. This would have to be a large simulation, perhaps 10^5 liters, and would require cooperative effort from many disciplines; but the cost would still be far less than even a single exobiology mission.

When I suggest such a natural environment to chemists they tend to dismiss it as far too complex and undefined for useful experiments. The biologists are more tolerant. Everyone wants to know exactly what to look for.

The point is, of course, that one must *begin* studying realistic chemical complexity with the eye of a naturalist, looking for significant chemical concentrations or exceptional chemical behavior or structures which may in turn suggest some regular underlying chemical constraints. Only then can analytical study begin. This is, after all, no different from the problem of initial complexity that all sciences have already faced in their historical development. The criterion for such experiments must be an honest acceptance of the complexity of the sterile earth in contrast to attempts to artificially simplify the initial conditions for ease of analysis.

A second, less natural, type of experiment on hierarchical origins may be done by computer simulation. Here the idea is to find an acceptable degree of program constraint on the computer while still leaving the computer's internal structure free enough to develop its own constraints. An interesting study of this nature has been done by Kauffman [3] in which a collection of switching elements with definite but randomly selected switching functions, connected by a fixed but random network, condenses out simple, stable cycles of behavior. It remains to be shown how this rather general process of simplification in a complex network could represent a real chemical reaction system on the primitive earth. Another computer study of the behavior of complex interactions by Conrad uses the basic constraints of an ecosystem to model primitive evolutionary dynamics [1]. The model assumes a population of coded, self-replicating units of the simplest imaginable type, in an environment with conservation of matter and movement of species within a closed, one dimensional space. Many more simulations of this type need to be done. Hopefully they will give us some intuition about the crucial conditions for originating hierarchical constraints, and how we should look for them in chemical systems.

While I have no doubts that experiments will continue to clarify many facets of the origin of life problem, there are still many questions where I believe that more theoretical understanding will be needed before we can even design the right experiment. Most of these questions are related to the simplification, integration, and optimization of a complex environment which is the physical basis of function and description. Why are there so few basic monomers in all living matter? How do coding or language constraints originate? How can codes and grammars evolve? Is the present code a frozen accident? Could it evolve further? Could there be any superior code in the sense of optimizing rates of evolution? What is the simplest code or description? We shall have no idea of what type of experiment to perform until we have some theoretical models which help us decide what questions are empirically testable.

Acknowledgement

This study is a part of work sponsored by the National Science Foundation, General Ecology Program, Grant number GB 16563.

References

[1] Conrad M. and H.H. Pattee, J. Theoret. Biol. (1970) in press.
[2] Dobzhansky T., Genetics and the Origin of Species (Columbia University Press, New York, (1951) p. 20.

[3] Kauffman S.A., J. Theoret. Biol. 22 (1969) 437.
[4] Von Neumann J., in: Theory of Self-Reproducing Automata, ed. A.W. Burks, (University Illinois Press, Urbana, Fifth Lecture, 1966) p. 74.
[5] Orgel L.E., J. Mol. Biol. 38 (1968) 381.
[6] Pattee H.H., in: The Origin of Prebiological Systems ed. S. Fox (Academic Press, New York, 1965) p. 400.
[7] Pattee H.H., in: Hierarchical Structures eds. L.L. Whyte, A.G. Wilson and D. Wilson (American Elsevier, New York, 1969) p. 179.
[8] Pattee H.H., in: Biological Hierarchies, ed. H. Pattee (Gordon and Breach, New York, 1970) in press.
[9] Pattee H.H., in: Textbook of Mathematical Biology, ed. R. Rosen (Academic Press New York, 1970) in press.

Chemical Evolution and the Origin of Life, eds. R. Buvet and C. Ponnamperuma

ENERGETICAL CONTINUITY BETWEEN PRESENT-DAY AND PRIMEVAL SYNTHESES OF BIOLOGICAL COMPOUNDS

René BUVET, Elisabeth ETAIX, Francine GODIN,
Philippe LEDUC and Louis LE PORT

Laboratoire d'Energétique Electrochimique de la Faculté des Sciences de Paris, 10, rue Vauquelin, Paris 5ème, France

Many experimental studies have shown that the formation of biological compounds could have proceeded spontaneously under different kinds of circumstances frequently encountered in the primeval terrestrial environment. But in fact, most published reports of these experiments mention hardly anything about the origin of "life" itself, because the essential character of life does not only reflect the kind of molecules from which living systems are built, but the way of energy utilization, mainly electromagnetic, along which these biological molecules are formed. This essence of life is schematically represented by the different tables of metabolic pathways which have recently been assembled [18]. As for investigations on the origin of life, the major problem is therefore to define first how syntheses of biologically interesting compounds, *obeying the same principles as present-day biological syntheses – but feasible with the only reagents, intermediates and, if necessary, structures available in primeval environment* – have come into existence. In other words, what are the original and simplest forms, the archetypes, of the metabolic processes. Afterwards, it becomes necessary to understand how and why these archetypes have evolved to present-day status.

It is only when archetypes of some present-day biosyntheses require the presence of a given component whose formation cannot be accounted for by an archetype of its own present-day biosynthesis that hypotheses concerning the occurrence of volcanoes, shock waves, radioactivity, lightning, etc., have to be taken into consideration for justifying the presence of this component at the time life originated.

With this view-point in mind, this paper deals with a search for the most evident conditions which govern the occurrence of syntheses as closely akin to present-day biosyntheses as possible, but as far as possible excluding the presence of elaborated reagents, catalysts or structures, whose formation should be justified separately if they are deemed necessary.

Since one of the most general features of biosyntheses lies in their thermodynamic character, mainly the fact that they proceed endergonically through energy transfer, and since some general thermodynamic laws of evolution seem to govern the evolution of energetically open systems [9], this discussion will be entirely developed on energetical bases. Actually, it will consist mainly of an analysis of energy balances of elementary actions occurring on simplified templates of present-day biosyntheses. Although this kind of treatment may be generalized to other kinds of processes, we shall limit its presentation here to the very widely represented bioreactions involving endergonic condensations in aqueous media, as schematically represented below:

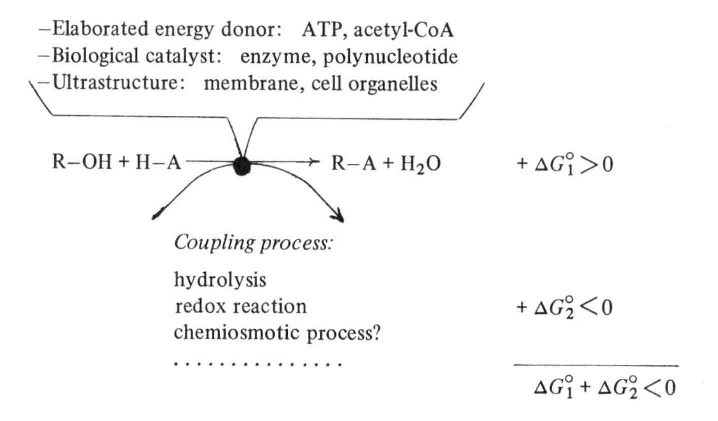

Such reactions correspond mainly to the formation of bio-oligomers and polymers; the coupling reaction most frequently met is the hydrolysis of an energy-rich condensate: e.g. in peptide syntheses, transphosphorylations, transacylations. Sometimes, the coupling process is a redox reaction, e.g. in oxidative phosphorylations or thiolesterifications. We shall not deal with these oxidative couplings in the main part of this paper but will devote it essentially to a thermodynamic discussion of the conditions driving the transdehydration processes.

As regards the definition of archetypes of such biochemical transdehydrations, we shall examine the extent to which the same kind of energy transfer process is possible between the same kinds of active chemical groups, without: elaborated energy donors, i.e. with inorganic polyphosphates instead of ATP, or with simple thiolesters instead of acyl-coenzyme A; polymeric catalysts, i.e. without any catalyst or in the presence of the only metallic cations available under primeval conditions; more or less elaborated structures, i.e. in homogeneous aqueous solutions or in the presence of simple precipitates of spontaneously occurring materials, as for instance suspensions of minerals [20].

We shall first recall [4] how the energy balance of any condensation reaction in dilute aqueous solutions at buffered pH can be graphically represented. In fig. 1, the free enthalpy change for the formation of ethyl acetate from acetic acid and ethyl alcohol is plotted as a function of pH. Below pH 4.8, i.e. pK_a for acetic acid, the hydrolysis of the ester yields protonized acetic acid and the material balance of the reaction does not involve protons:

$$CH_3COOH + HOC_2H_5 \rightleftharpoons CH_3COOC_2H_5 + H_2O \qquad (1)$$

consequently the energy balance does not depend on pH. Above pH 4.8, the

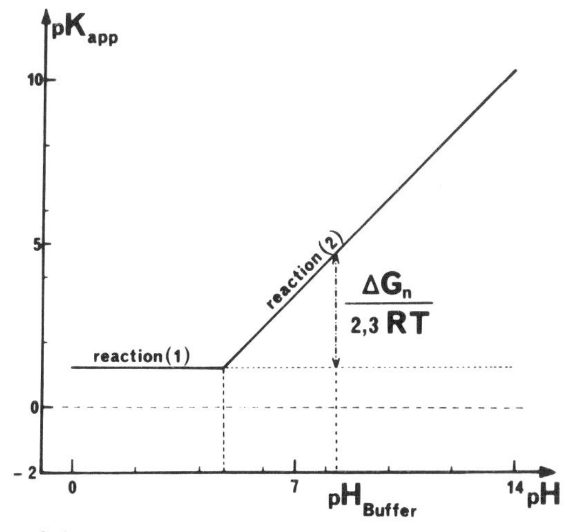

Fig. 1. Variation of the apparent equilibrium constant for the formation of ethyl acetate in aqueous solution from acetic acid and ethanol:

$$CH_3COOH + C_2H_5OH \rightleftharpoons CH_3COOC_2H_5 + H_2O\ H_2O \qquad (1)$$

$$CH_3COO^- + H^+ + C_2H_5OH \rightleftharpoons CH_3COOC_2H_5 + H_2O \qquad (2)$$

ΔG_n = free enthalpy change involved in the complete neutralization of 1 mole of molar aqueous solution of acetic acid in buffered solution at pH = pH_{buffer}.

predominant species of acetic acid is the acetate ion and the balance must be written:

$$CH_3COO^- + H^+ + HOC_2H_5 \rightleftharpoons CH_3COOC_2H_5 + H_2O \qquad (2)$$

in this case, the energy balance depends on the chemical potential of the proton, i.e. the pH. At any pH, the esterification equilibrium obeys the mass action law:

$$\frac{c_{CH_3COOC_2H_5}}{c_{CH_3COOH} \times c_{C_2H_5OH}} = K_{\text{apparent}}$$

where the c represents the *total* molarity of each component, in any of its protonation state, respectively:

$$c_{CH_3COOC_2H_5} = |CH_3COOC_2H_5|$$

$$c_{C_2H_5OH} = |C_2H_5OH|$$

$$c_{CH_3COOH} = |CH_3COOH| + |CH_3COO^-|$$

(where the bars refer to the *real* concentrations of species), since ethyl acetate and ethyl alcohol have no acid-base properties in dilute aqueous media. Consequently, below pH 4.8 the law of mass action becomes :

$$\frac{|CH_3COOC_2H_5|}{|CH_3COOH| \times |C_2H_5OH|} = K_{\text{app}} = K_c$$

where K_c is the actual pH-independent equilibrium constant, valid here only in the pH range 0–4.8. The corresponding standard free enthalpy change of condensation is therefore pH-independent:

$$\Delta G^\circ = 2.3\, RT\, \mathrm{p}K_c$$

Above pH 4.8, due to the protolysis of acetic acid, K_{app} varies with pH according to:

$$\frac{|CH_3COOC_2H_5|}{|CH_3COO^-| \times |C_2H_5OH|} = K_{\text{app}} = \frac{K_c}{K_a} |H^+|$$

where K_a is the acidity constant of acetic acid. The over-all apparent free enthalpy change for the ester linkage formation involves the neutralization standard free enthalpy change of acetic acid, in this pH range, and increases by $2.3RT$ per unit pH:

$$\Delta G^{\circ\prime} = 2.3RT\, \mathrm{p}K_{\text{app}} = \Delta G^\circ + 2.3RT(\mathrm{pH} - \mathrm{p}K_a)$$

Fig. 2, compiled from [14], gives the variations of these apparent standard

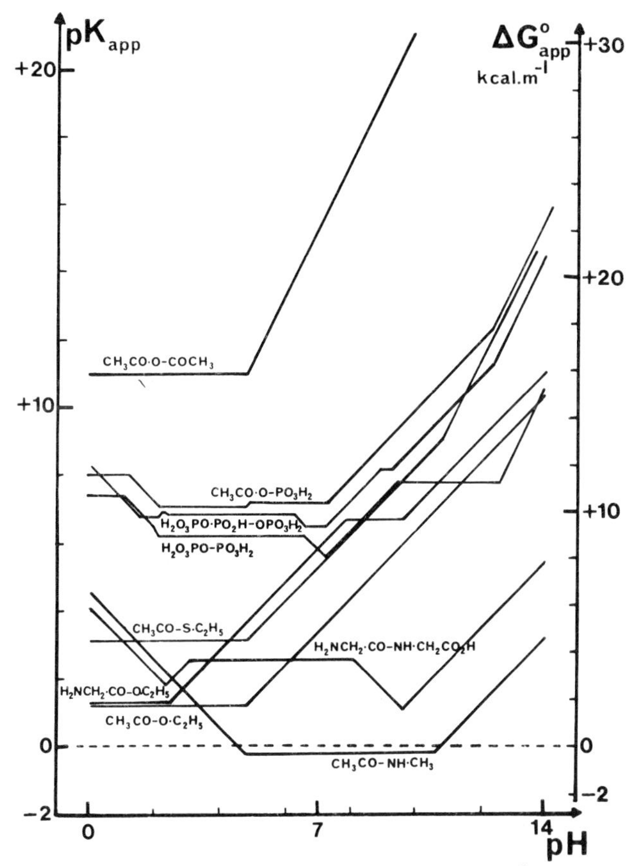

Fig. 2. pK_{app} and apparent standard free enthalpy changes ΔG^{o}_{app} vs. pH for the formation by condensation in aqueous solution from their hydrolysed components of: acetic anhydride, acetyl-phosphate, tripolyphosphate, pyrophosphate, thiolethylacetate, diglycine, ethyl glycinate, ethyl acetate and *N*-methylacetamide. Neutral forms of each condensate are mentioned on corresponding curves.

free enthalpy changes with pH, for a number of reactions in which condensates are formed from their corresponding hydrolysed components. Each curve possesses the same number of slope changes as the total number of acid functions of all the species involved in the reaction. All molecules considered in this diagram are simple enough to be compatible with the primeval conditions, and they represent the active groups occurring in different kinds of biological condensations–hydrolysis.

With the help of these diagrams, let us now review some more or less obvious conditions which govern the feasibility of transdehydration processes.

First, fig. 3 visualises the possibility of achieving the "simple" transfer, to a receptor, of one moiety of a condensed donor after hydrolysis, e.g. the formation of a peptide bond from the energy of hydrolysis of esters:

$$CH_3COOC_2H_5 + H_2O \rightleftharpoons \text{acetic acid} + C_2H_5OH \qquad -\Delta G^{\circ\prime}_{\text{ester}}$$

$$\text{acetic acid} + \text{methylamine} \rightleftharpoons CH_3CONHCH_3 + H_2O \qquad +\Delta G^{\circ\prime}_{\text{amide}}$$

The requisite condition for this transfer is that the algebraic sum $(\Delta G^{\circ\prime}_{\text{amide}} - \Delta G^{\circ\prime}_{\text{ester}})$ of the over-all transfer reaction must be negative. As a rule, we shall always consider that $\Delta G^{\circ\prime}$ represents the energy balance, mostly positive, of reactions proceeding towards condensation, thus affecting the minus sign for the reverse hydrolysis reactions. The above acetyl transfer is possible, from these data, with a good yield only above pH 4, which is approximately the abscissa of the intersection of both curves. The main point is that, if such transfer is possible, it is due to the integration of the neutralisation energy of acetic acid into the energy balance. The same

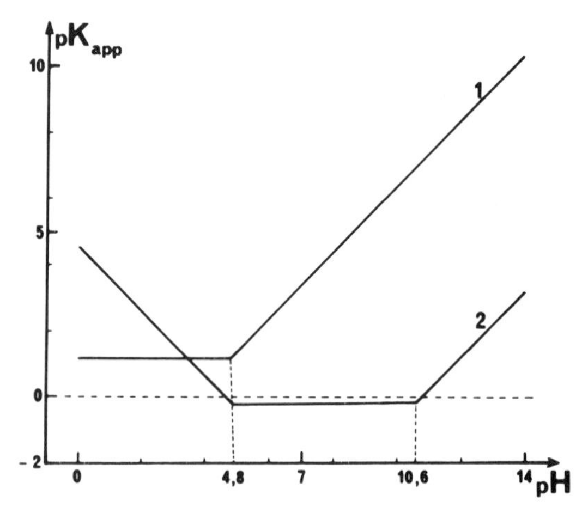

Fig. 3. Graphical representation of pH conditions required for negative energy balance of simple transfer:

$$CH_3COOC_2H_5 + \text{methylamine} \rightleftharpoons CH_3CONHCH_3 + C_2H_5OH$$

(1) Variation of pK_{app} for the formation of ethyl acetate from its hydrolysed components; (2) variation of pK_{app} for the formation of *N*-methylacetamide from its hydrolysed components.

considerations apply to the formation of oligopeptides from the corresponding esters or thiolesters.

Fig. 4 refers to the case of "double" transfers; reactions such as:

$$AB + A' + B' \rightleftharpoons A'B' + A + B \qquad (3)$$

where AB and A′B′ are more or less energy-rich, condensed molecules, can never be performed in a one-step mechanism because of their trimolecular character. So, reaction (3) has to be split into two simple transfers, i.e.:

$$AB + B' \rightleftharpoons AB' + B$$

$$AB' + A' \rightleftharpoons A'B' + A$$

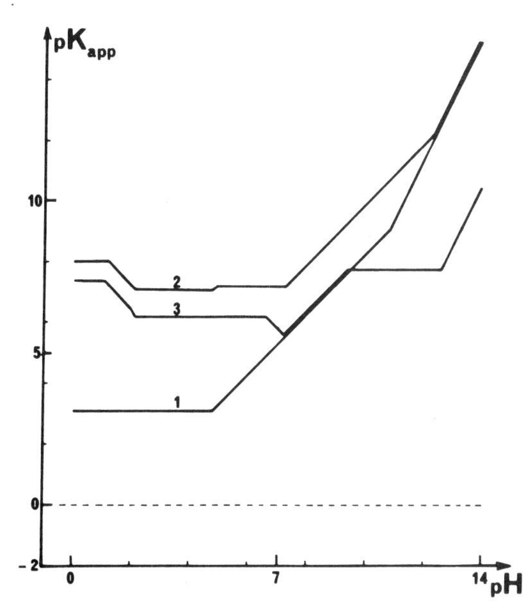

Fig. 4. Graphical representation of the pH conditions required for negative or null energy balance of the two steps of the double transfer:

$$\text{thiolethylacetate} + 2\,P_i \rightleftharpoons P_iP_i + \text{acetic acid} + \text{ethanethiol}.$$

(1) variation of pK_{app} for the formation of thiolethylacetate from its hydrolysed components; (2) variation of pK_{app} for the formation of acetylphosphate from its hydrolysed components; (3) variation of pK_{app} for the formation of pyrophosphate from its hydrolysed components. The first step is possible with null energy balance, only above pH 12 (curves 1 and 2). The second step has a negative energy balance at any pH (curves 2 and 3). Consequently the double transfer is possible only above pH 12, although the comparison of curves 1 and 3 gives an overall null or negative energy balance above pH 7.

and the above conditions must be valid for both steps of this mechanism. Such double transfer reactions may correspond to the formation of polyphosphoric bonds from thiolesters in alkaline aqueous media, e.g.

$$AcSC_2H_5 + P_i \rightleftharpoons AcP_i + HSC_2H_5$$

$$AcP_i + P_i \rightleftharpoons P_iP_i + AcOH$$

The first step occurs, above pH 12, with a global standard free enthalpy change near zero, which corresponds to a low-yield transfer.

A second set of pH-dependent effects is relative to conditions which might facilitate transfer reactions. They correspond to the possible presence of components, for example cations, which can form complexes with anionic donors and substrates. This new possibility entails two effects: readiness of contacts between reagents involved in transfer reactions, due to the association of two or three anionic ligands on one single cation, which implies changes in the kinetics of the possible transfers between these ligands; alteration of the energetics of transfer reactions which then take into account the standard free enthalpy changes for the formation of the complexes involved.

It is well known, qualitatively, [6] that the basic properties of anionic ligands and the acidic properties of metal cations limit the formation of complexes between both to a pH range more or less extended towards low and high pH values respectively.

As for the effects related to the energy balance of transfer processes, in the acid range where the complexes are formed, the relationships between the apparent free enthalpy changes of hydrolysis-condensation reactions and pH are modified. Unfortunately, experimental data do not permit quantitative computation of such apparent standard free enthalpy changes involving both pH and concentration of cations, e.g. Mg^{2+}, Ca^{2+}, Mn^{2+}, which occur in transdehydration processes.

Thus, in the case of a double transfer from thiolester leading to a polyphosphoric linkage, one can only qualitatively surmise that the reaction yield must change markedly in the presence of cations such as Mg^{2+} or Ca^{2+}. Since these cations give no stable complexes with the species involved in the thiolester condensation, they have no effect on the energy balance of this condensation. On the other hand, complexes of these cations are formed with the phosphate species and they are fairly stable for polyphosphoric species; the energy requirement for the formation of pyrophosphate from two orthophosphates is strongly decreased in the presence of an excess of Mg^{2+} ions [13]. Unfortunately, nothing can be inferred about the energy balance of the formation of acylphosphate, the intermediate step in the double transfer.

A third set of conditions directing transfer processes involving pH is related to the hydrolysis of the energy donor. The hydrolysis of almost any condensed species is an exergonic process, but the kinetics of such hydrolysis is generally infinitely slow in pure water. The half-life of condensates is often considerably shortened either by proton catalysis in acidic solutions, or by hydroxyl catalysis in basic solutions.

The pH-dependence of the hydrolysis of simple esters or polyphosphoric species, in aqueous solutions containing no other hydrolysis catalyst, has been studied quantitatively [2, 11, 21] and some hypotheses have been suggested for the mechanism of catalysis involving H^+ or OH^- ions [10]. Other simple hydrolytic catalysts are known, such as certain cations or metal hydroxides, not to mention the enzymes.

The results schematically given in fig. 5 allow us to discuss some possible interferences between the simple hydrolyses of esters and amides and the transfer reaction in the specific case of peptide bond formation from esters. A first criterion, which governs the feasibility of the transfer, is that the condensate produced must be metastable in the medium where it is synthe-

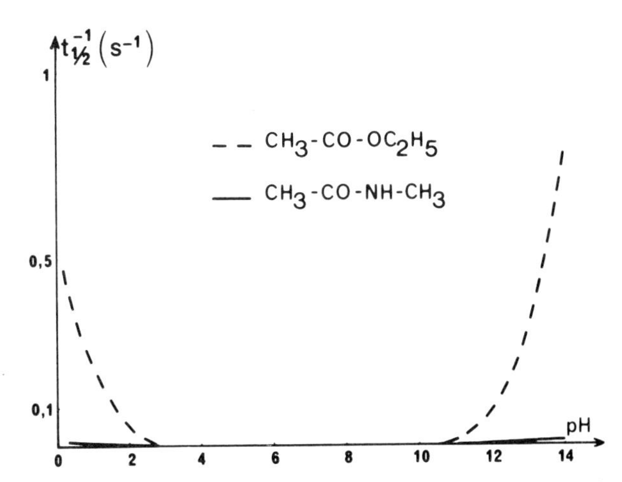

Fig. 5. Analysis of the kinetic conditions arising from simple hydrolyses in the case of the transfer reaction:

$$CH_3COOC_2H_5 + \text{methylamine} \rightleftharpoons CH_3CONHCH_3 + C_2H_5OH$$

The curves schematically give half-life variations of the two reacting condensates against pH. *1st condition:* $CH_3CONHCH_3$ metastable at any pH; *2nd condition:* $CH_3COOC_2H_5$ metastable at $3 < pH < 10$ or (?) $CH_3COOC_2H_5$ rapidly hydrolysed at $pH < 3$ or $pH > 10$. The energy balance permits the transfer only above pH 4.

sized. In the case of peptide bond formation, the half-life of the amide is sufficiently long at any pH to permit its observation as a metastable compound. Other conditions must be defined relative to the metastability of energy donors.

According to classical chemistry, we may consider trying to avoid any side-reaction leading to direct hydrolysis of the donor, i.e. by choosing experimental conditions in which the donor is metastable, that is between pH 3 and 10 and in the absence of any other hydrolytic catalysts when an ester linkage is used as a donor. Nevertheless, such a choice must be more closely examined.

Different kinds of qualitative arguments suggest that, contrary to such apparently obvious conclusion, in order to obtain transfer it may be necessary to implement the conditions in which the energy donor is rapidly hydrolysed. As far as mechanisms are concerned, this is the case if, during hydrolysis, an intermediate is produced which can attack the acceptor molecule, by any classical elementary step, more rapidly than it can be converted into the hydrolysed products. But such a situation may be clarified by considering the general thermodynamic evolution principles [9] which imply qualitatively that the relaxation of chemical energy in such systems should be as slow as possible. Consequently if hydrolysis is carried out in the presence of an acceptor which may, through simple transfer, give a metastable new condensate, this reaction path must be used, because it generally corresponds to a much slower relaxation of chemical energy, i.e. it involves storage of a part of the energy supplied by the donor.

At present, neither systematic nor dispersed experimental study enables us to choose between these two possibilities for the kinetics of the donor hydrolyses. We shall present an experimental contribution to this discussion in another paper during this conference [17].

To conclude this search for general criteria which govern the feasibility of transdehydration processes, it seems that without involving the pre-existence of complex biological donors, catalysts or structures, some directing conditions of such processes may be found. Taking as clues the selection against pH, many transdehydrations involving simple molecules assembled only with the active groups of biological condensates or substrates, seem to be feasible under conditions compatible with the primeval environment. Nevertheless no systematic attempt has been made to define experimentally on these grounds to what extent these criteria are sufficient.

Should the case arise – even for only a part of present day metabolic pathways sufficient to set up a complete chart of energy relaxation paths in primeval conditions – the problem of the origin of life would appear on renewed bases. The archetypic syntheses of prototypes of present-day biomolecules should be considered as the very beginning of life processes, the complexity and the efficiency of the primordial metabolic chart being further increased through a re-insertion, as intermediates, catalysts or structure components, of some of the compounds first produced.

Within such a frame, the major problem which now arises is the primeval production of the most energy-rich donors using redox energy. Some hypotheses consider as prototypes of energy donors, unsaturated compounds such as cyanate, cyanamide or dicyandiamide, which may have been produced in the first steps of atmospheric reactions [25,20]. C.N. Matthews proposes a mechanism for the formation of proteins, intermediate between the latter and the present-day one, short-circuiting the formation of free amino acids [19]. A. Katchalsky and coworkers obtained the same result from artificially synthesized aminoacyl phosphates or adenylates [15,16]. Other authors have assumed that inorganic polyphosphates obtained through heating minerals on primitive earth may have played the role of energy donors [22]. S.W. Fox proposed to build proteinoids directly from free amino acids [8] under similar conditions involving the use of thermal energy to produce endergonic condensations at ambiant temperatures. But these hypotheses will never remove the need for archetypes of present-day processes effecting the endergonic elimination of water between condensable substrates from redox energy in aqueous solutions. Experimental indications have been put forward [1,3] but we observed that in fact the proposed oxidation of ferrous iron [3] or manganous ion by hydrogen peroxide in the presence of orthophosphates produces no detectable quantities of pyrophosphate at any pH [7]. The discussion presented here for transdehydration may be extended to oxidative dehydrations, taking into account recent data connected to electron-proton electrochemical mechanisms in redox chemistry [12] and graphical representations of variations of redox potentials with acidity, similar to Carpenter diagrams [5,23,26]. But as yet there are no experiments which allow this theoretical frame to be exploited.

On the contrary, some of the results we have cited [15,16,20,25] and those we will present elsewhere during this conference [17] together with the theoretical considerations we have just reported, indicate, mainly when taking into account the basicity of primeval terrestrial periphery, that a number of transdehydration processes must have proceeded under primitive conditions before any compounds or biological structures now participating in such processes were formed.

References

[1] Barltrop J.A., P.W. Grubb and B. Hesp, Nature 199 (1963) 759.
[2] Bunton C.A., C.O'Connor and T.A. Turney, Chem. Ind. (1967) 1835.
[3] Calvin M., Chemical Evolution (Oxford University Press, 1969) p. 165.
[4] Carpenter F.H., J. Am. Chem. Soc. 82 (1960) 1111.
[5] Charlot G., Chimie Analytique Générale, Vol. 1 (Masson, Paris, 1967) chapter V, pp. 60–70.
[6] Charlot G., Chimie Analytique Générale, Vol. 1 (Masson, Paris, 1967), chapter VII, pp. 75–81.
[7] Etaix E., unpublished data.
[8] Fox S.W., Nature 205 (1965) 328.
[9] Glansdorff P. and I. Prigogine, Physica 30 (1964) 351.
[10] Gould E.S., Mechanism and Structure in Organic Chemistry (Holt, Reinehart and Winston 1969) pp. 315–332.
[11] Griffith E.J., Ind. Engr. Chem. 51 (1959) 240.
[12] Jacq J., Electrochem. Acta 12 (1967) 1345.
[13] Jencks W.P., in: Handbook of Biochemistry, ed. H.A. Sober (Chemical Rubber Co., Ohio) pp. J-144, J-149.
[14] Jencks W.P. and J. Regenstein, in: Handbook of Biochemistry, ed. H.A. Sober (Chemical Rubber Co., Ohio) pp. J-150, J-189.
[15] Katchalsky A. and M. Paecht, J. Am. Chem. Soc. 76 (1954) 6042.
[16] Katchalsky A., Paper presented at the Third International Biophysics Congress, Cambridge (Mass.), 1969.
[17] Le Port L., E. Etaix, F. Godin, P. Leduc and R. Buvet, this volume, p. 197.
[18] Marre E., G. Forti, S. Cocucci, B. Ferrini, B. Elviri and G. Michal, in: Handbook of Biochemistry, ed. H.A. Sober (1968) back cover.
[19] Matthews C.N. and R.E. Moser, Proc. Natl. Acad. Sci. U.S. 56 (1966) 1087.
[20] Miller S.L. and M. Parris, Nature 204 (1964) 1248.
[21] O'Connor C. and T.A. Turney, Chem. Ind. (1968) 1697.
[22] Ponnamperuma C. and A. Schwartz, Nature 218 (1968) 443.
[23] Pourbaix M., Atlas d'Equilibres Thermodynamiques (Gauthiers-Villars, 1963).
[24] Rabinowitz J., J. Flores, R. Krebsbach and G. Rogers, Nature 224 (1969) 795.
[25] Steinman G., D.H. Kenyon and M. Calvin, Nature 206 (1965) 706.
[26] Wurmser R., Traité Biochimie Générale (Masson, Paris, 1959).

Chemical Evolution and the Origin of Life, eds. R. Buvet and C. Ponnamperuma

UNSUCCESSFUL ATTEMPTS OF ASYMMETRIC SYNTHESIS UNDER THE INFLUENCE OF OPTICALLY ACTIVE QUARTZ. SOME COMMENTS ABOUT THE POSSIBLE ORIGIN OF THE DISSYMMETRY OF LIFE

Annie and Henri AMARIGLIO
Centre de 1er Cycle, Bd des Aiguillettes,
54 - Nancy, France

Some authors thought that the dissymmetry of organic substance might have arisen primarily, before life appeared on the earth, under the influence of physical inorganic factors. Bernal [8] thus put forward the hypothesis that optically active quartz might have been such an asymmetric agent.

Optically active quartz as a dissymmetric catalyst

Ostromisslensky [39] was the first (in 1908) to suggest that dissymmetric crystals might be used as stereospecific catalysts.

In 1932 and 1934, Schwab and Rudolph [43,44], published the first experimental results obtained in such a way. Their experiments have been resumed and extended by Terentjew and Klabunowski [27,50,51] since 1947. According to these authors, optically active quartz used as a catalyst (bare or metal covered) would be able to induce dissymmetry in a chemical reaction. Other instances are known [7,10,11,25,27,29,35,41,46,48,52,53] – particularly, asymmetric adsorptions – showing the stereospecific ability of this mineral. But several authors failed in attempting new experiments [23,26,35,47,53] or trying to reproduce previous work [36,45]. In a review already published about this subject [2], it was shown that this mineral, even with regard to the successful results claimed, never induced more than a very small optical activation of the reaction products. In fact, we thought the experimental conditions used by the previous authors could be improved in several ways, particularly with respect to the catalyst preparation. This remark prompted us to resume some of the previous work – with the hope of improving the results. We are now going to give some information about our experiments, the main details of which have already been published [3,6].

We used a photoelectric polarimeter, characterized by its high sensitivity ($10^{-3\circ}$), its direct reading of rotatory powers on a galvanometer-scale and its recording allowance. Thus we could exclude some sources of errors and non-reproductiveness, able to disturb the measurements performed with classical polarimeters. Optically active quartz was first crushed into minute pieces and then, for some experiments, carefully washed with HF. Quartz was used as a catalyst either pure or metal-covered. Mixed catalysts were prepared by impregnating quartz with a solution of a Ni, Cu or Pt salt and then dry-freezing, according to a method we have already been using in our laboratory and which leads to very active catalysts [1]. We could thus study, at fairly low temperatures, three reactions: the hydrogenation of methylethyl ketone and the dehydration or dehydrogenation of 2-butanol [3,I]. On the other hand, we attempted to resolve various racemic modifications by liquid–solid [3,II] or gas–solid chromatography [6] on optically active quartz columns.

Numerous attempts under carefully controlled conditions always led to negative results but, in some experiments, we could observe *apparent* optical rotations. This extraneous effect was due to minute quartz particles carried away with the reaction products, but had nothing to do with the rotatory power of quartz. In fact, a suspension of particles of an optically inactive matter (such as crushed glass, for instance) becomes doubly refracting under the influence of an orientation factor (here, gravity). This effect was recognized and studied at the beginning of the century [11,34] but was at once forgotten. However, in 1935, it vitiated conclusions of experimental studies [30–33] and it was again given attention. Nevertheless, only Heller [22], in 1942, ever talked about this bias. The failure to observe any asymmetric effect in our experiments led us to a critical analysis of previous work where small, but positive, effects were claimed [3,III]. We concluded that all the measurements might have been vitiated by this extraneous effect of which the authors were not aware. This artefact might have been all the more important as all the measurements were performed at the limit of sensitivity of the polarimeters used. We shall not talk about some important comments which make a number of results inconsistent or unlike. Let us say only that some of them, equally claimed successful, contradict themselves as to the sign of the enantiomer preferentially adsorbed by a given-handed quartz.

The analysis of these failures led us to try and determine clearly the conditions required for a reaction to be stereospecific [4] and we were able to show why optically active quartz is unsuccessful in transmitting its dissymmetry [5]. So, we were prompted to doubt the ability of an asymmetric adsorption on optically active quartz surface, except perhaps for complicated enough molecules such as B_{12} vitamin used by Seifert [44] for his epitaxy experiments. Therefore asymmetric reactions, despite the fact they are talked about in many books and treatises * – only qualitatively and without any critical point of view – are indeed a wager.

Failures of the various hypotheses about the abiogenic origin of the dissymmetry of life

Bernal's hypothesis seems unconvincing in explaining the origin of the dissymmetry of life. The other hypotheses about this subject fall into two categories, according to whether they consider the primary emergence of dissymetry before that of life, or they make life itself accountable for dissymetry. On the one hand, if there had been, during the prebiological epoch, everywhere on the earth's surface, a dissymmetric factor, one cannot conceive why it does not still manifest itself; this does not agree with the fact that laboratory syntheses always produce a racemic modification whether they start from symmetric molecules or a racemic modification and use no optically active reagents or catalysts and no asymmetric physical influence. On the other hand, the only hypothesis of this sort that cannot entirely be dismissed is that of Byk [9] which explains life's dissymmetry by the influence of circularly-polarized light. According to this author, the partially polarized light of the sky becomes partially right-circularly-polarized when reflected by water, under the influence of the terrestrial magnetic field. However, laboratory attempts to perform asymmetric syntheses under the influence of circularly-polarized light were disappointing and led only to very low optical activations in even the most favourable instances. ** Nevertheless, the experimental conditions were chosen as favourable as possible for obtaining stereospecificity: the initial compound was dichroic for the wavelength of the monochromatic light used which was also situated inside the photochemical activation fields of the reaction considered.

The weakness of these guesses led authors to suppose that dissymmetry might have appeared, not during the prebiological period, but during the biological period, and that life itself was responsible for setting up dissymmetry. It has been pointed out that, even in symmetric conditions which ought to lead only to a racemic modification, there may be, in a transient fashion and in a given place, a slight excess of molecules of one sign due to

* For some samples, see [2] page 21.
** For more details, see [2].

fluctuations. Effectively, in the limited case where only one asymmetric molecule appears this molecule must be right or left. For instance, in some crystallization experiments where there were only a few seeds, the sign of the first determined the sign of the whole crystal [21]. The characteristics of living organisms (self-reproduction, adaptation to life conditions, self-production of substances the organism needs, natural selection) must then be used to explain how dissymmetry evolves.

We should like to define this way of tackling the problem by a few comments which show that dissymmetry could have emerged during the biological period.

Life itself accountable for the dissymmetry of living matter

As early as 1924, Oparin [37] put forward the hypothesis, now universally accepted as valid, that the first living beings were heterotrophs whose appearance and growth used the ready-made organic substances that had gathered in the primary oceans during the prebiological period. First, we must notice the important fact – which does not seem to have been clearly expressed till now – that the natural synthesis of the macromolecules necessary for the emergence of the first living being (for instance, proteins) might have happened in initially symmetric surroundings. Let us indeed consider such an organic storehouse, constituted only of symmetric molecules and racemic modifications. If we imagine the condensation of two molecules of an amino acid possessing an asymmetric C-atom, the dipeptide LL will not have the same probability of appearance as the dipeptide LD because LL and LD are no longer enantiomers, but diastereoisomers. Ellenbogen [14] especially studied the effectively different properties of such diastereomers. Thus one conceives that peptide chains which form in such symmetric surroundings may be optically pure – that is wholly constituted by L monomeric units (or D monomeric units) – rather than mixed. A mechanism was proposed [12,38] which explains why, once the polymer starts out as active, it continues to grow by the addition of monomer units of only one sign, and some experimental results support this point of view. So, in 1956, when Price and Osgan [42] polymerized (±)propylene oxide with ferric chloride as a catalyst, the resulting polymer could be separated in two fractions, one partly crystalline (isotactic) and the other partly amorphous (atactic). The crystalline fraction included macromolecules constituted only by either a succession of D monomeric units or of L monomeric units. (Of course, since there is an equal chance of starting a polymer chain either from a (+)oxide or a (–)oxide molecule, there will be statistically equal numbers of chains of opposite configurations and the resulting polymer will be racemic). Likewise, in 1960, when Pino, Lorenzi and Lardicci [40] polymerized (±)4-methyl hexene, they obtained macromolecules, each of which were made up of one-handed monomer units. Besides, the experiments of Idelson and Blout [24], and of Farina, Natta and Bressan [15] show that some stereospecific polymerizations may be autocatalytic.

It is thus feasible to think that surroundings which were on the whole symmetric did not hinder the first living germs from appearing; so we can imagine, for instance, that there were coacervates [37] or microspheres [18,19] of both signs. Let us now consider the appearance of the first living germ, or at the least, of a primary structure able to catalyse its own formation from the surrounding materials. Therefore, as soon as this germ appears, a way of formation of similar germs, the much more efficient than that of spontaneous generation, also arises. Now, if we consider the growth of the population of all the germs, those formed by self-reproduction become far more numerous than those due to the spontaneous generation. This is all the more true because self-reproduction is more efficient. But since there are only a few parent-germs, one sign must predominate because the formation of a racemic modification results statistically where large numbers of molecules are involved and is an exception when there are only a few units synthesized. The absence of any dissymmetric influence only induces for the preferential sign to be merely stochastic. As soon as a primary dissymmetry appears, it must increase in very large proportions because of the autocatalytic production of germs. However the storehouse, initially racemic, is able to induce the formation of enantiomeric germs; it may be asked why an opposite-handed life would not have developed (even later or more slowly).

This question can be answered by two ways which are not exclusive:

1) Racemization is a thermodynamically favoured transformation. We may then conceive that racemization factors (for instance, catalysts, heating, radiations) exerted themselves upon the organic storehouse so that the enantiomers preferentially used for the life processes could be, even partly, regenerated. Such an explanation has already been used for Havinga's experiments [21] which concern the crystallization of a solution, in chloroform, of racemic methyl-ethyl-allyl-aniline iodide. Because this compound easily racemized in solution and also because there were only a very few germs, the author could obtain a one-handed crystal.

2) The living-germ development being autocatalytic, we may consider that, very quickly, at every place, it was limited by the speed with which the organic molecules could reappear (after a chemical reaction at the same place,

or after a migration from a richer one). Consequently, the later appearance of enantiomeric germs took place in surroundings which were poorer in organic matter. For growth, spread and reproduction, germs of both signs compete since they both need some symmetric substance (such as glycocolle or succinic acid), they will then use them up in proportion to their respective populations. So, the first species to appear will keep (and enlarge) its preponderance, even if the absolute speed of its population-growth decreases.

The dissymmetry thus obtained by the heterotrophic living germs would then have been transmitted to the autotrophic germs that followed them and the predominance of the preferential kind of life would have grown in an autocatalytic way, and so be established decisively.

Conclusion

The comments just made about the possible origin of the dissymmetry of life, are only mere hypotheses. But their advantage lies in the fact they do not suppose the preexistence of any dissymmetric factor and consider the emergence of the dissymmetry as dependent on life itself. These hypotheses allow us to consider (as was already suggested by Fox [17] for instance) that D-amino acids, which only occur in some bacteria, should be a vestige of the early evolutionary stages of life. Let us notice that in this scheme, chance is accountable only for the choice of the preferential configuration of the first germ(s) and it is difficult to imagine anything else if no dissymmetric factor is involved. But from this scheme, it follows that one form of life will necessarily predominate over the other.

References

[1] Amariglio A., H. Amariglio and X. Duval, Compt. Rend. Acad. Sci. 262C (1966) 1227.
[2] Amariglio A., H. Amariglio and X. Duval, Ann. Chim. 3 (1968) 5.
[3] Amariglio A., H. Amariglio and X. Duval, Helv. Chim. Acta 51 (1968) 2110.
[4] Amariglio A., H. Amariglio and X. Duval, Bull. Soc. Chim. France 5 (1969) 1539.
[5] Amariglio A., H. Amariglio and X. Duval, Bull. Soc. Chim. France 5 (1969) 1546.
[6] Amariglio A., Compt. Rend. Acad. Sci. 268C (1969) 1981.
[7] Bailar J.C. and D.F. Peppard, J. Am. Chem. Soc. 62 (1940) 105.
[8] Bernal, J.D., The Physical Basis of Life (Routledge and Kegan Paul, London 1951).
[9] Byk A., Z. Physiol. Chem. 49 (1904) 641.
[10] Busch D.H. and J.C. Bailar, J. Am. Chem. Soc. 75 (1953) 4574; 76 (1954) 5352.

[11] Chaudier J., Compt. Rend. Acad. Sci. 137 (1903) 248; 142 (1906) 201.
[12] Corey E.J., Tetrahedron Letters 2 (1959) 1.
[13] Das Sarma B. and J.C. Bailar, J. Am. Chem. Soc. 77 (1955) 5476.
[14] Ellenbogen E., J. Cellular Comp. Physiol. 47 (1956) Suppl. 1 151.
[15] Farina M., G. Natta and G. Bressan, J. Polymer Sci. PtC 4 (1964) 141.
[16] Fox S.W., J.E. Johnson and A. Vegotsky, Science 124 (1957) 923.
[17] Fox S.W., J. Chem. Ed. 34 (1957) 472.
[18] Fox S.W., G. Krampitz and T.V. Waehneldt, Bild Wiss. 12 (1967) 1014.
[19] Fox S.W., Atomes 23 (1968) (225) 348.
[20] Frank F.C., Biochim. Biophys. Acta 11 (1953) 459.
[21] Havinga E., Chem. Weekblad 38 (1941) 642; Biochim. Biophys. Acta 13 (1954) 171.
[22] Heller W., Rev. Mod. Phys. 14 (1942) 401.
[23] Henderson G.M. and H. Gordon Rule, J. Chem. Soc. 2 (1939) 1568.
[24] Idelson M. and E.R. Blout, J. Am. Chem. Soc. 80 (1958) 2387.
[25] Karagounis G. and G. Coumoulos, Prakt. Akad. Athenon 13 (1938) 414; Nature 142 (1938) 162; Atti. Congr. Intern. Chim. Rome 2 (1938) 278.
[26] Karagounis G., Helv. Chim. Acta 32 (1949) 1840.
[27] Klabunowski Je. I., in: L'Origine de la Vie; Quelques Aspects du Probleme, ed. M. Florkin (Gauthiers-Villars, Paris, 1962), p. 152; Asymmetrische Synthese (VEB deutscher Verlag der Wissenschaften, Berlin, 1963).
[28] Klabunowski Je. I. and W.W. Patrikejew, Dokl. Akad. Nauk SSSR. 78 (1951) 485.
[29] Kuebler J.R. and J.C. Bailar, J. Am. Chem. Soc. 74 (1952) 3535.
[30] Kunz J. and A. Mac Lean, Nature 136 (1935) 795.
[31] Kunz J. and S.H. Babcock, Phil. Mag. 22 (1936) 616.
[32] Kunz J. and R.G. La Baw, Nature 140 (1937) 194.
[33] Ludlam E.B., A.W. Pryde and H. Gordon Rule, Nature 140 (1937) 194.
[34] Meslin G., Compt. Rend. Acad. Sci. 136 (1903) 930; 136 (1903) 1059; 136 (1903) 1305; 136 (1903) 1641.
[35] Nakahara A. and R. Tsuchida, J. Am. Chem. Soc. 76 (1954) 3103.
[36] Ohara M., I. Fujita and T. Kwan, Bull. Chem. Soc. Japan 35 (1962) 2049.
[37] Oparin A.I., The Origin of Life on the Earth (Oliver and Boyd, London, 1957).
[38] Osgan M. and C.C. Price, J. Polymer Sci. 34 (1959) 153.
[39] Ostromisslensky I., Chem. Ber. 41 (1908) 3035.
[40] Pino P., G.P. Lorenzi and L. Lardicci, Chim. Ind. 42 (1960) 712.
[41] Ponomarjew A.A. and W.W. Selenkowa, J. Gen. Chem. USSR (English translation of Z. Obshch. Khim.) 23 (1953) 1544; Dokl. Akad. Nauk SSSR 87 (1952) 423 (Chem. Abstr. 48 (1954) 663).
[42] Price C.C. and M. Osgan, J. Am. Chem. Soc. 78 (1956) 4787.
[43] Schwab G.M. and L. Rudolph, Naturwissenschaften 20 (1932) 363.
[44] Schwab G.M., F. Rost and L. Rudolph, Kolloid-Z. 68 (1934) 157.
[45] Schwab G.M. and B. Wahl, Naturwissenschaften 43 (1956) 513.
[46] Schweitzer G.K. and C.K. Talbott, J. Tenn. Acad. Sci. 25 (1950) 143.
[47] Seifert H., Naturwissenschaften 42 (1955) 13; Z. Elektrochem. 59 (1955) 409; Actes du 2e Congres International de Catalyse, Paris, 1960, Vol. 2 (Editions Technip, Paris, 1961) p. 1897.
[48] Seifert H. and W. Borchardt-Ott, Z. Kristallogr. 122 (1965) 206.

[49] Stankiewicz A., Thesis Köningsberg (1938); results reported by Je. I. Klabunowski in: Asymmetrische Synthese (VEB deutscher Verlag der Wissenschaften, Berlin, 1963) p. 121.

[50] Terentjew A.P., Je.I. Klabunowski and W.W. Patrikejew, Dokl. Akad. Nauk SSSR 74 (1950) 947; Chem. Abstr. 45 (1951) 3798.

[51] Terentjew A.P. and Je.I. Klabunowski, Sbornik Statei po Obshch. Khim. 2 (1953) 1521 (Chem. Abstr. 49 (1955)5262 f); ibid. 2 (1953) 1598 (Chem. Abstr. 49 (1955) 5262 g); ibid. 2 (1953) 1605, 1612 (Chem. Abstr. 49 (1955) 5262 h, 5263 a); Sci. Rep. Moskow State Univ., No. 151; Org. Chem. 8 (1951) 145; in: L'Origine de la Vie; Quelques Aspects du Probleme, ed. M. Florkin (Gauthiers-Villar, Paris, 1962) p. 114.

[52] Tsuchida R., M. Kobayashi and A. Nakamura, Bull. Chem. Soc. Japan 11 (1936) 38; J. Chem. Soc. Japan 56 (1935) 1339 (Chem. Abstr. 30 (1936) 926^8.

[53] Tsuchida R., A. Nakamura and M. Kobayashi, J. Chem. Soc. Japan 56 (1935) 1335 (Chem. Abstr. 30 (1936) 963^5.

Chemical Evolution and the Origin of Life, eds. R. Buvet and C. Ponnamperuma

ORIGIN AND DEVELOPMENT OF OPTICAL ACTIVITY OF BIO-ORGANIC COMPOUNDS ON THE PRIMORDIAL EARTH *

Kaoru HARADA
Institute of Molecular Evolution and Department of Chemistry, University of Miami, Coral Gables, Florida 33134, USA

The configuration of amino acids within proteins is usually assumed to be all L. Some organisms low on the evolutionary scale may contain D-amino acids. However, the formation of these D-amino acids may not be genetically controlled but is probably derived from L-amino acids by epimerization reactions.

Most of the biologically important sugars found in nature are in the D-configuration.

The enzyme is a kind or organic catalyst composed of protein. Enzymes have specific secondary and tertiary structures. If the higher structure of an enzyme is destroyed, the enzyme shows less catalytic activity unless the higher structure is reconstructed. The dissymmetric building blocks of an enzyme usually confer stereospecificity on the reaction catalyzed. For example, fumaric acid is converted to L-malic acid or L-aspartic acid, as catalyzed by fumarase and aspartase, respectively. Such stereospecificity of reaction is the most characteristic in the chemistry of living bodies. This special character in the chemistry of living bodies could be considered a reflection of highly ordered organization which arose as a result of chemical and biological evolution.

Evolutionary view of chemical substances

The earth is believed to have been constructed from proto-planetary nebulae about 4.5 thousand million years ago. The oldest life or prelife is dated at 3.1 thousand million years. Therefore, organic compounds important

* A similar paper appeared in Naturwissenschaften 57 (1970) 114–119.

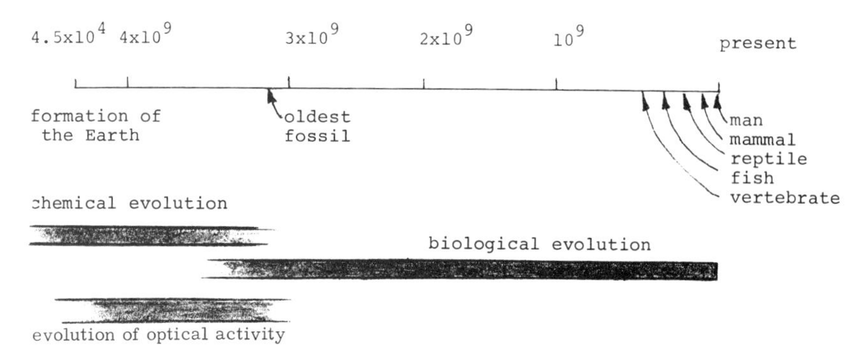

Fig. 1. Evolutionary view.

in the development of life would have formed abiotically on the primitive earth between 3.1 to 4.5 thousand million years ago, probably during the early part of this period.

The formation, development, and organization of organic compounds on the primitive earth can be viewed in the general context of the evolution of chemical substances. The origin of life would be the ultimate stage of chemical evolution (fig. 1). Therefore, the origin and development of optically active organic molecules on the earth is an important aspect of chemical evolution. The word "life" is very difficult to define; however, "life" is a highly organized and highly ordered system. "Racemic" life might exist in concept, but not in reality. In the later stages of chemical evolution, optically active organic molecules could have appeared on the earth. The appearance of optically active molecules would accelerate the organization of molecules and the further development of optical activity.

Definition of the "origin of optical activity" is difficult; however, let us define it here for convenience. The origin of optical activity could be "the appearance of some amount of optically active organic compounds on some limited area on the primitive earth." The origin of optical activity might not be a single process. Several individual or cooperative processes could constitute the origin of optical activity. I would like to discuss some of the experimental results which may be related to the origins of optical activity.

Circularly polarized light

The dissymmetric molecule could be synthesized under the influence of a dissymmetric moiety or of a dissymmetric physical force.

In the beginning of this century, Van't Hoff [34] suggested a role of circularly polarized light in the synthesis of optically active compounds. Kuhn and Braun [19] were the first to succeed in stereoselective photodestruction by the use of circulary polarized ultraviolet light. Similarly, Tsuchida [31] observed stereoselective photodestruction in the field of metal complex chemistry. Karagunis and Drikos [16] first accomplished an absolute asymmetric synthesis of organic compounds by the use of circularly polarized light. Davis et al. [28] first synthesized a natural organic compound, d-tartaric acid, from diethyl fumarate by oxidation with hydrogen peroxide under the influence of right-handed circularly polarized light.

It has been suggested that partially plane polarized light from the sky could be converted into partially right-handed circularly polarized light when reflected on the surface of the earth under the influence of the magnetic field of the earth [3]. If this is the case, it is interesting that the use of right-handed circulary polarized light resulted in the formation of natural d-(+)-tartaric acid [28]. Certainly, circularly polarized light poses interesting problems theoretically and experimentally in connection with the origin of optical activity.

Dissymmetric crystals

It has been known that there are many optically active crystals in nature. Upon dissolution or melting, some of these crystals lose their optical activity. Optical activity of this group is, therefore, due to the dissymmetric crystalline structure. A typical crystal of this group is quartz which has two crystalline forms, d- and l-quartz. Silica is a major component of the crust of the earth; therefore, it is natural to assume that the dissymmetric quartz crystals may have taken part in the origin of optical activity of organic compounds on the earth.

Tsuchida [31] first observed stereoselective adsorption of cobalt complex on l- and d-quartz. Karagunis and Coumoulos (1938) published the column chromatographic resolution of the chromium complex. Ferroni and Cini [4] succeeded in the partial resolution of beryllium complex by the use of optically active sodium chlorate crystals. Schwab and Rudolph [25] reported stereoselective dehydration of butan-2-ol by the use of a modified quartz catalyst which was coated with copper, platinum, or nickel. Terentev and Klabunovski [29] showed that the quartz-metal catalyst could also be used as an asymmetric hydrogenation catalyst.

Some of these reports might be necessary to reexamine the experiments because of the insensitivity of the polarimeter used in these days.

Bernal [2] emphasized the role of one-handed quartz on the origin of optical activity of organic compounds. Akabori [1] also discussed the role of quartz in his polyglycine theory. Recently Harada [13] observed the formation of optically active organic compounds under the influence of optically active quartz powder.

Localization of one-handed quartz on the earth is an important problem in connection with the origin of optical activity. The problem has not been settled yet. However, there could be a possiblity of localization of one-handed quartz on the primitive earth. If one-handed quartz is localized in some area of the primitive earth, it makes possible an origin of optically active organic compounds.

Preferential crystallization of racemic solutions

There are two types of resolution of racemic materials by preferential crystallization: a) by seeding with dissymmetric materials, b) by spontaneous resolution (without adding any material).

In each case, the formation of *racemic mixtures* (conglomerates) is preferred when racemic materials crystallize from supersaturated solutions. Some organic compounds which form *racemic compounds* under usual conditions might form *racemic mixtures* under some other suitable conditions. Few studies have been made on the relationship of *racemic compounds* to *racemic mixtures*.

Optical resolution of organic compounds by seeding has been reviewed by Secor [26]. Tartaric acid, malic acid, lactic acid, threonine, histidine, glutamic acid, and many others were resolved by this process. Ostromyslenskii [24] reported the resolution of DL-asparagine and also sodium ammonium DL-tartarate by addition of glycine crystal which has no asymmetric carbon in it. However, Iitaka [15] found a new form of glycine crystal, γ-glycine, which has dissymmetric helical crystalline structure. Therefore, Ostromyslenskii's glycine might be the one-handed γ-glycine. Harada and Fox [9] crystallized optically pure aspartic acid copper complex from supersaturated solution of racemic copper complex by seeding with optically active D- and L-copper complex. Similarly, when wool or cotton were added to the same supernatant racemic solution, D-rich aspartic acid copper complex crystallized [12]. Because of the supersaturation of the solution, stereoselectively absorbed D-aspartic acid copper complex might form microcrystals on the biopolymers for further preferential crystallization.

DL-Aspartic acid could not be resolved by seeding with D- or L-aspartic acid from the racemic supersaturated aqueous solution. However, we found that when ammonium formate solution was used for the solvent, the racemic aspartic acid was easily resolved upon crystallization by seeding with optically active aspartic acid [10,11]. Glutamic acid, asparagine, and glutamine were also resolved almost completely by the use of this method.

It has been postulated that many kinds of organic compounds accumulated in the primitive sea, before life evolved. In a lagoon or in a shallow pond, some organic compounds might have crystallized out by evaporation of water. On crystallization of the "primitive soup", some racemic organic compounds might have resolved into their optically active isomers. Ammonium formate is actually a hydrolyzate of hydrogen cyanide which was probably a key precursor of organic compounds on the primitive earth.

In addition to the seeding experiment with wool and cotton (1968), Harada [12] seeded l-quartz powder into a supersaturated solution of racemic aspartic acid copper complex. The resulting aspartic acid copper complex is L-rich. Quartz sand along the Gulf coast of Florida, which is crystalline, also resulted in L-rich aspartic acid copper complex. The facts could be explained by the stereospecific absorption of L-aspartic acid complex on l-quartz and subsequent preferential crystallization. However, this is not valid because D-quartz also did give L-rich crystals. Amorphous quartz powder, glass powder, also give L-rich crystals and finally, even without any seed, L-rich crystals were the result. These crystallization experiments were carried out using a millipore filter to eliminate possible dust, and in three different locations in our institute. Dr. Nakaminami (Osaka University) and Dr. Hayakawa (Shinshu University) kindly performed the same experiment in Japan, and obtained results similar to those we obtained in Miami.

One explanation could be the effect of optically active microparticles (dust). Dusts are everywhere and their size ranges from visibly resolvable down to molecular dimensions. It is practically impossible to eliminate all of these particles completely from the experimental system. Biopolymers such as cellulose and protein are the most likely dissymmetric dust on the earth. However, as mentioned earlier, cotton or wool resulted in the crystallization of D-aspartic acid copper complex. Therefore, the seed, if any, would not be cotton or wool in type, but must be some other type of material to account for the L-selectivity. Also, the dust, if any, must be distributed widely over the earth. The experimental results carried out in a dust-free room show lower optical activity and obtained scattered results. This fact might suggest the presence of some active dust particle in these experimental systems.

The cause of crystallization of L-isomer could be considered in another very speculative way which might be related to the problem discussed in the following.

Dissymmetric nature of matter

Several possible origins of optical activity mentioned above are mainly dependent on "chance". Localization of L-quartz or D-quartz in some area is a chance problem. On the other hand, some scientists believe that evolution of L-amino acids or L-amino acid organisms is a predetermined process in the Universe. This idea is based on the discovery of non-conservation of parity by Lee and Yang (1956). How can this breakdown of symmetry at the elementary particle level relate to the dissymmetry of the molecular level? Goldhaber et al. [8] showed that electrons emitted from ^{60}Co in the β-decay are polarized, and that the "Bremsstrahlung" from the polarized β-rays were found to be circularly polarized electromagnetic waves (γ-rays). Therefore, the dissymmetry in the elementary particle level could affect the molecular level with dissymmetric physical force. The relationship between optical activity of organic compounds and the fact of non-conservation of parity was first suggested by Fox [5], Haldane, and was discussed by Ulbricht [32].

Recently, Garay [7] irradiated alkaline L- and D-tyrosine solution with β-rays from ^{90}Sr for 18 months and claimed that D-tyrosine was found to be decomposed more than L-tyrosine. This could be a result of stereoselective destruction of tyrosine by polarized β-rays. Unfortunately, a similar experiment using DL-tyrosine was not carried out. Because of the theoretical importance of this type of experiment, similar experiments may be carried out more quantitatively by many investigators.

In addition to the breakdown of symmetry in weak interactions, Yamagata [36] proposed the possible dissymmetry of electromagnetic interaction. Therefore, molecular wave functions of D- and L-α-amino acids differ slightly. Therefore, D- and L-α-amino acids are slightly different energetically, and their reactivity would be different. If we define the ratio of reaction of L- and D-isomers in reactions 1, 2, 3 --- n as p_1, p_2, p_3 --- p_n, so N_L/N_D, ratio of L- and D-isomers after n times of reaction will be expressed as

$$N_L/N_D = (p_1)(p_2)(p_3) \cdots (p_n)$$

For simplicity, all p_1, p_2, --- p_n may be taken as equal to p. As stated already, p would not be equal to unity, but would be larger than it by an extremely small quantity, $\epsilon = f^2/\hbar c = 10^{-6} - 10^{-7}$, that is, $p = (1 + \epsilon)$. Thus N_L/N_D is as follows:

$$N_L/N_D = p^n = (1 + \epsilon)^n \approx e^{\epsilon n}$$

If the n is larger than $10^8 \sim 10^9$, N_L would be a very large number compared with N_D. According to this hypothesis, the optical activity of an organic compound is a result of "concentration" of the breakdown of symmetry in electromagnetic interaction by successive chemical reactions on the primitive earth.

According to the discussion mentioned above, the appearance of L-amino acid organisms on the earth or on some planet elsewhere was not by chance but was predetermined by the dissymmetric nature of matter. Future study on this subject and also future space probes might resolve this theoretically interesting problem. In such space probes, finding of optically active organic compounds could be a good sign of life; however, it could not be sufficient evidence for life, because optically active organic compounds could be formed abiotically under suitable conditions.

Development of optical activity

When some quantity of optically active organic compounds were formed in a limited area of the primordial earth, such a situation could be celebrated as the origin of optical activity on the earth. In the early developmental processes of optical activity, a continuous "fight" between L- and D-molecules would have taken place [35]. When the optically active molecule could act as a seed for preferential crystallization or could act as an asymmetric center in other asymmetric syntheses. One other type of propagation process of molecular asymmetry is a kind of chemical resolution reaction

$$DL + L^* \rightarrow L + DL^*$$

When a solution of DL-aspartic acid was added to an aqueous solution of L-alanine copper complex, D-aspartic acid copper complex which was almost optically pure crystallized out [10,11]. Therefore the solution became a mixture of L-aspartic acid and L-alanine.

Such developments of optical activity by chemical reactions are certainly important. However, during the course of chemical evolution, a more efficient process, the appearance of an asymmetric catalyst, must have taken place on the primitive earth. In the usual asymmetric synthesis (in organic chemistry), one asymmetric center could produce only a small amount of the new asymmetric center. If an asymmetric catalyst appeared, it could be possible to produce thousands of new asymmetric molecules by successive reactions. The appearance of an asymmetric catalyst, therefore, would have accelerated the development of optical activity of organic compounds on the primitive earth. Asymmetric catalysts would also have guided the direction of chemical evolution. This implies that the chemical reactions were not only controlled thermodynamically but were also controlled kinetically because of the appearance of the asymmetric catalyst. The more evolution of optical activity took place, the more efficient asymmetric catalysts would appear. The increase of optical purity of biochemical substances made the organized system more functional. After the successive development of pre-cellular systems, when a self-duplication mechanism had appeared, we could call it evolution of primitive life.

The self-duplication mechanism is an important part of genetic systems. One interesting problem is the relationship between the evolution of optical activity and the genetic mechanism. In order to account for the genetic storage and retrieval system even in the most primitive organism, it was necessary for asymmetric building blocks to be present.

Interactions between (poly)nucleotides and (poly)amino acids would be stereospecific because these two biopolymers have dissymmetric structure. Because of this, development of a genetic system could be considered as a configurational and conformational problem. The species of bases and also how these bases are arranged in space should be considered to determine their affinities to any given specific amino acid. Therefore, it is conceivable that when the genetic system was first established in the primitive organism, the development of optical activity of key biomolecules was already accomplished (fig. 1). The evolution of optical activity and the evolution of the genetic system are like two wheels of a car. They evolved simultaneously, interacting with each other. When the car arrived at the goal of evolution of primitive life, the two wheels also arrived at the goal at the same time. By the accomplishment of the genetic system, existence of the organism became more certain. At the same time, evolutionary directions of the organized systems were further determined by their own internal mechanisms which were developed in evolution. Organized systems created more highly ordered systems and there high ordered systems finally gained a new quality. The new levels of organization, primitive life, lived their lives towards the future, fighting with the environment, under control of inner mechanisms and still maintaining their own variability.

Acknowledgment

This work was aided by Grant no. 10-007-052 from the National Aeronautics and Space Administration. Contribution no. 155 from the Institute of Molecular Evolution.

References

[1] Akabori, S., Kagaku 25 (1955) 54.
[2] Bernal, J.D., Physical Basis of Life (Routledge and Kegan Paul, London, 1951).
[3] Byk, A., Z. Physiol. Chem. 49 (1904) 641; Chem. Ber. 42 (1909) 141; Naturwissenschaften 13 (1925) 17.
[4] Ferroni, E., and R. Cini, J. Am. Chem. Soc. 82 (1960) 2427.
[5] Fox, S.W., J. Chem. Educ. 34 (1957) 472.
[6] Fox, S.W., J.E. Johnson, and A. Vegotsky, Science 124 (1956) 923.
[7] Garay, A.S., Nature 219 (1968) 338.
[8] Goldhaber, M., L. Grodzins and A.W. Sunyar, Physiol. Rev. 106 (1957) 826.
[9] Harada, K. and S.W. Fox, Nature 194 (1962) 768.
[10] Harada, K., Nature 205 (1965) 590.
[11] Harada, K., Nature 206 (1965) 1354; Bull. Chem. Soc. Japan 38 (1965) 1552.
[12] Harada, K., Nature 218 (1968) 199.
[13] Harada, K., Naturwissenschaften 57 (1970) 114.
[14] Harada, K., unpublished experiment.
[15] Iitaka, Y., Acta Cryst. 11 (1958) 225.
[16] Karagunis, G. and G. Drikos, Naturwissenschaften 21 (1933) 697; Nature 132 (1933) 354.
[17] Klabunowski, Je.I., Asymmetrische Synthese (VEB Deutscher Verlag der Wissenschaften, Berlin, 1963).
[18] Karagunis, G. and G. Coumoulos, Nature 142 (1938) 162.
[19] Kuhn, W. and E. Braun, Naturwissenschaften 17 (1929) 227.
[20] Lee, T.D. and C.N. Yang, Physiol. Rev. 104 (1956) 254.
[21] Morowitz, H.J., J. Theoret. Biol. 25 (1969) 491.
[22] Oparin, A.I., The Origin of Life on the Earth, 3rd ed. (Oliver and Boyd, London, 1957).
[23] Oparin, A.I. et al. ed., The Origin of Life on the Earth, International Union of Biochemistry Series, Vol. 1. (Pergamon Press, London, 1959). Especially articles by A.P. Terentev and E.I. Klabunowskii (p. 95) and by E.I. Klabunowskii (p. 158).
[24] Ostromyslenskii, I., Chem. Ber. 41 (1903) 3035.
[25] Schwab, G.M. and L. Rudolph, Naturwissenschaften 20 (1932) 363.
[26] Secor, R.M., Chem. Rev. 63 (1963) 297.
[27] Stryer, L., Optical asymmetry, in: Biology and the Exploration of Mars, eds. C.S. Pittendrigh, W. Vishniac and J.P.T. Pearman (National Academy of Science, National Research Council, 1966) p. 141.
[28] Tenney, L., D. Davis and J. Ackerman, J. Am. Chem. Soc. 67 (1945) 486.
[29] Terentev, A.P. and E.I. Klabunowski, Sb. Statei po Obshch. Khim. 2(1953) 1521, 1598.
[30] Tsuchida, R., A. Nakamura and M. Kobayashi, J. Chem. Soc. Japan 56 (1935) 1335.
[31] Tsuchida, R., M. Kobayashi and A. Nakamufa, Bull. Chem. Soc. Japan 11 (1936) 38.
[32] Ulbricht, T.L.V., Quart. Rev. 13 (1959) 48.
[33] Ulbricht, T.L.V., Comp. Biochem. 4 (1962) 1.
[34] Van't Hoff, J.H., Die Lagerung der Atome in Raume (3rd ed.), Braunschweig (1908).
[35] Wald, G., Ann. N.Y. Acad. Sci. 69 (1957) 352.
[36] Yamagata, Y., J. Theoret. Biol. 11 (1966) 495.

PART III

SYNTHESES OF SMALL MOLECULES

Chemical Evolution and the Origin of Life, eds. R. Buvet and C. Ponnamperuma

PRIMARY TRANSFORMATION PROCESSES UNDER THE INFLUENCE OF ENERGY FOR MODELS OF PRIMORDIAL ATMOSPHERES IN THERMODYNAMIC EQUILIBRIUM

G. TOUPANCE, F. RAULIN and R. BUVET
Faculté des Sciences, Laboratoire d'Energétique Electrochimique,
10 rue Vauquelin, Paris 5°

For the last fifteen years, a great amount of work has been carried out on the evolution of models of primitive terrestrial environments and the sound results obtained support the theory of chemical evolution. At present, we must replace these results with respect to one another. This can be done in the context of an energetic approach to these processes.

As a matter of fact, in many different models of primitive atmospheres and under various energetic conditions of excitation, the syntheses of a number of simple molecules of biological interest such as amino acids, nucleic bases and sugars, have been observed.

A survey of all the results obtained shows that the methods used for excitation – i.e. sparks, silent electric discharges, ultraviolet radiations, ionizing radiations – do not seem to influence the very nature of the products obtained [5]. Nevertheless, the energy involved must be at a sufficient quantic level to be assimilable by the model of the atmosphere studied.

On the other hand, it appears that the oxidizing or reducing property of the atmosphere plays an essential part [4]. If positive results have always been obtained with a reductant environment, it has always been difficult to obtain the synthesis of significant amounts of biologically interesting products in an oxidant environment.

Finally, the analysis of the results shows that it is possible to decompose the building process of biological monomers into two successive steps [2,3]: the first corresponds to the building up of precursors such as HCN in the gas phase; the second corresponds to the aqueous phase evolution of these precursors towards the building up of molecules of biological interest.

Taking all these facts into account, it seems necessary to bring a final answer to the following question: what is the range of elementary atomic composition for the primeval atmosphere which allows the building up of unsaturated intermediates required for subsequent syntheses in the aqueous phase?

In order to estimate the amount of energy stored during the evolution of the studied atmospheres, it is of interest first to define for each of the possible atomic compositions, the molecular state of lower energy.

The general problem of finding the equilibrium composition of a medium with a well defined atomic constitution in C, H, O, N, can be solved mathematically. However, the whole set of equations does not allow us to express the solution under explicit algebraic expressions. The resort to an iterative solution with the help of a computer is inescapable.

Nevertheless, simplifications in the mathematical solution appear for some peculiar atmospheres. Such is the case when an important excess of one element exists, for example H or O. Then it is possible to determine the standard state of lower energy of such a medium and, therefore, to set up the list of the standard free enthalpies of the primitive syntheses of a number of molecules in this atmosphere ground state.

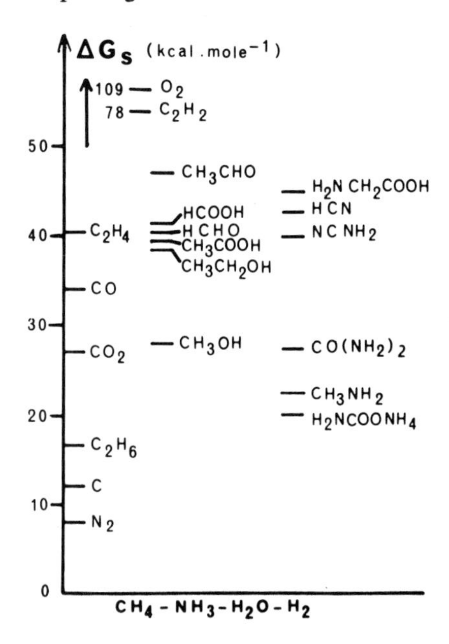

Fig. 1. Standard free enthalpies of syntheses of several molecules from a gaseous mixture CH_4, NH_3, H_2O, H_2.

Now, let us consider the case of an atmosphere rich in hydrogen. Thanks to numerical data tables, we can easily show that one system exists – that composed of CH_4, NH_3, H_2O, H_2, in their reference state – from which the calculated standard free enthalpies of synthesis of all other compounds are positive. It is, therefore, mainly in this form that the hydrogenated C, H, O, N systems are in their state of lower free energy. This form is precisely that most frequently considered as a model for primordial atmospheres.

In fig. 1 we have collected the values of the standard free enthalpies of some primitive syntheses. On the *Y* axis we have written the standard free enthalpies of synthesis of different compounds calculated from the previous medium of lower energy – CH_4, NH_3, H_2O, H_2 – which has been placed at zero level. These energies are expressed in kilocalories per mole of the compound formed. There is no unit on the *X* axis and the horizontal dimension is used to make the graph clear. We must note that the orders of magnitude of the energies stored during the synthesis of each molecule, range between 10 and 50 kcal.

Likewise, let us consider an atmosphere rich in O_2. With the help of numerical data tables we can easily show that the system CO_2, H_2O, N_2, O_2

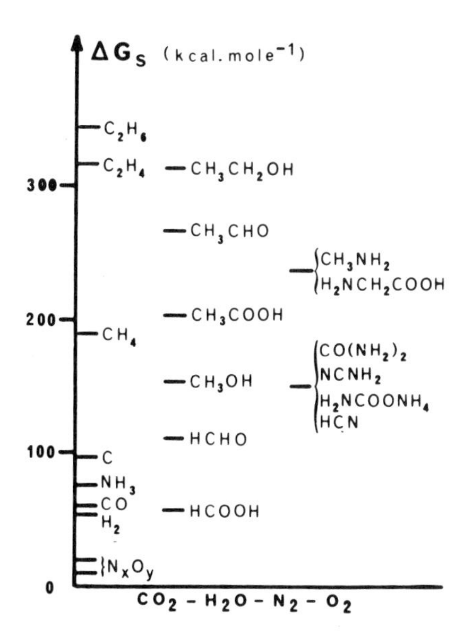

Fig. 2. Standard free enthalpies of syntheses of several molecules from a gaseous mixture CO_2, H_2O, N_2, O_2.

represents the standard state of lower energy. In fig. 2 we have gathered the characteristics of this medium as regards the formation of some kinds of biologically valuable compounds. On the *Y* axis we have drawn the standard free enthalpies of synthesis of these compounds computed from the medium of lower standard energy, i.e. CO_2, H_2O, N_2, O_2, placed at zero level.

We notice immediately that the energies required for the syntheses of organic molecules are far more important than in the previous hydrogen-rich case. The compounds which have the lowest free enthalpy of synthesis are here formic acid with 58 kcal/mole, then formaldehyde with 112 kcal/mole and urea with 156 kcal/mole. In the previous case of hydrogenated media we had formation of ethane as low as 16 kcal/mole and that of urea at 28 kcal/mole.

These differences are enhanced in fig. 3 where we have used the same scale for both media.

Though these considerations deal only with the free standard enthalpies and do not take into account concentration effects, we understand why it has always been more difficult to obtain syntheses of biologically interesting compounds in the case of oxygen-rich media.

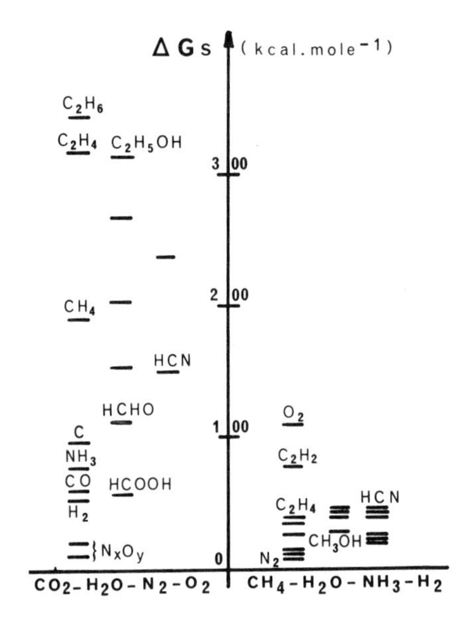

Fig. 3. Standard free enthalpies of syntheses of several molecules:
left: from a gaseous mixture CO_2, H_2O, N_2, O_2;
right: from a gaseous mixture CH_4, H_2O, NH_3, H_2.

Finally, we have tried to compare more precisely the conditions of syntheses of several molecules and to clearly foresee the possibilities of exergonic evolution from one kind of molecule to another. To that end, it is convenient to use a means of representation which, as first approximation, rules out the effect due to stoichiometric coefficients of the synthesis reaction. In the most frequent case of hydrogenated atmosphere, all free standard enthalpies of synthesis have been roughly calculated to the same mass number, by dividing each of them by the total number of all C, N and O atoms included in the molecules, that is all atoms except hydrogen atoms. In fact this convention roughly refers to bond energies.

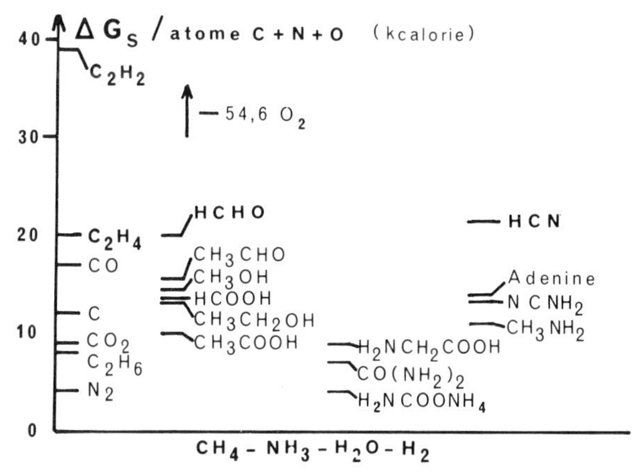

Fig. 4. Standard free enthalpies of syntheses of several molecules from a gaseous mixture CH_4, NH_3, H_2O, H_2. Each of them is divided by the total number of (all) C, N, O atoms constituting the considered molecule.

The results are gathered in fig. 4. The unsaturated intermediates known to appear during the first step of evolution in important quantity in gaseous phase such as HCN, HCHO, C_2H_4, are drawn in bold face. We note that these compounds are above the molecules of biological interest corresponding in this diagram to energies from 8 up to 15 kcal/atom save hydrogen. Then we understand why the former compounds can transform spontaneously into the latter in aqueous phase. Let us stress that the building energies of excited species obtained through electronic or photonic impact on molecules of the primeval atmosphere are very high, higher than 200 kcal [6].

On the experimental ground, and taking into account such previous theoretical data, we have started a detailed and exhaustive investigation on the nature of precursor compounds formed in gaseous phase when atmospheres of various compositions are subjected to energetic excitation. The atmospheres have been defined only by their elementary composition in C, H, O, N. For each composition, the initial molecular state chosen is always the molecular state of lower standard energy. This condition is necessary if we want to clearly develop the analysis of the primary processes of energetic accumulation. However, for the choice of the molecular composition of the initial mixture, we leave out species which occur at low concentration in the state of lower standard energy.

As regards energetic excitation, we have chosen silent electric discharge under low pressure. This type of discharge is known to occur under pressures ranging from 10^{-2} torr to a few dozen torrs and with voltages between 500 to 700 V. It proceeds by step ionizations. In the first step the charged particles spontaneously occurring at or emitted by the electrodes are accelerated under the influence of the potential difference. Then, when the accumulated kinetic energy is sufficient, it is transferred to another particle which ionizes in turn and takes part in the conduction of the medium. Consequently, it appears that the maximal energies which occur during these processes are, by their very nature, all very close to the minimal energies of excitation of the present molecules [1]. Accordingly, this type of discharge provides excitation energy with an excellent yield and prevents destruction of the formed products by too strong conditions of excitation. Thus, we can obtain important rates of syntheses which help experimental investigation.

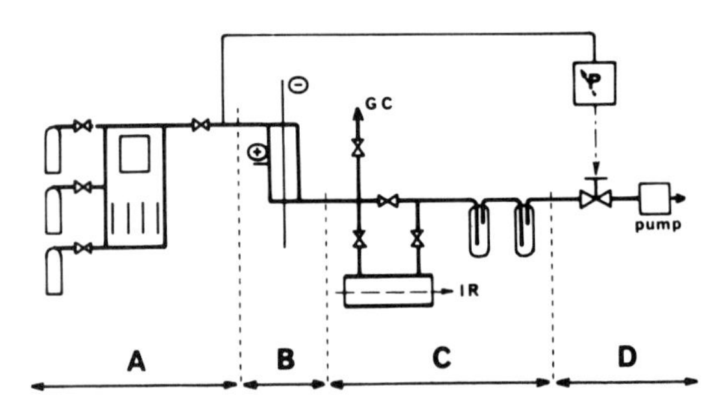

Fig. 5. Block diagram of the apparatus. A) atmosphere generator; B) discharge tube; C) analytical device; D) pumping device.

The block diagram of our apparatus is represented in fig. 5. A represents an atmosphere generator which allows us to achieve any possible mixture of water vapor with three compressed gases. The ratio of each constituent is adjustable from 1 to 100 percent. The reactor B mainly consists of a cylindrical open metal tube working as anode and a coaxial tungsten wire working as cathode. The gaseous mixture is flushed continuously into the reactor at a well known rate of flow. Working with reactors of various lengths and at various rates of flow, we have been able to investigate the state of evolution of the systems after an exposure time to the discharge varying between a few tenths of a second and several tens of seconds. C represents a device of valves allowing the sampling of gaseous samples which are directed either to the cell of a 337 Perkin Elmer infrared spectrophotometer or to the sample loop of a Variant Aerograph 1520 gas chromatograph which are both permanently connected. A group of traps cooled with liquid nitrogen allows the condensing of the effluents and occasionally their recovery. D represents a pumping device and a vacuum regulator which allows us to keep the pressure in the reactor at its reference value, generally from 10 to 20 torrs in spite of the continuous injection of fresh gaseous mixture. A pressure gauge is situated upstream of the reactor, bearing the indication of the reference pressure, and controls the opening of an automatic valve situated at the intake of the vacuum pump.

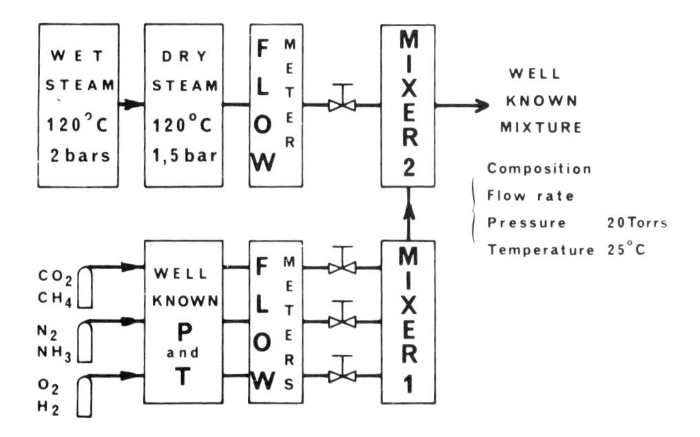

Fig. 6. Block diagram of the atmosphere generator.

The block diagram of the atmosphere generator is represented in fig. 6. Each of the compressed gases is expanded at well-known pressure and temperature (1.9 absolute bar and 25 °C); then it is injected at a well-known rate of flow into the first mixer. Water vapor is supplied by a generator at a temperature of 120 °C and then expanded in an oven at the same temperature. At that time, the vapor is used like the other gases, mixed with them at

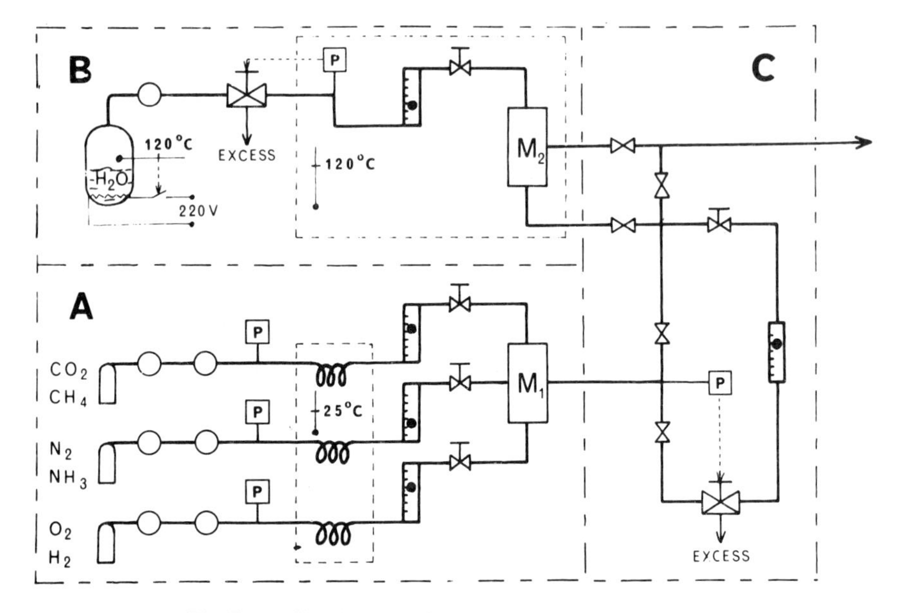

Fig. 7. Detailed diagram of the atmosphere generator.

a well-known rate of flow in the second mixer. The mixture is then at low pressure and can be used at room temperature. It is defined by its composition, its rate of flow, its pressure (about 20 torrs) and its temperature (room temperature).

A detailed diagram is given in fig. 7. In A the compressed gases are mixed, in B the vapor is supplied and mixed with the other gases and in C is a commutation device. The diagram of the discharge tube is given in fig. 8. It is made of an open brass cylinder whose inside diameter is 20 mm and which varies in length. We have often used a 10 cm long tube. This tube is enclosed in a glass envelope consisting of several elements fitting together with tight joints. The cathode is a tungsten wire which is 1 mm in diameter; it is positioned in the axis of the cylinder.

The evolution of several initial gaseous mixtures has so far been investigated. We have determined the nature and amount of many of the compounds formed. When only CH_4 occurs in the reactor, we observe the formation of all the possible hydrocarbons although they are present in quite different proportions. We have analyzed the complex mixtures obtained, mainly by gas chromatography. Fig. 9 gives the shape of a chromatogram

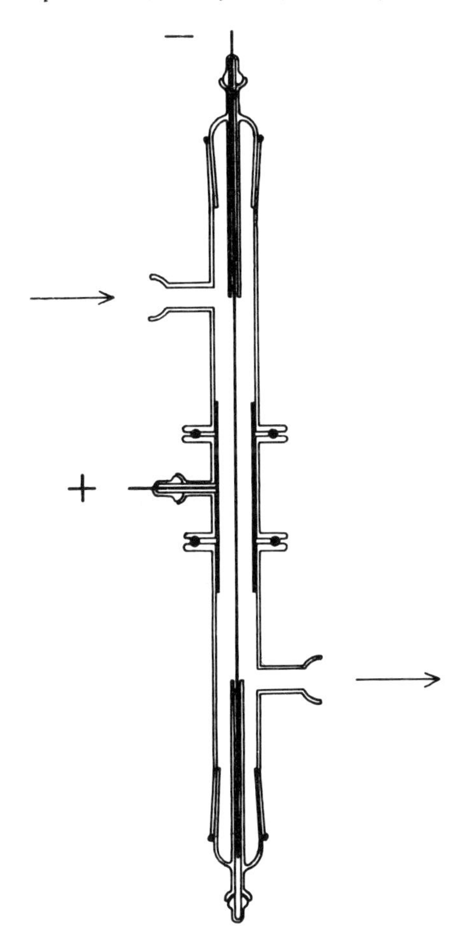

Fig. 8. Discharge tube.

performed with a Porapak Q column at 105°C with helium as carrier gas and a flame ionization detector. We clearly observe the residual methane, a peak which corresponds to C_2H_2 and C_2H_4, then other peaks for ethane, propylene, propane; another peak corresponds probably to the methyl acetylene, and others to hydrocarbons with four carbons including isobutane, butene and normal butane.

The resolution of ethylene and acetylene (fig. 10) has been performed with a silica gel column at 105 °C with He as the carrier gas. We have used the

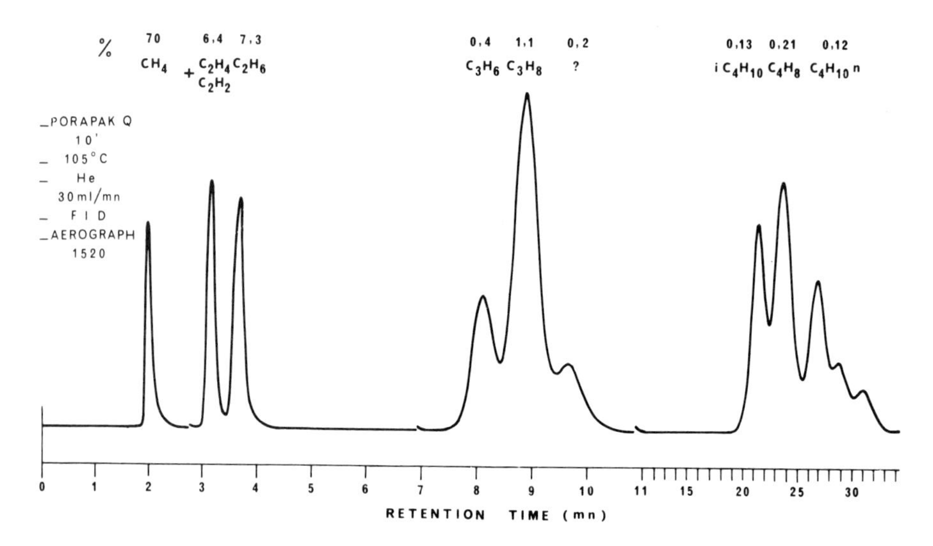

Fig. 9. Gas chromatogram of the products formed when CH_4 alone is treated.

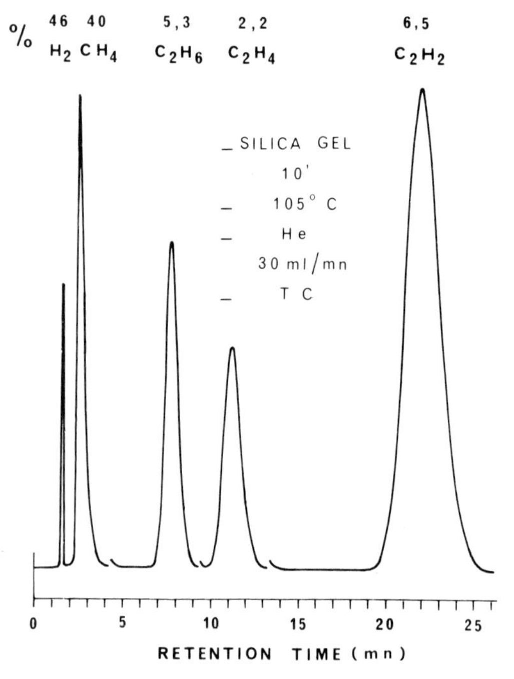

Fig. 10. Resolution of C_2H_4 and C_2H_2 in the products formed when CH_4 alone is treated.

thermal conductivity detector that has allowed us to determine simultaneously the amount of hydrogen produced. We can see H_2, CH_4, C_2H_6, C_2H_4, C_2H_2. The heavier hydrocarbons have higher retention times and are produced in amounts too small to be detected by the detector in use.

The behavior of such mixtures as a function of the exposure time to discharge is summarized in fig. 11. For a pressure of 20 torrs and a discharge current of 100 mA we can see that the maximal quantities produced are 2% for C_2H_4, 5% for C_2H_6 and 7% for C_2H_2. These quantities are expressed in mole per cent in the effluents and they appear after about one second of exposure time.

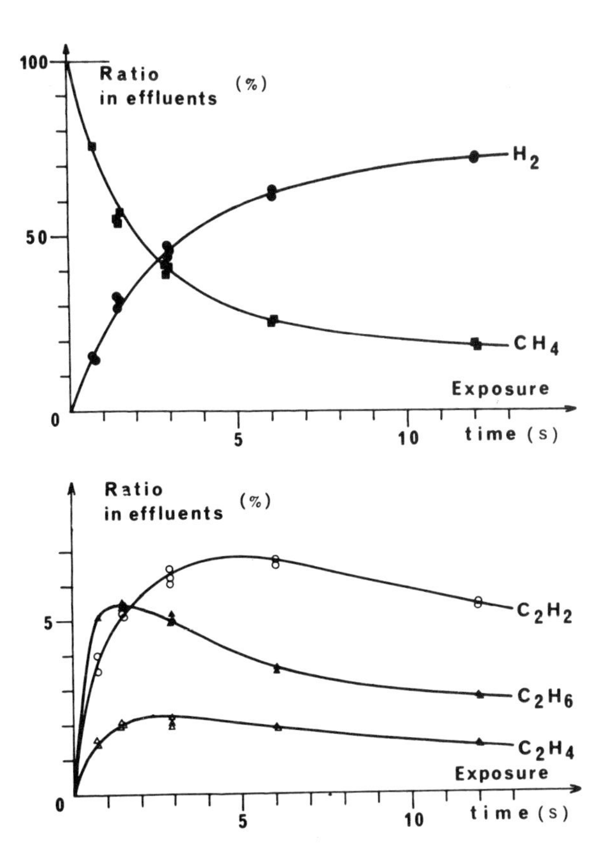

Fig. 11. Composition of effluents as a function of the exposure time in the discharge when CH_4 alone is treated.

The treatment of the mixture $CH_4 + NH_3$ has been completed and we obviously find the formation of very important quantities of HCN. The amount of this compound has been determined by IR spectrophotometry thanks to its intense absorption band at 712 cm^{-1}.

In fig. 12 we have plotted the quantity of HCN produced with a mixture of NH_3 and CH_4 as a function of the exposure time to the discharge. We note that HCN is very stable under the conditions existing in the reactor, since it only weakly decomposes even after long stays in the reactor.

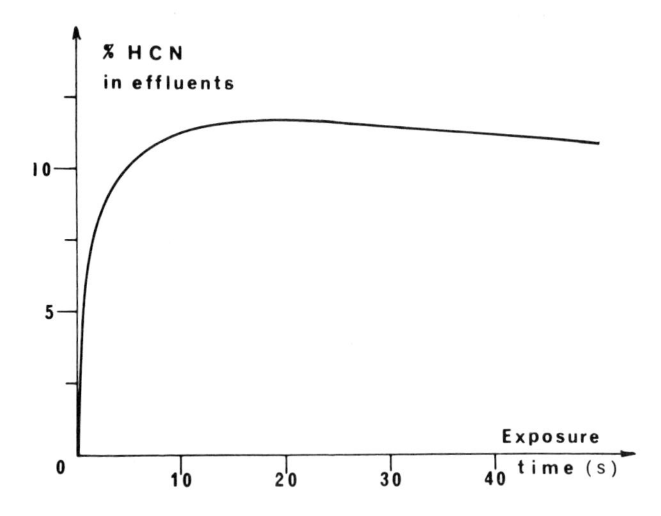

Fig. 12. Ratio of HCN in effluents as a function of the exposure time in the discharge when the mixture 50% NH_3–50% CH_4 is trated.

We have started the study of the compounds formed in the discharge as a function of the initial composition in CH_4 and NH_3. We can say that the amount of HCN formed is maximum when the ratio of NH_3 to CH_4 equals one or so, whereas the amount of C_2H_2 formed increases parabolically with the increase of CH_4.

The other hydrocarbons observed with CH_4 alone also appear but in smaller amounts. Moreover, we note some new peaks, which are now under study.

A study of water vapor and methane mixtures is being carried out. A gas chromatographic examination of the products shows the presence of other peaks, besides the hydrocarbon peaks already identified, which we have attributed to methanol, acetaldehyde and formic acid. We have not been able to detect formaldehyde through the two physical methods used. This compound is indeed difficult to identify in low quantities by means of these methods. We are trying to titrate it by chemical analysis – colorimetry – after trapping the condensates.

In the mixtures resulting from CH_4 and water vapor treatment, we observe that the amount of unsaturated hydrocarbons decreases faster than the amount of other hydrocarbons when the proportion of water vapor increases. This probably means that these unsaturated hydrocarbons play an important role in the synthesis of oxygenated compounds.

Finally, we have started a study of oxygenated atmospheres. In this particular case the amounts of compounds produced are very small and we are not yet able to give any quantitative results.

We can say as a conclusion that a very large number of products are formed after a few seconds of treatment in the reactor. The rate of synthesis of each is very different from one to the other. For the heavier compounds the rate of synthesis is very low but we must note that they are at the same time the most condensable. So they may have reached a significant concentration in aqueous solution. Taking into account all the published results, we can see that only very few of the synthesized products have been studied in their subsequent chemical evolution, and we think that it is possible to find new paths of evolution by taking into account the results of our study, and the ones that we will obtain from it in a near future.

References

[1] Boutry G.A., Physique Appliquée aux Industries du Vide et de l'Electronique (Masson, Paris, 1962).
[2] Ferris J.P. and L.E. Orgel, J. Am. Chem. Soc. 88 (1966) 3829.
[3] Miller S.L., Biochim. Biophys. Acta 23 (1957) 480.
[4] De Rosnay J., Ann. Chim. 2 (1967) 57.
[5] Thomas J.A., Biocytologia, Colloque sur les Systèmes Biologiques Elémentaires et la Biogénèse (Masson, Paris, 1968).
[6] Watanabe K., M. Zelikoff and E.C.Y. Inn, AFCRC Technical Report (June 1953), No. 53.23.

Chemical Evolution and the Origin of Life, eds. R. Buvet and C. Ponnamperuma

THE ROLE OF SHOCK WAVES IN THE FORMATION OF ORGANIC COMPOUNDS IN THE PRIMEVAL ATMOSPHERE

Adolf R. HOCHSTIM
Research Institute for Engineering Sciences,
Wayne State University, Detroit, Michigan 48202, USA

In the primeval atmosphere shock waves could have been produced by the following phenomena (see fig. 1 and Appendix 1) which might have lead to the formation of organic molecules:

At high altitudes (low pressures):
1) Micrometeorites and cometary meteors [12,18].
2) Meteorites [12,18].

At low altitudes:
1) Meteorites [12,18]. (see fig. 2).
2) Thunder from lightning [5].
3) * Sudden eruption of volcanoes into atmosphere – a possible source of shock waves.

High pressure regime: (gas-surface reactions)
1) Meteorite colliding with water [18] (radius, $R > 1$ meter)
2) Meteorite colliding with earth with explosive force ($R > 1$ meter)
3) * Possible presence of cavitations in the water due to the action of ocean surface waves [1].
4) * Shock waves in earth (resulting from meteoritic impact, shear waves or earthquakes) and passing through entrapped gas bubbles in rocks.

All of the above processes have the following time history in common: sudden heating (shock wave) and rather rapid cooling. Different regimes of shocks around a meteorite have different rates of cooling. The shock produced by lightning (thunder) is similar to the bow shock wave (e.g., see fig. 1 in [18]).

* No estimate available.

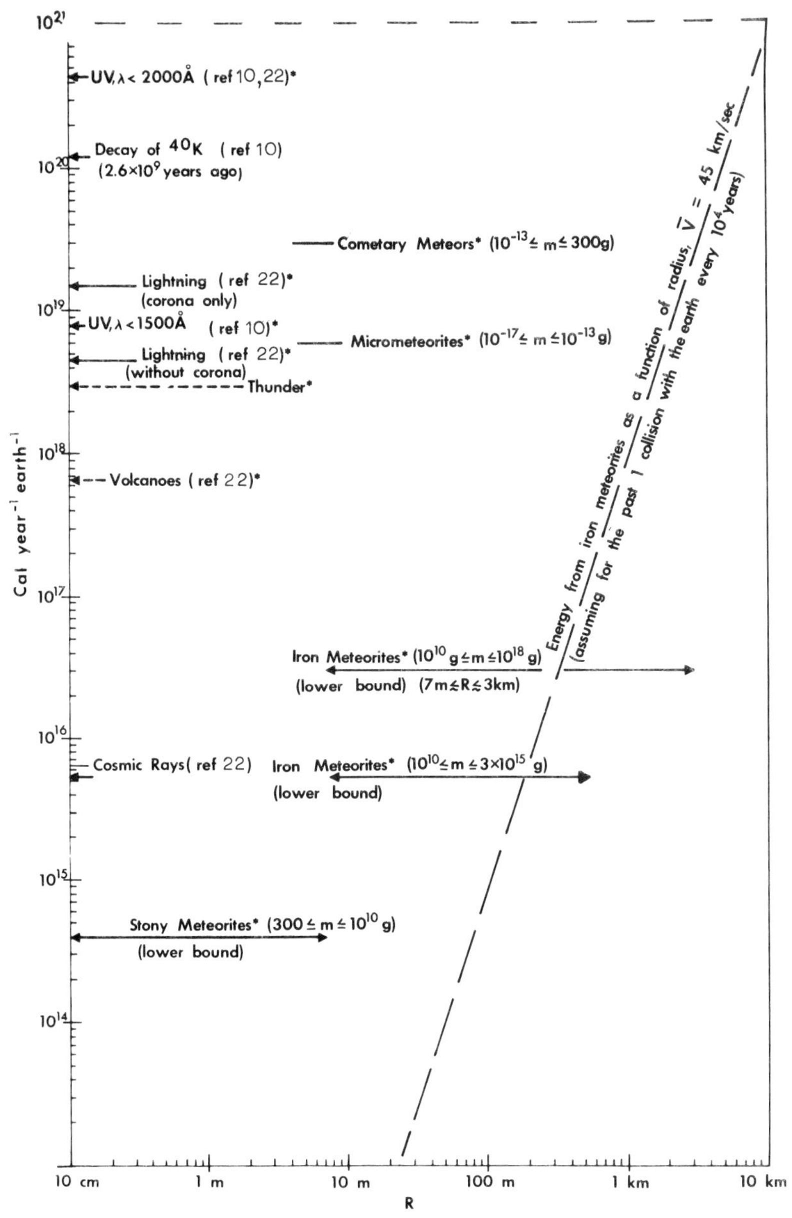

Fig. 1. Comparison of estimates of the average flux of energy from various processes. *Estimated for present times.

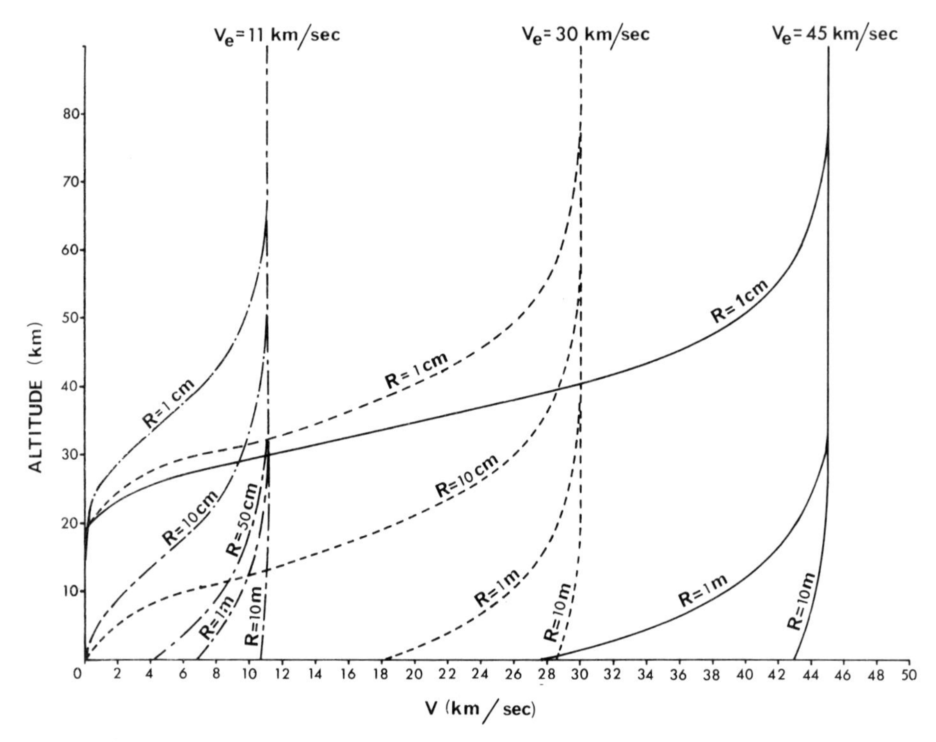

Fig. 2. An illustration of the slow down of various non-ablating spherical iron meteorites in the atmosphere having a scale height, $H = 8.4$ km (see Appendix II). Vertical fall.

At high altitudes the processes are not in thermodynamic equilibrium and the concentration of various species are determined by various chemical rates coupled to the hydrodynamic flow field [2]. There are many free radicals present and vibrational and electronic excitation plays an important role. Many complex species can be formed in the rapidly cooling regime, since different processes have different rates, and those are sensitive functions of translational and vibrational temperature. At the present time many of those rates are not known and a meaningful rate calculation can not be yet done. At lower altitudes the mean free path is small and the local thermodynamic equilibrium assumption is valid in the high temperature regime. One can therefore compute rather accurately the concentration of all species behind shocks. Such calculations were made for 26.3% CH_4, 26.3% NH_3, 13.2% H_2 and 34.2% H_2O by W.K. Park [25], using the numerical method of free energy minimization for 61 species. The results are shown in figs. 3–5. As

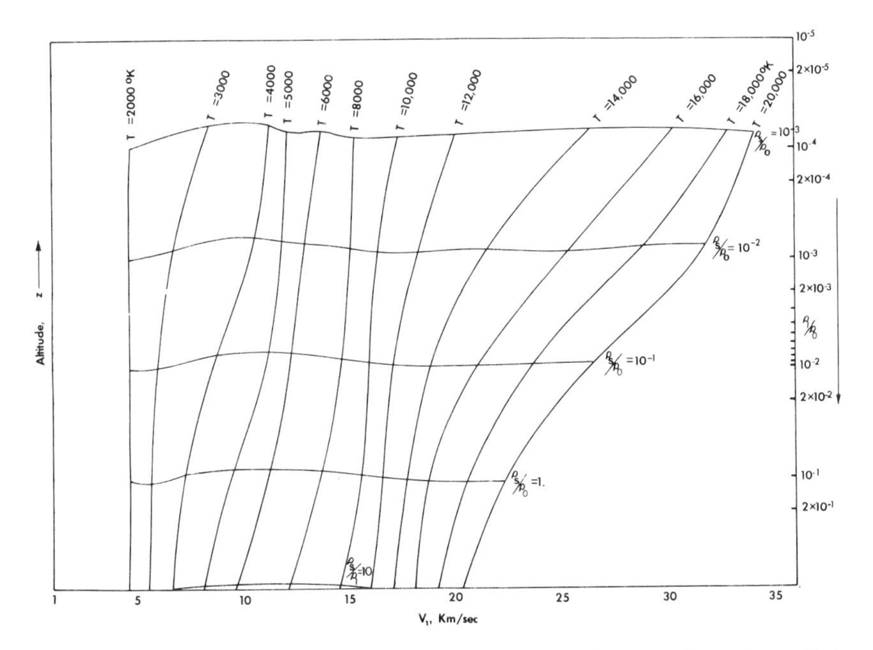

Fig. 3. Temperature and density behind normal shocks (ρ_S) in thermodynamic equilibrium as a function of velocity of meteorite and ambient density, ρ_1 (or altitude). The ambient atmosphere: 26.3% CH_4, 26.3% NH_3, 13.2% H_2 and 34.2% of H_2O. ρ_0 = the sea level density.

those species (e.g., see figs. 5A–5D) enter the expanding volume, the temperature drops and the recombination takes place. But the "memory" of the ambient composition is lost. Many chemical reactions may be frozen in the expanding shocks since some recombination processes are slower than others. The concentration of species may be determined by steady state which is quite different from a complete equilibrium. Furthermore, any interception of the shock wave with water level or water spray may cause a cascade of very fast quenching reactions [18]. Similarly, ablated meteorites may give rise to a multitude of surface reactions [18]. Fig. 1 lists various processes contributing to the total energy input and the results are the best estimates available. The straight line is only for the purpose of comparison, since we do not know the frequency or the sizes of meteorites striking the earth in primeval times. If we assume meteorites with a radius of 1 km striking the earth on the average every 10,000 years this would result in about 10^{18} cal/year. At present the best estimate is that every 10,000 years on the

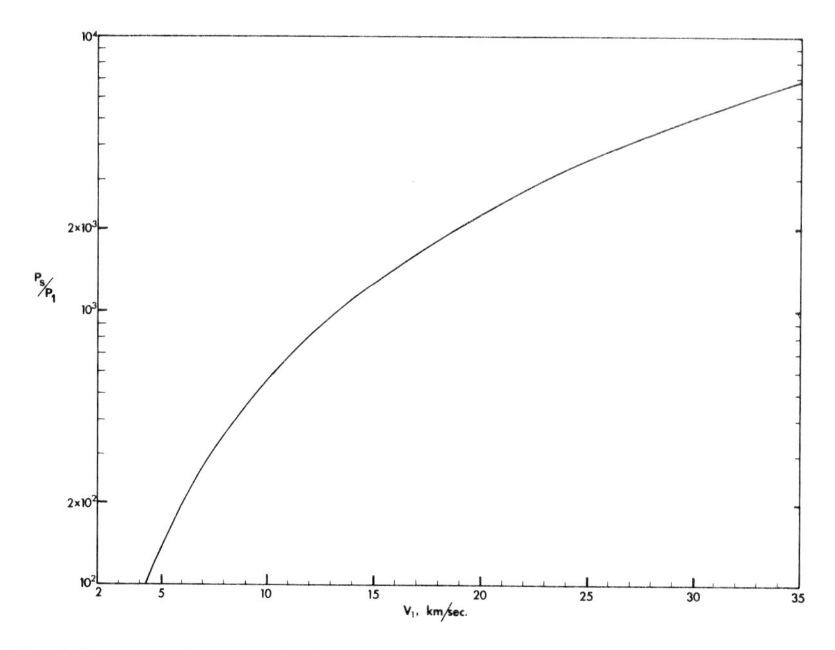

Fig. 4. Pressure ratio across shock (P_S, behind, P_1, in front) as a function of velocity of meteorite. The same ambient atmosphere as in fig. 3.

average 6 meteorites struck the earth with $R>10$ meters (see Appendix 1). The lower bound of energies from meteorites are derived in Appendix 1. If we assume that the results from shock tube by A. and N. Bar-Nuns, Bauer and Sagan [5] apply, the efficiency of production of organic compounds in a rapidly cooling (quench) reaction should yield up to six orders of magnitude greater yield of organic molecules than UV radiation. Thus, shock waves from micrometeorites, meteors, meteorites and thunder are of interest from the viewpoint of contributing significantly to the total accumulation of organic compounds in primeval times and also due to the importance of chemical kinetics in the multitudes of recombination reactions, leading possibly to more complex compounds [12,18].

There is no estimate available of the amount of transient cavitation (violent collapse of bubble) in the oceans due to the action of water impingement on water. It is known that cavitations can be produced hydrodynamically (in addition to acoustically), that shock waves result from the collapse, that the temperature inside of such a collapsing bubble may reach a few thousand degrees Kelvin and that chemical reactions do occur

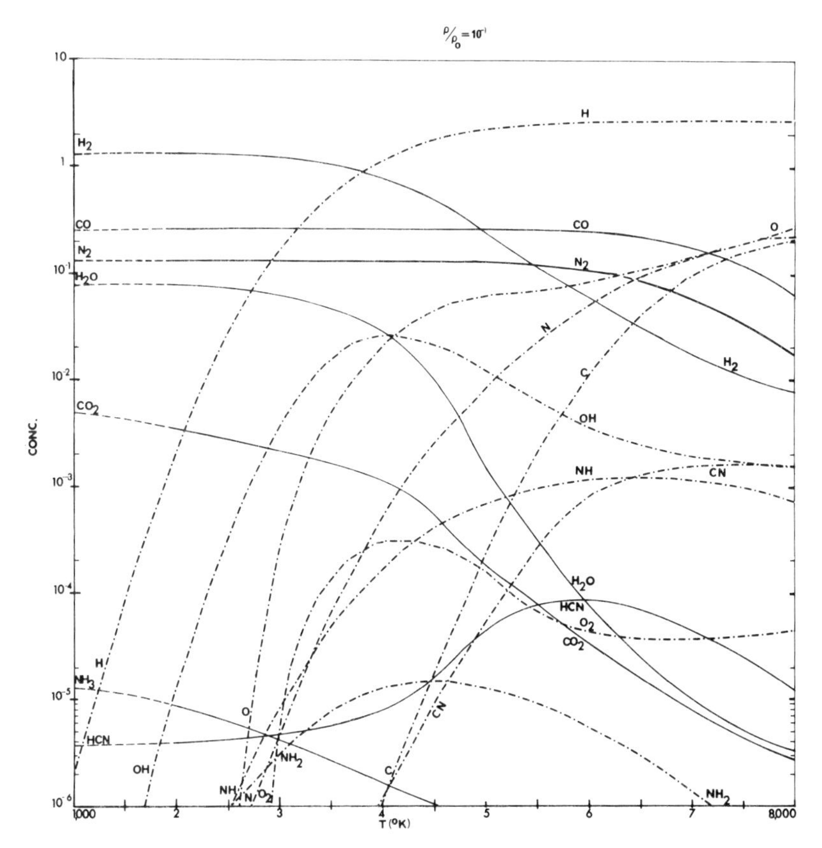

Fig. 5A. Concentration of various major species in thermodynamic equilibrium as a function of temperature and density ρ behind shocks. The scale if multiplied by $2.7 \times 10^{19}\ \rho/\rho_0$ gives number per cm^3. Same atmosphere as in figs. 3 and 4.

inside, emitting radiation (sonoluminescence) [28]. Anbar and Ponnamperuma found amino acids, etc. in ammoniacal water saturated with methane and subjected to ultrasonic cavitation [1].

"Meteroids" fall into the earth's orbit through the atmosphere, and then are called micrometeorites, meteors (if smaller than 300 g in mass) and meteorites (if larger), although the nomenclature used in the literature is more involved, and not very consistent [3,16,17,23,24,27].

In Appendix 1 we have calculated *lower bound* kinetic energies available to micrometeorites, cometary meteorites, stony and iron meteorites. These calculations are based on the observational data and various found meteorites

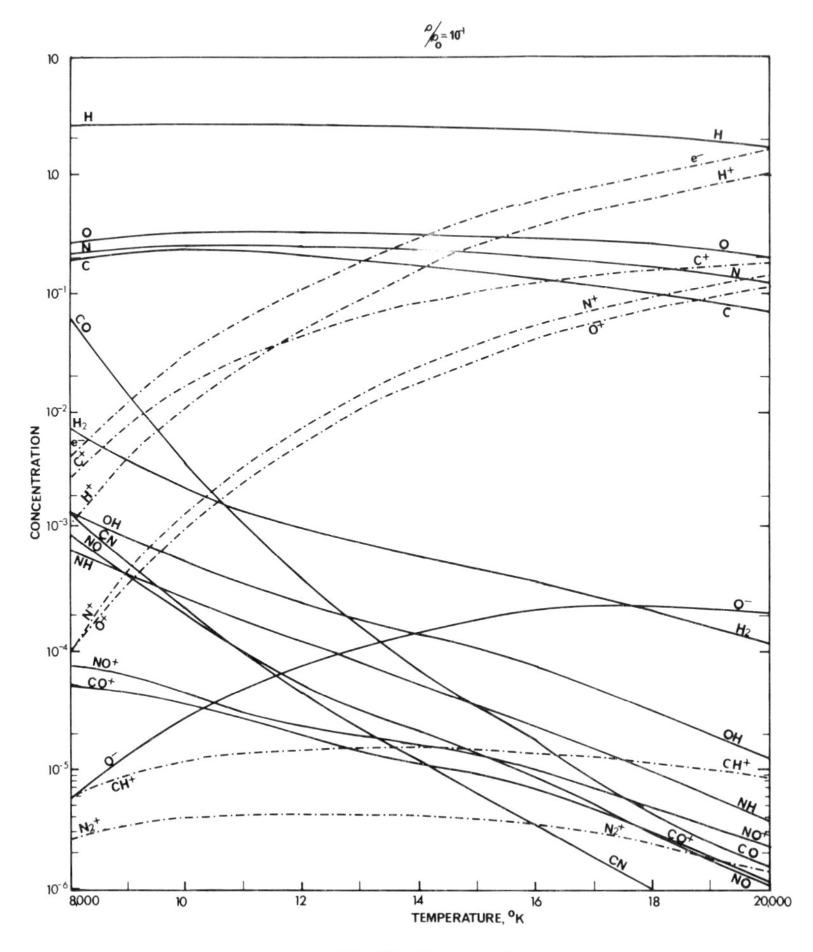

Fig. 5B. Continued.

correlated by Hawkins [17]. Obviously, the data is limited, since many meteorites must have fallen into the sea or in unpopulated areas and have not been counted. Furthermore we may assume that in the past larger numbers of meteorites were swept by the earth. It follows that the total energies indicated in fig. 1 from meteorites in the primeval times probably need to be multiplied by some factor of about 10 to 100, or more.

The efficiency of the energy transfer from the micrometeorites and smaller cometary meteors for the formation of lasting organic compounds is probably small because either they burn or slow down in high altitudes (60–80 km)

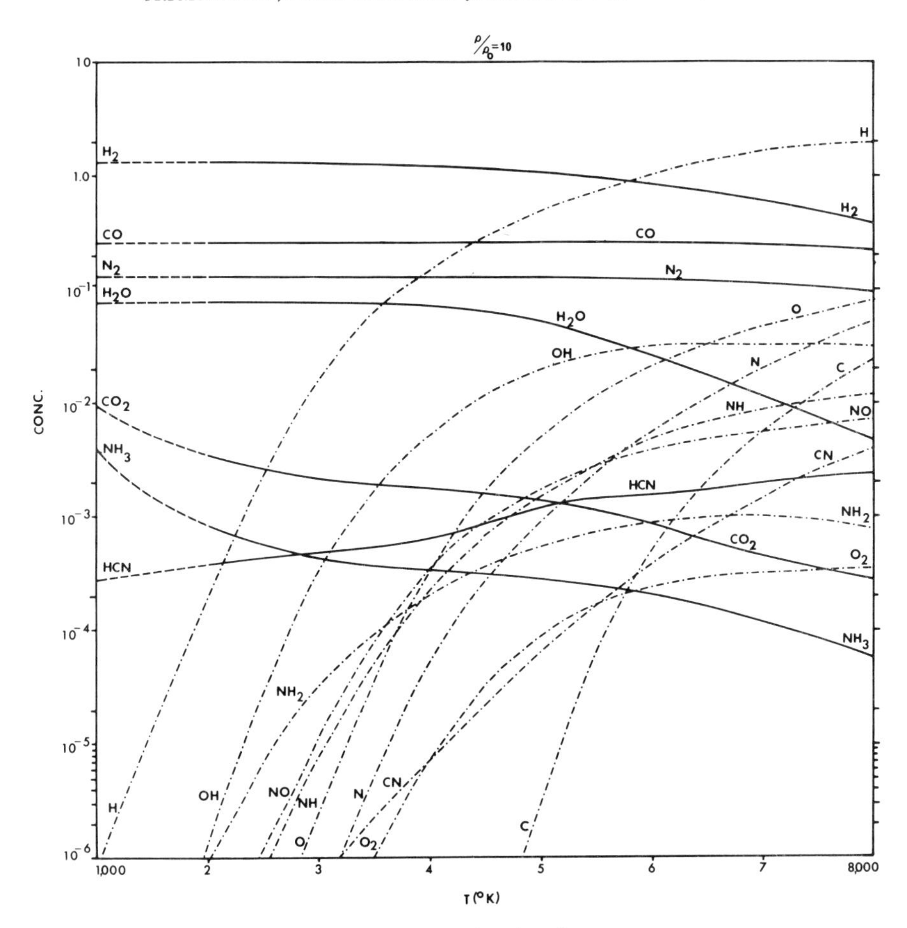

Fig. 5C. Continued.

and any organic molecules produced in the primeval atmosphere would have to travel a long way and a long time before they reached the surface of the earth, and may decompose on the way. On the other hand very large meteorites which reached the surface of the earth would have dissipated a considerable amount of energy in an explosion (similar to a nuclear explosion) on the surface of land or sea (energy of fragmentation, lifting of spray dome, etc.). The intermediate meteorites probably were most efficient.

Under continuous deceleration (see Appendix II, eqs. 2.9–2.13) very high pressures (see fig. 4) build up in front of the meteorite (between its surface and the shock wave) so that there is very high probability that a large

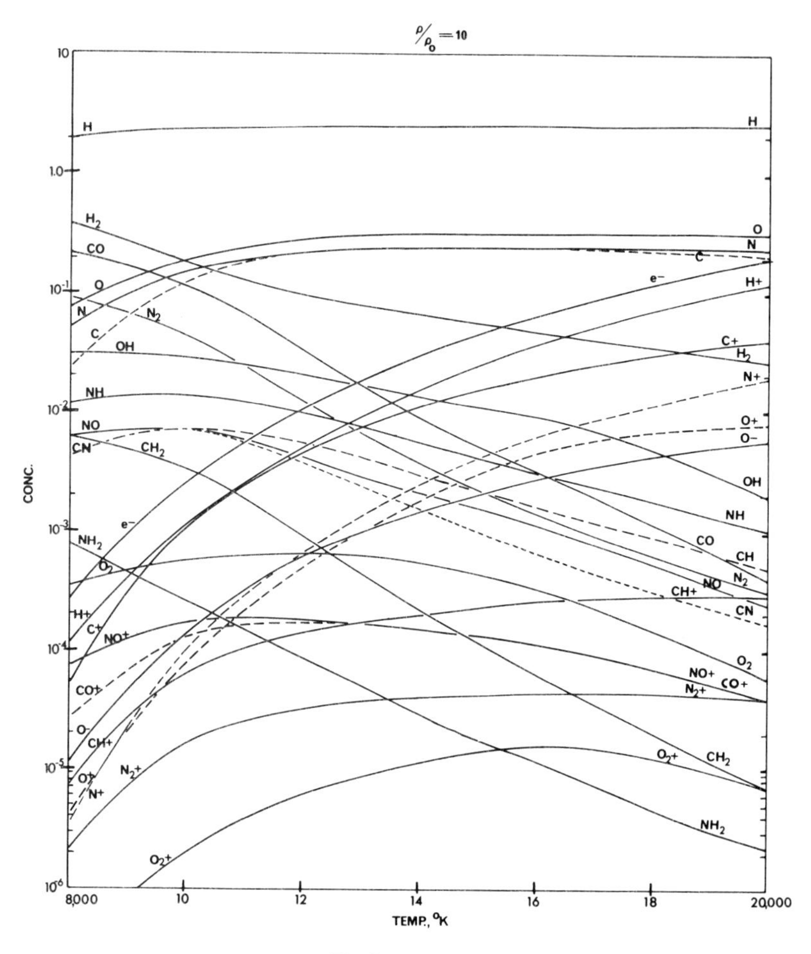

Fig. 5D. Continued.

meteorite may break up at high altitude ($h \sim 10$–40 km) into many fragments, each entering with its own shock wave.

Meteorites have an additional feature, absent in many other energy transfer mechanisms: collision with water, or even land, produces various high pressure processes suitable for polymerization and provides the environment for an efficient dispersal of the formed compound on sea and land through rain ("fallout" of organic molecules) [18].

The most complete simulation of meteoritic impact in the presence of various reaction regimes (heating, cooling, ablation, wake, gas bubble, effect of spray, etc.) is the impact of a bullet (preferably of meteoric material) fired from a hypersonic gun [18]. Such a preliminary experiment was performed in 1963 by Dr. W. Hooker [12,18]. This may be done inexpensively, since many experimental stations are usually arranged in the course of such experiments to measure various shock-properties, and an additional station at the end of a flight costs very little. To study a single regime at a time is also desirable. The regime of gas heating (with slow recombinations) can be satisfactorily reproduced in an ordinary shock tube. It was successfully used by many in pyrolysis of hydrocarbons [9], in decomposition of methane [19], ammonia [4,13], among many others (see also ch. X, [9]). The shock wave heating with subsequent rapid cooling can be made in a modified shock tube (so-called "single-pulse" shock tube) and was used successfully by S.H. Bauer and his coworkers (e.g. isomerization of butene [20], polymerization of C_2N_2 in presence of argon [8], and lately in the formation of amino acids from a mixture of C_2H_6, NH_3 and CH_4 [5,6].

We are performing a few experiments at NASA Ames Research Center for the study of the effects of shock waves on the formation of organic molecules, for which the enthusiastic cooperation of Dr. C. Ponnamperuma is here greatly acknowledged.

Professor Calvin's reservations in our discussion in 1967 [10] regarding the equilibrium calculations for dry air, which I used as a simple example [18], prompted the reporting here of new calculations using a more realistic model of the primeval atmosphere. Comparing the temperatures in [18] with fig. 3 here, we see that at velocity of meteorite of 11 km/sec at sea level the temperature in front of the meteorite will be 16,500°K, while in the primeval atmosphere it would have been only 6,000°K, with a high concentration (see fig. 5C) of H, H_2, CO, N_2, H_2O, O, OH, N, NH, NO, C, HCN, NH_2, CO_2, CN, O_2, NH_3, etc. All those species were predicted before in the wake of a meteorite [18].

Acknowledgement

This work was partially supported by NASA Grant NGR-05-007-215 to the University of California at Los Angeles, through the courtesy of Professor W.F. Libby.

References

[1] Anbar M. and C. Ponnamperuma, Am. Chem. Soc. Meeting, Atlantic City, September (1968) (Div. Biol. Chem., Abstract 100).
[2] Andersen H.C., in: Kinetic Processes in Gases and Plasmas, ed. A.R. Hochstim (Academic Press, New York, 1969).
[3] Astapovich T.S., Meteoric Phenomena in Earth's Atmosphere (GIFML, Moscow, 1958).
[4] Avery H.E. and J.N. Bradley, Trans. Faraday Soc. 60 (1964) 857.
[5] Bar-Nun A., N. Bar-Nun, S.H. Bauer and C. Sagan, Science 168 (1970) 470.
[6] Bar-Nun A., N. Bar-Nun, S.H. Bauer and C. Sagan, this volume, p. 114.
[7] Brown H., J. Geophys. Res. 65 (1960) 1679.
[8] Bauer S.H. and W.S. Watt, J. Chem. Phys. 44 (1966) 1327.
[9] Bradley J.N., Shock Waves in Chemistry and Physics (Methuen, London; Wiley, New York, 1962) chapter on: Chemical reactions in shock tube.
[10] Calvin M., Chemical Evolution (Oxford University Press, London, 1969).
[11] Davis J.G. and J.C. Gill, Monthly Notices Roy. Astron. Soc. 121 (1960) 437.
[12] Gilvarry J.J. and A.R. Hochstim, Nature 197 (1963) 624, 626.
[13] Guenebaut H. et al., Compt. Rend. Acad. Sci., France 251 (1960) 1166.
[14] Hartmann W.K., Icarus 4 (1965) 157.
[15] Hawkins G.S., Nature 197 (1963) 781.
[16] Hawkins G.S., Meteors, Comets Meteorites (McGraw-Hill, New York, 1964).
[17] Hawkins G.S., Ann. Rev. Astron. Astrophys. 2 (1964) 149.
[18] Hochstim A.R., Proc. Natl. Acad. Sci. U.S. 50 (1963) 200.
[19] Kozlov G.I. and V.G. Knorre, Combust. Flame 6 (1962) 253.
[20] Lifshitz A., S.H. Bauer and E.L. Resler jr., J. Chem. Phys. 38 (1963) 2056.
[21] Martin J.J., Atmospheric Entry (Prentice-Hall, Englewood Cliffs, N.J., 1966).
[22] Miller S.L. and H.C. Urey, Science 130 (1959) 245.
[23] Nazarova T.N., Space Sci. Rev. 8 (1967) 455.
[24] Öpik E.J., Physics of Meteor Flight in the Atmosphere (Interscience, New York, 1958).
[24a] Öpik E.J., Irish Astrophys. J. 5 (1958) 34.
[25] Park, Woo Kong, Ph.D. Thesis, College of Engineering, Wayne State University, Detroit, Michigan, in preparation.
[26] Schoemaker E.M., R.J. Hackman and R.E. Eggleton, Advan. Astronaut. Sci. 8 (1962) 70.
[27] Vedder J.F., Space Sci. Rev. 6 (1966) 365; Dohnanyi J.S., J. Geophys. Res. 75 (1970) 3468.
[28] West C.D. and R. Howlett, Nature 215 (1967) 727; M. Anbar, Science 161 (1968) 1343; V.A. Konstantinov, Dokl. Akad. Nauk SSSR 56 (1947) 259; P. Jarman, Proc. Phys. Soc. (London) 73 (1959) 628; E. Meyer and H. Kuttruff, Angew. Phys. 11 (1959) 25; C. Hunter, J. Fluid Mech. 8 (1960) 241; K. Negishi, J. Phys. Soc. Japan 16 (1961) 1450; P.D. Jarman and K.J. Taylor, Brit. J. Appl. Phys. 15 (1964) 321; 16 (1965) 657; F.B. Peterson, Ph.D. dissertation, 1966 (Northwestern University, Light Emission from Hydrodynamic Cavitation); F.B. Peterson and P.T. Anderson, Phys. Fluids 10 (1966) 874.

Appendix I. Energy estimates from the flux of micrometeorites, meteors and meteorites

If the total number of meteors (or meteorites) striking 1 km^2 area in a year on the earth, having mass between m and $m + dm$ and velocity between v and $v + dv$ is denoted by $\hat{N}(m)$, then the total number striking with mass equal to or greater than m is

$$N(m) = \int_{v_0}^{v'} \int_{m}^{m'} \hat{N}(m) \, f(m, v) \, dm \, dv \tag{1.1}$$

where m' is the upper bound of observed mass in a given meteoroid category, v_0 and v' are the lowest and highest observed velocity ($v_0 \cong 11$ km/sec, $v' \cong 72$ to 82 km/sec) respectively, $f(m, v)$ is the joint distribution function of mass and velocity.

The kinetic energy of meteors (or meteorites) falling on 1 km^2 in a year and having masses equal to or greater than m is

$$E(m) = \tfrac{1}{2} \int_{m}^{m'} dm \, m \int_{v_0}^{v'} dv \, v^2 \hat{N}(m) \, f(m, v) \tag{1.2}$$

$E(o)$ is the total kinetic energy of all meteorites (or meteors), bounded by m'. We assume that the speed of meteorite (or meteor) is independent of its mass. Then we can write $f(m, v) = f_1(m) \, f_2(v)$ and

$$E(m) = \tfrac{1}{2} \overline{v^2} \int_{m}^{m'} m \hat{N}(m) \, f_1(m) \, dm \tag{1.3}$$

where

$$\overline{v^2} = \int_{v_0}^{v'} v^2 \, f_2(v) \, dv \, . \tag{1.4}$$

Differentiating,

$$\frac{dE}{dm} = -\tfrac{1}{2} m \overline{v^2} \, \hat{N}(m) \, f_1(m) \tag{1.5}$$

$$\frac{dN(m)}{dm} = -\hat{N}(m) \, f_1(m) \, . \tag{1.6}$$

One can verify the expression below by appropriate differentiation and using eq. (1.4) and eq. (1.9).

$$E(m) = \frac{m\overline{v^2}}{2}\left[N(m) + \frac{1}{m}\int_m^{m'} N(m)\,\mathrm{d}m\right]. \tag{1.7}$$

Since Hawkins [17] gives for $m \leqslant m'$

$$\log_{10} N = a - b\log_{10} m \tag{1.8}$$

i.e.,

$$N = \frac{\exp(a\ln 10)}{m^b} \tag{1.9}$$

resulting from a survey of measured numbers of micrometeors, cometary meteors and a survey of all collected large stony and iron meteorites. The results are a significant lower bound for the present flux, and the flux was probably even larger in the past. Substituting eq. (1.9) into eq. (1.7) we obtain *average lower bound energies*:

$$E(m) = \tfrac{1}{2}\, m\overline{v^2}\,\frac{N(m)}{b-1}\left[b - \left(\frac{m}{m'}\right)^{b-1}\right], \quad b > 1 \tag{1.10a}$$

$$E(m) = \tfrac{1}{2} m\overline{v^2}\, N(m)\,[1 + \ln(m'/m)]\;, \quad b = 1 \tag{1.10b}$$

and

$$E(m) = \tfrac{1}{2}\, m'\overline{v^2}\,\frac{N(m')}{1-b}\left[1 - b\left(\frac{m}{m'}\right)^{1-b}\right], \quad b < 1 \tag{1.10c}$$

For the velocity distribution we took Davies and Gill [11] data with $\bar{v} = 39.2$ km/sec and obtained $\sqrt{\overline{v^2}} \approx 45$ km/sec, which we used.

A. *Micrometeorites*

$R \leqslant 2.5$ μm, $m \leqslant 10^{-13}$ g, $\bar{\rho}_m \approx 3$ g/cm³. From Hawkins [17]

$$\log_{10} N(m) = 12.43 - 0.39\log_{10} m\;, \tag{1.11}$$

where m is in grams.

m (g)	E (cal. year^{-1}/earth)
10^{-17}	6.2×10^{18}
10^{-16}	6.1×10^{18}
10^{-15}	6.0×10^{18}
10^{-14}	5.6×10^{18}
10^{-13}	3.8×10^{18}

i.e., micrometeorites with mass equal to or greater than 10^{-17} g carry on the average 6.2×10^{18} cal on earth per year, and with mass equal to a greater than 10^{-14} g, only 5.6×10^{18}.

B. *Cometary meteors* *.

10^{-13} g $\leqslant m \leqslant 300$ g, $\bar{\rho}_m \approx 0.4$ g/cm^3. From Hawkins [17]

$$\log_{10} N(m) = 0.41 - 1.34 \log_{10} m \tag{1.12}$$

m (g)	E (cal. year^{-1}/earth)
10^{-13}	3.2×10^{19}
10^{-12}	1.4×10^{19}
10^{-10}	3.0×10^{18}
10^{-8}	6.3×10^{17}
10^{-2}	1.2×10^{14}

C. *Meteorites*

1. *Stony meteorites*

$\bar{\rho}_m \approx 3.4$ g/cm^3, 300 g $\leqslant m \leqslant 10^{10}$ g (before ablation). From Hawkins [17]:

$$\log_{10} N(m) = -0.73 - \log_{10} m . \tag{1.13}$$

m (g)	E (cal. year^{-1}/earth)
300	4.1×10^{14}
10^5	2.8×10^{14}
10^{10}	2.2×10^{13}

* Nazarova [23] for $10^{-16} \leqslant m \leqslant 10^{-5}$ g gives $N(m)$ as compiled by Hemenway from various space rockets, satellites, balloons, etc. The data lie below Hawkins [17] $N(m)$ by 1 to 2 orders of magnitude. For more information see also review by Vedder [27] and Dohnanyi [27].

The uncertainties include the amount of ablated mass (i.e., the pre-entry mass was estimated), the flux over entire earth and the flux in the past. It is quite possible that these figures need to be multiplied by a factor of 10 to 100 for primeval times.

2. *Iron meteorites*

$\bar{\rho}_m \approx 7.8$ g/cm^3, 10^{10} g $\leqslant m \leqslant m'$. From Hawkins [17] :

$$\log_{10} N(m) = -3.51 - 0.7 \log_{10} m \,. \tag{1.14}$$

The upper bound m' is not well known, although estimates of 10^{19}–10^{20} g have been made [15].

m (g)	$m' = 3 \times 10^{15}$ g	$m' = 10^{18}$ g
	E (cal.year^{-1}/earth)	E (cal.year^{-1}/earth)
10^{10}	5.3×10^{15}	3.1×10^{16}
10^{12}	5.1×10^{15}	–
10^{14}	4.0×10^{15}	2.9×10^{16}

(Spherical meteorite with radius $R = 1$ km weighs about 3×10^{16} g; for asteroid Eros which passes close to earth $m = 3 \times 10^{18}$ g.)
Similarly to the stony meteorites, these figures should also be considered as lower bounds, possibly to be multiplied by a factor of 10 to 100 for primeval times. The estimate by Calvin [10] is consistent with the straight line in fig. 1, if $R \approx 1$ km, $\bar{v} = 45$ km/sec and for 1 meteorite every 10^4 years.

3. *The frequency of large meteorites*

From Hartman [14] the estimate for the number of craters on the earth having diameter greater than 1 km and resulting from impact per 10^4 years is:

a) Meteorite and asteroid observations: 10.5 (Öpik [24a]); 2.5 to 11.5 (Brown [7]); 55 (Hawkins [17]).
b) Astrobleme * counts in central USA: 0.5 to 5 (Schoemaker, Hackman and Eggleton [26]).
c) Crater counts in Canadian Shield: 0.5 to 7.5 (Hartman [14]).

Hartman's [14] best estimate is a fall of 6 meteorites per 10^4 years on earth, making crater greater than 1 km in diameter, which corresponds to iron meteorites having radius $R > 10$ meters (see footnote after eq. 2.8).

* "Star-wound", i.e. scars due to meteoritic impact, deduced from examination of shatter cones in the rocks.

Appendix II. The equation of motion of meteorites in the atmosphere

In this appendix we are going to derive the critical size necessary for a meteorite to reach the surface of the earth without appreciable slowdown in the atmosphere.

The equation of motion of a spherical meteorite for hypersonic velocity ($v \geqslant 1$ km/sec) is (e.g., see Martin [21])

$$m \frac{dV}{dt} = mg \sin\gamma - \tfrac{1}{2} \rho C_D A V^2 \tag{2.1}$$

where $m^* = \tfrac{4}{3} R^3 \pi \rho_m$, $A = \pi R^2$, (for a sphere), R is the radius of a meteorite, m its mass, V the velocity, γ the entry angle with the horizontal ($\gamma \geqslant 5^\circ$), ρ the atmospheric density, C_D is the aerodynamic drag coefficient ($C_D = 1$ for the sphere) and ρ_m is the density of meteorite.

Using for the density an exponential atmosphere:

$$\rho = \rho_0 \, e^{-z/H} \tag{2.2}$$

where the scale height $H=P_0/\rho_0 g=RT/Mg$, where M is the molecular weight ** of the atmosphere.

Since

$$\frac{dV}{dt} = \frac{dV}{dz}\frac{dz}{dt} = -V \sin\gamma \frac{dV}{dz} = -\tfrac{1}{2} \frac{dV^2}{dz} \sin\gamma \,, \tag{2.3}$$

and with $y = V^2$ we obtain

$$\frac{dy}{dz} = By\, e^{-z/H} - 2g \,, \tag{2.4}$$

where

$$B = \frac{C_D A}{m} \frac{\rho_0}{\sin\gamma} = \tfrac{3}{4} \frac{\rho_0}{\rho_m} \frac{1}{R \sin\gamma} \,.$$

The initial conditions are at $z=z_0$ (e.g. z_0=300 km), $v=v_E$. The solution in an isothermal atmosphere for non-ablating spheres is (H=const. and g=const.).

$$V^2 = e^{-BH(e^{-z/H} - e^{-z_0/H})} \times \left[V_E^2 + 2g \int_z^{z_0} e^{BH(e^{-z/H} - e^{-z_0/H})} \, dz \right] \tag{2.5}$$

* Assuming negligible fraction of ablated mass.

** In the present earth's atmosphere $H \approx 8.3$ km at sea level.

or with $z_0 \gg H$,

$$V^2 \approx e^{-BH e^{-z/H}} \left(V_E^2 + 2g \int_z^{z_0} e^{BH e^{-z/H}} dz \right)$$

$$\approx e^{-BH e^{-z/H}} \left[V_E^2 + 2g(z_0 - z) + 2gH \sum_{n=1}^{\infty} \frac{(BH)^n}{n \cdot n!} e^{-nz/H} \right]. \quad (2.6)$$

The terms with $2g$ can be neglected for $v_E \geqslant 10$ km/sec.

If $BH = \frac{3}{4}(\rho_0/\rho_m)(H/\sin\gamma)(1/R) \equiv R_{min}/R < 1$, i.e. if

$$R > R_{min} = \frac{3}{4} \frac{\rho_0}{\rho_m} \frac{H}{\sin\gamma} = \frac{3}{4} \frac{1}{\rho_m} \frac{P_0}{g \sin\gamma}, \quad (2.7)$$

$$V \approx V_E ,$$

i.e., there is no appreciable drag for $R \gg R_{min} \simeq 1$ meter. It is interesting that one can obtain this answer intuitively * by assuming that the mass of the meteorite must exceed the mass of the swept out atmosphere, i.e.

$$m = \frac{4}{3} R^3 \pi \rho_m > \pi R^2 \int_0^{\infty} \rho dz = \pi R^2 \left(\frac{H}{\sin\gamma}\right) \rho_0 ,$$

or

$$R > \frac{3}{4} \frac{\rho_0}{\rho_m} \frac{H}{\sin\gamma} = R_{min} ,$$

which is exactly the limit derived in eq. (2.7). Thus for a vertically incident iron meteorite in the atmosphere with P_0=1 atm there will be no appreciable (less than 50%) slow down if R>1 meter, or from fig. 2 less than 5% if

$$R > 10 \text{ meters} . \quad (2.8)$$

On impact, such meteorites will make craters greater than 60 m in diameter **. The deceleration is given by

$$D \equiv -\frac{dV}{dt} \approx \frac{1}{2} \frac{\sin\gamma}{H} \frac{R_{min}}{R} V^2 e^{-z/H} , \quad (2.9)$$

* Dr. J.A.L. Thomson, Wayne State University, private communication; also Hawkins [16].

** Diameter of a crater, $d \simeq 3 \times 10^{-6} (\frac{1}{2} m V^2)^{1/3}$, where m is in grams, V in cm/sec and d in cm [14,26].

where

$$V \approx V_E \, e^{-\frac{1}{2}(R_{min}/R)\, e^{-z/H}} \, . \tag{2.10}$$

For $R < R_{min}$, the deceleration is maximum ($dD/dz = 0$) at altitude

$$Z_{max} = H \ln (R_{min}/R) \, . \tag{2.11}$$

At this altitude the velocity is

$$V = V_E/\sqrt{e} = 0.607 \, V_E \, . \tag{2.12}$$

For $R \geqslant R_{min}$ the deceleration increases continuously with the fall to the earth, where the impact velocity is

$$V(z=0) \simeq V_E \, e^{-\frac{1}{2}R_{min}/R} \, . \tag{2.13}$$

Chemical Evolution and the Origin of Life, eds. R. Buvet and C. Ponnamperuma

SHOCK SYNTHESIS OF AMINO ACIDS IN SIMULATED PRIMITIVE ENVIRONMENTS †

Akiva BAR-NUN *
Department of Chemistry, Cornell University,
Ithaca, New York, USA

Nurit BAR-NUN **
Laboratory for Planetary Studies, Cornell University

S.H. BAUER
Dept. of Chemistry, Cornell University

and

Carl SAGAN
Laboratory for Planetary Studies, Cornell University

A wide variety of energy sources are effective in generating α-amino acids (among many other organic compounds) from simulated primitive environments; the atmosphere of precursor gases may apparently have any composition, provided only that it is reducing and that the atoms H, N, O, and C are present [1,4,11,12,15]. While electrical discharge, ultraviolet, thermal, hypervelocity impact, and fast elementary particle energy sources have been used, there has heretofore been no report of prebiological organic synthesis employing shock tubes. Such syntheses may be of importance because: (a) the composition and temperature of the gas mixture, the duration of activation, and the rate of quenching can be controlled; (b) the significant

† Part of the manuscript was published in Science, April 24, 1970.
* On leave of absence from Dept. of Chemistry, The Hebrew University, Jerusalem, Israel. Present address: NASA Ames Research Center, Moffett Field, California, USA.
** On leave of absence from Dept. of Botany, The Hebrew University, Jerusalem, Israel. Present address: Palo Alto Medical Research Foundation, Palo Alto, California, USA.

kinetic reactions occur under strictly homogeneous conditions, since the sample is heated gas dynamically while the tube walls are cold; and (c) as discussed below, the product of energy available from shock sources on the primitive earth and the efficiency of synthesis may make shock excitation a major source of primordial organic molecules.

Experimental

The single pulse shock tube used for these experiments consists of a uniform bore, 1 X 3 inch I.D. Pyrex low-pressure (driven) section, 69 X 3 inch long, and an equal diameter brass high-pressure (driver) section, 18 X 3 inch long. These were separated by 3 mil Mylar diaphragms which were burst (thus initiating the shock) when the helium driver gas pressure was raised to about 120 psi *. Pressure traces were recorded with a dual beam oscilloscope operating a total scan times of 100 μsec and 5 msec. The first trace gave the shock passage time between two very fast response quartz piezoelectric gauges and thus permitted calculation of the shock speed; the second trace recorded the duration of the reflected shock at the terminal end of the tube, at the region from which a sample was extracted for analysis subsequent to passage of the quench wave.

The reaction mixture, prepared in a large glass bulb, consisted of 3.3% CH_4, 11% C_2H_6, and 5.6% NH_3 (Matheson "pure" grade, used without further purification), diluted with Matheson ultra-pure argon. We do not believe the results would be significantly affected by many other choices of reduced gases [1,4,11,12,15]. A second glass bulb was partly filled with distilled water and degassed several times under vacuum with the water alternately frozen and heated.

The glass section of the shock tube was cleaned 25 times with toluene and ethanol. After each cleaning the paper towel was checked with ninhydrin spray for amino acids; all tests were negative. With a heating tape the temperature of the driven section was raised to an approximate 80°C and pumped to below 10^{-4} torr for 5 hr. Water vapor at pressures of 5–25 torr was then introduced into the *driven* section from the water reservoir bulb attached to one of the valves located at the terminal end of the test section.

* This tube is a streamlined modification of the first single pulse shock tube constructed at Cornell (A. Lifshitz, S.H. Bauer and E.L. Resler, Jr., J. Chem. Phys. 38 (1963) 2056). This early tube was used in 1962 (A. Lifshitz and S.H. Bauer, unpublished) in unsuccessful attempts at shock synthesis of organic compounds from $CH_4/NH_3/H_2O$ mixtures. A general introduction to shock-tube operation can be found in S.H. Bauer, Science 141 (1963) 867.

To the extent that these experiments simulate meteoroid impact into the earth's atmosphere, introduction of water vapor at the far end of the shock tube corresponds to the stratification of water vapor in the lower atmosphere. The reaction mixture was then introduced into the *driven* section from the diaphragm end *via* the pump port. Sample pressures ranged from 10 to 90 torr. Two minutes were allowed for partial mixing. The helium driver gas was then introduced slowly until the diaphragm burst, sending the shock wave down the tube. As successive sections of the sample gas are traversed by the incident shock front their temperature rises sharply, in 10^{-8} sec to 1000–2000 °K (depending on the shock speed). Due to momentum conservation, the wave reflected from the terminal plate consists of another compression shock, so that the sample gas is further heated (2000–4000 °K) and additionally compressed to a net density approximately six times the original. When the reflected shock wave meets the now somewhat diffused boundary region between the driven and the driver gases, there is a sharp drop in pressure and temperature. Quenching is augmented when the length of the *driver* section is adjusted so that the expansion wave reflected from the opposite end of the tube simultaneously arrives at this intersection. Cooling rates in the range of $0.5–1.0 \times 10^6$ °K sec^{-1} result, depending on the care taken with this "tuning". The dwell periods for the reaction during which the mixtures are subjected to the reflected shock temperature (for this tube) ranged from 500 to 1000 μsec, determined by the shock speed. The incident and reflected equilibrium shock temperatures are calculated from the conservation equations, the thermodynamic functions of the reaction mixture, and the measured shock speed.

As rapidly as one can move, after initiating the shock, the valve between the terminal end of the shock tube and the previously evacuated sample -collecting bulb is opened. For these experiments the latter was a two-liter glass vessel into which 10 ml of 0.1 N HCl (degassed) has been placed. After collecting the shocked and quenched gas sample, the bulb was detached from the tube, vigorously shaken and the walls washed by slowly rotating the bulb. It was then evacuated, reattached to the test section, and the next run initiated. A series of the three shock samples was obtained for each range of shock speeds and the product samples collected in one bulb. This solution was then analyzed for amino acids by paper chromatography, with *n*-butanol: methanol:water:ammonia (10:10:5:2) and by column chromatography in an amino acid analyzer.

Results

Ten shocks were made, covering the range of compositions and shock speeds summarized in table 1. All the dwell times were approximately 500 μsec. Three control runs (group D) were obtained with the reaction mixture used for groups B and C. For group D, the sample was introduced into the driven section in the usual manner, allowed to remain in the shock tube for five minutes and then slowly pushed into the sample bulb by a stream of helium. The mixture had a residence time of about five minutes at 380 °K. Group A shocks also constitute a control. No amino acids were detected in either groups A or D. The former demonstrates that the shock tube was "clean"; i.e., the (hydrocarbon/ammonia/water) mixture is necessary for amino acid production. The latter demonstrates that shock heating is essential; a five-minute exposure of the ingredients to 380 °K is not sufficient. In groups B and C amino acids were found in the following concentration (μmoles per 10 ml):
B) glycine, 73; alanine 34.5; valine 0.55; leucine 0.105;
C) glycine, 62.5; alanine 17.6; valine 3.25; leucine 0.27.

Table 1
Experimental conditions for shock synthesis [a]

Shock no.		H_2O pressure (torr)	Total initial pressure (torr)	Incident shock speed (mm/μsec)	Approximate [b] reflected shock temperature (°K)
(A)	1	25	50	1.18	2000
	2	22	47	1.18	2000
	3	12	32	1.33	3000
(B)	4	19	69	1.08	1600
	5	10	35	1.30	3000
	6	10	35	1.30	3000
(C)	7	20	110	0.95	1000
	8	7	30	1.30	3000
	9	5	20	1.42	3500
	10	3	15	1.42	3500

[a] Average dwell periods in reflected shock regime ≃500 μsec.
[b] Estimated on the basis of no reaction, assuming argon was the driven gas. Actual sample temperatures are several hundreds of degrees lower.
Group A: Test gas was argon plus water.
Groups B and C: Test gas was 3.3% CH_4, 11% C_2H_6, 5.6% NH_3 plus H_2O (as indicated); the major constituent was argon.

Yields calculated for group B were surprisingly high; 36% of the ammonia present was converted to amino acids. This demonstrates direct and highly efficient amino acid production initiated by shocks under strictly homogenous conditions, with all constituents in the gas phase.

Thermodynamic considerations

Because only a few experiments have been performed thus far, no rate laws can be determined. Nonetheless, thermochemical considerations place strict limits on the possible processes. For example, were glycine produced under conditions of constant temperature and pressure, its yield would be limited by $\Delta G^\circ_{\cdot}(T)$ for the equivalent single step reaction:

$$NH_{3(g)} + C_2H_{2+2n(g)} + 2H_2O_{(g)} = \underset{\displaystyle NH_2}{\underset{|}{H_2C}}-COOH_{(g)} + (2+n)\,H_{2(g)}$$

We have considered cases with $n = 0$, 1, 2. Regrettably only very rough estimates are available for the enthalpy and entropy of sublimation and the heat capacity of glycine. It is evident, however, that for the above reaction $\Delta S^\circ(T) > 0$ for $n = 0$, 1. If one assigns the reasonable value of 74 e.u. to gaseous glycine at 298 °K, $\Delta S^\circ{}_{298}$ $(n = 2) \simeq +8$ e.u.; $\Delta S^\circ{}_{298}$ $(n = 1) \simeq -21$ e.u. and $S^\circ{}_{298}$ $(n = 0) \simeq -48$ e.u. To counterbalance the net positive $T\Delta S^\circ$ term, the associated enthalpy change must be negative. For the above reaction, $\Delta H^\circ{}_{298}$ $(n = 2) \simeq +39$ kcal mole^{-1}, $\Delta H^\circ{}_{298}$ $(n = 1) \simeq +8$ kcal mole^{-1}, and $\Delta H^\circ{}_{298}$ $(n = 0) \simeq -34$ kcal mole^{-1}. However, the latter value is somewhat misleading since at shock temperatures the heat capacity correction is more important than the negative enthalpy of reaction. Clearly acetylene is the most favorable reagent, but the predicted yield at high temperatures is still very small. One is thus forced to the conclusion that the observed high-production efficiency is a consequence of the thermal history of the shock sample, i.e., the high-temperature pulse during which the reagents are partially fragmented *must be followed by a rapid quench*, during which the radicals react with the remaining reagents in a chain mechanism. The essential reunification steps occur with relatively low activation energies. The high temperature thermodynamic barrier is thus bypassed. It is convenient to introduce a name for this process; for the present, we propose kinetic thermoquench (ktq).

The equilibrium compositions of systems which consist of C,H,O,N,S . . . have been calculated for a wide range of conditions [2]. These show that while the mole fractions of low molecular weight radicals and molecules are appreciable at high temperatures, the relative concentrations of the complex species are very low indeed. Hence, reaction and condensation in the quench step are essential. The relative concentrations of the four amino acids found in this study are close to the mole fraction ratios of the radicals H, CH_3, $(CH_3)_2CH$, and $(CH_3)_2CH\text{-}CH_2$ calculated at equilibrium, around 2000 °K [3]. The corresponding ratios are: $H/HC_3 \simeq 1.5$, $CH_3/C_3H_7 \simeq 40$, and $CH_3/C_4H_9 \simeq 800$. In experiment B the ratios of the amino acids are: Gly/Ala = 2.1, Ala/Val = 63, and Ala/Leu = 350. These rough comparisons indicate that while thermodynamic relations do impose a barrier for α-amino acid production and condensation if one insists on isothermal processes, no such restriction is present in ktq processes. Quantitative demonstrations of feasibility require measurement of specific rate constants in the quench period – a difficult but not impossible task for current shock-tube technology.

The above-described shock-tube techniques can be applied to the study of fragmentation and reassembly reactions in the HCN/H_2O system, as demonstrated by one brief experiment which we have performed with a mixture of $C_2H_6/C_2N_2/H_2O$. A product with a broad UV absorption spectrum was generated, indicating condensation reactions had taken place, but no specific α-amino acids were found. There is also the interesting possibility that during the last stages of the quench period the concentrations of amino acids, polymers, pyrimidines, etc., could vastly exceed their equilibrium sublimation pressures at the corresponding ambient gas temperatures, so that nucleation and crystallization from the gas phase may occur. If such crystallizations take place, it is possible that concurrent separation of D- and L-isomers occurs.

Conversion efficiency and comparison with other energy sources

To place the prebiological synthetic role of shock waves in context, we consider the following rough estimates of relative efficiencies. For the shocks used we estimate the energy input to the test sample to have been 10–12 kcal $mole^{-1}$, or $4\text{–}5 \times 10^{11}$ ergs $mole^{-1}$. Since only 5×10^{-4} moles of reagents were present, the rest being argon, these absorbed $\simeq 2 \times 10^8$ ergs, which produced 3.7×10^{-5} moles of amino acids, that is, 2.2×10^{19} molecules, implying $\simeq 9 \times 10^{10}$ molecules per erg of shock-injected energy. In contrast, the quantum yield for ultraviolet photoproduction of α-amino acids under simulated primitive terrestrial conditions is $\sim 10^{-6}$, or less if Hg-sensitization is not employed [6]. This corresponds to $\sim 10^5$ molecules per erg of absorbed long-wavelength ultraviolet light. The efficiency advantage of shock over ultraviolet excitation thus appears to be some six orders of magnitude.

To compare the overall significance of shock and ultraviolet primitive energy sources, we must compare the absolute energies available. One obvious source of shock waves on the primitive earth, comparable in many respects to those produced in the experiments reported above, is hypervelocity impact by cometary meteors and micrometeorites, which dissipate their kinetic energy at altitudes above the earth's surface. Thus both the experiments and the interpretation reported here must be distinguished from the proposal of Hochstim [8], which invokes the shock formation around large meteorites and the subsequent collision of the shock-heated gaseous envelopes and the meteorites themselves with liquid water at hypersonic velocities. There are many more orders of magnitude of kinetic energy available in the smaller objects which do not reach the surface at hypersonic velocities; in addition, the quenching of a shock-heated gas mixture by liquids differs fundamentally from that produced by gas-dynamic expansion *. Above an altitude of some 90 km, most cometary meteors and micrometeorites have been either completely ablated or thermalized so the remnants float down to the surface like a fine rain (see e.g. [14]). The distribution function of incoming particles is a power law strongly weighted towards the smallest particles. Some uncertainty still exists on values of this distribution function for the contemporary earth. On the primitive earth a larger flux, particularly of cometary debris, may be expected. A representative and conservative mass flux for the primitive earth is 2×10^{-14} g cm^{-2} sec^{-1} with arrival velocities of some 35 km sec^{-1} [9, cf. 7]. The corresponding energy flux is $\sim 10^{-1}$ erg cm^{-2} sec^{-1} $\sim 10^{-1}$ cal cm^{-2} yr^{-1} **.

An additional source of shocks on the primitive earth is thunder; we wish clearly to distinguish between synthesis in the lightning column and synthesis in the resulting shock wave. The latter can be approximated by a pressure pulse propagating away from a cylindrical energy source. At the center the temperature is approximately 2×10^4 °K, where no complex chemical reactions can occur. Behind the propagating pressure pulse the temperature drops by expansion and radiative cooling, reaching 4000 °K at a distance 10–30 cm from the center [17]. At this point, temperatures are low enough for

* Although the emphasis in Hochstim's theory is on a collision of the hot gas with water, he also mentions the possibility of formation of quite complex species at the extreme end of the wake that follows an incoming meteorite.

** H.R. Hulett, by using other data, reaches an erroneous value of 4×10^{-5} cal cm^{-2} yr^{-1} [10].

chemical processes of the type reported in the present paper to occur. Further cooling, due to expansion alone, freezes the reaction mixture, at a rate comparable to the present experiment. If we assume a rate of electrical discharge on the primitive earth similar to that on the contemporary earth [13], and knowing that ~ 75 percent of the electrical energy in such a discharge appears as the accompanying shock wave, an additional energy source of several times 10^{-1} cal cm^{-2} yr^{-1} appears.

Thus, a total shock flux on the primitive earth ~ 1 cal cm^{-2} yr^{-1} does not seem unreasonable. By comparison, the total energy available as ultraviolet light on the primitive earth, at wavelength shorter than about 3000 Å where some organic photochemistry can be expected, is computed [5,16] from models of the evolution of the sun to be ~ 10^3 cal cm^{-2} yr^{-1}. Accordingly, ultraviolet light is a thousand times more abundant than shock waves on the primitive earth; but shocks are a million times more efficient in producing amino acids. The only other energy source of any magnitude [13] appears to be volcanic thermal activity, in which the synthesized molecules are left for a significant time in the high-energy region, leading to significantly lower efficiencies than in the case for rapidly quenched shocks. This is probably also the principal reason for the low ultraviolet quantum yields. We are thus led to the unexpected conclusion that the impact of cometary meteors and micrometeorites, and thunder, were the principal energy sources leading to building blocks essential for the origin of life, by a factor of perhaps 10^3.

The bulk of the impacting objects dissipate their energy at a pressure corresponding to the 100 km altitude level in the contemporary terrestrial atmosphere. If the primitive atmosphere had a base pressure ~ 1 bar, then most of the organic compounds synthesized by impact shock would reach the surface only by slow diffusive processes and run considerable risk of ultraviolet photo-dissociation on the way. If the base pressure were less, the situation is improved. In any case, thunder and larger meteors dissipate their energy in the lower troposphere; a significant fraction of shock dissipation in the primitive earth occurred near surface water. The high-pressure shocks of meteors which reach the sea provide an interesting environment for anhydro-copolymerization of amino acids and other monomers which do not readily polymerize in water in the absence of special condensing agents; some other scenarios for anhydrocopolymerization and subsequent solution in liquid water have been criticized on grounds of geophysical implausibility (see e.g. [4]).

With the numbers presented above, it follows that ~ 10^{10} organic molecules cm^{-2} sec^{-1} were produced by shocks on the primitive earth. If the mean molecular weight of synthesized molecules is ~ 100, the production flux is $\sim 10^{-12}$ g cm^{-2} sec^{-1}. For a primitive reducing atmosphere lasting 10^9 years a total production of organic molecules ~ 30 kg cm^{-2} is implied. This quantity of carbon is quite comparable to the quantity present as carbonate sediments in the earth's crust, suggesting that a major fraction of the present crustal carbon on the earth was once in the form of organic compounds. Were all this organic matter to have survived and been dissolved in oceans of contemporary depth and extent, a 10 percent solution of organic compound would result. Even with efficient loss mechanisms it appears that shock production of organic molecules may have played a significant role in the origin of life.

Acknowledgements

This research was supported in part by the U.S. Air Force Office of Scientific Research Contract No. AF49(638)-1448 at the Dept. of Chemistry, and by NASA Grant No. NGR-33-010-101 at the Laboratory for Planetary Studies.

References

[1] Abelson P.H., Ann. N.Y. Acad. Sci. 69 (1957) 274.
[2] Dayhoff M.O., E.R. Lippincott and R.V. Eck, Science 153 (1966) 628; see also NASA Special Report SP-3040 (1967).
[3] Duff R.E. and S.H. Bauer, J. Chem. Phys. 36 (1962) 1754.
[4] Fox S.W., ed., The Origins of Prebiological Systems (Academic Press, New York, 1965).
[5] Fox S.W., ed., ibid. p. 238.
[6] Groth W.E. and H. von Weyssenhoff, Planetary Space Sci. 2 (1960) 79.
[7] Hawkins G.S., Ann. Rev. Astron. Appl. 2 (1964) 149.
[8] Hochstim A.R., Proc. Natl. Acad. Sci., U.S. 51 (1963) 200; Nature 197 (1963) 624.
[9] Hodge P.W. and F.W. Wright, Icarus 10 (1969) 214.
[10] Hulett H.R., J. Theoret. Biol. 24 (1969) 56.
[11] Kenyon D.H. and G. Steinman, Biochemical Predestination (McGraw Hill, New York, 1969).
[12] Miller S.L., J. Am. Chem. Soc. 77 (1955) 2351.
[13] Miller S.L. and H.C. Urey, Science 130 (1959) 245.
[14] Öpik E.J., Physics of Meteor Flight in the Atmosphere (Interscience, New York, 1958).
[15] Ponnamperuma C. and N.W. Gabel, Space Life Sci. 1 (1968) 65.
[16] Sagan C., Organic Matter and the Moon, Natl. Acad. Sci. Publ. 757 (1961).
[17] Troutman W.W., J. Geophys. Res. 74 (1969) 4595.

Chemical Evolution and the Origin of Life, eds. R. Buvet and C. Ponnamperuma

PREBIOTIC SYNTHESIS OF THE AROMATIC AND OTHER AMINO ACIDS

Nadav FRIEDMANN, William J. HAVERLAND
and Stanley L. MILLER
Department of Chemistry, University of California,
San Diego, La Jolla, California 92037, USA

There are so far two basically different processes for the synthesis of amino acids under primitive earth conditions – electric discharge reactions in a reduced atmosphere and analogous syntheses using other sources of energy, and amino acids obtained from ammonium cyanide polymerizations. There have been a number of claims for the synthesis of valine, isoleucine, leucine, phenylalanine, tyrosine, and tryptophan by such processes, but these compounds were identified only by paper chromatography or only by the position of a peak on the amino acid analyzer. It is evident that a single R_f value, either on paper or on an ion exchange column, is insufficient evidence to identify an amino acid.

Valine and isoleucine

An example of this problem are the small amounts of valine, isoleucine and leucine reported to be synthesized during an ammonium cyanide polymerization [21,22]. We have been unable to detect these amino acids from such a polymerization reaction [12].

A litre of 1.1 M HCN and 1.3 M NH_3 was heated at 40 °C for 21 days. All reagents were distilled before use, and the HCN and NH_3 were added as gases from a vacuum system. The dark brown solution and black polymer were hydrolyzed in 3 M HCl for 20 hr at 100 °C. The black polymer was discarded and the hydrolysate was desalted by absorbing on Dowex 50 (H^+) and eluting the amino acids off with 2 M NH_4OH. The NH_4OH fraction was evaporated to dryness and chromatographed on Dowex 50 (H^+) [47]. The tubes where valine, alloisoleucine, isoleucine and leucine appear were evaporated to dryness and chromatographed on paper (butanol:acetic acid:water, 4:1:5). A number of unidentified ninhydrın spots were found, but none with the R_f value of valine, alloisoleucine, isoleucine or leucine. Chromatography of the appropriate fractions from the Dowex 50 (H^+) chromatography on the buffered columns of a modified Beckman amino acid analyzer [6] showed less than 2×10^{-9} moles of valine, alloisoleucine, isoleucine or leucine. The yield of these amino acids was less than 1×10^{-6} percent based on cyanide carbon.

A modification of the cyanide polymerization can, however, yield valine. Acetone (0.09 M) was added to a mixture of hydrogen cyanide (1.3 M) and NH_3 (2.9 M), and kept at 40 °C for 21 days. The isolation of the valine was similar to the above procedure. The yield was 0.3 percent based on the acetone.

Acetone is a reasonable compound to occur in the primitive ocean. Indirect evidence for its synthesis by electric discharges has been obtained [27], and it can be synthesized by ultraviolet light with ethane and carbon monoxide [8]. It could also be synthesized by the hydration of methyl acetylene, which can be synthesized by the pyrolysis of C_1 to C_4 hydrocarbons [10] and by the action of electric discharges on C_1 to C_4 hydrocarbons [11]. The polymerization of 0.01 M HCN, 0.02 M NH_3 and 0.001 M acetone gave a valine yield of 2×10^{-3} percent based on acetone. This valine synthesis would therefore be effective only in the concentrated HCN solutions of the HCN eutectic needed for purine synthesis [37].

A plausible mechanism of this synthesis, based on the fact that cyanide polymerizations give substantial yields of glycine nitrile [21,22,33], would be:

$$CH_3-\overset{\overset{O}{\|}}{C}-CH_3 + \underset{\underset{NH_2}{|}}{CH_2}-C{\equiv}N \rightarrow (CH_3)_2\underset{\underset{OH}{|}}{C}-\underset{\underset{NH_2}{|}}{CH}-C{\equiv}N \rightarrow$$

$$(CH_3)_2C{=}\underset{\underset{NH_2}{|}}{C}-C{\equiv}N \xrightarrow[\text{(2) hydrolysis}]{\text{(1) reduction}} (CH_3)_2CH-\underset{\underset{NH_2}{|}}{CH}-COOH$$

Evidence for this mechanism was obtained by incubating 0.1 M HCN, 0.1 M NH_3, 0.025 M H_2CO (equivalent to 0.025 M glycine nitrile) and 0.05 M acetone for 20 days at 40 °C. The yield of valine with these more dilute solutions was 0.2 percent based on acetone. In addition to substantial quantities of α-amino-isobutyric acid, β-hydroxyvaline was found. This compound can be considered analogous to α,β-dihydroxyisovaleric acid, an intermediate in the biosynthesis of valine.

A similar set of experiments with methyl ethyl ketone instead of acetone gave a 0.01 percent yield of isoleucine and alloisoleucine (alloisoleucine/isoleucine = 1.1) from the concentrated NH_4CN and methyl ethyl ketone polymerization, and 0.02 percent from the more dilute glycine nitrile and methyl ethyl ketone experiment.

It is not clear why the yields are substantially lower for isoleucine than for valine. It should be noted that nearly equal amounts of isoleucine and alloisoleucine were obtained. Other syntheses might give a different ratio of these diastereomers, but any chemical synthesis of isoleucine should give alloisoleucine as well.

These experiments suggest a plausible mechanism for the synthesis of valine and isoleucine on the primitive earth, in addition to the proposed Strecker synthesis with the appropriate aldehyde [26,32] and the radical addition of propene and butene to polyglycine [1].

Phenylalanine and tyrosine

The synthesis of the aromatic amino acids, phenylalanine and tyrosine, have been reported by a number of workers, but the identifications were based only on paper chromatography [20,34,43], and in one case on a single peak on the amino acid analyzer [15]. The lack of phenylalanine and tyrosine in such experiments combined with the apparent difficulty of synthesizing them results in their absence from some lists of "primitive" amino acids [5,7,17].

We propose that phenylalanine was synthesized on the primitive earth from phenylacetylene, followed by hydration to phenylacetaldehyde and a Strecker synthesis with the HCN and NH_3, which are generally assumed to have been present in the primitive oceans [12]:

$$C_6H_5{-}C{\equiv}CH \longrightarrow C_6H_5{-}CH_2{-}CHO \xrightarrow[HCN]{NH_3}$$

$$C_6H_5{-}CH_2{-}CH(NH_2){-}C{\equiv}N \xrightarrow{\text{Hydrolysis}} C_6H_5{-}CH_2{-}CH(NH_2){-}COOH$$

Phenylacetylene can be synthesized from a variety of mixtures of simple hydrocarbons by high temperatures, electric discharges, and ultraviolet light (table 1). The synthesis of phenylacetylene from various hydrocarbons by an electric arc [3,48], by electro cracking [28], and by pyrolysis [42] has been reported. The phenylacetylene was identified by its gas chromatography retention time on three columns, by gas chromatograph–mass spectrometer combination, and by its ultraviolet and infrared spectra. Benzene, styrene, and naphthalene are also synthesized in most of these experiments. Success in these experiments is dependent on removing the products from the region of the hot wire or electric discharge by cooling the walls of the flask to approximately −20°C [36]. This is analogous to reactions in the atmosphere where the phenylacetylene would have time to diffuse to the ocean before being broken up by various energy sources in the atmosphere. The yields of phenylacetylene in these experiments are between 0.001 to 5 percent. The rate of production of phenylacetylene in the primitive atmosphere is difficult to estimate, but the phenylacetylene yields in these experiments suggest that the rate could be substantial.

The hydration of phenylacetylene is difficult to achieve under primitive

Table 1
Formation of phenylacetylene from simple hydrocarbons. The hot wire source was a tungsten wire at approximately 1300°C in a 1-liter flask. The spark source was a spark-type discharge from a Tesla coil in a 3-liter flask. The ultraviolet source was a Hanovia medium-pressure Hg ultraviolet lamp at 22 °C.

Gas	Energy source	Time (hr)	Yield of ϕ–C≡CH (%)
CH_4 (450 mm)	Hot wire	3	0.05
	Spark	12	0.001
C_2H_6 (450 mm)	Hot wire	1.5	4.7 [a]
	Spark	12	0.003
C_2H_4 (50 mm) + CH_4 (400 mm)	Hot wire	2	1.5 [b]
	Spark	12	0.04 [b]
C_2H_2 (5 mm) + CH_4 (445 mm)	Hot wire	3	0.40 [b]
	Spark	12	0.02 [b]
C_2H_2 (10 mm) + CH_4 (360 mm)	Ultraviolet	0.7	0.12 [b]

[a] The yield is not affected substantially on the addition of 200 mm of nitrogen or 50 mm of ammonia to the reaction flask.
[b] Yield based on either C_2H_4 or C_2H_2.

earth conditions. The hydration can take place in two ways, electrophilic (acid catalyzed) to give acetophenone [31] and nucleophilic (base catalyzed) to give phenylacetaldehyde [25]. A Strecker synthesis with acetophenone gives α-amino-α-phenylpropionic acid, a compound which is presumed to have played no role in the origin of life. Heating phenylacetylene in aqueous solutions buffered between pH 7 and 10 results in very small yields of phenylacetaldehyde and acetophenone. The hydration products were detected by converting them to the amino acids with NH_3 and HCN, followed by hydrolysis. It is important to deaerate these solutions; otherwise the phenylacetylene is oxidized to benzaldehyde, which gives phenylglycine as a Strecker product. The phenylacetylene, H_2S, HCl, and NH_4OH were distilled immediately before use. The mixture was desalted, and the amino acids were analyzed on the modified Beckman-Spinco amino acid analyzer and by paper chromatography. Heating aqueous ammonia solutions buffered between pH 7 and 10 also gives only small yields of phenylacetaldehyde. The half life is approximately 500 years at 100 °C and pH 8. Increasing the pH to 14 does not increase the yield because of decomposition of the phenylacetaldehyde produced. At lower temperatures the rate of addition of OH or NH_3 to phenylacetylene would probably be too slow to have been effective in the primitive ocean.

The nucleophilic addition of HCN to phenylacetylene gives cinnamonitrile, which upon addition of NH_3 and hydrolysis yields β-aminohydrocinnamic acid rather than phenylalanine. The addition of NH_3 to cinnamic acid gives exclusively β-aminohydrocinnamic acid at pH values between 5 and 14 (Bada and Miller, unpublished results). Furthermore, the equilibrium constant for phenylalanine synthesis from cinnamic acid, a reaction catalyzed by phenylalanine–ammonia lyase, is unfavorable [16].

The nucleophilic addition of H_2S to phenylacetylene at pH's between 7 and 12 results in good yields of phenylacetaldehyde. The conditions and yields are given in table 2. The second-order rate constant for HS^- as the nucleophile is about 10^4 times as great as that for the addition of OH^- and is comparable to the rate of addition of *p*-toluene thiolate anion to phenylacetylene [45]. The phenylalanine was identified by its R_f value in three paper chromatography solvents, and its position on the amino acid analyzer. This identification was confirmed by preparing the phenylhydantoin derivative. Its melting point was 171 to 172 °C, the same as that of authentic 3-phenyl-5-benzylhydantoin, and the mixed melting point was not depressed.

Phenylacetylene can also be hydrated to phenylacetaldehyde by way of radical addition of H_2S.

The radical addition of mercaptans to phenylacetylene is well known [41], but the product $C_6H_5CH{=}CH{-}SR$, cannot rearrange and hydrolyze to phenylacetaldehyde. An aqueous mixture of H_2S and phenylacetylene, buffered at pH 7.5 in the absence of air, was irradiated for 5 min with a Hanovia medium-pressure Hg ultraviolet lamp. Ammonia and HCN were added, and the mixture was hydrolyzed and desalted. The yields of phenylalanine are comparable to the nucleophilic addition of H_2S (table 2). The HS

Table 2
Formation of phenylalanine from phenylacetylene.

H_2S (M)	pH	Temp (°C)	Time (hr)	Yield based on [a] phenylacetylene added	Yield based on [b] phenylacetylene used
3×10^{-4}	7.8	60	84	0.03	
6×10^{-3}	7.8	60	67	0.5	3.8
6×10^{-3}	8.3	80	44	6.8	20.4
6×10^{-3}	7.8	100	67	5.1	17.0
1.5×10^{-2}	7.5	22	0.08 (UV) [c]	1.7	10.5
1.3×10^{-3}	7.5	22	0.16 (UV) [c]	0.1	

[a] Based on initial amount of phenylacetylene added. The phenylacetylene concentration was 3×10^{-3} M in these experiments.

[b] Based on the amount of phenylacetylene reacted which was followed spectrophotometrically.

[c] Irradiation with a Hanovia medium-pressure Hg ultraviolet lamp.

radicals generated from $H_2O_2 + H_2S$ and from the action of visible light on yellow solutions of ammonium polysulfide are also effective in synthesizing phenylacetaldehyde.

In addition to the phenylalanine obtained from radical reaction of H_2S, a small yield of tyrosine was found. No tyrosine was observed in the nucleophilic addition of H_2S to phenylacetylene. The tyrosine was isolated by elution from Dowex 50 (H^+) with increasing concentrations of HCl [47]. The tyrosine peak was evaporated to dryness. Re-chromatography of this fraction on the buffered columns of the amino acid analyzer showed a peak corresponding to tyrosine. The yield of tyrosine was 2 percent of the phenylalanine yield. Both *o*- and *m*-tyrosine, which may also have been synthesized along with the tyrosine, were not isolated, but peaks corresponding to these amino acids were observed on chromatograms from the amino acid analyzer.

Tyrosine would also have been synthesized on the primitive earth by hydroxylation of phenylalanine. Solutions of phenylalanine are hydroxylated by ultraviolet light and γ-rays to tyrosine and dihydroxyphenylalanine [9,46; 18,23,39,44]. Various hydrogen peroxide reagents also will hydroxylate benzene rings. Molecular oxygen combined with reducing agents, such as ascorbic acid [30,38], also hydroxylate phenylalanine, but molecular oxygen was probably absent from the primitive earth except for traces.

These results suggest that the presence of H_2S in the primitive ocean is necessary for the synthesis of phenylalanine. Although H_2S is the stable form of sulfur under reducing conditions, an excess of Fe^{2+} would precipitate all the sulfide. These results imply that the sulfide may have been more abundant than iron in the primitive ocean. The presence of HS^- in the ocean would be accompanied by H_2S in the ocean and in the atmosphere. Hydrogen sulfide absorbs ultraviolet light at wavelengths above 2000 Å (10^{-4} atm of H_2S has an optical density of 1 at 2600 Å). Therefore, H_2S can protect NH_3 from photodecomposition in the atmosphere and can protect organic compounds absorbing below about 2600 Å in the oceans.

Tryptophan

By almost any standard tryptophan would be considered the most complex of the twenty amino acids in proteins. It is frequently held that tryptophan was introduced into proteins after the development of the genetic code [5,7,17]. One reason is that tryptophan has only one codon.

Tryptophan is also considered to be very unstable, as indeed it is to acid hydrolysis. However, it is stable to basic hydrolysis. We have heated solutions of tryptophan at 100 °C for 20 days at pH values between 5 and 9. The tryptophan in the pH 5 solution was almost completely destroyed, but the tryptophan in the pH 9 solution was almost completely recovered. These results imply that tryptophan might be stable for long periods in the primitive ocean at pH 8 and low temperature.

A reasonable precursor for a primitive synthesis of tryptophan is indole, which can be synthesized by pyrolysis and electric discharges on mixtures of hydrocarbons and ammonia (table 3). The yields vary from 8×10^{-5} to 3.1 percent, with the highest yields from the pyrolysis of acetylene and ammonia.

The 3-carbon of indole is nucleophilic and could undergo Michael type addition to an activated double bond. An example would be dehydroalanine [40]. The enzyme tryptophan synthetase may convert serine to dehydroalanine and then catalyze the addition of indole to the dehydroalanine. However, dehydroalanine is unstable, tautomerizing to the imine and then hydrolyzing to pyruvic acid and ammonia. Metzler et al. [24] have synthesized small amounts of tryptophan from indole and serine using pyridoxyl and aluminium as catalysts. However, the same system also catalyzed the decomposition of tryptophan to indole. Furthermore, pyridoxyl has not yet been established as a primitive earth compound.

Dehydroalanine is relatively stable in a peptide [14]. We have used *N*-acetyl serine amide as a peptide model. Heating 6.8×10^{-3} M indole and 2.2×10^{-3} M *N*-acetyl serine amide for 192 hr at 100 °C at pH 8.74, followed by basic hydrolysis gave 0.05 percent yield of tryptophan based on the

Table 3
Formation of indole from simple hydrocarbons and ammonia.
The apparatus are described in table 1.

Hydro-carbon	(mm)	Ammonia (mm)	Energy source	Time (hr)	Yield of indole (%) [a]
CH_4	426	51	Hot wire	2.0	4.0×10^{-3}
CH_4	407	100	Spark	1.0	$<1.0 \times 10^{-5}$
C_2H_6	355	55	Hot wire	1.3	1.0×10^{-1}
C_2H_6	340	46	Spark	1.5	3.3×10^{-3}
C_2H_4	432	45	Hot wire	1.0	1.0×10^{-1}
C_2H_4	350	100	Spark	1.5	6.4×10^{-3}
C_2H_2	210	46	Hot wire	1.0	3.1
C_2H_2	110	60	Spark	1.0	1.4×10^{-2}

[a] Yield based on the initial amount of carbon present.

N-acetyl serine amide. This is a rather slow reaction. It is analogous to Akabori's [1] "fore-protein" synthesis of amino acids. A similar reaction was tried with polydehydroalanine, but no tryptophan was obtained [5], possibly because the addition is slow.

The addition of indole is more rapid to dehydroalanine hydantoin. Dehydroalanine hydantoin is a reasonable primitive earth compound. It can be formed by the addition of cyanate to serine, both of which are primitive earth compounds, to give serine hydantoic acid. The hydantoic acid would be in equilibrium with serine hydantoin, which in turn would be in equilibrium with dehydroalanine hydantoin. The equilibrium constants and rates for these reactions are not known except that the maximum rate of dehydration of serine hydantoin is at pH 8.5. Serine hydantoin can also be synthesized from glycine hydantoin by addition of formaldehyde to the 5 position. Although dehydroalanine hydantoin was probably present in the primitive ocean, there are not enough data available to estimate its concentration.

$$HO-CH_2-CH(NH_2)-COOH \xrightarrow{^-NCO} HO-CH_2-CH(NH-CO-NH_2)-COOH$$

$$\rightleftharpoons \text{HO}-CH_2-CH\text{ (serine hydantoin)} \rightleftharpoons CH_2{=}C\text{ (dehydroalanine hydantoin)}$$

$$\xrightarrow{\text{Indole}} \text{indolyl}-CH_2-CH\text{ (hydantoin)} \xrightarrow{\text{Hydrolysis}} \text{indolyl}-CH_2-CH(NH_2)-COOH$$

Heating 1×10^{-2} M indole and 1×10^{-2} M serine hydantoin for eight days at 100 °C and pH 8.55 gives 0.01% yield of tryptophan. The yield after 53 days at 60 °C is 0.5%. It appears that lower temperatures give increased yields of tryptophan because of the decomposition of the serine and dehydroalanine hydantoin at high temperatures.

The rate of these reactions at lower temperatures remains to be determined. These reactions will probably be relatively slow at lower temperatures although small yields of tryptophan could be expected under primitive earth conditions. Other possible analogs of dehydroalanine which would react more rapidly with indole at lower temperatures are being investigated.

Serine

Serine has been synthesized from the cyanide polymerization and electric discharge reactions. An additional serine synthesis would be the reaction of glycine with formaldehyde:

$$\underset{\displaystyle NH_3^+}{\underset{|}{CH_2}}-COO^- + H_2CO \rightleftharpoons \underset{\displaystyle OH}{\underset{|}{CH_2}}-\underset{\displaystyle NH_3^+}{\underset{|}{CH}}-COO^-$$

This is an equilibrium reaction catalyzed by the enzyme serine hydroxymethyl transferase [2,4]. The equilibrium constant is reported to be 3×10^3 to form L-serine (6×10^3 to form DL-serine), although there are indications that the reaction may be even more favorable than this.

Both glycine and serine are primitive earth compounds. Glycine is the most abundant amino acid in almost all primitive synthetic reactions, and formaldehyde is synthesized efficiently by electric discharges and ultraviolet light, and it is needed for sugar synthesis.

Solutions of 1×10^{-3} M of glycine and 5×10^{-3} M formaldehyde were heated in phosphate buffered solutions at 60, 80 and 101 °C. The serine produced was measured by the amino acid analyzer. The data are plotted in fig. 1 as pseudo first order rate constants for 5×10^{-3} M H_2CO. The effect of pH on the rate of serine synthesis can be calculated on the basis that H_2CO adds to glycine carbanion ($H_3^+N-CH-COO^-$), which in turn is formed by the reaction of OH^- with the dipolar ion of glycine ($H_3^+N-CH_2-COO^-$). This gives the rate equation:

$$\frac{d(\text{ser})}{dt} = k_1(H_2CO)(\text{gly}^{+-})(OH^-) = k_1(H_2CO)\frac{(\Sigma\text{gly})K_w}{\left(1+\frac{K_2}{H^+}+\frac{H^+}{K_1}\right)(H^+)} \quad (1)$$

where k_1 is the second order rate constant, Σgly is sum of the concentrations of the various ionic forms of glycine, K_w is the dissociation constant of water, K_1 and K_2 are the first and second dissociation constants for glycine.

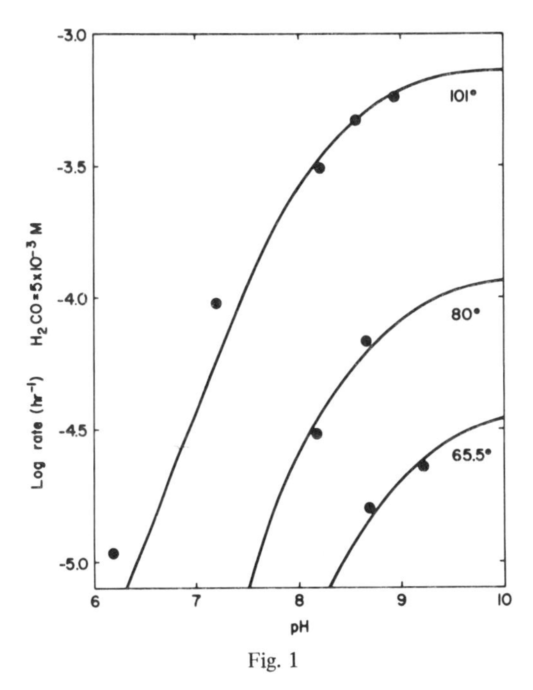

Fig. 1

The solid lines in fig. 1 are the calculated curves from the above equation using k as an adjustable parameter.

The experimental pH dependence agrees rather well with the calculated dependence. Experimental points for solutions of pH > 10 are not in accord with this equation because of decomposition of the serine that is synthesized.

The reaction appears to be more complicated than indicated by eq. 1. The rate of serine synthesis has more than a first order dependence on the formaldehyde concentration. This can be interpreted on the basis of two reactions:

$$\frac{\mathrm{d(ser)}}{\mathrm{d}t} = k_1(\mathrm{H_2CO})(\mathrm{gly}^{+-})(\mathrm{OH}^-) + k_2(\mathrm{H_2CO})(\mathrm{gly \cdot H_2CO})(\mathrm{HO}^-)$$

where $\mathrm{gly \cdot H_2CO} = \mathrm{HO{-}CH_2{-}NH{-}CH_2{-}COO^-}$.

This complication makes it impossible to extrapolate the rate of serine synthesis at lower temperatures with the above data, although it seems likely that the rate would be substantial at even at 0 °C and at formaldehyde concentrations needed to synthesize sugars (~ 10^{-2} M).

Pyridoxyl has been shown to be a catalyst for the synthesis of serine from glycine and formaldehyde [24]. The above data show that formaldehyde is also a catalyst, and it seems likely that other aldehydes would catalyze this reaction. Since the equilibrium constant is so favorable for serine synthesis it is possible that all the glycine would be converted to serine at concentrations of H_2CO needed for sugar synthesis.

Conclusion

The fact that simple electric discharge experiments or cyanide polymerizations do not give all of the twenty amino acids that occur in proteins does not mean that they did not occur in the primitive oceans. A variety of conditions would have occurred on the primitive earth, and it is reasonable to expect that organic compounds synthesized under one set of conditions would react with products synthesized in another place. Organic compounds that have been previously shown to have been synthesized can be considered acceptable reagents for reactions with compounds directly synthesized by primitive energy sources, provided the concentrations and conditions are not extreme. When viewed in this light, we expect that plausible pathways will be worked out for the synthesis of all twenty amino acids.

References

[1] Akabori S., in: The Origin of Life on the Earth, ed. A.I. Oparin (Pergamon Press, New York, 1959).
[2] Alexander N. and D.M. Greenberg, J. Biol. Chem. 220 (1956) 775.
[3] Baumann P., Chem. Ing. -Tech. 20 (1948) 257.
[4] Blakley R.L., Biochem. J. 65 (1957) 342.
[5] Crick F.H.C., J. Mol. Biol. 38 (1968) 367.
[6] Dus K., S. Lindroth, R. Pabst and R.A. Smith, Anal. Biochem. 18 (1967) 532.
[7] Eck R.V. and M.O. Dayhoff, Science 152 (1966) 363.
[8] Faltings K., Chem. Ber. 72 (1939) 1207.
[9] Friedberg F. and G.A. Hayden, Science 140 (1963) 825.
[10] Friedmann N., H.H. Bovee and S.L. Miller, J. Org. Chem. 35 (1970) 3230.
[11] Friedmann N., H.H. Bovee and S.L. Miller, in preparation.
[12] Friedmann N. and S.L. Miller, Science 166 (1969) 766.
[13] Friedmann N. and S.L. Miller, Nature 221 (1969) 1152.
[14] Gross E. and J.L. Morell, J. Am. Chem. Soc. 89 (1967) 2791.

[15] Harada K. and S.W. Fox, Nature 201 (1964) 336.
[16] Havir E.A. and K.R. Hanson, Biochemistry 7 (1968) 1904.
[17] Jukes T.H., Molecules and Evolution (Columbia Univ. Press, New York, 1966) pp. 68.
[18] Kenyon D.H. and M.S. Blois, Photochem. Photobiol., 4 (1965) 335.
[19] Klinger W., Naturwissenschaften 9 (1959) 545.
[20] Kolomiychemko M.A., Biokhim. Zh. 36 (1964) 216 (Chem. Abstr. 61 (1964) 2084).
[21] Lowe C.U., M.W. Rees and R. Markham, Nature 199 (1963) 219.
[22] Mathews C.N. and R.E. Moser, Nature 215 (1967) 1230.
[23] Matsuda G., T. Maekawa, C. Ohkawa and M. Yamada, Nagasaki Igakkai Zassi 29 (1954) 10.
[24] Metzler D.E., M. Ikawa and E.E. Snell, J. Am. Chem. Soc. 76 (1954) 648.
[25] Miller S.I. and G. Shkapenko, J. Am. Chem. Soc. 77 (1955) 5038.
[26] Miller S.L., Ann. N.Y. Acad. Sci. 69 (1957) 260.
[27] Miller S.L. J. Am. Chem. Soc. 77 (1955) 2351.
[28] Nevmerzhitskaya E.A., A.N. Belyaeva, V.A. Poprotskaya and N.A. Kudryavtseva, Khim. Prom. 41 (1965) 895; Chem. Abstr. 64 (1965) 10992.
[29] Nofre C., A. Cier and C. Michou-Saucet, Compt. Rend. Acad. Sci. Paris 251 (1960) 811.
[30] Norman R. and J. Smith, in: Oxidases and Related Redox Systems, ed. T.E. King (Wiley, New York, 1965) p. 131.
[31] Noyce D.S. and M.D. Schiavelli, J. Am. Chem. Soc. 90 (1968) 1020.
[32] Oro J., in: The Origin of Prebiological Systems, ed. S.W. Fox (Academic Press, New York, 1965) p. 137.
[33] Oro J. and S.S. Kamat, Nature 190 (1961) 442.
[34] Pavlovskaya T.E., A.G. Pasynskii and A.I. Grebenikova, Dokl. Akad. Nauk SSSR 135 (1960) 743; Chem. Abstr. 55 (1960) 10531.
[35] Sakakibara S., Bull. Japan. Chem. Soc. 34 (1961) 205.
[36] Sanchez R.A., U.S. Patent (1969) 3,410,922; Chem. Abstr. 70 (1969) 47166.
[37] Sanchez R., J.P. Ferris and L.E. Orgel, Science 153 (1966) 72.
[38] Schmidt F. and W. Klinger, Z. Physiol. Chem. 310 (1958) 31.
[39] Shocken K., Exptl. Med. Surg. 9 (1951) 465.
[40] Snyder H.R. and J.A. MacDonald, J. Am. Chem. Soc. 77 (1955) 1257.
[41] Sosnovsky G., Free Radical Reactions in Preparative Organic Chemistry (MacMillan, New York, 1964).
[42] Studier M.H., R. Hayatsu and E. Anders, Geochim. Cosmochim. Acta. 32 (1968) 151.
[43] Taube M., S.Z. Zdrojewski, K. Samochocka and K. Jezierska, Angew. Chem. 79 (1967) 239.
[44] Tominaga F., Nagasaki Igakkai Zassi 33 (1958) 753.
[45] Truce W.E. and R.F. Heine, J. Am. Chem. Soc. 81 (1959) 512.
[46] Vermeil C. and M. Lefort, Compt. Rend. Acad. Sci. Paris 244 (1957) 889.
[47] Wall J.S., Anal. Chem. 25 (1953) 950.
[48] Zobel F., Chem. Ingr.-Tech. 20 (1948) 260.

Chemical Evolution and the Origin of Life, eds. R. Buvet and C. Ponnamperuma

CONTRIBUTION FROM LOWER ALDEHYDES TO ABIOGENESIS OF BIOCHEMICALLY IMPORTANT COMPOUNDS

T.E. PAVLOVSKAYA

A.N. Bakh Institute of Biochemistry,
Academy of Sciences of the USSR, Moscow, USSR

Prebiological formation of substances indispensable for life seems to have been a step in the origin of life on earth [15].

Simple, relatively stable molecular intermediates played a decisive part in abiogenic synthesis of various biochemically important compounds.

Contribution from hydrocyanic acid, particularly from its polymeric forms, to formation of peptides, amino acids, and purine bases, have been reported [8,10,11,16,24].

The data reported and those obtained by us seem to show that compounds of sufficient reactivity, such as lower aldehydes, took part in abiogenic synthesis.

These compounds could arise under the conditions of the primitive earth and atmosphere. Formaldehyde was shown to be synthesized by shortwave irradiation of gaseous mixtures of $CO + H_2O$, $CH_4 + H_2O$ [3] and $CO + CH_4$ [7]. It was also formed by irradiation of a mixture of $CO_2 + H_2O$ with a beam of helium ions [6], as well as by the action of electric discharges [12] or β-rays [19] on a mixture of methane, ammonia and water.

The recent radioastronomic discovery of organic molecules, of formaldehyde, in the Galaxy [27] is an indication that it could have been present in the primitive atmosphere of the earth.

The high reactivity of lower aldehydes is due to their structure, first of all to the double bond of carbonyl. Electron displacement with formation of a positive, electrophilic, center on the carbon atom and a negative, nucleophilic, center on the oxygen atom

$$\rangle C : \; : \ddot{O} : \; \leftrightarrow \; \rangle C^+ : \ddot{\underset{..}{O}}^- :$$

is characteristic just of the carbonyl of aldehydes.

Complete electron displacement occurs in chemical reactions, under the action of the electric field of a reactant (for example of HCN), whereas non-reacting aldehydes show but partial displacement responsible for their potential chemical activity.

The electrophilic center at the carbon atom readily adding nucleophilic reactants, for example HCN to produce cyanhydrins, NH_2OH to produce aldoximes, plays a decisive part in these reactions. The great diversity of reactions of addition at the double carbonyl bonds makes carbonyl one of the most reactive organic functions.

Along with the many reactions connected with polar forms of lower aldehydes, of great importance under certain conditions is the transition of carbonyl to a biradical state, for example by action of light

$$\rangle C{=}O \xrightarrow{h\nu} -\dot{C}-\dot{O}-$$

The primary steps of photodissociation of lower aldehydes in the gas phase studied in the range of 3300–2500 Å [2] follow a simple mechanism involving the appearance of formyl radicals and radicals R$^{\cdot}$, and formation of stable hydrocarbon and carbon monoxide molecules

$$R.CHO \xrightarrow{h\nu} CHO + {}^{\cdot}R \qquad (1)$$

$$R.CHO \xrightarrow{h\nu} RH + CO \qquad (2)$$

Quantum yields φ_1 and φ_2 are characteristic of the two reactions, respectively. The ratio of φ_2/φ_1 for various simple aldehydes is approximately the same at any wavelength. Reaction (1) is important for all wavelengths, whereas reaction (2) is unimportant at 3130 Å.

Å	φ_2/φ_1
3130	0.001–0.05
2804	0.34
2654	0.95
2537	1.22

The formation and accumulation of formyl radicals in aqueous formaldehyde glass at 77 °K and its mixtures with ammonium nitrate on irradiation at 2537 Å was established in our laboratory by the ESR technique [13].

This suggests that photodissociation of aldehyde yielding formyl radicals may be one of the primary steps in photochemical synthesis of biochemically important compounds with participation of lower aldehydes.

An important conversion of formyl radicals may be their dimerization to form a dialdehyde: glyoxal

$$CHO + CHO \rightarrow \underset{\text{glyoxal}}{HOC + CHO}$$

The rate constant of this process was determined [29] and appeared to be 10^{11} cm^3 sec^{-1} $mole^{-1}$.

Glyoxal and its simple derivative, methylglyoxal (pyruvic aldehyde) contain an open chain of conjugated double bonds, which is responsible for their exclusively high reactivity. Activation of carbonyl both by displacement to the polar state and by transition to a biradical state may be conceived for dialdehydes. This refers particularly to photochemical reactions proceeding via radical forms of carbonyl. The reactivity of dialdehydes, compared to that of aldehydes, is high. Glyoxal was found in all cases of formaldehyde generation from simple gases by various energy sources [7,28]. We have identified glyoxal in an aqueous solution of formaldehyde irradiated at 2537 Å [14]. The ready formation, relative stability, and at the same time high reactivity of lower aldehydes and dialdehydes could make them important reactive intermediates in prebiological synthesis.

This suggestion is supported by the well known abiogenic formation of amino acids and peptides in heated formaldehyde-hydroxylamine solutions [18], by the formation of 2-deoxyribose and 2-deoxyxylose with condensation of acetaldehyde and glyceroaldehyde in aqueous solutions in the presence of basic catalysts, such as Ca^{2+}, Mg^{2+} [17]. UV irradiation of aqueous formaldehyde solution to produce ribose, deoxyribose and other sugars [23], as well as production of hexoses, trioses and tetroses by heating formaldehyde solutions in the presence of kaolinite [5] and the contribution of formaldehyde to photochemical formation of porphins [9], also supports the above suggestion. Formaldehyde was found to be an important intermediate, for example in the Strecker synthesis of amino acids by passing electric discharge through a mixture of methane, ammonia, water and hydrogen [12].

It was found earlier [20] that amino acids were synthesized in the formaldehyde-ammonium salt system. It seemed of interest to find out the conditions for synthesis of amino acids most important for the enzymic activity of proteins in evolution, such as: serine, histidine, tyrosine, tryptophan, cysteine, etc., i.e. the amino acids which are now contained in active centers of enzymes.

It may be suggested that selection of these amino acids in the formation of a catalytically active center took place at the stage of formation of high-molecular peptides that could have appeared on absorbent surfaces [1] or by simple heating of amino acids [4] formed by other plausible methods.

Synthesis of some of the amino acids mentioned was conducted by irradiation at 2537 Å of an aqueous solution of CH_2O (2.5%) + NH_4NO_3 (1.5%). Serine and threonine were synthesized under these conditions, along with glycine, alanine and basic amino acids [21]. It was found that accumulation of amino acids occurs with increase in irradiation time from 20 to 350 hr, at a rate characteristic of every acid and always attaining a stationary value [25].

Replacement of formaldehyde by acetaldehyde resulted in additional formation of aspartic acid, glutamic acid, valine and leucine. Neither histidine nor tryptophan were found in the irradiated solutions of lower aldehydes and ammonium salt.

Synthesis of peptides displaying neutral and basic properties was observed in the same system of acetaldehyde and ammonium nitrate. Analysis of the amino acid fraction by means of an analyzer has shown that the peak amplitudes increased after hydrolysis. Fifty-seven percent of fixed nitrogen fell to peptides. Five amino acid residues represented the average peptide length. This study is still in progress.

Besides the amino acids diazotizable compounds identified as imidazole, 4-hydroxymethylimidazole, 4-methylimidazole, 4-oxomethylimidazole and various amine derivatives were found to form in the irradiated system (CH_2O + NH_4NO_3). The qualitative content of the compounds formed was virtually independent of variations in time from 3 to 350 hr [26].

The rate of accumulation of certain imidazole compounds sometimes passed through a maximum, and in other cases tended to a stationary level. This seems to show possible conversion of certain imidazole compounds from one form to another. In particular, 4-oxomethylimidazole might be formed from 4-hydroxymethylimidazole, and amine derivatives might appear by amination of imidazole, 4-hydroxymethylimidazole and 4-oxomethylimidazole.

The formation of imidazole compounds in this case might occur by the Radzievskii mechanism:

$$\begin{array}{l} R\text{-}C\text{=}O \\ \quad | \\ H\text{-}C\text{=}O \end{array} + \begin{array}{l} NH_3 \\ \\ NH_3 \end{array} + O\text{=}CH_2 \rightarrow \begin{array}{c} R\text{-}C\text{-}N \\ \| \quad \| \\ H\text{-}C \quad C\text{-}H \\ \backslash / \\ N \\ | \\ H \end{array} + 3\,H_2O$$

4,5-Dimethylimidazole and 2-methylimidazole are synthesized, along with imidazole, 4-hydroxymethylimidazole and certain unidentified imidazole derivatives, in the system containing acetaldehyde instead of CH_2O [21].

Possible synthesis of these compounds in UV-irradiated mixtures of simple aldehydes and ammonium salts is of interest in connection with the fact that various metabolites, including histidine, the most important amino acid, contain the imidazole ring. Moreover, 4,5-disubstituted imidazole derivatives are intermediates in the synthesis of purine bases.

The initial step of photochemical synthesis of amino acids in aqueous solution of formaldehyde and ammonium nitrate was investigated [14]. Of importance was the ESR detection of formyl radicals on irradiation at 2537 Å in aqueous formaldehyde glass at 77 °K, and the identification of glyoxal in irradiated formaldehyde solutions. The conversion of glyoxal and methylglyoxal, kept in solution with ammonium nitrate or ammonia, to glycine and alanine, respectively, as well as to imidazole compounds, is evidence of their high reactivity. Further oxidation of glyoxal to glyoxalic acid or of relevant ketoaldehydes to ketoacids is suggested to represent an intermediate step in the photochemical formation of α-amino acids. The UV-induced interaction of these compounds with ammonia and hydrogen atoms or alkyl radicals

$$CH_2O \xrightarrow{UV} \cdot CHO + CHO \rightarrow CHO{-}CHO \xrightarrow{OH} CHO{-}COOH$$

$$\xrightarrow[NH_3H(R)]{UV} CH_2NH_2COOH$$

might be one of the routes of glyoxalic acid and ketoacid conversions to α-amino acids under these conditions.

Such a postulated mechanism of abiogenic synthesis of amino acids, via highly reactive intermediates, somewhat resembles that of amino acid biosynthesis in the contemporary and ancient organisms.

Some aliphatic amines, such as methylamine, ethanolamine, and isopropylamine, were also synthesized by UV irradiation of the formaldehyde-ammonium nitrate system. Urea and hexamethylenetetramine were obtained under the same conditions.

Irradiation of systems containing acetaldehyde and ammonium salt produced indole, tryptamine, indolealdehyde and hydroxymethylindole (identified by thin-layer chromatography), but no amines or amides [22]. The formation of these compounds is of interest as the indole ring enters into such an important amino acid as tryptophan, and also because it permits suggesting the possibility of pyrrole ring synthesis in the system studied.

This is an incomplete enumeration of all the research on contribution of lower aldehydes to biochemically important compounds in abiogenic synthesis.

Finally, it may be stated that the role of formaldehyde and other lower aldehydes in prebiological chemistry could be very great, and detection of formaldehyde in intergalactic space adds a new dimension to the idea that lower aldehydes were precursors of primitive compounds useful in prebiological synthesis.

References

[1] Akabori S., in: The Origin of Life on the Earth, Vol. 1, ed. Oparin et al. (Pergamon Press, London, New York, 1959) p. 189.
[2] Calvert J.G. and J.H. Pitts, Photochemistry (John Wiley, New York, 1966).
[3] Dodonova N.A. and A.I. Sidorova, Biofizika 6 (1961) 149.
[4] Fox S. and K. Harada, Science 128 (1958) 1214.
[5] Gabel N.W. and C. Ponnamperuma, Nature 216 (1967) 453.
[6] Garrison W.M., J.G. Hamilton, D.C. Morrison, A.A. Benson and M. Calvin, Science 114 (1951) 416.
[7] Groth W., Z. Physiol. Chem. B 37 (1937) 315.
[8] Kliss R.M. and C.N. Matthews, Proc. Natl. Acad. Sci. U.S. 48 (1962) 1300.
[9] Krasnovskii A.A. and A.V. Umrikhina, Dokl. Akad. Nauk SSSR 155 (1964) 691.
[10] Matthews C.N. and R.E. Moser, Proc. Natl. Acad. Sci. U.S. 56 (1966) 1087.
[11] Matthews C.N. and R.E. Moser, Nature 215 (1967) 1230.
[12] Miller S., Biochim. Biophys. Acta 23 (1957) 48.
[13] Niskanen R.A. and T.A. Telegina, Vtoroi Vsesoyuznii Biokhimicheskii S'ezd, Tezisy Sektsii 1 (1969) 105, Tashkent, FAN Akad. Nauk Uzbek. SSSR.
[14] Niskanen R.A., T.E. Pavlovskaya, V.S. Sidorov, T.A. Telegina and V.A. Sharpatyi, Izv. Akad. Nauk SSSR, Ser. Biol. no. 2 (1971) in press.
[15] Oparin, A.I., Vozniknovenie Zhizni na Zemle (Izd. Nauk, Moscow, 1957).
[16] Oró J., Arch Biochem. Biophys. 94 (1961) 217.
[17] Oró J., in: Problemy Evolutsionnoi i Tekhnicheskoi Biokhimii (Izd. Nauka Moscow, 1964) p. 63.
[18] Oró J., A. Kimball, R. Fritz and F. Master, Arch. Biochem. Biophys. 85 (1959) 115.
[19] Palm C. and M. Calvin, J. Am. Chem. Soc. 84 (1962) 2115.
[20] Pavlovskaya T.E. and A.G. Pasynskii, in: The Origin of Life on the Earth, ed. Oparin et al. (Pergamon Press, London, New York, 1959) p. 151.
[21] Pavlovskaya T.E. and A.G. Pasynskii, in: Problemy Evolutsionnoi i Tekhnicheskoi Biokhimii (Izd. Nauka, Moscow, 1964) p. 70.
[22] Pavlovskaya T.E., A.G. Pasynskii, V.S. Sidorov and A.I. Ladyzhenskaya, in: Abiogenez i Nachalnii Stadii Evolutsii Zhizni, (Izd. Nauka, Moscow, 1968) p. 41.
[23] Ponnamperuma C. and R. Mariner, Radiation Res. 19 (1963) 183.
[24] Sanchez R.A., J.P. Ferris and L.E. Orgel, J. Mol. Biol. 30 (1967) 223.

[25] Sidorov V.S., Dokl. Akad. Nauk SSSR 164 (1965) 692.
[26] Sidorov V.S., T.E. Pavlovskaya and A.G. Pasynskii, Zh. Evolut. Biokhim. Fiziol. 2 (1966) 293.
[27] Snyder L.E., D. Buhl, B. Zuckerman and P. Palmer, Phys. Rev. Letters 22 (1969) 679.
[28] Terenin A.N., in: The Origin of Life on the Earth, ed. Oparin et al. (Pergamon Press, London, New York, 1959) p. 136.
[29] Yee Quee M.J. and J.C. Thynne, Trans. Faraday Soc. 63 (1967) 1656.

Chemical Evolution and the Origin of Life, eds. R. Buvet and C. Ponnamperuma

EXPERIMENTS IN JOVIAN ATMOSPHERE

Mohindra S. CHADHA *, James G. LAWLESS, Jose J. FLORES
and Cyril PONNAMPERUMA
*Exobiology Division, Ames Research Center,
Moffett Field, California, USA*

1. Introduction

The present atmosphere of Jupiter consisting of methane, ammonia and hydrogen may be considered to resemble the primitive atmospheres of the earth [3,5,8]. Experiments to explore the action of various forms of energy on simulated Jupiter atmospheres have, therefore, a direct relevance to our understanding of the origin of life. In the present study, the nature of products formed, when an electric discharge is passed through a mixture of methane and ammonia has been explored.

A number of experiments in which methane and ammonia mixtures were subjected to semicorona and electric discharges have already been reported from this laboratory [6,9]. In the semicorona experiments, when 0.3 mA current was passed through a mixture of methane and ammonia (2:1), the consumption of methane and ammonia was found to be accompanied with the formation of hydrogen and nitrogen (fig. 1) and traces of HCN. In the 0.5 mA arc experiments (fig. 2), from an equimolar mixture of methane and ammonia, methane was used up much faster than in the previous case and the yield of HCN was somewhat greater. In contrast to the above two cases, in a 5 mA experiment, the utilization of methane was very rapid and the yield of HCN quite sizeable (fig. 3). The gradual increase in methane level after a rapid drop within the first 1 hr was quite remarkable and may eventually throw some light on the mechanisms of the product synthesis in this experiment.

The volatile fraction from the 0.5 mA arc experiment was analyzed by gas chromatography (fig. 4), and the products were shown to be ammonium

* Senior Research Associate, National Academy of Sciences, Washington, D.C. On leave from: Biology Division, Bhabha Atomic Research Centre, Bombay-85, India.

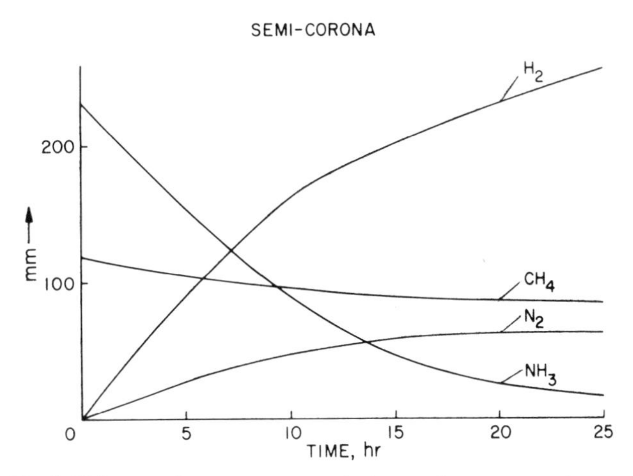

Fig. 1. Major gas phase components during semicorona experiment.

cyanide, acetonitrile, propionitrile, aminoacetonitrile and its *C*-methyl and *N*-methyl homologs [6]. The identification of the nitriles was carried out in a definitive manner by the mass spectral and NMR spectral examination of each of the fractions separated by gas-liquid chromatography. The nonvolatile material in this experiment was a translucent ruby red material. The present study describes further examination of the colorless material entrapped in a cold finger of a modified reaction vessel and investigation of the ruby red material deposited on the upper walls of the reaction vessel.

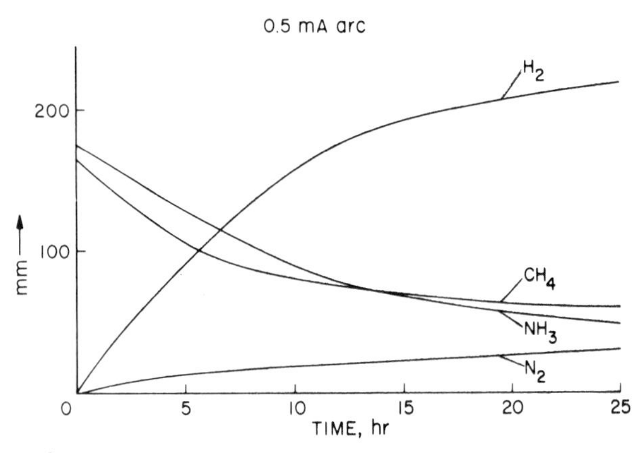

Fig. 2. Main gas phase components during 0.5 mA arc experiment.

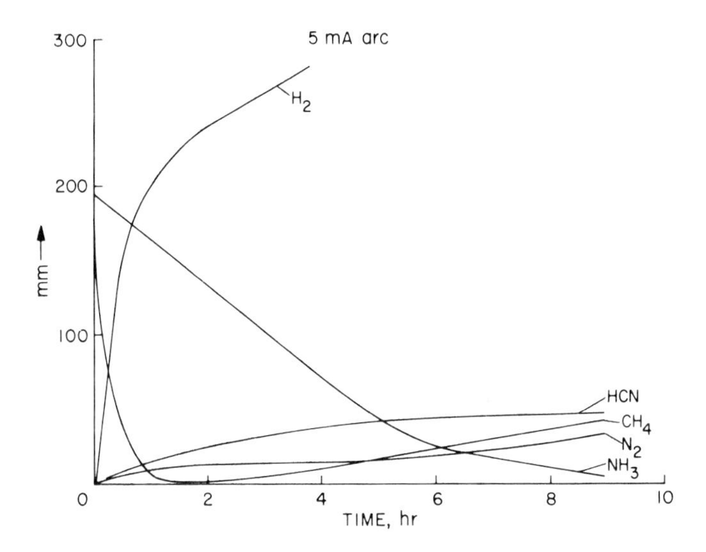

Fig. 3. Course of the 5 mA arc process, byproducts (acetylene, ethylene, benzene) not shown.

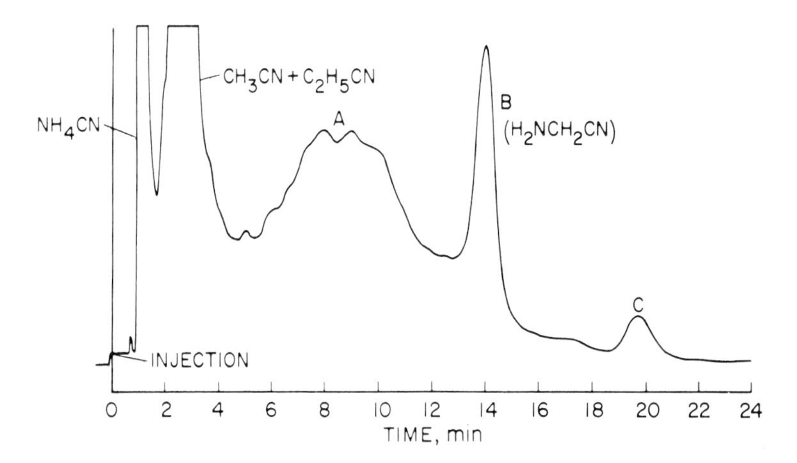

Fig. 4. Gas chromatogram of 0.5 mA arc, colorless product. A represents *C*-methyl and *N*-methyl homologs of aminoacetonitrile.

Experimental

A 1000 ml reaction vessel, fitted with a vacuum sealed stopcock, two 7/25 standard tapered joints for the electrodes, and a T29/40 connecting joint was used for the purpose. The flask which has a 12 cm × 0.8 cm cold finger fitted at the bottom, after evacuating to 10 cm pressure, was filled with an equimolar mixture of anhydrous methane and ammonia. It was then rested on a narrow mouthed Dewar flask containing dry ice, so that the lower end of the cold finger was about a centimeter above the dry ice level. The electrodes were adjusted to provide a gap of about 0.8 cm between them so that the current flowing through the electrodes was approximately 0.5 mA and a continuous spark could be observed. This was continued for 20 hr, at which time a colorless liquid had condensed in the cold finger and a reddish brown viscous liquid had condensed on the inside of the connecting joint and the upper walls of the reaction vessel. After the passage of spark was discontinued, the reaction products were allowed to stand for 3 days with the cold finger at −78°C and the rest of the flask at room temperature.

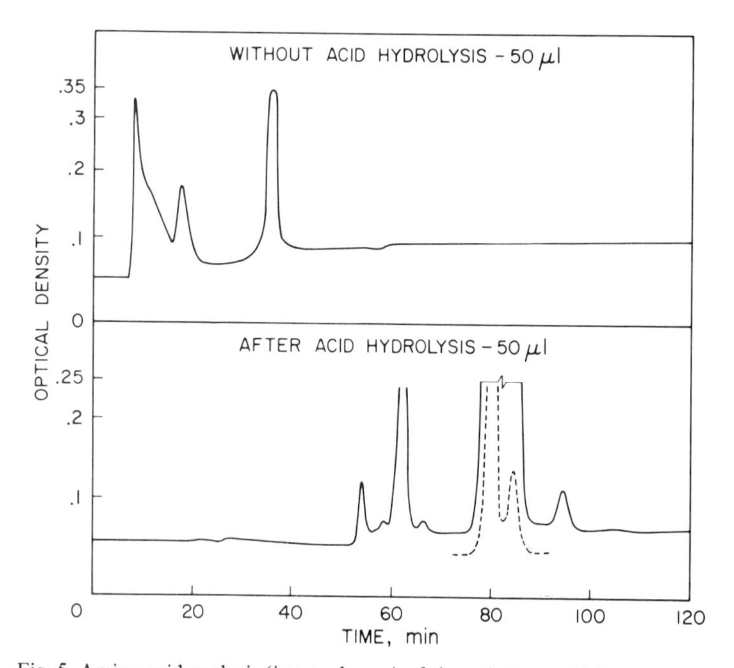

Fig. 5. Amino acid analysis (ion exchange) of the colorless cold finger material.

The colorless cold finger material (A) was dissolved in 10 ml of dry CH_2Cl_2 while the reddish brown material (B) was dissolved in 10 ml of water. The pH of the aqueous solution was 10.2.

A portion of the CH_2Cl_2 solution was analyzed on a Beckman model 120 C automatic amino acid analyzer before and after acid hydrolysis (6 N HCl, 100 °C, 12 hr) (fig. 5). Likewise, a portion of the colored aqueous solution after standing at room temperature for 30 min at pH 10.2 was analyzed on the amino acid analyzer before and after acid hydrolysis (6 N HCl, 100 °C, 12 hr) (fig. 6).

For a rigorous proof of structure of the hydrolysis products of B, an acid hydrolyzed sample was evaporated to dryness. The residue was esterified with 3 N HCl in butanol (1 hr) and then the butylated material after removal of butanol was acylated with trifluoroacetic anhydride in CH_2Cl_2. The resulting

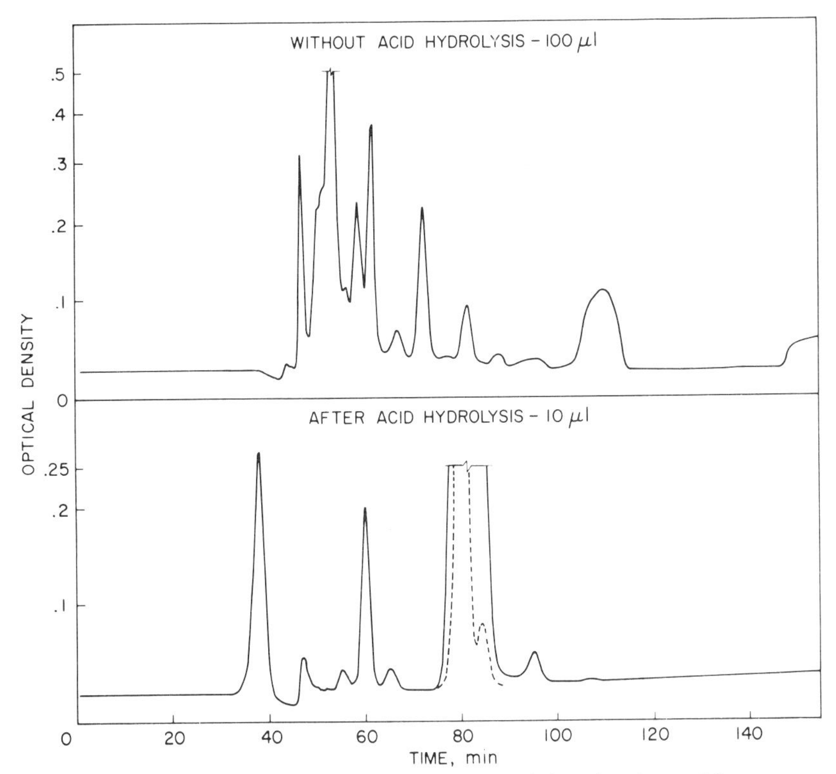

Fig. 6. Amino acid analysis (ion exchange) of the colored material.

product was examined for *N*-TFA-*n*-butyl esters of amino acids on an OV-17 column (0.125 inch × 4.5 ft , stainless steel 1.5% on 80/100 mesh H.P. chromosorb G) [1].

The derived mixture was also examined by combined gas chromatography-mass spectrometry [2] using a Loenco 160X gas chromatograph and a CEC 21-491 mass spectrometer. For chromatography, a Perkin Elmer 50 ft × 0.02 inch stainless steel support coated open tubular (SCOT) column coated with OV-17 was used. The gas chromatograph was connected to the mass spectrometer by means of a single stage Llewellyn separator (Varian V-5620). Each component was detected by a Carle 100 dual thermistor microdetector – as it emerged from the GLC column and recorded on one pen of the dual pen Honeywell Electronik 194 recorder. The sample then passed into the source of the mass spectrometer where it was ionized at 70 eV. The beam monitor provided the second trace on the recorder. The mass spectra were recorded at the rate of 4 sec/decade on a Honeywell 1508 A Visicorder.

Results

The ion exchange amino acid analysis of the material A (CH_2Cl_2 solution) showed the absence of easily recognizable amino acids e.g. glycine and alanine, whereas the material B (aqueous solution, pH 10.2) showed the presence of small proportions of glycine and alanine. Obviously, in the basic solution resulting from the dissolution of the spark material, a finite degree of hydrolysis had resulted.

The products of acid hydrolysis in both cases, gave by ion exchange recognizable amino acids (glycine-alanine, etc.) in significant amounts (fig. 7).

The combined gas chromatography–mass spectrometry of the acid hydrolyzed sample B (fig. 8) showed the presence of a large number of peaks whose spectra were recorded. Of these, 9–10 peaks represented a major portion of the material, of which five have been characterized positively and three tentatively. On the basis of molecular ions and fragmentation patterns, and comparisons with known samples, the first three major peaks were identified as alanine, glycine and sarcosine.

The fourth compound to be identified showed a molecular ion at mass 341. However, the retention time and mass spectrum ruled out the possibility of the compound being derivatized aspartic acid. Analysis of the mass spectrum indicated that the compound in question could be the *N*-TFA-*n*-butyl ester derivative of iminodiacetic acid. This was indeed found to be the case, when the G.C. retention time and mass spectral fragmentation pattern were compared to an authentic sample of the derivatized iminodiacetic acid.

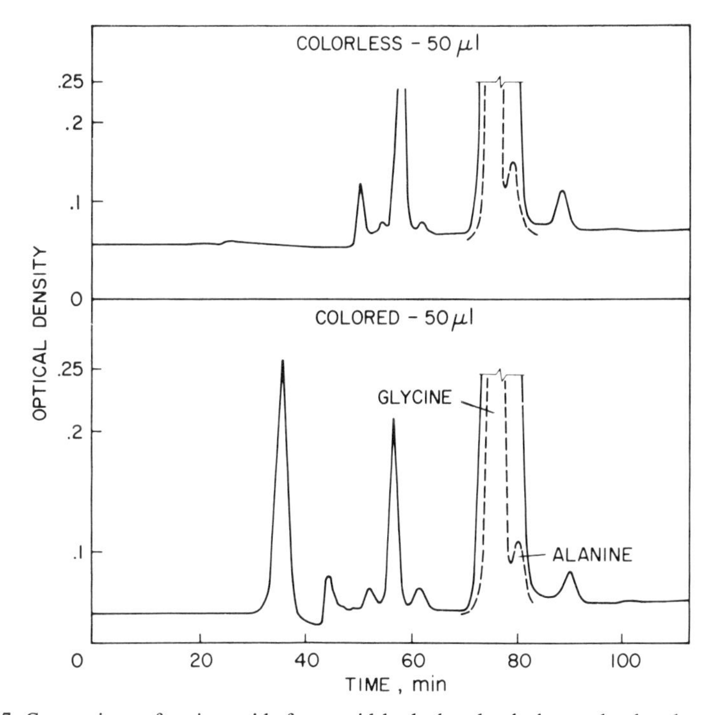

Fig. 7. Comparison of amino acids from acid hydrolysed colorless and colored material.

The next compound to be identified gave a molecular ion at 355 and a mass spectrum indicative of the *N*-TFA-*n*-butyl ester of iminoacetic-β-propionic acid. By comparison of the G.C. retention time and mass spectral fragmentation pattern of the unknown with an authentic sample, the compound was characterized as *N*-TFA-*n*-butyl ester of iminoacetic-β-propionic acid.

Some of the other peaks, for which the molecular ions were discernible and the fragmentation patterns rationalizable, enable us to make the following tentative proposals. The G.C. peak closely following the derivatized iminodiacetic acid was found to show a molecular ion at 355 and the mass spectrum suggested it to be due to iminoacetic-α-propionic acid. Two of the G.C. peaks gave mass spectra with molecular ions at 255, arising possibly from β-amino-iso-butyric acid and from N-methylalanine derivatives, and one of the G.C. peaks giving a mass spectrum with a molecular ion at 259 could be accounted for by the dibutyl ester of *N'N*-dimethylaminomalonic acid.

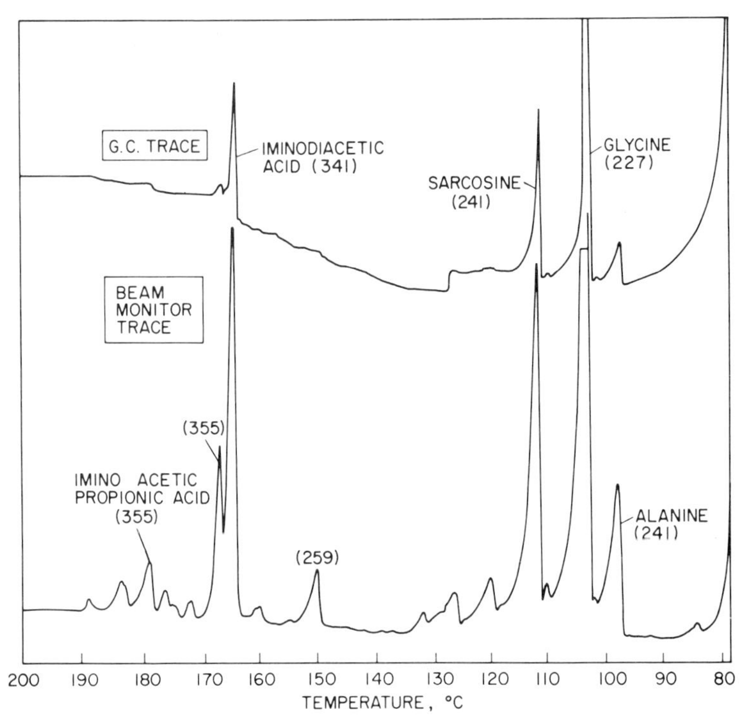

Fig. 8. Combined gas chromatographic-mass spectrometric examination of derivatized acid hydrolysed colored material. Gas chromatographic separation on OV-17 SCOT column, temperature program, 6°/min (80°–200°).

A few G.C. peaks give molecular ions at 355 and 369 and suggest the presence of *N*- and/or *C*-methyl homologs of aspartic, glutamic, iminodiacetic or iminoacetic propionic acid. Out of these one of the products is suspected strongly to be *N*-TFA-*n*-butyl derivative of α,α'-iminodipropionic acid.

Discussion

The formation of amino nitriles during electric discharge of a mixture of methane and ammonia has been further substantiated. The acid hydrolysis of the brown material leading to the formation of the amino and imino acids discussed above demonstrates the significance of the corresponding nitriles which are formed under the conditions of electric discharge. Work in this laboratory has shown that when electric discharge is passed through methane and ammonia, polymeric products having molecular weights in the 2–3,000 range are formed [4]. It is not unlikely that the polymerization is resulting through the formation of an amidine linkage by the interaction of cyano and amino groups, which is a facile reaction. Once formed, the amidine groups can easily hydrolyse to the amide groups. The hydrolysis of the cyano groups is somewhat more difficult, but as can be expected and is also indicated from the present study, a slow hydrolysis of cyano groups at pH around 10 is possible. It is noteworthy that in the dry methane and ammonia experiments, the amounts of aspartic and glutamic acids formed, if any, are minimal; whereas, in the wet experiments (methane, ammonia and water), aspartic acid is formed in siginificant amounts [7].

Acknowlegement

Thanks are due to Dr. A. Duffield of Stanford University for the interpretation of some of the mass spectra.

References

[1] Gehrke C.W., R.W. Zumwalt and L.L. Wall, J. Chromatog. 37 (1968) 398.
[2] Gelpi E., W.A. Koenig, J. Gilbert and J. Oro, J. Chromatog. Sci. 1 (1969) 604.
[3] Mc. Elroy M.B., J. Atmospheric Sci. 26 (1969) 798.
[4] Noda H. and C. Ponnamperuma, in: Proceedings of 3rd International Conference on Origin of Life, Pont-a-Mousson, France, 1970).
[5] Opik E.J., Icarus 1 (1962) 200.
[6] Ponnamperuma C., and F.H. Woeller, Currents Mod. Biol. 1 (1967) 156.
[7] Ponnamperuma C., F. Woeller, J. Flores, M. Romiez and W. Allen, Advan. Chem. Ser. 80 (1969) 280.
[8] Rasool S.I., Astronautics Aeronautics 6 (1968) 24.
[9] Woeller F. and C. Ponnamperuma, Icarus 10 (1969) 386.

Chemical Evolution and the Origin of Life, eds. R. Buvet and C. Ponnamperuma

SYNTHESIS OF ADENINE, GUANINE, CYTOSINE, AND OTHER NITROGEN ORGANIC COMPOUNDS BY A FISCHER–TROPSCH-LIKE PROCESS

C.C. YANG and J. ORÓ

Departments of Chemistry and Biophysical Sciences, University of Houston, Houston, Texas 77004, USA

The Fischer–Tropsch synthesis is a catalytic non-equilibrium process which converts CO and H_2 mixtures into hydrocarbons and related substances. It has been suggested that this process may have been involved in the prebiotic synthesis of the alkyl chains of biochemical compounds, such as amino acids and fatty acids [8], the hydrocarbons found in carbonaceous chondrites [6,11] and the nitrogenous compounds, e.g. adenine, guanine, etc. [5] reported to be present in these meteorites [4].

The work described here is not concerned with the controversial aspects of the hydrocarbons and other organic compounds detected in meteorites [9]. It is only directed to a study of the formation of purines, pyrimidines and other bases from CO, H_2, and NH_3 under similar conditions to those used in the Fischer–Tropsch process. This is done in an attempt to substantiate and possibly extend the results obtained by Hayatsu et al. [5] where insufficient data was reported for the unequivocal identification of the products.

In our experiments we have used industrial nickel–iron catalyst powders of defined composition, instead of iron meteorite filings, and we have put the emphasis on accumulating enough product material to be able to isolate the purine and pyrimidine derivatives as solid compounds and identify them by infrared spectrophotometry. Variations in the catalyst composition and reaction conditions are described below.

Experimental methods and results

The experimental conditions are summarized in table 1. In a typical run a quartz reaction vessel was rapidly heated to a peak temperature of 650 °C, followed by step-wise cooling to room temperature. According to theoretical

Table 1
Reaction conditions and products identified.

Run	Temperature (°C)	Catalysts	Total pressure at room temperature	Compounds identified		
1	600, 1 hr 140, 3 hr 410, 8 hr 100, 1 hr 310, 24 hr 200, 6 hr	Fe–Ni alloy (58–42%) 1.474 g SiO_2 0.650 g Al_2O_3 0.580 g	25 psig	adenine guanine cytosine	urea biuret guanylurea	melamine cyanuric acid
2	600, 1 hr 280, 6 hr 340, 7 hr 160, 20 hr 300, 20 hr 74, 4 hr	Fe–Ni alloy (58–42%) 2.140 g	26 psig	adenine guanine cytosine	urea biuret guanylurea	melamine cyanuric acid
3	650, 1.2 hr 125, 19 hr 309, 9 hr 75, 4 hr 220, 7 hr	SiO_2 1.047 g Al_2O_3 0.916 g	26 psig		urea biuret guanylurea	melamine cyanuric acid
4	950, 0.5 hr 120, 5.5 hr 310, 9 hr 75, 14 hr 210, 28 hr	no catalyst	1 atm	adenine	urea biuret guanylurea	melamine cyanuric acid
5	450, 1.5 hr 105, 7 hr 300, 24 hr 70, 10 hr 220, 15 hr	Fe–Ni alloy (58–42%) 1.370 g SiO_2 0.56 g Al_2O_3 0.47 g	26 psig		urea biuret guanylurea	melamine cyanuric acid
6	Repeat run 1 ten times in order to obtain enough sample for infrared (in KBr) spectrophotometry.					

[a] Gas composition: $CO : NH_3 : H_2 = 1 : 0.4 : 2$ (mole ratio).

considerations these temperature variations were intended to simulate conditions in the solar nebula: brief thermal spikes superimposed on a general downward trend [1,5,6,11].

The total pressure of the reactant gases varied from 1 to 3 atm (at room temperature); the temperature varied from room temperature to 650 °C in most of the experiments (950 °C in one case). No higher pressures or temperatures were tried because of the limited strength of the reaction vessel. However, it is known that relatively high temperatures are sometimes required for the synthesis of some precursors of biologically significant compounds. For instance, only at a temperature of 600 °C or higher can HCN be synthesized from a carbon monoxide and ammonia mixture in considerable yield [12].

Reaction conditions

The reaction vessel was a quartz tube (about 38 cm in length, 2.5 cm i.d.) having a volume of 244 ml, sealed at one end and joined at the other to a stainless steel control valve. The sealed end was inserted to a depth of 12 cm in the electric furnace which was heated to the desired temperature. The rest of the vessel was left outside the furnace. The system was not strictly isothermal, however, as one end was kept close to room temperature. In this arrangement, volatile compounds could distil away from the high temperature zone, thus remaining protected to some extent from further reaction or destruction.

The total reaction time varied from 48 hr to 72 hr. At the end of the reaction, the reaction vessel was cooled to room temperature and as expected the inside pressure was in partial vacuum. After the reaction the color of the catalysts changed from gray to deep black, near the cold end, some liquid formed on the bottom of the reaction vessel and some white powder deposited on the upper inner surface of the reaction vessel.

All the catalysts (58%–42% Fe–Ni alloy, Al_2O_3 SiO_2) used in these experiments were reduced in H_2 flow at 400 °C for 2 hr then baked out under vacuum in a torch for 2 hr.

Since NH_3 was proportionally lowest in the reactant mixture ($CO:H_2:NH_3$ = 1:2:0.4), it was filled into the evacuated reaction vessel first to the desired pressure, followed by the CO and H_2 mixture (1:2 mole ratio) to the desired total pressure.

Fractionation procedures

The separation procedures used in this work were designed for the nitrogen compounds only. They are summarized in the scheme of fig. 1.

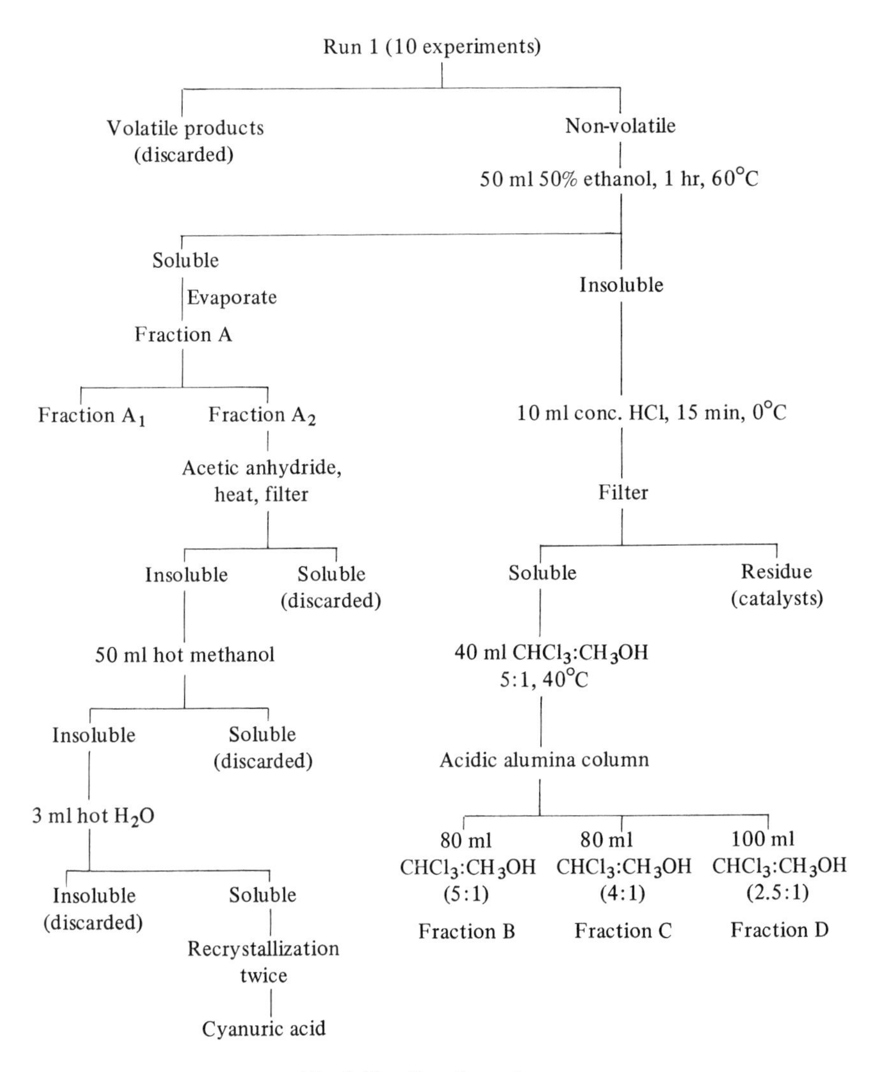

Fig. 1. Fractionation scheme.

Volatile products (at room temperature, 1 atm) were not analyzed. The non-volatile products were extracted by a one-hour treatment with 50 ml of 50% ethanol and 50% distilled water at 60 °C (made acidic, 0.05 N–0.1 N, by adding required amounts of dilute HCl). After the solvent mixture was evaporated under vacuum at 35 °C, a white residue remained (fraction A).

Four compounds were detected in one half of this residue (fraction A_1) by two dimensional paper chromatography: urea, biuret, melamine, and adenine. Four unidentified spots were also present.

Although the presence of cyanuric acid was suspected in this fraction, we consistently failed to identify it by paper chromatography because it shows no characteristic color reaction. It was finally isolated by a solubility method. The other half of the residue (fraction A_2), was acetylated with acetic anhydride according to the method developed by Cason [2]. It was refluxed with 75 ml of acetic anhydride for one hour; under these conditions cyanuric acid was not acetylated (melamine formed diacetylmelamine, 98% conversion). The resulting slurry was filtered and the residue washed with water three times, dried, and extracted with 5 ml hot methanol (at 60 °C). The methanol-insoluble fraction was dissolved in 3 ml of hot water and filtered rapidly. A white powder soon separated from the filtrate. After two recrystallizations from water, a residue of 0.6 mg remained which was set aside for infrared spectrophotometry and comparison with cyanuric acid standard.

Three further fractions (B, C and D) were obtained from the insoluble residue remaining after removal of fraction A with slightly acidic 50% ethanol. The residue, consisting largely of catalyst and free carbon, was extracted with 10 ml concentrated HCl at 0 °C for 15 min. The solution was diluted with 20 ml H_2O, filtered, and evaporated to dryness under vacuum below 30 °C. The pale greenish yellow residue was dissolved in 40 ml of a 5:1 mixture of chloroform and methanol at 40 °C, and the opaque solution was passed through a 20×0.8 cm alumina column pretreated with 1 N HCl [13]. Fraction B, eluted with 80 ml $CHCl_3$:CH_3OH (5:1), contained adenine, urea, biuret, and guanylurea, along with 3 unidentified spots. Fraction C, eluted with 80 ml of 4:1 mixture of $CHCl_3$:CH_3OH, gave an especially large variety of compounds: adenine, guanine, cytosine, melamine, urea, guanylurea, and 6 unidentified spots on the paper chromatogram. The last fraction, D, was eluted with 100 ml of 2.5:1 mixture of $CHCl_3$:CH_3OH. It contained adenine, melamine, guanine, urea, and 3 unidentified spots.

In separating and detecting the products, the acidic alumina column step is necessary to remove $FeCl_2$, $FeCl_3$, and $NiCl_2$, derived from the reactions of catalysts with concentrated HCl during extraction. These metallic chlorides can cause misidentification on paper chromatograms. All three compounds showed up as dark spots under short wavelength ultraviolet light (λ = 253.7 nm) and the first also gave blue spots with alkaline ferricyanide-nitroprusside.

Detection by paper chromatography

Both one-dimensional and two-dimensional paper chromatography were used for the detection and preliminary identification of the synthesized compounds. One-dimensional paper chromatography was unable to separate these compounds with good resolution. Even in two-dimensional paper chromatography (on Whatman 3 MM paper, 21×21 cm) guanine and melamine were not separated in the solvent mixtures used: First dimension, glacial acetic acid:butanol:water (20:60:20); second dimension, concentrated HCl:isopropanol:H_2O (16.6:65:18.4).

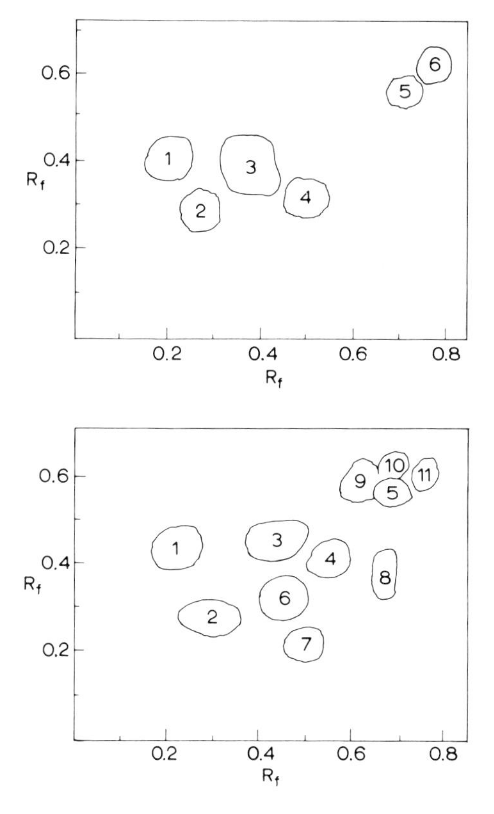

Fig. 2. Two-dimensional paper chromatograms of standards and reaction product (experiment 48, fraction C). First dimension: butanol:acetic acid:H_2O (60:20:20) vertical direction; second dimension: isopropanol:conc. HCl:H_2O (65:16.6:18.4), horizontal direction. Upper figure: standards. Lower figure: reaction product; (1) guanylurea, (2) guanine + melamine, (3) adenine, (4) cytosine, (5) urea, and (6–11) unidentified.

After the paper was developed in two dimensions (about 10 and 32 hr, respectively) and dried, the ultraviolet light absorbing compounds were detected by means of an ultraviolet light (λ = 253.7 nm) lamp. The ultraviolet absorbing spots were circled in pencil and the R_f values of synthesized products were compared with those of authentic standards. A two-dimensional chromatogram of fraction C, compared with that of a mixture of

Table 2
Identification of synthetic compounds and yield.
(Product from experiment 1 repeated 10 times.)

Compound	Reagent [a] and spectro-photometry	R_f X 100 in two dimensional paper chromatogram [b]		Amount (mg)	% yield [c]
		HAC: butanol H_2O (1st)	HCl: isopropanol H_2O (2nd)		
Urea	DAB	54 (53)	75 (73)	0.49	0.475
Biuret	DAB	59 (55)	82 (80)	0.21	0.178
Melamine	DAB, UV, IR	28 (29)	29 (29)	0.88	1.220
Guanylurea	FCNP	37 (37)	21 (20)	0.30	0.353
Adenine	UV, IR	38 (35)	40 (38)	0.47	0.507
Guanine	UV, IR	28 (29)	29 (29)	0.19	0.158
Cytosine	UV, IR	31 (29)	52 (48)	0.16	0.127
Cyanuric acid	IR			1.20	0.810

Non-volatile synthetic nitrogen compounds in different fractions

Fraction	
A	Urea, biuret, melamine, adenine, cyanuric acid
B,	Urea, biuret, granylurea, adenine (trace)
C	Urea, melamine, guanylurea, adenine, cytosine, gyanine
D	Urea, melamine, adenine, guanine

[a] DAB = *p*-dimethylaminobenzaldehyde
FCNP = alkaline ferricyanide–nitroprusside

[b] Values in parentheses refer to standards. 1st dimension: glacial acetic acid : butanol : water (20:60:20). 2nd dimension: isopropyl alcohol : conc. HCl : water (65:16.6:18.4) on Whatman No. 3 MM paper. R_f values for melamine and guanine are the same. These two compounds have been separated in 1.25% phenolic solution (in water) with R_f values of 0.79 and 0.43, respectively. Cyanuric acid was identified by its infrared spectrum alone.

[c] Gas mixture: CO:H_2:NH_3 = 1:2:0.4; 244 ml at 3 atm. Yields are calculated based on conversion of NH_3.

standards is shown in fig. 2, where all the standard compounds are separated with the exception of the guanine–melamine pair which gave a single spot. After extraction of this spot from the two-dimensional paper chromatogram, a third chromatographic procedure in a 1.25% phenolic (in water) solution could separate guanine and melamine in one dimension with R_f values of 0.43 and 0.79 respectively.

Each ultraviolet-absorbing spot was cut off and set aside for extraction and identification of individual compounds. The paper chromatograms were then used for the spraying of DAB solution (*p*-dimethylaminobenzaldehyde), or FCNP solution (alkaline ferricyanide–nitroprusside), to detect the presence of products not detectable by ultraviolet light absorption [7]. Urea and biuret (also melamine) gave yellow color spots immediately after spraying [7]. The quantities of biuret and urea synthesized were determined by extracting the yellow paper chromatogram spots with pyridine and measuring the extracts at 449 nm. Guanylurea gave a very strong purple spot immediately after spraying with the FCNP reagent.

A summary of the results obtained by fractionation and paper chromatography on the preliminary identification of the compounds synthesized is shown in table 2.

Identification by ultraviolet and infrared spectrophotometry

Infrared and ultraviolet spectrophotometric identification of the individual compounds was carried out as follows: each ultraviolet-absorbing spot and a spot from the blank paper chromatogram (paper developed in the same solvent mixtures, both spots having the same size) were cut off in small pieces, and extracted in two test tubes with about 3 ml of 0.1 N HCl for at least one day and the ultraviolet absorption spectra were obtained. Figs. 3, 4 and 5 give the spectra corresponding to three of the compounds synthesized in comparison with authentic adenine, guanine and cytosine standards.

When the ultraviolet spectrum showed the characteristic absorption and without too much absorption from impurities, the samples were accumulated in order to obtain enough product to make a KBr pellet. We considered the infrared spectrum as the most unequivocal and final identification of the products.

The standard infrared spectra of free adenine, adenine monohydrochloride, and adenine dihydrochloride are different. Therefore, it was necessary to remove the HCl from the hydrochlorides before preparing the KBr pellets. Concentrated ammonium hydroxide was added to the sample solution (in 0.1 N HCl), until the solution was slightly basic. Due to the relatively low solubilities of these nitrogen compounds in alkaline solution, adenine, guanine and

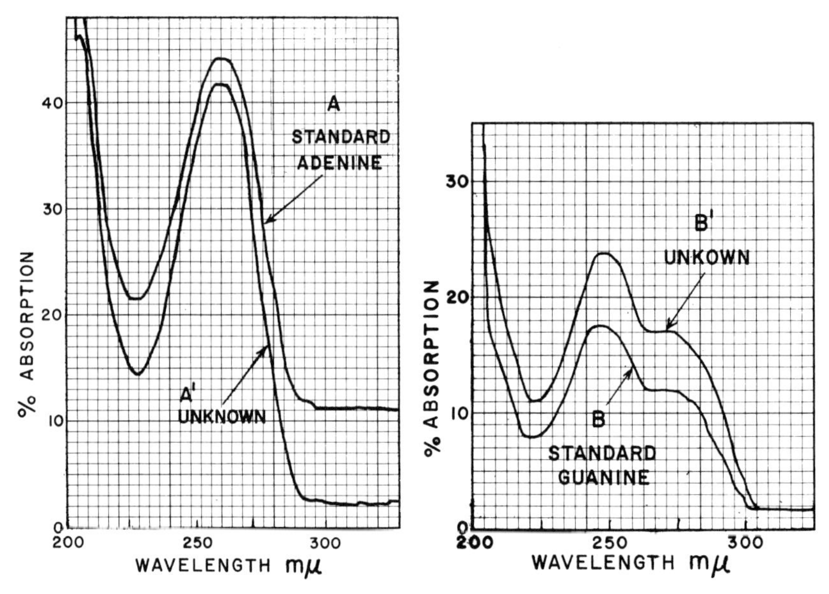

Fig. 3. Ultraviolet spectra of adenine and unknown in 0.1 N HCl.

Fig. 4. Ultraviolet spectra of guanine and unknown in 0.1 N HCl.

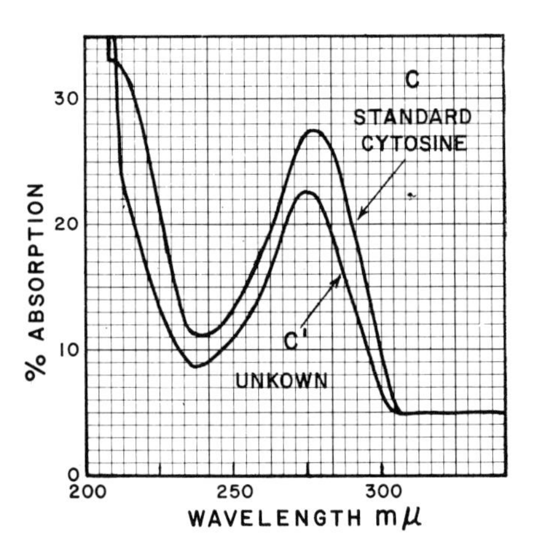

Fig. 5. Ultraviolet spectra of cytosine and unknown in 0.1 N HCl.

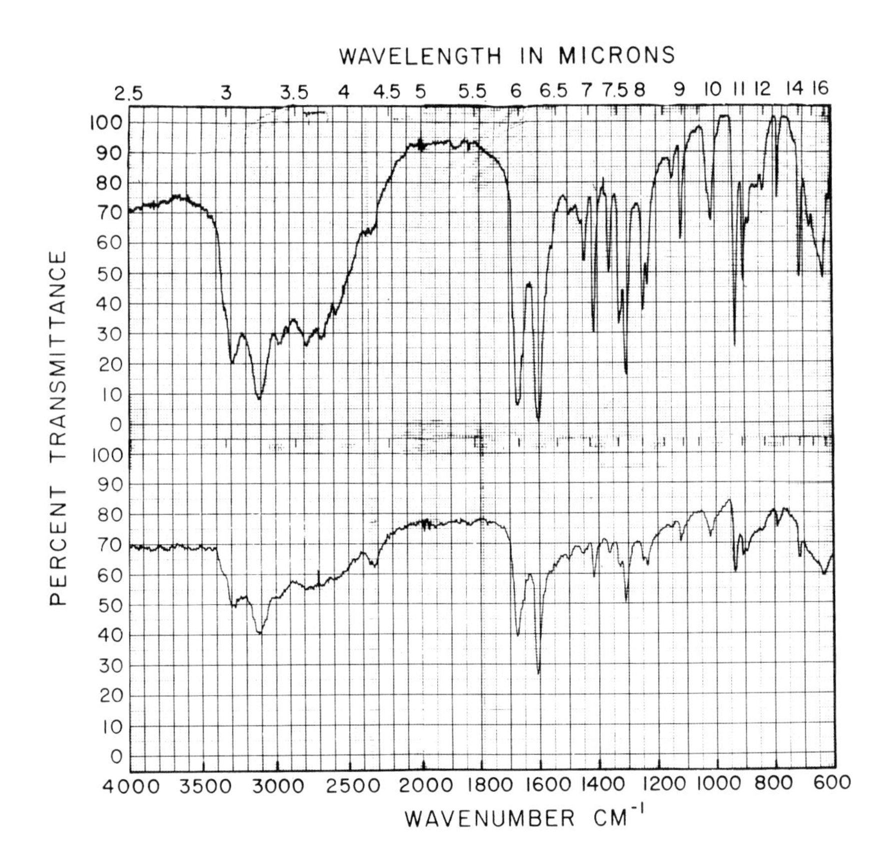

Fig. 6. Infrared spectra of standard adenine and unknown. (Both in KBr). Upper spectrum: standard adenine; lower spectrum: unknown.

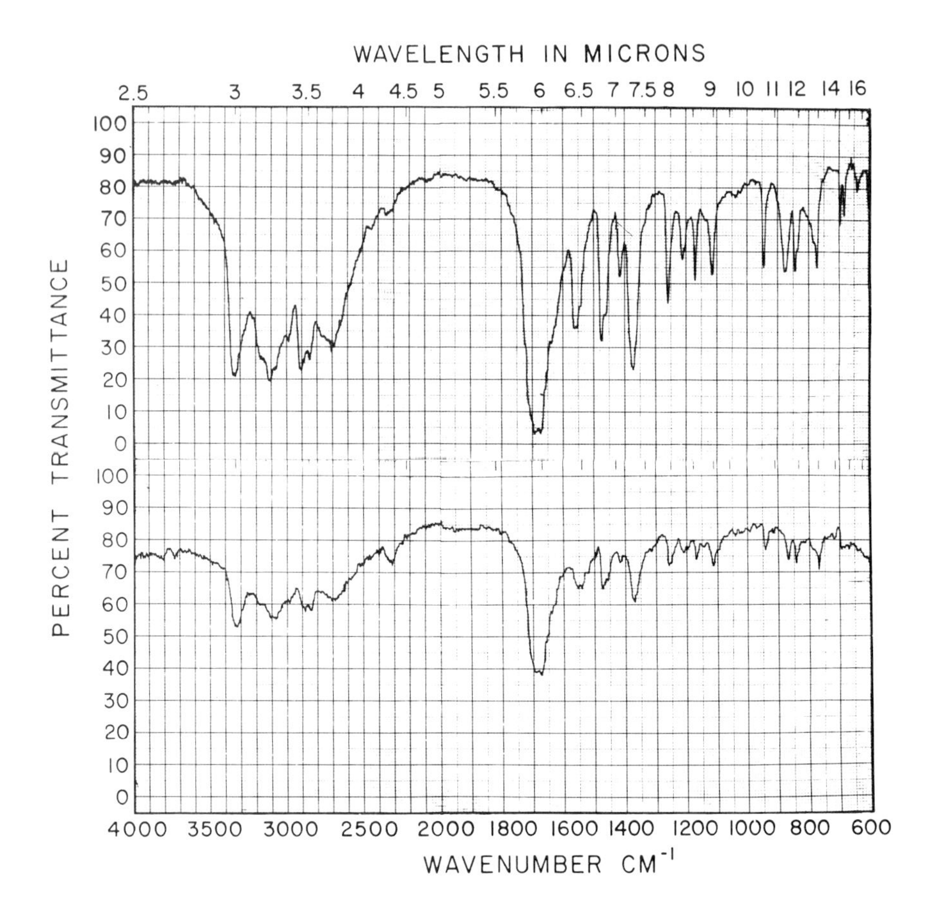

Fig. 7. Infrared spectra of standard guanine and unknown. (Both in KBr.) Upper spectrum: standard guanine; lower spectrum: unknown.

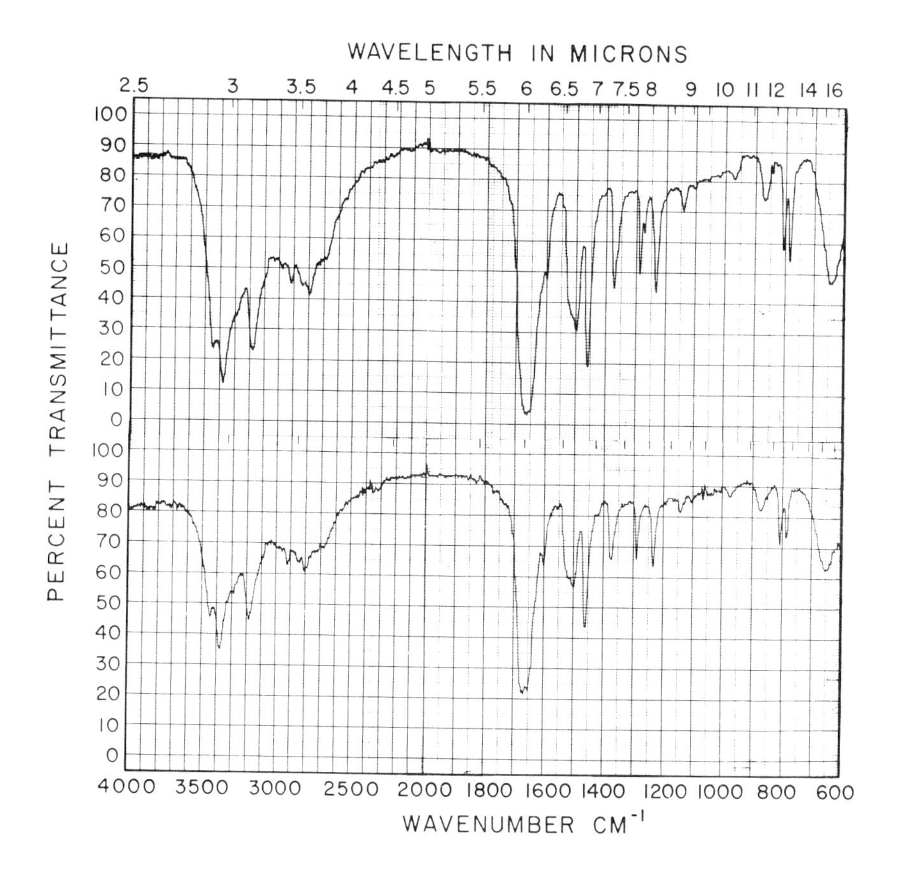

Fig. 8. Infrared spectra of standard cytosine and unknown. (Both in KBr.) Upper spectrum: standard cytosine; lower spectrum: unknown.

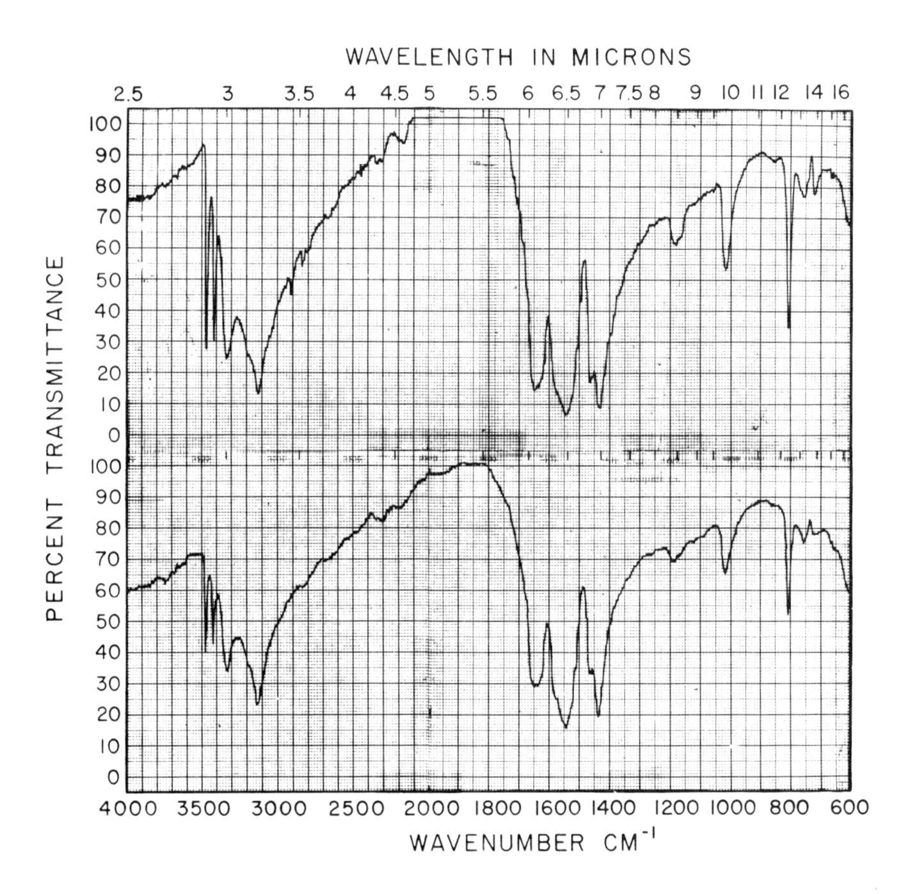

Fig. 9. Infrared spectra of standard melamine and unknown. (Both in KBr.) Upper spectrum: standard melamine; lower spectrum: unknown.

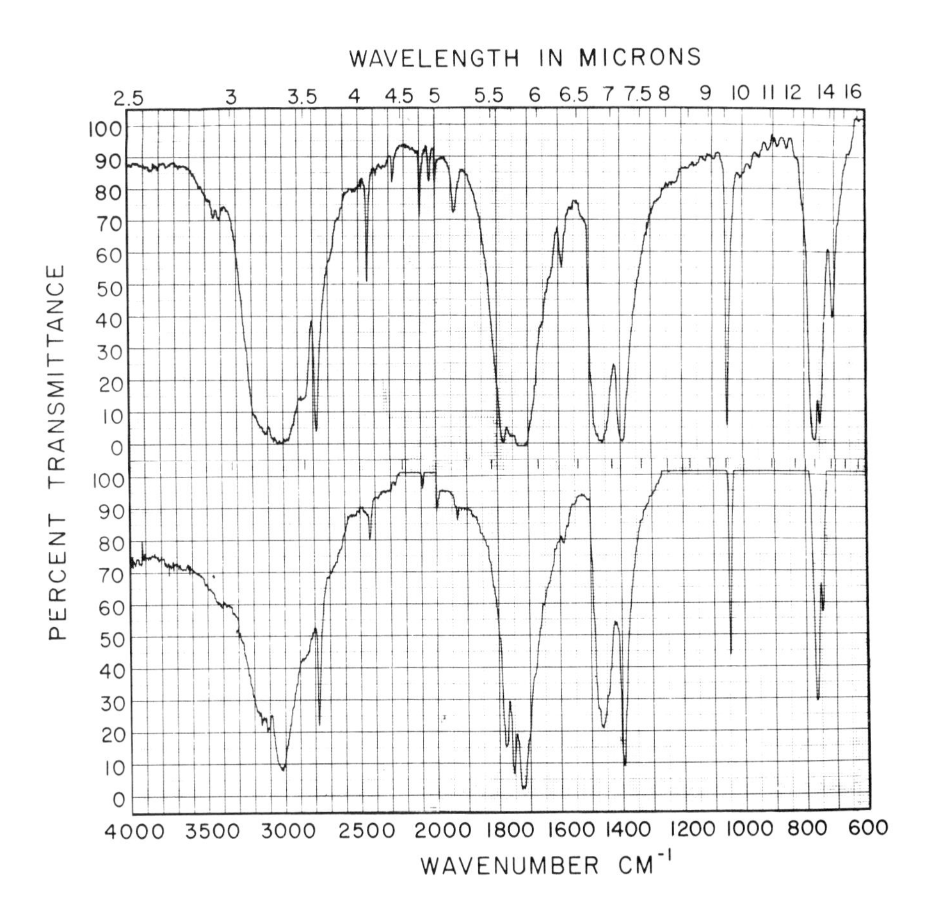

Fig. 10. Infrared spectra of standard cyanuric acid and unknown. (Both in KBr.) Upper s;petrum: standard cyanuric acid; lower spectrum: unknown.

cytosine (also melamine) were precipitated out by centrifuging (NH_4Cl dissolved in the solvent). The precipitates were then washed twice with distilled water. After obtaining enough sample of each compound, these samples and the spectroscopic grade KBr were dried in the oven for 24 hr at 100°C. A pellet was made out of well-ground mixture of KBr and sample under a pressure of 20,000 psi for 2 min.

The infrared spectrum was measured for each sample, and then compared with a standard free base spectrum. Figs. 6,7,8 and 9 show the infrared spectra of adenine, guanine, cytosine and melamine. Cyanuric acid was also identified in these experiments by its infrared spectrum (fig. 10) even though it did not show any color reaction with the reagents used on the paper chromatograms. The quantitative determination of adenine, guanine, cytosine, and melamine was done by ultraviolet light spectrophotometry. Comparisons of the absorption coefficients at their characteristic peaks were always used. In the case of cyanuric acid it was quantitatively determined by weighing the isolated crystalline product. As described earlier the amounts of other compounds were measured by color reactions.

Discussion and conclusions

Industrial nickel–iron alloy catalyzes the synthesis of adenine, guanine, cytosine and other nitrogenous compounds (including urea, biuret, guanylurea, melamine and cyanuric acid) from mixtures of CO, H_2 and NH_3 at temperatures of about 600 °C. Sufficient sample was accumulated to isolate as solid products adenine, guanine, and cytosine which were identified by infrared spectrophotometry. In the absence of nickel–iron catalyst, at 650 °C, or in the presence of this catalyst, at 450 °C, no purines or pyrimidines were synthesized. These results confirm and extend some of the work reported by Hayatsu et al. [5]. The mechanisms of synthesis of the purines and the pyrimidines can probably be interpreted in terms of hydrogen cyanide and cyanoacetylene formation; and followed by condensation of these compounds with urea, cyanate or other reactive molecules in accordance with the work reported from this [8,10] and other laboratories [3].

Acknowledgement

This work was supported by NASA Grant NGR-44–005–002.

References

[1] Cameron A.G.W., Icarus 1 (1962) 13.
[2] Cason J., J. Am. Chem. Soc. 69 (1947) 495.
[3] Ferris J.P., R.A. Sanchez and L.E. Orgel, Science 154 (1966) 784.
[4] Hayatsu R., Science 146 (1964) 1291.
[5] Hayatsu R., M.H. Studier, A. Oda, K. Fuse and E. Anders, Geochim. Cosmochim. Acta 32 (1968) 175.
[6] Larimer J.W. and E. Anders, Geochim. Cosmochim. Acta 31 (1967) 1239.
[7] Milks J.E. and R.H. Janes, Anal. Chem. 28 (1956) 846.
[8] Oró J., in: The Origins of Prebiological Systems, ed. S.W. Fox (Academic Press, New York, 1965) p. 137.
[9] Oró J., E. Gelpi and D.W. Nooner, J. Brit. Interplan. Soc. 21 (1968) 83.
[10] Oró J. and A.P. Kimball, Arch. Biochem. Biophys. 94 (1961) 217; 96 (1962) 293.
[11] Studier M.H., R. Hayatsu and E. Anders, Geochim. Cosmochim. Acta 32 (1968) 151.
[12] Tanaka K., J. Res. Inst., Hokkaido Univ. 11 (1964) 185.
[13] Wieland T., Z. Physiol. Chem. 273 (1942) 23.

PART IV

OLIGOMERS AND POLYMERS

Chemical Evolution and the Origin of Life, eds. R. Buvet and C. Ponnamperuma

THE EFFECT OF IMIDAZOLE, CYANAMIDE, AND POLYORNITHINE ON THE CONDENSATION OF NUCLEOTIDES IN AQUEOUS SYSTEMS

J. IBANEZ, A.P. KIMBALL and J. ORÓ

Department of Biophysical Sciences, University of Houston, Houston, Texas 77004, USA

Introduction

The polymerization of mononucleotides in possible primitive environments has already received the attention of several investigators [5,6,8]. In these reports, agents such as ethyl metaphosphate, polyphosphoric acid, or water-soluble carbodiimides were used to promote the condensations. Recently, the polymerization of thymidine-5-monophosphate (TMP) was carried out in a non-aqueous solvent (dimethyl formamide) using β-imidazolyl-4(5)-propanoic acid as the catalyst [4]. This type of reaction has more significance as a possible model for the abiogenic condensation of mononucleotides if carried out in aqueous solutions. We report here the polymerization of TMP in aqueous solutions using imidazole to promote this reaction. In addition, other experiments employing a polycationic polyamino acid as a prototemplate and cyanamide as the condensing agent are reported.

Experimental and results

In typical experiments, 2-^{14}C-TMP, 1×10^{-2} μmole, 0.5 μCi, and imidazole, 1.5×10^{-5} mole, were allowed to react in 0.20 ml of distilled water in stoppered glass vials at 90 °C for varying lengths of time. The reaction mixtures (and suitable control mixtures) were spotted on DEAE-cellulose paper and developed in a 0.5 M ammonium bicarbonate solvent. Detection of products was carried out by cutting the strips of paper and measuring their radioactivity in a Packard scintillation counter. Alternatively, the product of another reaction (to which 2.3 μmoles of non-radioactive TMP was also

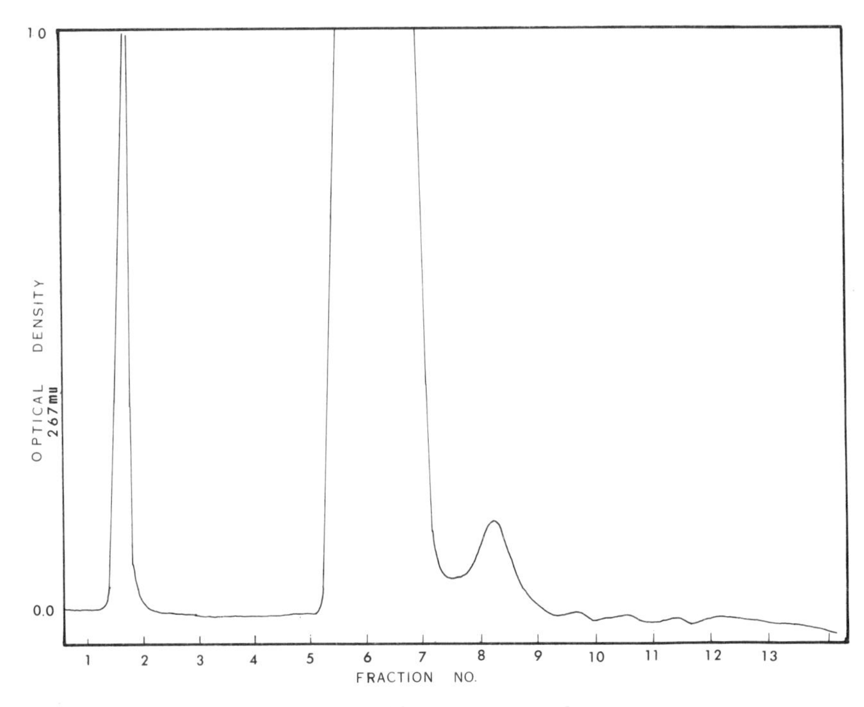

Fig. 1. The polymerization of 2-^{14}C-TMP, 1×10^{-2} μmole, 0.5 μCi; imidazole, 1.5×10^{-5} mole, in 0.2 ml of water. The reaction was carried out in sealed tubes at 90 °C for 24 hr. The material was passed through a column (10 cm × 9 cm) of DEAE-cellulose (BioRad) which had been equilibrated with 0.04 M lithium acetate buffer, pH 5.0. The material was eluted with a linear gradient of 0.04–0.40 M LiCl. Fractions were collected and counted by liquid scintillation. The effluent was monitored continuously at 267 nm.

added) was subjected to column chromatography using DEAE-cellulose columns (previously equilibrated with 0.04 M lithium acetate buffer, pH 5.0) and linear gradient elution with lithium chloride; 0.04 to 0.40 M. Enzymic characterization of the isolated products was carried out by coupled alkaline phosphatase–snake venom phosphodiesterase degradation or by snake venom phosphodiesterase degradation alone.

The profile of the TMP products off a DEAE-cellulose column is shown in fig. 1. The major peak is TMP (fraction 6) which is followed by several peaks of polymeric material that decreases in quantity with increasing size.

The peak at fraction 8 was collected and rechromatographed on DEAE-cellulose paper. Counting the radioactivity on the chromatogram gave one

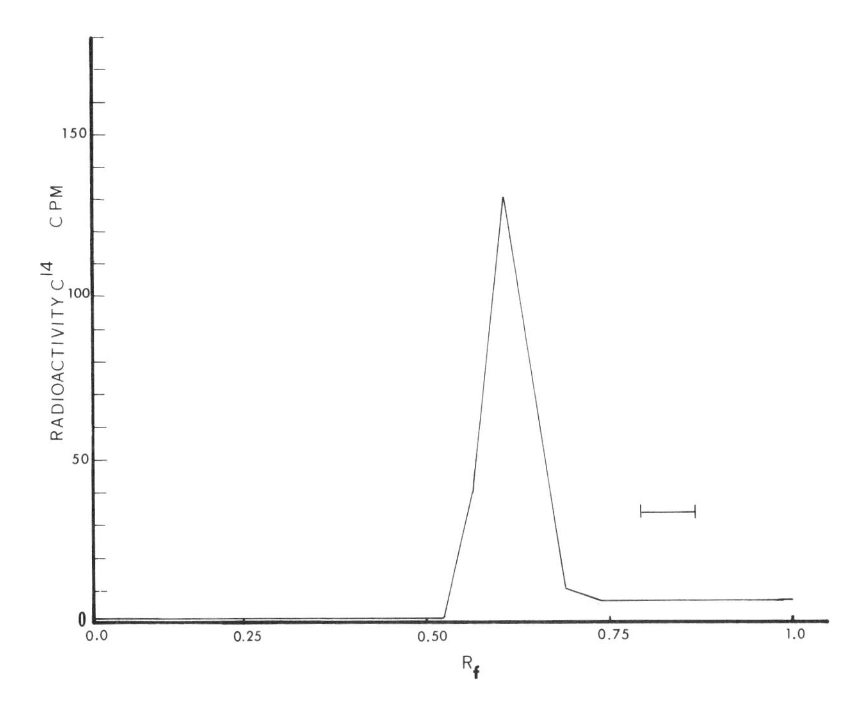

Fig. 2. Small oligomers of TMP. DEAE-cellulose paper chromatography and radioactive profile of fraction 8 (see fig. 1). The bar at approximately R_f 0.80 represents the position of standard TMP. The chromatogram was developed in a 0.5 M ammonium bicarbonate solvent. The dried and developed chromatogram was stripped and counted in a Packard (model 3380) scintillation counter.

peak at approximately R_f 0.60 (Standard TMP, R_f 0.80), fig. 2. This peak represents oligomers of TMP (2 to 3 units) [1].

Since the imidazole ring is a constituent of the amino acid, histidine, and the purine precursor, 4(or 5) amino 5(or 4)-imidazole carboxamide (AICA), we tried these as TMP polymerization catalysts. The results of this study are shown in fig. 3 which represents radioactivity on a DEAE-cellulose paper chromatogram. Neither histidine nor AICA showed catalytic activity under the experimental conditions employed as compared to the imidazole base. These imidazole derivatives shown in fig. 4 are substituted at positions 4(or 5) or 5(or 4). This data indicated that these substituents at these positions decrease catalytic activity. However, the imidazole side group of histidine is a well-known participant in some enzyme reactions. The amount of polymeric

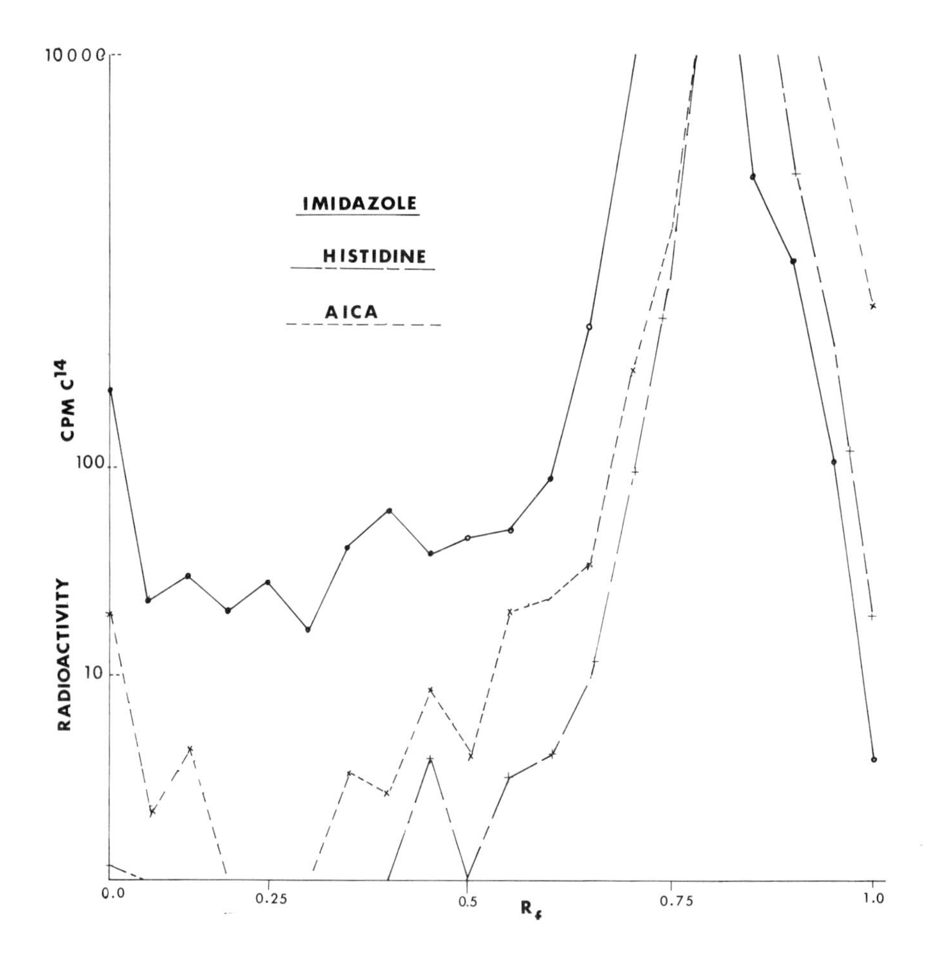

Fig. 3. Comparisons of imidazole compounds as catalysts in the polymerization of TMP. The reaction mixtures contained: a) 2-^{14}C-TMP, 2×10^{-3} μmole, 0.10 μCi; imidazole 1.5×10^{-5} moles, in 0.10 ml of water; (b) 2-^{14}C-TMP, 2×10^{-3} μmole, 0.10 μCi; histidine 5.3×10^{-6} mole in 0.10 ml of water; (c) 2-^{14}C-TMP, 2×10^{-3} μmole, 0.10 μCi; AICA 7.8×10^{-6} mole in 0.10 ml of water. Reaction time, 15 hr, at 90 °C. Chromatographed on DEAE-cellulose paper in 0.5 M ammonium bicarbonate.

material at the chromatogram origin (fig. 3) in the reaction where imidazole was used as the catalyst, gave 0.2% of the total radioactivity. The large peak at R_f 0.75 is TMP.

This peak at the origin (fig. 3, imidazole as catalyst) was eluted off and

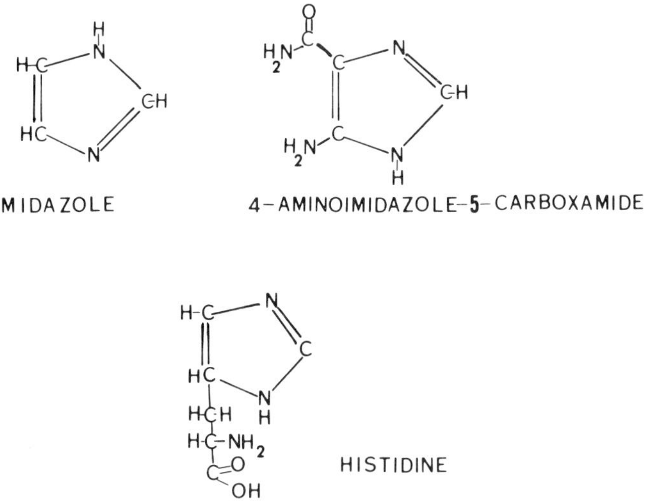

IMIDAZOLE

4-AMINOIMIDAZOLE-5-CARBOXAMIDE

HISTIDINE

Fig. 4. The structure of 3 imidazole compounds tried as condensing agents.

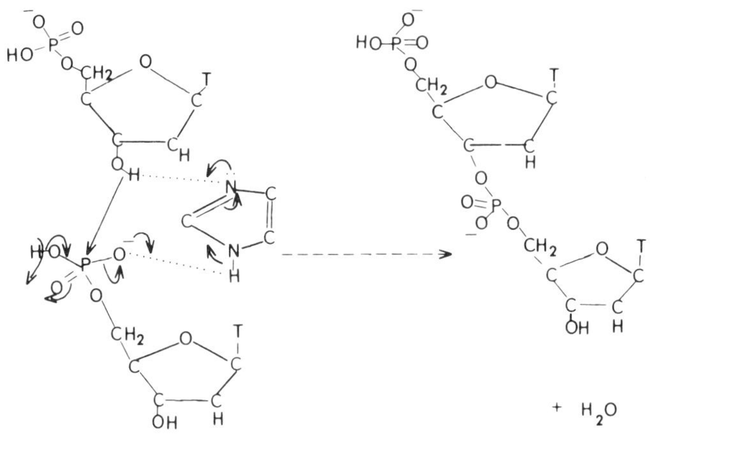

$+ H_2O$

Fig. 5. The postulated mechanism of TMP polymerization catalyzed by imidazole.

treated with snake venom phosphodiesterase. Rechromatography gave one radioactive peak corresponding to TMP.

One tentative mechanism for the role of imidazole catalysis in phosphodiester bond formation between adjacent TMP molecules is shown in fig. 5. Briefly, this involves electron shifts around an 8-membered ring formed by

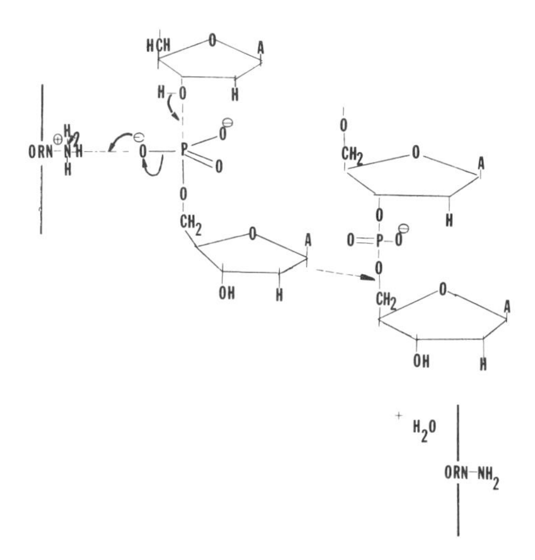

Fig. 6. The postulated role of an ornithine residue in binding and polymerization of nucleotides.

interactions of the nitrogens on imidazole with the hydrogen of the 3′-OH group of deoxyribose of one TMP and the negatively charged oxygen of the 5′-phosphate group of an adjacent TMP molecule. The oxygen of the 3′-OH group is attracted to the phosphorus completing the 8-membered ring. Water is eliminated, the phosphodiester bond is formed, and imidazole is left unaffected for a "tautomeric" shift of the proton. This mechanism implies that imidazole with alkyl or other type substituents on nitrogen would not catalyze this condensation reaction. Also, substituents on the 4 or 5 position, which would tend to make unavailable the unpaired electrons or the tertiary nitrogen of imidazole for bond formation with the hydrogen of the 3′-OH group, would decrease the catalytic ability.

The imidazole ring is a basic group as is the ϵ-amino group of lysine and the δ-amino group of ornithine. Both of these amino groups are protonated at pH 7 and should form ionic bonds with a negatively charged oxygen of the phosphate group of a nucleotide. This is shown schematically in fig. 6. Binding of this nature has been reported [2] for polylysine. It might be possible for such a basic group in a primitive polypeptide to participate in the formation of the phosphodiester bond (fig. 6). In this connection, we have recently found that RNA polymerase has a lysine residue in its active center [7].

We carried out experiments on the possible polymerization of dAMP-

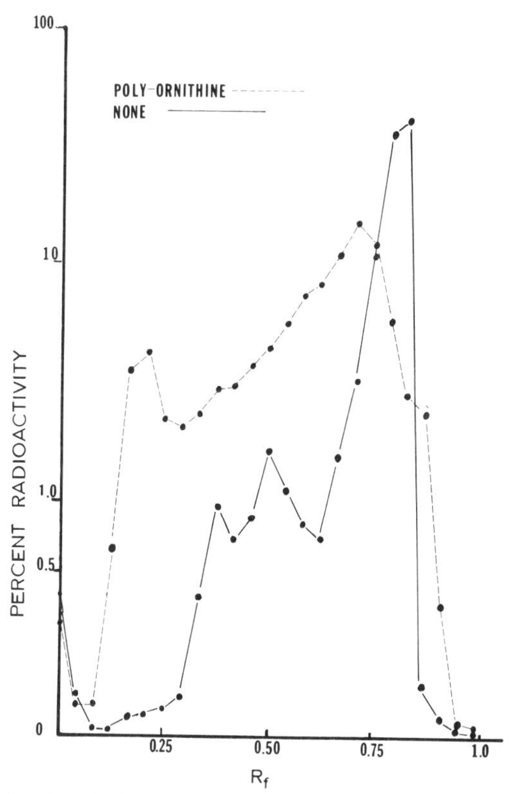

Fig. 7. Effects of polyornithine on the polymerization or complexation of dAMP. The reaction mixture contained in 0.75 ml; dAMP-8-^{14}C, 1 × 10^{-2} μmole, 1 μCi; 0.1 ml of a 50 percent solution of cyanamide; acetate buffer, pH 7.0, 0.01 mmoles; polyornithine, 0.5 mg, M.W. 200,000; and distilled water to 0.75 ml. The reaction mixtures with (– – –) or without (——) polyornithine were reacted in sealed tubes at 40°C for 24 hr. The products were analyzed by methods outlined in the legend for fig. 2.

8-^{14}C with cyanamide in the presence and absence of polyornithine. The mixtures after 24 hr reaction time were chromatographed on DEAE-cellulose paper in 0.5 M ammonium bicarbonate and counted on the liquid scintillation counter (fig. 7). The chromatogram indicated that a high molecular weight material was present in about 4 percent yield at an R_f of approximately 0.20 in the reaction mixture containing polyornithine. No such material occurred at R_f 0.20 in the reaction lacking polyornithine. Further work is required to characterize this possible polymer of dAMP or complex.

In similar experiments with dGMP and polyornithine, pronase degradation of the "high molecular weight" material followed by chromatography gave back dGMP only. Our conclusions were that dGMP was not polymerized but only very tightly bound to the polycationic amino acid polymer. The apparent stability of the dGMP-polyornithine complex opens up the possibility that these type of complexes could, of themselves, serve as prototemplates if their conformations allowed this. We mean by prototemplate that the polycationic amino acid polymer could order in nucleic acid precursors in stacked arrays by ionic bonding.

Summary and conclusions

This report thus outlines two models for the condensation of nucleotides under possibly prebiotic conditions. The first was the promotion of this type of reaction in the presence of imidazole and substituted imidazole compounds. The second is the condensation of mononucleotides with cyanamide in the presence and absence of a prototemplate such as polyornithine. The yields of products remaining at the origin of a paper chromatogram (presumably, oligomers larger than 8–10 units) are quite small in these condensation reactions. In general, the average yield was between 0.3–0.8% of the starting material (calculated from several experiments). In addition, oligomers of 2–7 units in length decreased in amounts with increasing size. The enzymatic degradation of the products at the origin of a paper chromatogram (R_f 0.0) with snake venom phosphodiesterase resulted in the formation of TMP (97%). This indicates that the principal type of bonding in the product is 3′–5′ phosphodiester linkage. Further degradation with alkaline phosphatase results in the formation of thymidine.

Thus, although relatively small yields of condensation products are obtained when carried out in aqueous systems, their formation is significant. Accumulation of these products might have occurred later on biopolymers such as polyornithine, or may have been concentrated under evaporation conditions. In addition, inorganic surfaces might have accumulated these products. In fact, Paecht-Horowitz, Berger, and Katchalsky [3] have recently shown the formation of polypeptides from amino acid adenylates in aqueous systems in the presence of montmorillonite. They report that the presence of clay particles in the polymerization medium strongly inhibits competitive hydrolysis of the active groups and favors rapid formation of peptide bonds. In a similar manner, the condensation of nucleotides could possibly be facilitated after their adsorption on clays.

Acknowledgement

This work was supported by NASA Grant NGR-44–005–002.

References

[1] Bollum F.J., J. Biol. Chem. 237 (1962) 1945.
[2] Lacey J. and K. Pruitt, Nature 224 (1969) 799.
[3] Paecht-Horowitz M., J. Berger and A. Katchalsky, to be published.
[4] Pongs O. and Ts'o, P.O.P., Biochem. Biophys, Res. Comm. 36 (1969) 475.
[5] Schramm G., H. Grotsch and W. Pollmann, Agnew Chem. Intl. Ed. Engl. 1 (1962) 1.
[6] Schwartz A.W. and S.W. Fox, Biochem. Biophys. Acta. 134 (1967) 9.
[7] Spoor T., F. Persico, J. Evans and A.P. Kimball, Nature 227 (1970) 57.
[8] Sulston J., R. Lohrmann, H. Miles and L.E. Orgel, Proc. Natl. Acad. Sci. U.S. 60 (1968) 409.

Chemical Evolution and the Origin of Life, eds. R. Buvet and C. Ponnamperuma

THE POSSIBLE PARTICIPATION OF ESTERS AS WELL AS AMIDES IN PREBIOTIC POLYMERS

Alexander RICH
Department of Biology, Massachusetts Institute of Technology, Cambridge, Massachusetts, USA

One of the first problems to be faced in any formulation of prebiological chemical evolution is the development of polymers. Polymers are the materials which nature uses in the development of biological structures and they serve as the framework for the major classes of biological macromolecules. In the prebiological era, it is likely that polymers formed abiogenically and later were ultimately formed by biological processes . In both circumstances, these built up large molecules. We would like to know which polymers were biologically important and how they were formed. The polymers which are used in contemporary biological systems are well characterized. These include polyamides (proteins), polynucleotides and polysaccharides. Enzymatic activities are found in the polyamide protein molecules. It is possible that prebiological polymers included the polyamides as well as the sugar polymers and the sugar–phosphate polymers which form the backbone of the nucleic acids. However, could other polymers have been important in the prebiological era? Two considerations are important in this regard. The monomeric material had to be available in the prebiological period so that they could be incorporated into polymers. Furthermore, if they were incorporated into polymers, they might have been biologically significant if they could serve as a basis for the development of the early biological systems. For the former consideration, it is important that bifunctional molecules form in the prebiological environment, i.e., molecules which have two chemical functions which are capable of being utilized in polymer formation. This is illustrated clearly in the formation of α-amino acids in the prebiological atmosphere experiments of Miller [8]. These experiments yielded significant amounts of α-amino acids. These are bifunctional molecules which contain both a carboxyl group and an amino group. Because of these two functions they can form a polymer by making a peptide bond between the carboxyl group of one amino acid and the amino group of the next. The purpose of this communication is to point out the fact that α-hydroxy acids may have participated in the formation of prebiological polymers in a manner analogous to the participation of the α-amino acids. The α-hydroxyl acids could be incorporated to make polyesters or, more probably, polymers with a mixture of amide and ester linkages. In order to make this plausible, it would be of interest to know whether it was likely that α-hydroxyl acids were present in the prebiological environment. In addition, we would like to know whether contemporary biological systems are able to form ester linkages as well as amide linkages, and form polyesters or polymers containing both amide and ester linkages.

It is easy to answer the first of these questions. Analysis of the products of the Miller spark experiments in a reducing atmosphere shows that there is an appreciable yield of α-hydroxyl acids. In some experiments the total yield of various α-hydroxyl acids is almost as great as the yield of α-amino acids, depending upon the conditions of the spark experiments. Thus, it is likely that α-hydroxyl and α-amino acids were present in the period during which abiogenic polymerization took place. It is possible that many of the polymerization mechanisms may not have differentiated between them since amides and esters are somewhat similar in their reactivity.

In addition, the amide ($\overset{\diagdown}{\underset{O^{\swarrow}}{C}}-\overset{H}{N}\diagdown$) and the ester bond ($\overset{\diagdown}{\underset{O^{\swarrow}}{C}}-O\diagdown$) have very similar geometry.

We may ask whether esters and amides are handled in similar ways in contemporary biological systems. More specifically, can the peptide bond forming machinery of the cell be active in the formation of esters and polyesters? Also, can the peptide cleaving machinery cleave ester linkages? If this were the case it might lend additional credence to their possible participation in prebiological systems. Accordingly, we describe here experiments which indicate that the system for forming peptide bonds in present day biological organisms is equally competent in forming ester and polyester bonds.

Ribosome catalyzed ester bond formation

Peptide bonds are formed in ribosomes. This process is carried out through the formation of an intermediate complex containing messenger RNA and transfer RNA which has been activated so that an amino acid is attached to it. The enzyme which is responsible for forming the peptide bond in the ribosome is called peptidyl transferase. During normal peptide bond formation, two transfer RNAs are involved. One of them has attached to the hydroxyl of its 3′-adenosine terminus the nascent or growing polypeptide chain. Next to this on the aminoacyl site of the ribosome is another transfer RNA which has an amino acid attached to its 3′-terminal adenosine. Both the peptide chain and the amino acid are attached by an ester bond to ribose hydroxyl groups probably the 3′-position. Peptidyl transferase acts by transferring the peptidyl group from one tRNA. Thus the ester bond which normally attaches the nascent polypeptide chain to one adenosine is cleaved and the peptidyl group is transferred to the α-amino group of the adjoining aminoacyl-tRNA. A peptide bond is then formed and the polypeptide chain has been elongated by one residue. This process is repeated in a cyclic fashion during the polymerization of the polypeptide chain. We might ask whether the ribosomal peptidyl transferase is also capable of forming an ester bond? One way of carrying out this experiment is to use the antibiotic puromycin which is an analog of the aminoacyl-adenosine end of tRNA. The puromycin molecule (fig. 1) finds its way specifically to the aminoacyl site of the ribosome and there it acts as an acceptor for the nascent polypeptide chain. The polypeptide chain is transferred to the puromycin residue with the formation of a peptide link onto the α-amino group of puromycin. The peptidyl puromycin molecule is subsequently released from the ribosome. The activity of puromycin as an antibiotic is thus due to its ability to abort the growing polypeptide chain and prevent the completion of the entire peptide needed for the intact protein [12].

In the experiments described here [3] analogs of puromycin were made in which the α-amino group was removed and an α-hydroxyl group was substituted for this. This is accomplished by two methods as shown in fig. 1. In one of these the α-amino group of puromycin was removed by reaction with nitrous acid and the α-hydroxyl puromycin derivative was obtained. In another set of experiments an α-hydroxyl analog of puromycin was synthesized directly by reacting phenyllactic acid with puromycin amino-nucleoside. This resulted in the formation of a product ψ-hydroxy-puromycin which is analogous to the α-hydroxy-puromycin except for the absence of a paramethoxy group on the phenylalanyl side chain of puromycin. It has already been shown that the parامethoxy group does not modify the reactivity of puromycin for the release of the peptide chain [11].

The activity of puromycin can be assayed most simply by using *N*-formylmethionine as a peptidyl group. *N*-Formylmethionine-tRNA is normally used as the initiator of protein synthesis in *E. coli*; due to its blocked α-amino group, it is likely that initiation takes place by having this group enter the

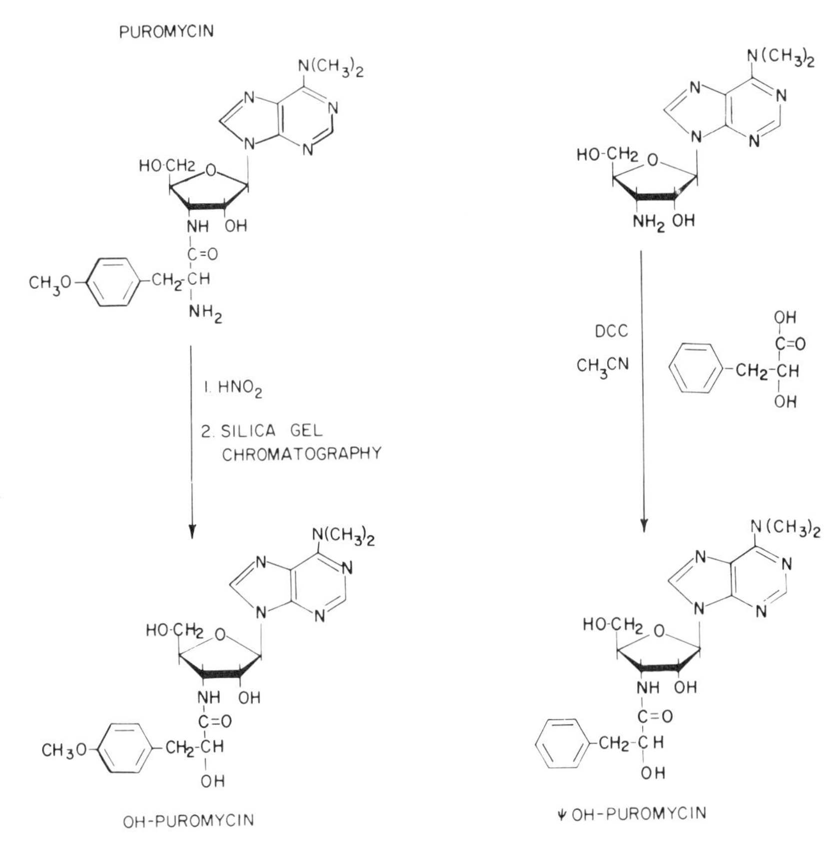

Fig. 1. Chemical synthesis of puromycin derivatives. Hydroxy(OH)-puromycin is made by nitrous acid deamination of puromycin, followed by chromatography. During this reaction there is a molecular rearrangement so that the proper isomer must be isolated. The analogue ψ-hydroxy-puromycin is synthesized by condensation in dicyclohexyl-carbodiamide (DDC) of puromycin amino nucleoside and L-lactic acid. ψ refers to the fact that the methoxy group is not attached to the phenyl ring.

peptidyl site of the ribosome. We can thus use the α-hydroxy-puromycin molecule to see whether ribosomal peptidyl transferase will catalyze the formation of the ester analogue of formylmethionyl-puromycin (Fmet-puromycin). This can be tested in the fragment reaction, so called because it only uses a T_1-ribonuclease digested hexanucleotide fragment of fMet-tRNA plus ribosomes and puromycin [10]. The kinetics of the reaction in which α-hydroxy-puromycin and puromycin itself were used is shown in fig. 2.

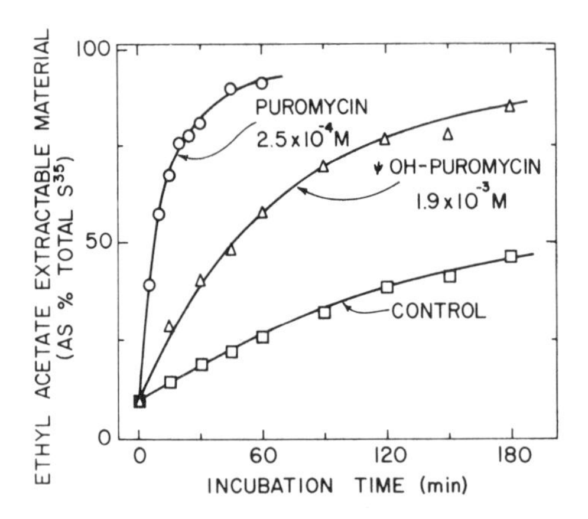

Fig. 2. Time course of the formylmethionyl-T_1 oligonucleotide fragment reaction with ψOH-puromycin and puromycin. ^{35}S-fMet-T_1 oligonucleotide fragment was prepared from ^{35}S-formylmethionyl tRNA. Prior to methanol addition, the reaction mixture contained 0.06 M tris-HCl buffer (pH 8.1, measured at 0 °C), 0.4 M KCl, 0.02 M magnesium acetate, 0.006 M β-mercaptoethanol, 13.8 A_{260} units per ml ribosomes, and the formylmethionyl-T_1 fragment (50,000 dpm/ml). The reaction was initiated with an equal volume of methanol (control) or a methanolic solution of puromycin (5×10^{-4} M) or ψOH-puromycin (3.7×10^{-3} M). Concentrations of puromycin and ψOH-puromycin were determined spectrophotometrically in 0.1 N HCl solution (λ_{max} = 267.5 nm; $\epsilon = 2.0 \times 10^4$ for both compounds). Incubation was at 0 °C. At various times, 0.1 ml aliquots were mixed with 25 μl of 0.1 M $BeCl_2$, and 0.1 ml of 0.3 M sodium acetate (pH 5.5) saturated with $MgSO_4$, and then 1.5 ml of ethyl acetate were added. The mixture was shaken at room temperature for 15 sec and centrifuged briefly. One ml of the ethyl acetate layer was counted in a liquid scintillation spectrometer.

Formylmethionyl-puromycin is soluble in ethyl acetate. It can be seen that both of these materials react to yield products which are soluble in ethyl acetate. Furthermore, the reactions go to completion. The control reaction shown in fig. 2 illustrates the formation of the methyl ester of formylmethionine, because the solvent for the fragment reaction contains methanol.

The product of these reactions were characterized in several ways. In paper elecrophoresis it was found that the ^{35}S labeled product of the puromycin and the α-hydroxy-puromycin reaction both migrate together over the pH range 3.0 to 4.2 (fig.3). The mobility of fMet-puromycin in this pH range is determined largely by the dimethyl amino group of adenine which has a pK of 3.7. Since both fMet-puromycin and the fMet-α-hydroxyl product have the same mobility on either side of this p*K* it is clear that the product contains the puromycin analog as well as the ^{35}S-formylmethionine. There is a

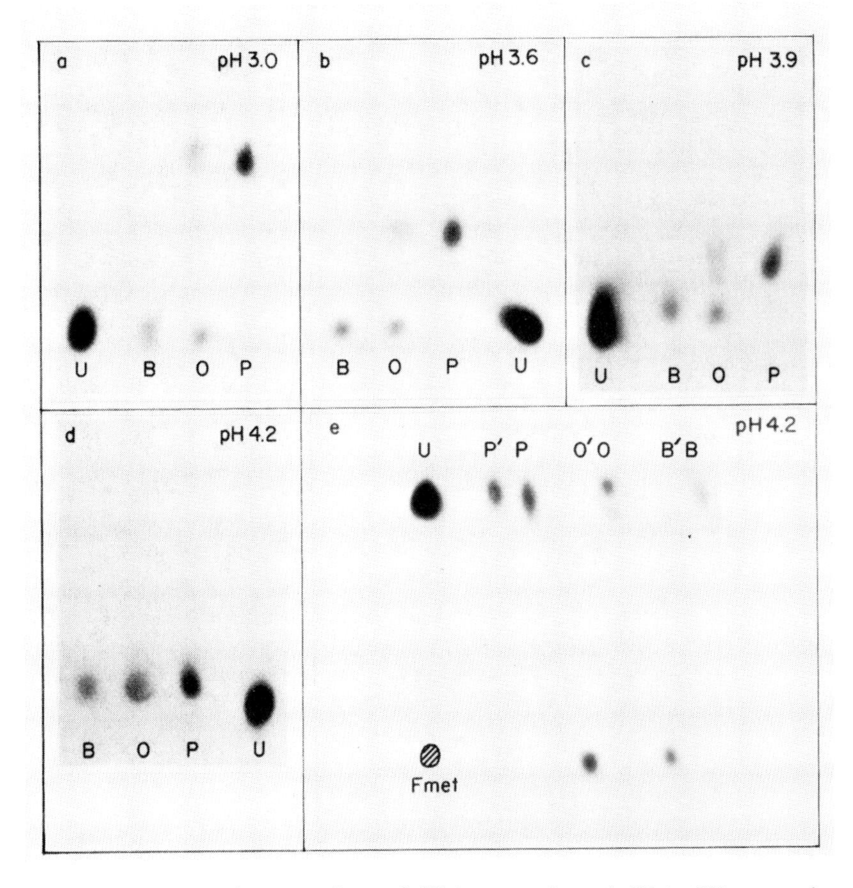

Fig. 3. Electrophoretic comparison of fMet-puromycin and fMet-ψOH-puromycin. Products were formed from ^{35}S-fMet T_1 oligonucleotide under conditions as described in fig. 2. Samples labeled *P* contained a final concentration of 5×10^{-4} M puromycin, those labeled *O* contained 5×10^{-4} M ψOH-puromycin, and those labeled *B*(blank) contained no puromycin or ψOH-puromycin. Samples were incubated 2–½ hr at 0 °C, then the reaction was terminated and the samples extracted with ethyl acetate as described in fig. 2. The ethyl acetate layer was dried under reduced pressure and the residue was dissolved in ethanol. In (e) aliquots of each of these ethanol solutions were incubated in 1 M triethylamine for 30 min at 37 °C, dried under reduced pressure, and the residue redissolved in ethanol. These triethylamine-treated samples are designated by *P'*, *O'* and *B'*, respectively. Aliquots of the products were subjected to electrophoresis on Brinkmann cellulose thin layer plates, at an ionic strength of 0.2. Radioactivity was located by autoradiography. Uridine-^{14}C was used to locate the origins (U). The direction of the cathode is up in all cases.

marked contrast however, in the lability of the products when exposed alkali as shown in fig. 3 (e). The products of the puromycin reaction (*P*) and the α-hydroxy-puromycin reaction (*O*) were exposed to 1 M triethylamine, and the electrophoresed at pH 4.2. It can be seen that the product of the α-hydroxy-puromycin reaction was unstable, and liberated labeled formyl-methionine while the puromycin product is stable. It is well known that esters are labile in alkali while amides are stabile. The alkaline lability of the products is shown in more detail in fig. 4.

While fMet-puromycin is unaffected at pH 9.0, fMet-ψOH-puromycin starts to be hydrolyzed appreciably at a pH just over 8, and is destroyed rapidly at pH 9.0. At pH 9.0, 30 °C in 0.04 M tris-HCl, fMet-ψOH-puromycin has a half life of 12 min. For comparison, the Fmet-methylester formed in the fragment system has a half life of 130 min and formyl-methionyl-adenosine, formed by pancreatic RNase digestion of Fmet-tRNA, has a half life of 3 min under these conditions.

We have concluded that this ester bond involves the α-hydroxyl group of hydroxy-puromycin rather than the 2′- or 5′-hydroxyls. Other analogs of puromycin in which the α-amino group is *N*-acetylated or replaced by a chlorine atom or methoxy group are not reactive under these conditions even

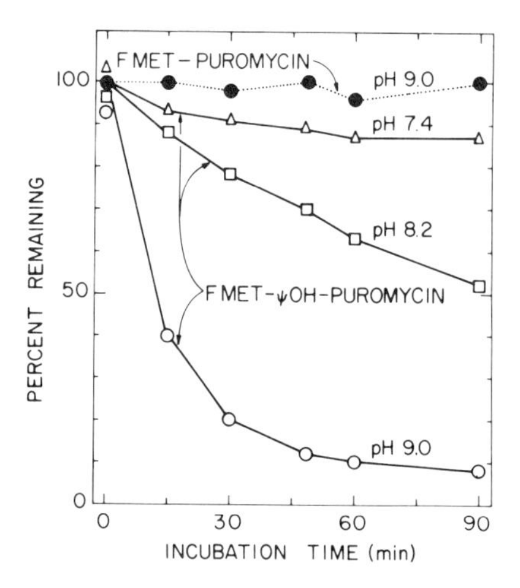

Fig. 4. Alkaline hydrolysis of fMet-ψOH-puromycin. The ethylacetate layer was dried under reduced pressure and the material redissolved in ethanol. Aliquots of these ethanol solutions were mixed with 0.04 M tris-HCl buffer at the pH given and incubated at 30 °C. Aliquots were removed at various times for assay.

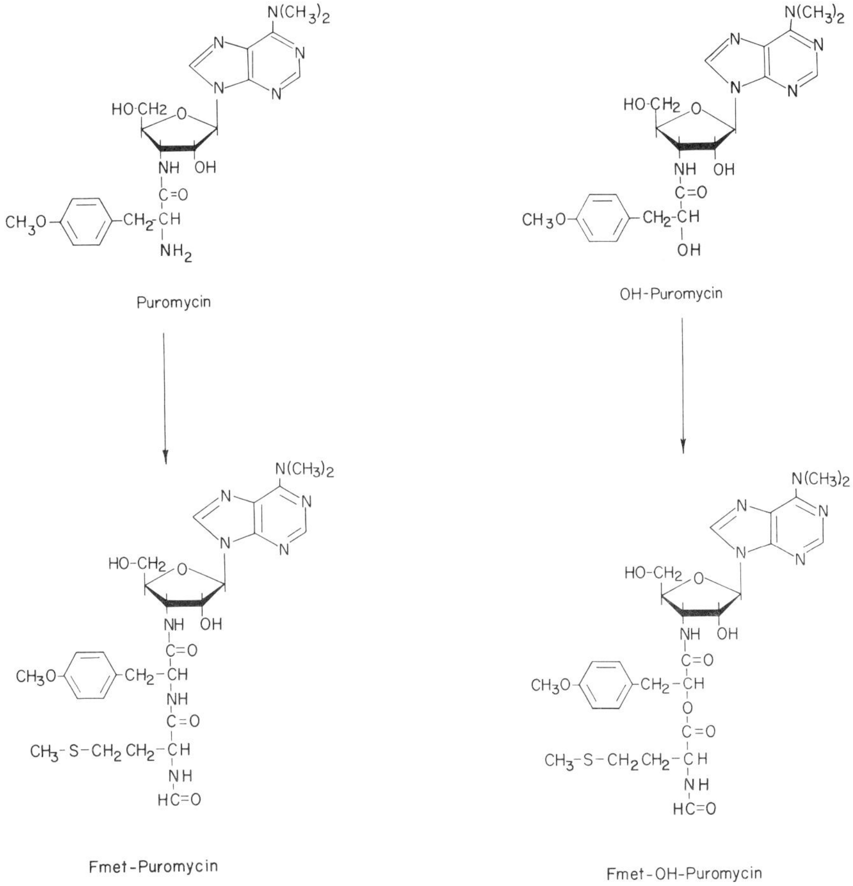

Puromycin

OH-Puromycin

Fmet-Puromycin

Fmet-OH-Puromycin

Fig. 5. Formulas of the products of the reaction of the formylmethionyl-tRNA fragment with either puromycin or hydroxy-puromycin.

though the 2′- and 5′-hydroxyls are unaltered in these compounds [3]. The products of the two reactions are as shown in fig. 5.

Participation of the ribosome

The reactions of the fMet-tRNA fragment with both puromycin and α-hydroxy-puromycin require ribosomes, and are inhibited by both chloramphenicol and gougerotin, known inhibitors of protein synthesis [3]. This suggests that they are both catalyzed by the ribosomal peptidyl transferase. Further similarity of these reactions is shown by the heat inactivation date of fig. 6. Ribosomes are preincubated for 5 min at various temperatures, then tested for their activity in a subsequent incubation in the fragment system at 0 °C. Preincubation of the ribosomes at temperatures about 60 °C results in a parallel loss of their ability to catalyze Fmet-puromycin and fMet-ψOH-puromycin formation. At temperatures below 60 °C a slight activation of the ribosomes is observed.

It has been shown that the activity of the ribosomal peptidyl transferase increases with increasing pH between 7.0 and 8.5 [7]. A similar pH dependence is observed for the reaction with ψOH-puromycin, as shown in fig. 7. This suggests that a functional group with a pK_a value in the range 7.5 to 8.0 is involved in the catalysis. However, the pH dependence for the puromycin reaction might also be due to protonation of the α-amino group of puromycin if its pK is altered slightly on complex formation. This ambiguity is now eliminated because the similar pH dependence of the ψOH-puromycin reaction clearly cannot be explained by substrate protona-

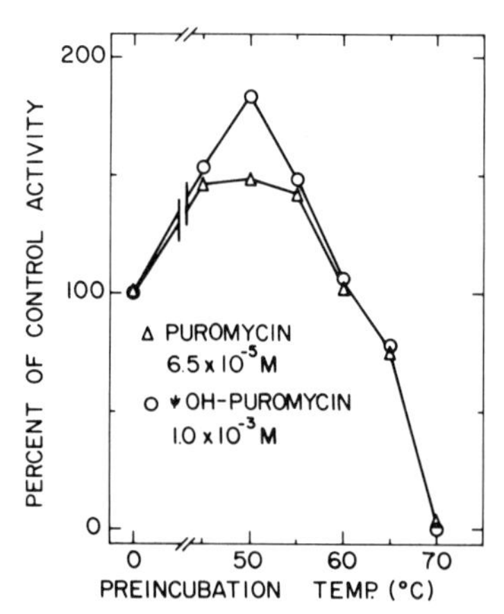

Fig. 6. Heat inactivation of ribosomes. Ribosomes were incubated in 0.4 M KCl, 0.02 M magnesium acetate, 0.06 M tris-HCl (pH 8.1 at 0 °C) for 5 min at various temperatures, then kept at 0°C for 15 min. ^{35}S-fMet-T_1 oligonucleotide (50,000 dpm/ml) was added, together with methanol (blank) or a methanol solution of ψOH-puromycin (2.0×10^{-3} M) or puromycin (1.0×10^{-4} M). After 15 min at 0 °C the reaction was terminated and product assayed as described in fig. 2. The values obtained reflected initial rates of the reaction. The data are expressed as percentage of the value obtained using ribosomes kept at 0 °C.

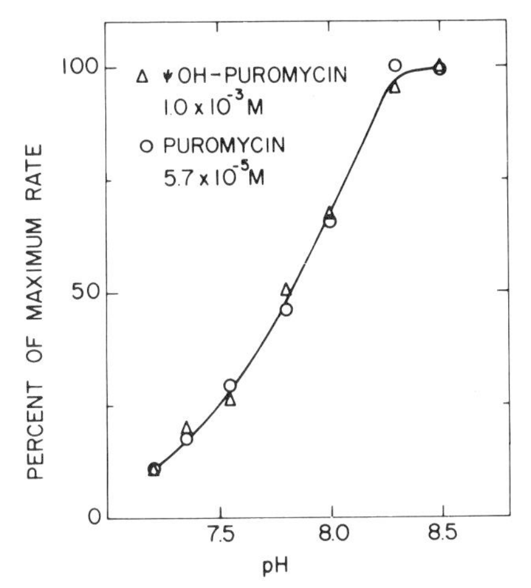

Fig. 7. Effect of pH on the rate of reaction between fMet-T_1 oligonucleotide and ψOH-puromycin or puromycin. After 15 min at 0 °C, the reaction was terminated and assayed as described in fig. 2. Data are expressed as percent of the pH 8.5 values, which were 368 dpm for puromycin and 175 dpm for ψ(H-puromycin.

tion. This is consistent with the idea that an imidazole residue or an *N*-terminal α-amino group might be involved in the catalysis of the reaction.

The hydroxyl analogues of puromycin are also active in bringing about the release of polypeptide chains. In an *E.coli* in vitro system synthesizing polyphenylalanine directed by polyuridylic acid, hydroxy-puromycin is able to bring about the release of oligophenylalanine into the supernatant. Like puromycin, hydroxy-puromycin is also able to stop the incorporation of radioactive amino acids into a culture of *E. coli* although higher concentrations are required. The activity of hydroxy-puromycin is thus similar to puromycin even though it is somewhat less effective, due to a reduced binding constant.

Experiments with messenger RNA

The experiments described above are directed towards answering questions regarding the action of peptidyl transferase in ester formation. However, will a complete protein synthetic system operate with transfer RNA molecules which have α-hydroxyl acids attached to them instead of α-amino acids? We can answer this question in two ways, using both synthetic and natural

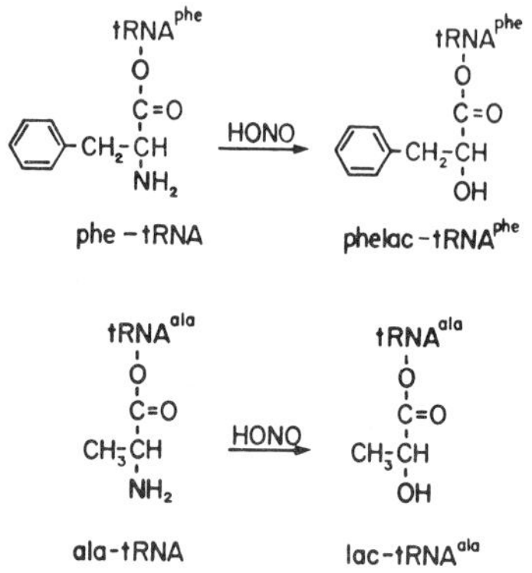

$tRNA^{phe}$–O–C(=O)–CH(NH_2)–CH_2–C_6H_5 $\xrightarrow{HONO}$ $tRNA^{phe}$–O–C(=O)–CH(OH)–CH_2–C_6H_5

phe – tRNA phelac – $tRNA^{phe}$

$tRNA^{ala}$–O–C(=O)–CH(NH_2)–CH_3 $\xrightarrow{HONO}$ $tRNA^{ala}$–O–C(=O)–CH(OH)–CH_3

ala-tRNA lac-$tRNA^{ala}$

Fig. 8. Preparation of α-hydroxyacyl tRNAs: tRNA was prepared from *E. coli* B and charged with alanine or phenylalanine. The RNA was dissolved in 2 ml 0.25 M sodium acetate, pH 4.3, 0.01 M magnesium acetate, 1 M $NaNO_2$, and incubated at room temperature (23 °C) on a pH-stat at pH 4.3 for 30 min. The products of the nitrous acid treatment of alanyl- and phenylalanyl-tRNA are lactyl- and phenyllactyl-tRNA, respectively. This was verified by paper chromatography in *n*-butanol : acetic acid : water (78:5:17) of alkaline digests of similar preparations made with the ^{14}C-amino acids. Nitrous acid treatment under these conditions results in minimal damage to the RNA.

mRNA. For these experiments we must create a form of tRNA activated with α-hydroxyl acids. These can be most conveniently made by treating aminoacyl-tRNA with nitrous acid, which deaminates them to the α-hydroxyl analogs. As shown in fig. 8, the alanyl- and phenylalanyl-tRNA were converted to lactic- and phenyllactic-tRNA. The products of the reaction were characterized by chromatographic analysis following alkaline digestion. Under the conditions used, there was minimal damage to the tRNA [1].

When polyuridylic acid is used as a synthetic mRNA, polyphenylalanine is obtained as a product. If this incorporation is carried out with phenyllactic-tRNA, polyphenyllactic ester is obtained [5]. This product is acid-precipitable in a manner similar to polypeptides, but when it is exposed to alkali, it is rapidly hydrolyzed. In initial experiments, about 80% of the residues in the polymer were phenyllactic acid and the remaining 20% were phenylalanine residues. A cleaner system would undoubtedly yield a pure polymer of phenyllactic acid. These experiments thus show that ribosomes can synthesize polyesters when given a synthetic mRNA.

Experiments in protein synthesis were carried out using the naturally occurring mRNA found in the bacteriophage virus R17. This RNA is widely used in in vitro protein synthesis and the major product is the viral coat protein [13] A mutant of bacteriophage R17 (am B_2) contains an amber codon at the seventh position of the coat protein cistron; this signals the ribosome to release the polypeptide chain. In in vitro systems, the RNA of this virus therefore directs the synthesis of the *N*-terminal hexapeptide of the coat protein with an amino acid sequence fMet–Ala–Ser–Asn–Phe–Thr. The synthesis of the hexapeptide product requires correct initiation and termination in the in vitro system. In the experiments described here [4] instead of using one of the required amino acids, the analogous α-hydroxyl acid was supplied to an α-hydroxyacyl-tRNA. The resulting "hexapeptide" product contains an alkali labile ester bond which is found at a specific position determined by the amino acid replaced by the α-hydroxyl acid.

The protocol is outlined in fig. 9. Three different types of experiments were carried out, those containing ^{35}S-fMet-tRNA and five aminoacyl-tRNAs, or four aminoacyl-tRNAs with either Lac-tRNA or Phelac-tRNA. The incorporating system also used ribosomes from *E. coli*, enzymes from an *E. coli* supernatant together with RNA from the amber mutant (am B_2) of bacteriophage R17. The results are shown in the autoradiogram of fig. 10. It contains the products of the reaction, together with mobility standards.

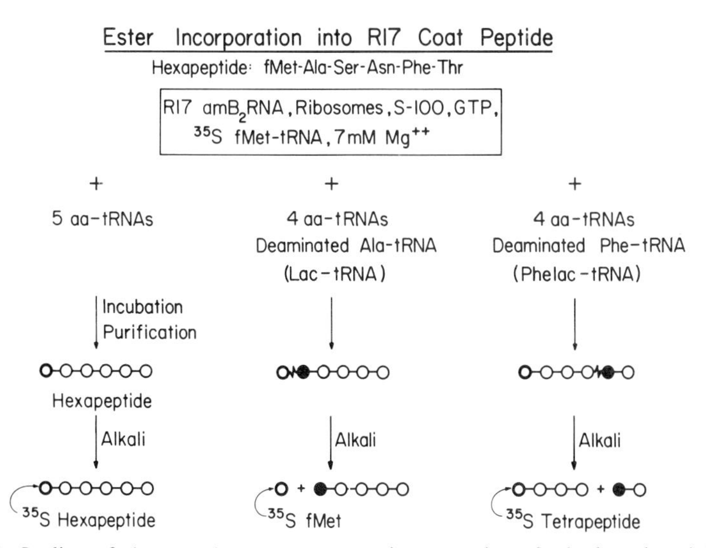

Fig. 9. Outline of the experiment to measure incorporation of α-hydroxyl acyl-tRNA into R17 viral coat peptide. Circles are used to indicate amino acids. The heavy circles represent ^{35}S-fMet, the only radioactive amino acid. The solid circles represent α-hydroxy acids.

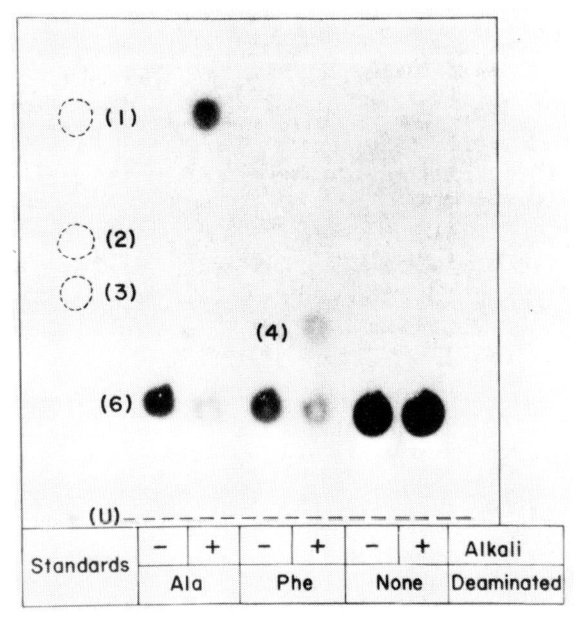

Fig. 10. Characterization of the products. After 30 min of incubation, the *N*-blocked peptides were isolated. The hexapeptide was further purified by electrophoresis on paper. An aliquot of this material was treated with 1 M triethylamine (pH about 12.5) at 35 °C for 20 min. Triethylamine treated and untreated samples were then subjected to electrophoresis at 4 °C on a cellulose-thin layer in a pyridine–acetate buffer, pH 4.8. The radioactivity was located by autoradiography. The standards are: (1) fMet; (2) fMet–Ala; (3) fMet–Ala–Ser; (U) uridine [^{14}C]. Uridine marks the origin corrected for endosmotic buffer flow. (6) Hexapeptide, (4) tetrapeptide.

Under the conditions of this electrophoresis there is a good separation of *N*-blocked peptides with a mobility which varied inversely with the molecular weight of the peptide [6]. There is good separation between all the peptides containing one to six amino acids. In fig. 10, the origin is marked by radioactive uridine (U). The numbers indicate the number of amino acids in the various peptides, with the dotted circles indicating the position of the non-radioactive standards.

The reaction products of the incubation containing only aminoacyl-tRNAs is a hexapeptide. Treatment with 1 M triethylamine which hydrolyzes ester bonds (as shown in fig. 3) leaves the products unaffected. Peptide bonds are known to be stable under these conditions. However, the hexapeptide made in the presence of deaminated Ala-tRNA contained an alkali labile bond between the first and second positions in the peptides since alkaline treatment resulted in the release of labeled fMet. Similarly, the hexapeptide made in the presence of deaminated Phe-tRNA contained an alkali labile bond between the fourth and fifth positions as the alkali treatment hydrolyzed the product to the labeled tetrapeptide, marked [4].

The results shown in fig. 10 are most simply interpreted as indicating the replacement of phenyllactic acid for phenylalanine or lactic acid for alanine in the viral *N*-terminal coat protein peptide. Furthermore, this occurs in the incubation under ionic conditions which allows accurate translation of natural mRNAs, i.e., at 7 mM magnesium [4]. In the work described above with the α-hydroxyl analogue of puromycin, we showed that the *E. coli* ribosome could catalyze the formation of one ester bond. The present results demonstrate that the peptidyl transferase of a single ribosome can catalyze interchangeably peptide and ester bond formation and that α-hydroxyl acids can be incorporated from α-hydroxyacyl-tRNAs into internal positions in a polypeptide under the direction of a natural messenger RNA. The possibility that peptide and ester formation are catalyzed by two different subsets of the ribosome preparation is clearly eliminated by these experiments. Here a polymer is synthesized which contains both peptide and ester bonds in the same chain. Since the polypeptide is assembled on a single ribosome and is not released from the ribosome until a terminator codon is encountered, it is clear that a single ribosome can incorporate both α-hydroxy acids and α-amino acids interchangeably. This shows that peptidyl transferase of a single ribosome can use either a nitrogen or an oxygen atom as a nucleophilic acceptor during polymer formation.

Relation to early biological processes

The capability of peptidyl transferase for ester formation reflects a flexibility in the catalytic mechanism; however, it does not necessarily indicate a special function of the enzyme in vivo in contemporary living systems. Nevertheless, the fact that the ribosome has this capacity opens the possibility that ester bonds could be inserted into polypeptides or may have been used in the past. Whether or not this occurs in vivo depends upon whether the appropriate substrates exist in the cell.

The ribosomal peptidyl transferase is a central enzyme in molecular biology: it is responsible for the synthesis of all proteins. It is therefore of considerable interest to understand the mechanism of its action. The work described here demonstrates its ability to catalyze the formation of an ester bond as well as its normal product, the peptide bond. However, it is well known that in the reverse enzymatic reaction, the hydrolysis of peptide bonds, most peptidases are also esterases [2]. This may indicate that the transition states for both the synthetic reaction and the hydrolytic reaction are the same. If this is true, it may be of considerable consequence in working out the mechanism of action of the enzyme. However, the fact that both esters and amides can be used as substrates for both the synthetic reactions as well as the degradative reactions makes it possible that these reactions could have been used in early stages of biological evolution.

Our best estimate of the composition and mode of formation of molecules in the primitive reducing atmosphere comes from experiments similar to those of Miller and Urey [8,9]. The large amount of α-hydroxyl acids produced in these experiments means that it is distinctly possible that early polymerizing mechanisms may have resulted in polymers containing both amides and esters. It is possible that the ability of contemporary ribosomal systems to polymerize both amides and esters and the ability of proteases to cleave both amides and esters reflects thc participation of both of these molecular species in early polymerizing mechanisms. Of course we do not know the basis of early polymerization systems. Nonetheless this may be looked upon as an attractive possibility, perhaps an example in which the contemporary system for building and destroying polymers retains an ability which may have been utilized at a more primitive stage of evolution. In any case, it is clear that further work will have to be done to determine the possible role of esters in the formation of biological polymers in the early history of life on this planet.

Polyesters in living systems?

Is it possible that a type of life could be developed around a system employing polyesters instead of polyamides? It is not easy to answer questions of this sort. There are certain advantages which accrue by using polyamides. For example, the peptide bond is more stable than the ester bond, especially in an alkaline medium. In addition, the presence of a proton attached to the nitrogen atom makes it possible to fold polypeptides into a number of tertiary configurations which would not be possible in polyesters. The alpha helix found in many proteins involves the participation through hydrogen bonding of the N–H proton in forming the helix. This helix would not be possible with polyesters because the ester link lacks the proton. This suggests that some features of polyesters are less desireable than those of polyamides. However, they cannot be ruled out entirely as a polymer which could be used to form an information-containing molecule analogous to proteins. The limited stability of polyesters means that such systems would have to be stabilized at a lower pH than contemporary terrestrial systems. In addition different three-dimensional structures would be assumed by polyesters. However, this would not in itself be a significant constraint. For example, if one looks in the field of synthetic textile polymers, we find that Nylon is a widely used polyamide which forms a useful and stable fiber. However, an equally useful commercially important fiber is the polyester Dacron. It has properties as a textile fiber which compare favorably with the polyamide, Nylon. Of course, it has quite a different three-dimensional structure but nonetheless it can serve equally well as a basis for a textile fabric. In the same way it may be that a polyester could be used as an information-containing polymer in biological systems although not necessarily those of terrestrial origin.

In this regard, it is interesting to note that our current investigations have revealed a very small amount of nitrogen present on the surface of Mars. At first one might think that this would make it impossible for Mars to have developed a living system. However, it is not impossible that a system of life could have developed on Mars which uses ester polymers rather than amide polymers. In such a system, nitrogen might be used as a slightly rare element, perhaps useful for carrying out important catalytic functions. In this respect, it might have an abundance in such a hypothetical Martian living system perhaps comparable to that of sulphur or phosphorus in terrestrial biological systems. At the present time, however, this must be regarded as purely a speculative possibility.

Acknowledgements

This work was supported by grants from the U.S. National Aeronautics and Space Administration, and the National Institutes of Health.

References

[1] Carbon J.A, Biochim. Biophys. Acta 95 (1965) 550.
[2] Dixon M. and E.C. Webb, Enzymes (Academic Press, New York, 1964) p. 225.
[3] Fahnestock S., H. Neumann, V. Shashoua and A. Rich, Biochemistry 9 (1970) 2477.

[4] Fahnestock S. and A. Rich, Nature, in press.
[5] Fahnestock S. and A. Rich, Science, in press.
[6] Kuechler E. and A. Rich, Nature 225 (1970) 920.
[7] Maden B.E.H. and R.E. Monro, European J. Biochem. 6 (1968) 309.
[8] Miller S.L., J. Am. Chem. Soc. 77 (1955) 2351; Science 117 (1953) 528.
[9] Miller S.L. and H.C. Urey, Science 130 (1959) 245.
[10] Monro R.E. and K.E. Marcker, J. Mol. Biol. 25 (1967) 347.
[11] Nathans D. and A. Neidle, Nature 197 (1963) 1076.
[12] Yarmolinski M.B. and G.L. de la Haba, Proc. Natl. Acad. Sci. U.S. 45 (1959) 1721.
[13] Zinder N.D., D.L. Englehardt and R.E. Webster, Cold Spring Harbor Symp. Quant. Biol. 31 (1966) 251.

Chemical Evolution and the Origin of Life, eds. R. Buvet and C. Ponnamperuma

ARCHETYPES OF PRESENT-DAY PROCESSES OF TRANSPHOSPHORYLATION, TRANSACYLATION AND PEPTIDE SYNTHESIS

Louis LE PORT, Elisabeth ETAIX, Francine GODIN,
Philippe LEDUC and René BUVET
Laboratoire d'Energétique Electrochimique de la Faculté des Sciences de Paris, 10, rue Vauquelin, Paris 5°, France

We have previously shown [2] that a thermodynamic treatment of said biological energy-transfer processes does permit us, to put into evidence driving conditions which do not imply the pre-existence of present-day biological substrates, energy-donors, catalysts and structures (table 1). From such treatment energy-transfer processes involving very simple molecules, built up with only the active groups of biological substrates and energy donors, possibly with such simple catalysts as metal ions, appear to be theoretically possible. Nevertheless, no systematic experimental investigations have been developed so far in order to examine the extent to which such

Table 1
Driving criteria of energy transfer processes summarized from [2]. ($\Delta G_d^{o\prime}$ is the apparent standard free enthalpies in aqueous solutions at given pH's, expressed for all reactions proceeding towards dehydration. The dehydrated component of the considered reaction is mentioned between brackets.

$\Delta G_d^{o\prime}$ (donor) + $\Delta G_d^{o\prime}$ (dehydrated product) $\leqslant 0$
Only bimolecular processes (simple transfers or additions to double bonds) are possible.
If double transfer is considered: $\Delta G_d^{o\prime}$ (donor) $\geqslant \Delta G_d^{o\prime}$ (intermediate) $\geqslant \Delta G_d^{o\prime}$ (dehydrated product)
Reactions may be facilitated by complex formations (e.g. with metal cations) as regards: location of reagents, influence on $\Delta G_d^{o\prime}$.
Kinetic conditions, at the considered pH: condensate obtained must be metastable; donor, if alone, may be either metastable or rapidly hydrolysed.

theoretical criteria are experimentally sufficient. However, if they are, this would be pertinent to studies on the origin of life, since many of the processes which are set into action in the metabolic pathways should appear compatible with the primordial terrestrial environment. Life should be considered as spontaneously originating from so-called non-living systems suitably supplied with energy.

The aim of this paper is to present a set of experimental data related to transdehydration reactions of biological importance, leading in particular to protein-like macromolecules, but proceeding in simple aqueous media from reagents and under conditions compatible with primeval terrestrial environment.

Some experimental results, reported from different sources in recent years, have been taken into account for guiding this systematic study for transfer processes compatible with primeval conditions. Some investigations connected with the study of the mechanisms of enzymic reactions have led us to put into evidence non-enzymic models of such processes, involving especially metal cations as catalysts. Thus, Lowenstein studied the transphosphorylation of the terminal phosphate group of ATP to labelled orthophosphate [12,14], and the formation of hydroxamates by double transfer through the action of ATP on free carboxylic acids in the presence of hydroxylamine [13,15] and divalent and monovalent metal cations. To go from these results to possible archetypes of transphosphorylations, the extent to which ATP may be replaced by inorganic tripolyphosphate under the same experimental conditions remains to be examined.

In more direct connection with the study of the origin of life, several authors have shown that the sugar moiety of nucleosides may be phosphorylated with low yields by polyphosphates [e.g. 17,19,20] and that the action of polyphosphates on free amino acids in aqueous solutions in the presence of metal cations may likely result in an overall process of peptide synthesis [e.g. 18].

Data related to the terminal step of such peptide syntheses were obtained by studies of transacylation and peptide synthesis from energy-rich amino acid derivatives, by Jencks et al. [3,6], Davidson et al. [5], Lipmann and Tuttle [11] and A. Katchalsky et al. [8,10]. From these contributions it appears that acylating simple transfers are possible in aqueous solutions from acylphosphates, esters and thiolesters even without metal catalysts.

Lastly, some authors have proposed that the primordial energy donors which have led to the overall formation of polyphosphates or peptides may have been different kinds of unsaturated compounds such as cyanate [1,16], cyanamide [21], dicyandiamide [1,21,22] or cyanoacetylene [4] in the presence of precipitated divalent metal cations. Such reactions may be considered as double transfers involving as the first step an addition to the double bond of the unsaturated donors [e.g. 7].

First of all we have examined the extent to which results similar to those of Lowenstein may be obtained when inorganic tripolyphosphate is used as energy and phosphoryl donor. In fig. 1 we have jointly reported Lowenstein's data relative to the transphosphorylating simple transfer:

$$\text{ATP} + \text{P}^* \rightarrow \text{P–P}^* + \text{ADP}$$

in the presence of Mn^{2+} and data obtained under similar conditions with tripolyphosphates (TP) in the presence of Ca^{2+} ions. In both cases the yields are comparable and depend on pH in similar ways. The order of catalytic activities of the different cations is:

$$Ca^{2+} > Cd^{2+} > Zn^{2+} > Mn^{2+} > Fe^{2+} > Ni^{2+} > Mg^{2+}$$

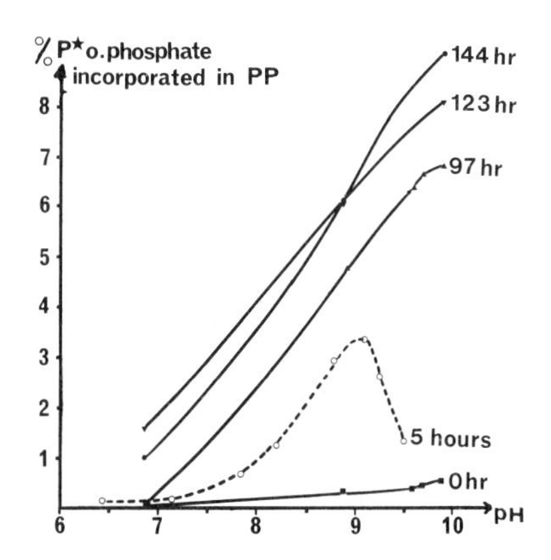

Fig. 1. Percentage of transphosphorylation (labelled orthophosphate in pyrophosphate) as a function of pH; incubation times on curves; – – – –, Lowenstein's results [12] for incubation of 50 μmoles/ml ATP, 5 μmoles/ml P* and 45 μmoles/ml Mn^{2+} at 38 °C. ——, our results for incubation of 50 μmoles/ml inorganic tripolyphosphate (TP), 5 μmoles/ml P* and 50 μmoles/ml Ca^{2+} at 0 °C.

i.e. very close to that noticed by Lowenstein when ATP is used:

$$Ca^{2+} > Mn^{2+} > Cd^{2+} > Zn^{2+} > Co^{2+} > Fe^{2+} > Cu^{2+} > Ni^{2+} > Be^{2+} > Mg^{2+}$$

Nevertheless the yields remain relatively low even for long incubation times. Similarly we have qualitatively observed that in the presence of acetic acid, hydroxylamine and Mn^{2+}, tripolyphosphate gives acetohydroxamate according to:

$$TP + CH_3CO{-}OH \rightarrow CH_3CO{-}O{-}P + P{-}P$$

$$CH_3{-}CO{-}OP + HNHOH \rightarrow CH_3{-}CO{-}NHOH + P$$

(P = orthophosphate group)

in the pH range 0–6 under similar conditions to those used by Lowenstein [13] with ATP in the presence of Mn^{2+}.

Such results tend to support the hypothesis that present-day transphosphorylation with nucleotide polyphosphates present no fundamental differences with reactions occurring simply with inorganic tripolyphosphates as donors and metal cations as catalysts, and which may be considered as archetypes of biological transphosphorylations.

A much more extended set of results has been obtained in connection with the study of transacylation processes. By extension of well-known methods of preparation of hydroxamates from esters [5], acyl phosphates [11] or other acyl donors [23], we have studied the acylation of different kinds of compounds with mobile hydrogen atoms. The simplest of such reactions is the acetylation of methylamine by simple energy transfer in aqueous solutions from ethyl acetate:

$$CH_3{-}CO{-}O{-}Et + HNHCH_3 \rightarrow CH_3{-}CO{-}NHCH_3 + EtOH$$

Fig. 3 in [2] shows that the free enthalpy balance of this process is negative only above approximately pH 4. In other respects, *N*-methylacetamide is practically metastable at any pH in this range, whereas ethyl acetate, which is metastable in neutral solutions, undergoes rapid hydrolysis through hydroxylic catalysis in alkaline solutions.

In fact, it is experimentally observed that any aqueous solutions of CH_3COOEt and NH_2CH_3 undergo no appreciable transformation at any pH below 9, at ambient temperature. On the contrary, when the alkaline hydrolysis of the ester occurs, very different evolutions of the reaction mixture are noticeable whether NH_2CH_3 is present or not at the time the hydrolysis proceeds. Indeed, the composition of the final mixture, assayed by pH metric titration, depends on the order in which the three reagents CH_3COOEt, NH_2CH_3 and KOH are added. For example, when the system containing 5 mmoles CH_3COOEt, 5 mmoles NH_2CH_3 and 10 mmoles KOH in 5 ml of aqueous mixture is titrated (fig. 2), the following observations are made:

(a) if the reagents are added in the order KOH, CH_3COOEt and after only one minute NH_2CH_3, i.e. NH_2CH_3 is added some time after the rapid and total hydrolysis of ester has proceeded, the final solution contains 5 mmoles of KOH, NH_2CH_3 and CH_3COO^- (fig. 2, curve 1);

(b) on the other hand, when the addition is performed in the order KOH, NH_2CH_3, then CH_3COOEt, i.e. NH_2CH_3 is present in the solution when the energy is released from the ester hydrolysis, the titration curve 2 in fig. 2 shows that stoichiometric amounts of acetate ion and amine have disappeared to give rise to a compound without acid–base properties. This compound, after extraction from the reaction mixture, was identified as *N*-methylacetamide (NMA) by infrared spectrometry.

The molar yield of the acetyl transfer can be defined as the proportion of methylamine which is acylated. It reaches 58% when an aqueous mixture of 2 M KOH, 1 M NH_2CH_3 and 1 M CH_3COOEt is used. The influence of

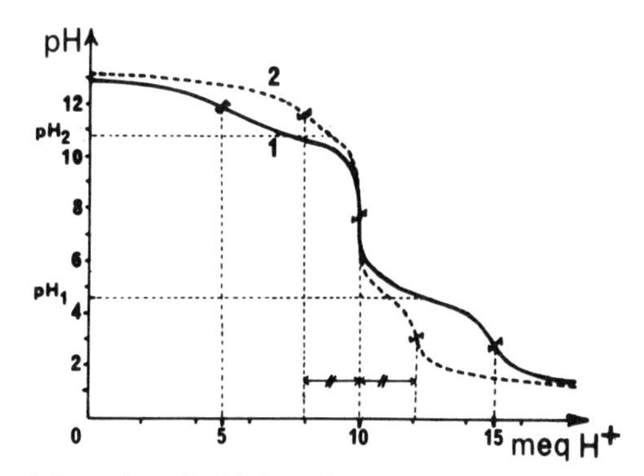

Fig. 2. Influence of the order of addition of reagents on the shape of the titration curve of 5 ml aqueous mixture of 10 mmoles KOH, 5 mmoles CH_3COOEt and 5 mmoles NH_2CH_3 at 25 °C. *Curve 1*: For comparison: titration curve when AcOEt is first mixed to KOH, then only after total hydrolysis of the ester is completed (after one minute) NH_2CH_3 is added; *curve 2*: titration curve when ester AcOEt is added last to the mixture KOH and NH_2CH_3. Equal amounts of NH_2CH_3 and CH_3COO^- disappear during reaction. $pH_1 \simeq 4.7$: acetic acid pK_a; $pH_2 \simeq 10.7$: methylammonium pK_a.

dilution on the acetyl transfer is given in fig. 3 for a mixture involving the three reagents with same initial molar quantities. This yield reaches a maximum of about 60% when the initial concentration is about 1 M.

When different molar quantities of the three reagents are taken, higher yields can be obtained [9].

Table 2 gives the transfer yields for different acyl donors, esters, thiolesters and anhydrides of simple carboxylic acids, malonic acid and glycine and different acceptors with amino groups including glycinate and β-alaninate anions. From these data it appears that: (1) the transfer yield does not depend strongly on the nature of the alcohol group of the ester; (2) the transfer depends strongly on the nature of the substituent group bonded to the N atom of the acceptor. For instance, when ethyl acetate acetylations of NH_2CH_3 through glycinate ion are compared, the yields observed under our conditions range from 57.5% to 9%; (3) the transfer yield depends also on the nature of the acyl group of the donor. When different esters are considered, the transfer yields range from 57.5% to 34% when ethyl acetate is replaced by ethyl glycinate; (4) the transfer yields are strongly increased when thiolesters are used as donors instead of esters. The replacement of -OC_2H_5 residue in ethyl acetate by -SC_2H_5 residue increases the yield of acylation of NH_2CH_3 from 57.5% to 89%. The glycinate ion is 65% acylated by ethyl thiolacetate. From these results it may be inferred that the yield of formation of peptidic linkages which can be expected from the action of aminoacyl thiolester on the amine group of α-amino acids or on terminal NH_2 of poly α-amino acids

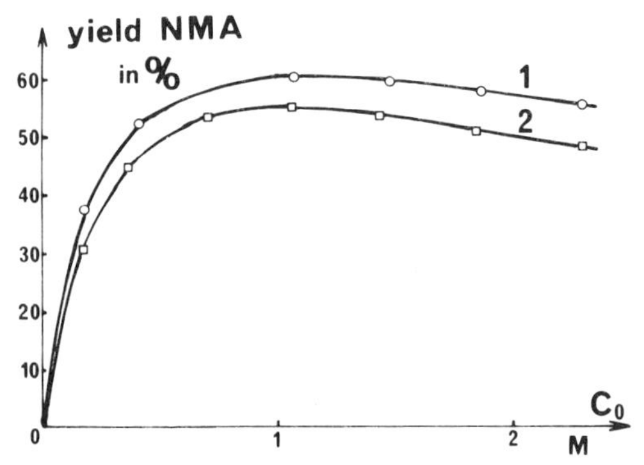

Fig. 3. Influence of dilution on the acetylation yield of *N*-methylacetamide (NMA) as a function of molar concentration Co of reagents, when equal molar quantities of KOH, NH_2CH_3 and AcOEt are mixed. *Curve 1*: reaction performed in aqueous solution; *curve 2*: reaction performed in water:ethanol solvent mixture (70:30, v/v).

Table 2
Yields of acyl transfer in aqueous alkaline medium at ambient temperature, for a number of donors and nitrogenated acceptors.

Acceptors \ Donors	$CH_3{-}CO{-}O{-}C_2H_5$	$CH_3{-}CO{-}O{-}CH_3$	$CH_3{-}CO{-}O{-}CH(CH_3)_2$	$CH_3{-}CO{-}O{-}CH_2{-}CH_2{-}OCH_3$	$CH_2{-}CO{-}O{-}CH_2{-}CH_2{-}OC_2H_5$	$CH_3{-}CO{-}SC_2H_5$	$CH_3{-}CO{-}O{-}OCH_3$	$NH_2{-}CH_2{-}CO{-}OC_2H_5$	$CH_2(CO{-}OC_2H_5)_2$
$H{-}NH_2$	2					10	53		
$H{-}NH{-}CH_3$	57.5 (70) [a]	59.1	50.8	53.3	56.6	89	69	34	67
$H{-}NH{-}C_2H_5$	24.5					76			
$N{-}NH{-}CH_2{-}CH_2OH$	19 (32) [a]					60	59.5	15	
$H{-}N(CH_3)_2$	20					92		19	
$H{-}NHOH$	83					93	90	67	97.5
$H{-}NH{-}CH_2{-}CO_2^-$	9 (12) [a]					65	58.5	3	
$H{-}NH{-}CH_2{-}CH_2{-}CO_2^-$	12								

Initial concentration of reagents: KOH 2 M, acceptor 1 M, donor 1 M. Given values have been determined from the titration curves of the mixture after reaction.

[a] Values between brackets are the yields at 0 °C.

must be sufficiently high for obtaining small polypeptides from such simple reagents.

Reactive H atoms bonded to atoms other than nitrogen may also be acylated under similar conditions. Moreover similar kinds of transfer reactions can be performed at pH's other than alkaline. But, here again such reactions are observed only at pH's where the donor can undergo simple hydrolysis in aqueous solution. Such is the case when $(CH_3CO)_2O$ is dissolved into aqueous solutions containing inorganic phosphate ions: large quantities of acetyl phosphate are obtained whatever the extent of protonation of the ortho or pyrophosphoric species used (fig. 4).

This last result tends to show that the observed acyl transfers are not

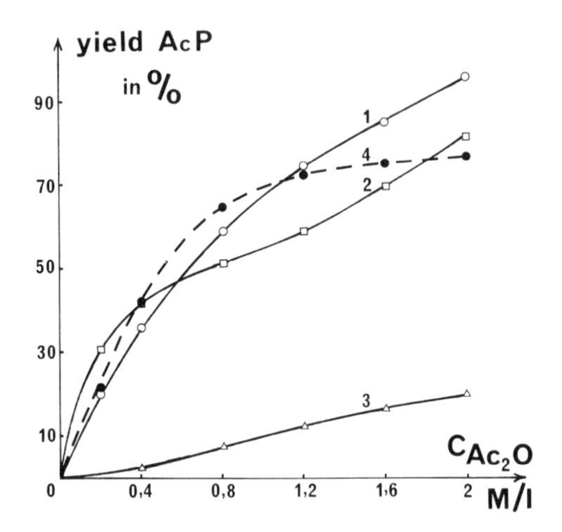

Fig. 4. Acetylphosphate (AcP) yield as a function of concentration of acetic anhydride added to non-buffered aqueous solutions of K_3PO_4 (*curve 1*), K_2HPO_4 (*curve 2*), KH_2PO_4 (*curve 3*) and $K_4P_2O_7$ (*curve 4*), respectively 0.4 M. Yields are calculated with respect to initial inorganic phosphate from pH metric titrations of the reaction mixtures after one hour at room temperature (under all experimental conditions here involved, acetic anhydride hydrolysis is practically complete in less than 45 min).

pH at the end of transfer process

	$[Ac_2O]$ = 0.4 M	$[Ac_2O]$ = 2 M
1)	6.1	4.2
2)	4.6	3.4
3)	3.2	3
4)	6	4.10

dependent upon the particular mechanism of an alkaline hydrolysis of the studied donors, but are probably fairly general provided that any kind of hydrolytic process of a donor be brought about in the presence of an acceptor convertible into a metastable condensate by simple transfer of a moiety of the energy-donor.

To conclude: One of the main consequences drawn from our previous theoretical analysis of energy transfer processes is that the neutralization energies involved in transdehydration reactions play a role of paramount importance in the free enthalpy balances. This conclusion is supported here by the experimental fact that simple esters and thiolesters, relatively energy-poor as regards their simple hydrolysis in acidic media, behave as relatively energy-rich donors in alkaline solutions. Consequently the alkalinity of the primeval aqueous media must have played an essential part in the development of the archaic biosphere, together with simple esters and thiolesters which occurred as prototypes of their present-day biological homologs.

Some of the experimental results reported here offer a new alternative as for the primeval syntheses of protein-like materials and more generally for primeval transacylation processes. In opposition to Fox's hypotheses – which take for driving conditions of such syntheses some problematical conditions without relation to present-day biological ones – our results concerning the feasibility in aqueous media of peptide syntheses from energy-rich acylating derivatives together with those of Jencks and A. Katchalsky et al., show that the pathways of primeval syntheses of proteins could have been archetypes of the present-day ones. The problem which now arises is to define how such energy-rich compounds, either acylphosphates, or esters and thiolesters, formed on the primitive earth from free acids through either transdehydration or direct oxidative dehydration.

As regards the theoretical criteria governing energy transfer processes, our experimental results lead to the conclusion that the transfers might be achieved only when the donor is capable of undergoing simple hydrolysis in the absence of condensable substrates in the aqueous medium considered. So, the breakage of the energy reserves should govern the energy transfer.

References

[1] Beck A. and L.E. Orgel, Proc. Natl. Acad. Sci. U.S. 54 (1965) 664.
[2] Buvet R., E. Etaix, F. Godin, P. Leduc and L. Le Port, this volume, p. 51.
[3] Di Sabato G. and W.P. Jencks, J. Am. Chem. Soc. 83 (1961) 4393, 4400.
[4] Ferris J.P., Science 161 (1968) 53.
[5] Fishbein W.N., T.S. Winter and J.D. Davidson, J. Biol. Chem. 240 (1965) 2402.
[6] Gerstein J. and W.P. Jencks, J. Am. Chem. Soc. 86 (1964) 4655.
[7] Jones M.E. and F. Lipmann, Proc. Natl. Acad. Sci. U.S. 46 (1960) 1194.
[8] Katchalsky A. and M. Paecht, J. Am. Chem. Soc. 76 (1954) 6042.
[9] Le Port L. and R. Buvet, Compt. Rend. Acad. Sci. Paris 270 (1970) 1753.
[10] Lewinsohn R., M. Paecht-Horowitz and A. Katchalsky, Biochim. Biophys. Acta 140 (1967) 24.
[11] Lipmann F. and L.C. Tuttle, J. Biol. Chem. 159 (1945) 21.
[12] Lowenstein J.M., Biochem. J. 70 (1958) 222.
[13] Lowenstein J.M., Biochim. Biophys. Acta 28 (1958) 206.
[14] Lowenstein J.M., Biochem. J. 75 (1960) 269.
[15] Lowenstein J.M. and M.N. Schatz, J. Biol. Chem. 236 (1961) 305.
[16] Miller S.L. and M. Parris, Nature 204 (1964) 1248.

[17] Ponnamperuma C. and R. Mack, Science 148 (1965) 1221.
[18] Rabinowitz J., Nature 224 (1969) 795.
[19] Rabinowitz J., S. Chang and C. Ponnamperuma 218 (1968) 442.
[20] Schwartz A. and C. Ponnamperuma, Nature 218 (1968) 443.
[21] Steinman G.R., R.M. Lemmon and M. Calvin, Proc. Natl. Acad. Sci. U.S. 52 (1964) 27.
[22] Steinman G.R., R.M. Lemmon and M. Calvin, Science 147 (1965) 1574.
[23] Vogel A.I., A Text-Book of Practical Organic Chemistry (Longmans, Green, London, 1961) p. 1062.

Chemical Evolution and the Origin of Life, eds. R. Buvet and C. Ponnamperuma

PHOSPHATE: SOLUBILIZATION AND ACTIVATION ON THE PRIMITIVE EARTH

Alan W. SCHWARTZ
Department of Exobiology, University of Nijmegen, Nijmegen, The Netherlands

Of the six common non-metallic elements which form the basis for all life on earth – carbon, hydrogen, oxygen, nitrogen, sulfur and phosphorus – only phosphorus occurs as neither a volatile nor a water-soluble compound. The other five elements share both of these desirable properties. As was pointed out by Gulick some years ago, this fact represents a theoretical difficulty which must be overcome in any general theory of the origin of life on the earth [10]. The problem is further compounded by the relatively unreactive nature of phosphoric acid with respect to esterification. This latter difficulty can be by-passed through the condensation or addition of phosphate to an activating agent. Thus, cyanamide and its dimer [30], cyanate [19], cyanogen [15] and cyanoacetylene have been proposed as having possible roles in prebiological phosphorylation reactions [6]. The direct thermal synthesis of condensed phosphate is another possible route to the activation of phosphate and eventual synthesis of phosphate esters [21,22,27]. The basic problem, however, must still be solved. Before there could have been significant activation of phosphate through any of these mechanisms, the solubilization of phosphate was required. Although heterogeneous processes might have played an important role on the primitive earth, only one such reaction has so far been demonstrated – the reaction of cyanate with the hydroxylapatite crystal surface to produce pyrophosphate [19]. In the absence of a solubilization mechanism, the crystal pyrophosphate would only be available for reaction through a further heterogeneous process. An interesting possibility presents itself here, however. If trimetaphosphate were to be formed through a similar surface-catalyzed reaction, the process might be driven by the removal of the product, since trimetaphosphate is the only member of the entire family of condensed phosphates to be completely soluble in the presence of excess calcium. This compound has interesting properties as a phosphorylating agent [26].

The purpose of this paper, however, is to suggest that a heterogeneous mechanism would not have been necessary, as a process for the solubilization of phosphate did, in fact, exist on the primitive earth.

Phosphorus is distributed throughout the igneous rocks in the form of fluoroapatite. Data on the average mineralogical composition of the igneous rocks reveal that apatite comprises about 0.6% of the total mineral mass [16]. It is found rather uniformly scattered as small, discrete crystals, although there are rare examples of the crystallization of fluoroapatite in large masses.

Table 1 shows the usual formula for the apatites. Although hydroxylapatite may form in the precipitation of phosphate from pure aqueous solution, in the oceans a complex carbonate-fluoroapatite is deposited. The mineral of marine phosphorites is basically of this type. As the carbonate substitution is low, it is essentially a fluoroapatite [9]. Chloroapatite appears to be the common phosphate mineral in meteorites, although it may also have some minor distribution in terrestrial rocks [7,8]. It is apparent from the solubility product constants that a very severe problem is to be overcome in the solubilization of phosphate.

To approach the problem first on a theoretical basis, the solubility of apatite was computed as a function of pH. As can be seen in fig. 1, only very low pH values result in a significant solubility of fluoroapatite, although the case for hydroxylapatite is somewhat better. At pH 5.0, for example, the solubility of hydroxylapatite is greater than 10^{-3} M in phosphorus, but fluoroapatite produces only 10^{-4} M phosphate. It is important at this point to consider the question of the existence of acid locals in primitive times, as it is clear that this offers a possible mechanism for phosphate solubilization. It is often stated that the lack of early Precambrian carbonate deposits, and the extensive limestone formations in the Paleozoic, indicate that a massive accumulation of outgassed compounds could not have been present at an early period of the development of the oceans [13]. It is therefore argued that the

Table 1
Structures of the apatites: $Ca_5(PO_4)_3X$.

X	Compound	pK_{sp}
OH	Hydroxylapatite	54 [a]
F	Fluoroapatite	60 [b]
Cl	Chloroapatite	–

[a] Lindsay and Moreno [14].
[b] Farr and Elmore [5].

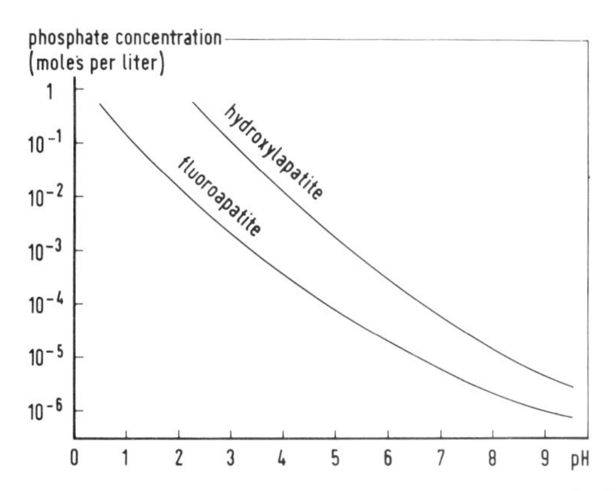

Fig. 1. Concentration of phosphate in solutions saturated with hydroxylapatite and fluoroapatite at various pH values.

primitive oceans could not have been acidic. These arguments were originally put forward by Rubey to demonstrate that the atmosphere and hydrosphere of the earth have developed gradually over geological time. An initial atmosphere containing up to 6% of the total mass of volatile components now present at the earth's surface, however, would not be incompatible with the existing geological data [23]. The sudden condensation of water from such atmosphere – containing carbon dioxide, hydrogen halides, and hydrogen sulfide – would have produced an extremely low initial pH. Of course, neutralization would have proceeded rapidly, but because apatite is one of the most rapidly weathered minerals, temporarily high levels of phosphate could have been present [2]. There are, in fact, local increases of phosphate in ocean waters adjacent to areas of volcanic activity even under the present highly buffered conditions [18]. Reaction with any carbohydrate or nucleoside then would have fixed the phosphate in the form of a highly soluble phosphate ester. On the other hand, the earliest stages of outgassing and condensation could have occurred at a slow enough rate for equilibrium to have been maintained between volatile acids and rocks, and acid localities may never have resulted [31]. I wish only to emphasize that we have insufficient information to rule out a possible role for reduced pH in the solubilization of phosphate. The critical issue being the relationship between the initial rate of outgassing and the temperatures of surface rocks.

There is, however, another possible mechanism which requires no assumptions that are not inherent in the Oparin-Haldane hypothesis. A casual glance at a metabolic chart reveals that many biologically important compounds, which are presumed to have been components of the primitive hydrosphere, are also excellent complexing agents for calcium. Prominent among these compounds are di- and tricarboxylic acids, including many members of the citric acid cycle [25]. That this is no new observation can best be illustrated by the fact that a widely used analytical method for the analysis of rock phosphate depends upon the solubilization of apatite with a large excess of citrate [20]. In order to assess the feasibility of such a process having had a significant prebiological role, it was necessary to compute, in a general way, the influence of a complexing agent on the apparent solubility of apatite. The approach was similar to that used by Peirs et al. for calcium phosphate, except that a series of general expressions were derived which are applicable to any apatite for which solubility data are available, and to any complexing agent at any pH.

Citric and oxalic acids were chosen as model complexing agents for numerical treatment. Fig. 2 summarizes the equations which were used for the system fluoroapatite-oxalate. Citric acid was treated in a similar way. Values for the various constants were chosen which are valid for an ionic strength of 0.5 [20,28]. In carrying out the computations with the aid of a computer, an iterative method was employed which will be described elsewhere. The relationship between the concentration of phosphate in a solution saturated with fluoroapatite, and the concentration of citrate, is illustrated in fig. 3 for several values of pH.

The results for citrate are not particularly impressive, taken by themselves. In no case is the increase in solubility in even molar citrate greater than about 100 fold, and even at pH 5, the phosphate concentration is only 0.0025 M. Now, of the various calcium-complexing compounds considered (table 2), citric acid certainly has the highest stability constant, about 10^3. However, oxalic acid, although slightly less efficient than citric acid, has another interesting property, and that is the low solubility of the calcium salt. The solubility product of calcium oxalate is usually given as about 10^{-9} [24]. This figure is based on the assumption that the salt is ionized in solution. When the stability constant of the calcium oxalate complex is taken into account, it is apparent that the concentration of free calcium in solution is very low indeed. It seemed quite probable that in a system where apatite was going into solution in the presence of oxalate, the beginning of precipitation of calcium oxalate would tend to drive the equilibria toward higher values for dissolved phosphate.

That this is indeed the case can be seen in fig. 4, which presents some of the results for fluoroapatite-derived phosphate at equilibrium with various levels of oxalate.

In this case, the value for oxalate represents either an initial concentration, solution unsaturated with CaOx $\left([\mathrm{CaOx}] < S_{\mathrm{CaOx}}\right)$:

$$K_{ST} = \frac{\left[\frac{5}{3}P_T - \left(3K_{SP}\Phi_{Ca}^5\Phi_P^3\Phi_F\right)^{\frac{1}{5}}P_T^{-\frac{4}{5}}\right]\Phi_{Ca}\Phi_{Ox}}{\left[\left(3K_{SP}\Phi_{Ca}^5\Phi_P^3\Phi_F\right)^{\frac{1}{5}}P_T^{-\frac{4}{5}}\right]\left[\mathrm{Ox}_T - \frac{5}{3}P_T + \left(3K_{SP}\Phi_{Ca}^5\Phi_P^3\Phi_F\right)^{\frac{1}{5}}P_T^{-\frac{4}{5}}\right]}$$

solution saturated with CaOx $\left([\mathrm{CaOx}] = S_{\mathrm{CaOx}}\right)$:

$$K_{ST} = \frac{S_{\mathrm{CaOx}}\Phi_{Ca}\Phi_{Ox}}{\left[\left(3K_{SP}\Phi_{Ca}^5\Phi_P^3\Phi_F\right)^{\frac{1}{5}}P_T^{-\frac{4}{5}}\right]\left[\mathrm{Ox}_T - \frac{5}{3}P_T + \left(3K_{SP}\Phi_{Ca}^5\Phi_P^3\Phi_F\right)^{\frac{1}{5}}P_T^{-\frac{4}{5}}\right]}$$

$$K_{ST} = \frac{[\mathrm{CaOx}]}{[\mathrm{Ca}^{2+}][\mathrm{Ox}^{2-}]} \qquad K_{SP} = [\mathrm{Ca}^{2+}]^5[\mathrm{PO}_4^{3-}]^3[\mathrm{F}^-]$$

$$P_T = [\mathrm{PO}_4^{3-}] + [\mathrm{HPO}_4^{2-}] + [\mathrm{H}_2\mathrm{PO}_4^-] + [\mathrm{H}_3\mathrm{PO}_4]$$

$$\mathrm{Ox}_T = [\mathrm{Ox}^{2-}] + [\mathrm{HOx}^-] + [\mathrm{H}_2\mathrm{Ox}] + [\mathrm{CaOx}] + [\mathrm{CaOx}]_S$$

$$\Phi_P = 1 + k_{\mathrm{PO}_4^{3-}(1)}[\mathrm{H}^+] + k_{\mathrm{PO}_4^{3-}(1)}k_{\mathrm{PO}_4^{3-}(2)}[\mathrm{H}^+]^2 + k_{\mathrm{PO}_4^{3-}(1)}k_{\mathrm{PO}_4^{3-}(2)}k_{\mathrm{PO}_4^{3-}(3)}[\mathrm{H}^+]^3$$

$$\Phi_F = 1 + k_{F^-}[\mathrm{H}^+]$$

$$\Phi_{Ox} = 1 + k_{\mathrm{Ox}^{2-}(1)}[\mathrm{H}^+] + k_{\mathrm{Ox}^{2-}(1)}k_{\mathrm{Ox}^{2-}(2)}[\mathrm{H}^+]^2$$

$$\Phi_{Ca} = 1 + k_{\mathrm{Ca}^{2+}(1)}[\mathrm{H}^+]^{-1} + k_{\mathrm{Ca}^{2+}(1)}k_{\mathrm{Ca}^{2+}(2)}[\mathrm{H}^+]^{-2}$$

Fig. 2. Summary of equations for the fluoroapatite-oxalate case.

before apatite is introduced, or the total quantity of oxalate which must be added to the system to bring the phosphate concentration up to the indicated level. Here we can finally see some rather respectable phosphate levels around pH 6 or a little lower. At pH 5, and 0.2 M oxalate, the concentration of phosphate is nearly 0.1 M. Even at 0.01 M oxalate, phosphate is about 0.003 M in concentration. It should be noted that these quantities of oxalate need not actually be added to the solution to obtain the corresponding concentrations of phosphate. It is only necessary to evaporate a dilute solution to

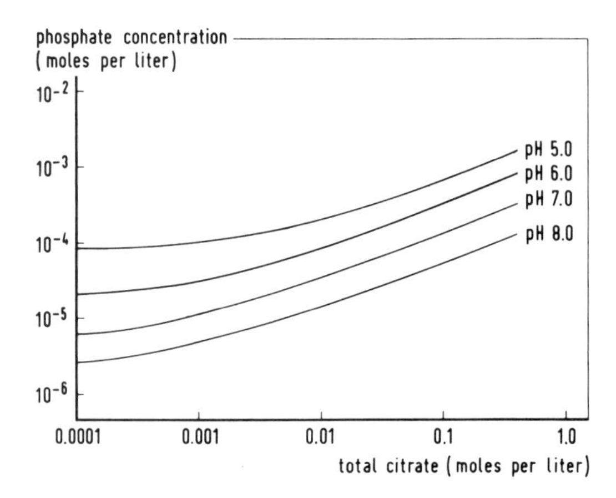

Fig. 3. Solubility of fluoroapatite at various total citrate concentrations.

obtain one that is concentrated both in oxalate and phosphate. Under most sets of conditions this can be accomplished without loss of phosphate by precipitation. This can be seen more clearly in table 3. The only limiting factor in concentration by evaporation is the solubility of the oxalate salt. This limit is roughly indicated for ammonium oxalate by the last figure in each column. Aş an example; a 1 liter sample at pH 5.0, containing

Table 2
Stability constants of calcium complexes.

Parent acid	$\log K_{st}$ (u = 0.16) [a]
Oxalic	3.0 [b]
Malonic	1.36
Succinic	1.2
Glutaric	1.06
Citric	3.15
Isocitric	2.47
Malic	2.06
Tartaric	1.78
Oxalacetic	1.60
Aconitic	1.50
α-Ketoglutaric	1.29

[a] From Schubert and Lindenbaum [25].
[b] From Sillen and Martell [28].

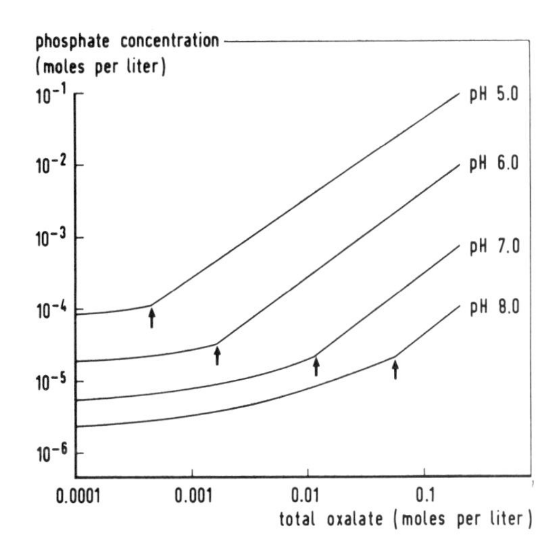

Fig. 4. Solubility of fluoroapatite in the presence of various quantities of total oxalate. The arrows indicate the point at which saturation with calcium oxalate is reached for each pH.

0.005 moles of oxalate and 0.0014 M in phosphate may be concentrated by evaporation 100 fold to produce a sample containing 0.5 moles per liter of oxalate and which is now 0.14 M in phosphate. As this level of oxalate is capable of maintaining a phosphate concentration of 0.213 M, no precipitation of apatite will occur. The pH range of 5 to 6 is interesting because this

Table 3
Phosphate concentration in solutions saturated with fluoroapatite as a function of total oxalate added.

Oxalate (M)	Phosphate (M)			
	pH 5.00	pH 6.00	pH 7.00	pH 8.00
0	8.4×10^{-5}	1.9×10^{-5}	5.5×10^{-6}	2.3×10^{-6}
0.001	2.5×10^{-4}	2.8×10^{-5}	8.1×10^{-6}	3.3×10^{-6}
0.005	1.4×10^{-3}	1.2×10^{-4}	1.5×10^{-5}	6.1×10^{-6}
0.05	1.8×10^{-2}	2.1×10^{-3}	1.4×10^{-4}	2.0×10^{-5}
0.1	3.8×10^{-2}	4.8×10^{-3}	3.3×10^{-4}	4.5×10^{-5}
0.3	0.123	1.8×10^{-2}	1.3×10^{-3}	1.8×10^{-4}
0.5	0.213			
1.0	0.443			

can easily be reached in the presence of carbon dioxide. However, this is a moot point, as the presence of ligands for calcium would shift the pH of the system. In making these calculations, the pH has been held constant. I should point out here that these values probably represent minima, as there is some evidence – for hydroxylapatite, at least – that the product of the ions in solution does not correspond to the product in the solid. The solubility, in fact, is somewhat higher than is calculated from thermodynamic data. In the case of hydroxylapatite, this is apparently due to the formation of a surface complex by hydrolysis, which then has a higher solubility than the bulk of the apatite. It is not known whether a similar process can occur with fluoroapatite [3]. The limitation of this approach is the necessity of treating apatite as an isolated calcium source, since the simultaneous weathering of other rocks would, of course, greatly affect the final state reached. This oversimplified approximation does, at least, offer encouragement to the idea that calcium complexing may have played a role in the primitive oceans, and we are investigating the weathering of apatite-bearing rocks of various compositions.

Oxalic acid has been synthesized by the decomposition of glycine, and from formic acid [11,12]. One of the more interesting pathways by which oxalate might have been synthesized is through the hydrolysis of cyanogen [4]. Not only is cyanogen a reasonable prebiological reagent, but as a highly water soluble gas, this mechanism could have provided a constant supply of oxalate at every rock surface exposed to weathering.

I do not wish to suggest that the key to the problem of apatite solubility lies in oxalic acid alone. Many other compounds would have had some contribution to make in complexing calcium. Although no other single compound may be as efficient in removing calcium from circulation as oxalate, others would have been in great abundance at some stage of chemical evolution, and their combined capacity for calcium may have been decisive. Amino acids and proteins may also have contributed significantly to this mechanism. It is perhaps more than coincidence that within contemporary cells, calcium is never in a free state, but is bound [1].

Acknowledgement

The assistance of Hans Deuss in developing a method of solution and programming the operations is gratefully acknowledged.

References

[1] Bianchi C.P., Cell Calcium (Butterworth, London, 1968) p. 5.
[2] Burger D., Can. J. Soil Sci. 49 (1968) 11.
[3] Chaverri J.G. and C.A. Black, Iowa State J. Sci. 41 (1966) 77.
[4] Dalgliesh C.E., A.W. Johnson and C. Buchanan, in: Chemistry of Carbon Compounds, ed. E.H. Rodd, Vol. 1B (Elsevier, Amsterdam, 1952).
[5] Farr K.L. and T.D. Elmore, J. Phys. Chem. 66 (1962) 315.
[6] Ferris J.P., Science 161 (1968) 53.
[7] Fuchs L.H., Meteorite Research, ed. P.M. Millman (Reidel, Dordrecht, 1969) p. 683.
[8] Goldschmidt V.M., Geochemistry (Clarendon, Oxford, 1958) p. 587.
[9] Gulbrandsen R.A., Econ. Geol. 64 (1969) 365.
[10] Gulick A., Am. Scientist 43 (1955) 479.
[11] Haissinsky M. and R. Klein, J. Chim. Phys. Physicochim. Biol. 65 (1968) 336.
[12] Heynes K. and K. Pavel, Z. Naturforsch. 12b (1957) 97.
[13] Johnson F.S., Space Sci. Rev. 9 (1969) 303.
[14] Lindsay W.L. and E.C. Moreno, Soil Sci. Soc. Am. Proc. 24 (1960) 177.
[15] Lohrmann R. and L.E. Orgel, Science 161 (1968) 64.
[16] Mason B., Principles of Geochemistry (John Wiley, New York, 1966) p. 100.
[17] McComas W.H. and W. Rieman, J. Am. Chem. Soc. 64 (1942) 2948.
[18] Mikhailov A.S., Litol. Polez. Iskop. (1968) (2) 43.
[19] Miller S.L. and M. Parris, Nature 204 (1964) 1248.
[20] Peirs S., J. Nicole and G. Tridot, Chim. Anal., Paris 51 (1969) 3.
[21] Ponnamperuma C. and R. Mack, Science 148 (1965) 1221.
[22] Rabinowitz J., S. Chang and C. Ponnamperuma, Nature 218 (1968) 442.
[23] Rubey W.W., Geol. Soc. Am. Bull. 62 (1951) 1111.
[24] Ruff O., Z. Anorg. Chem. 185 (1929) 387.
[25] Schubert J. and A. Lindenbaum, J. Am. Chem. Soc. 74 (1952) 3529.
[26] Schwartz A., J. Chem. Soc. D 23 (1969) 1393.
[27] Schwartz A. and C. Ponnamperuma, Nature 218 (1968) 443.
[28] Sillen L.G. and A.E. Martell, Stability Constants of Metal-Ion Complexes, Vol. 11 (Chem. Soc. London, Spec. Pub. 17, 1964).
[30] Steinman G., R.M. Lemmon and M. Calvin, Proc. Natl. Acad. Sci. U.S. 52 (1964) 27.
[31] Vinogradov A.P., Lithos 1 (1968) 169.

Chemical Evolution and the Origin of Life, eds. R. Buvet and C. Ponnamperuma

THE ROLE OF PHOSPHATES IN CHEMICAL EVOLUTION

Cyril PONNAMPERUMA
Exobiology Division, NASA–Ames Research Center,
Moffett Field, California, USA
and
Department of Biological Sciences, Stanford University,
Stanford, California
and
Sherwood CHANG
Exobiology Division, NASA–Ames Research Center,
Moffett Field, California, USA

In our study of chemical evolution, the main endeavor has been to reconstruct the path by which the constituents of the nucleic acid molecule may have arisen on the primordial earth before the appearance of life. In this context the search for prebiotic phosphorylating agents and the understanding of the mechanism of their action constitute integral parts of our research into the chemical origin of life.

This paper surveys some of the results which have been obtained in our laboratory in the phosphorylation of nucleosides. There appear to be two general approaches to the study of the problem: the hypohydrous thermal reaction between inorganic phosphates and nucleoside and the aqueous phosphorylation involving reactive intermediates formed by the addition of inorganic orthophosphates to potential prebiotic condensing agents. Both approaches have been used in our laboratory, but we shall discuss only the hypohydrous thermal reactions between nucleosides and inorganic phosphates. The products of the reactions have been identified, and an attempt has been made to elucidate the mechanism.

We have shown that fair yields of nucleoside phosphates are obtained when a nucleoside is heated with sodium dihydrogen orthophosphate, NaH_2PO_4 [4]. In a typical experiment 100 μl of solution containing 2 μmoles of a nucleoside and 2 μmoles of a phosphate were placed in a 5 ml pyrex tube and lyophilized. The tube was then sealed and heated to 160 °C for 2hr. The reaction products were dissolved in 200 μl of water for analysis. Two parallel sets of experiments were performed in which a labelled reactant was included: a ^{14}C-labelled nucleoside or ^{32}P-labelled phosphate. The analytical techniques used were thin layer and paper chromatography, and ion exchange chromatography. The last column of table 1 summarizes the results so obtained.

Marked decreases in yields of nucleotides were observed with monohydrogen phosphates, M_2HPO_4 except when one M was ammonium. In this case thermal dissociation to ammonia and dihydrogen phosphate was possible. Practically no phosphorylation took place with tribasic phosphates.

An examination of the chromatograms revealed that a large number of compounds had been synthesized. Among these could be identified some of the monophosphates (fig. 1).

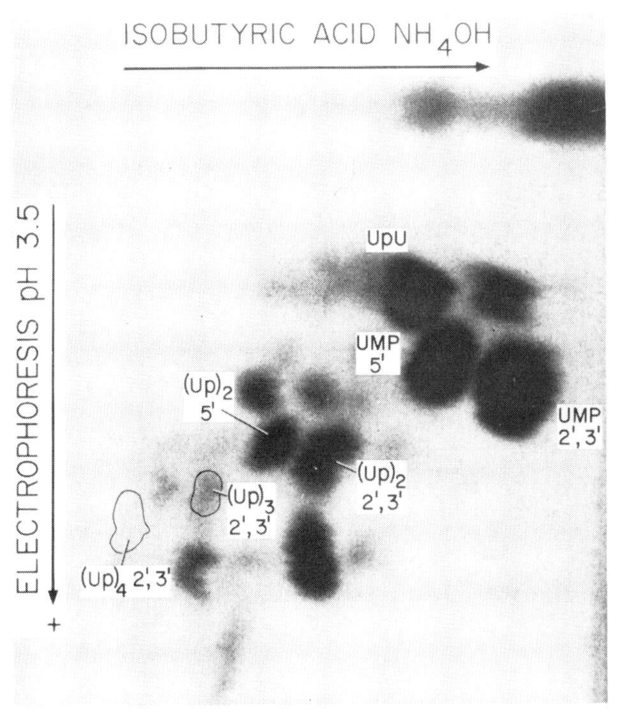

Fig. 1. Autoradiograph of products obtained by heating ^{14}C-labelled uridine with sodium dihydrogen phosphate.

Different nucleosides yielded different results with the sodium dihydrogen phosphate: adenosine, 3.1%; guanosine, 9.8%; cytidine, 13.7%; uridine, 20.6%; and thymidine, 6.3%. Since uridine gave the largest yields of monophosphates we examined the reaction between uridine and the orthophosphates more closely.

Generally, phosphorylation of organic compounds with the aid of a phosphoryl reagent containing a labile P–X bond involves the displacement of X or its conjugate acid at tetrahedral phosphorus by nucleophilic center [1].

$$N{:} + (RO)_2P(=O)\text{–}X \longrightarrow N\text{–}P(=O)(OR)_2 + {:}X$$

$$X = -O\overset{\overset{O}{\|}}{P}(OR)_2,\ -O\overset{\overset{NR}{\|}}{C}R,\ -O\overset{\overset{O}{\|}}{C}R,\ -O\overset{\overset{CH_2}{\|}}{C}OR,\ -NR_2,\ \text{Halogen, etc. . .}$$

(R may be H, aryl, aralkyl, or alkyl)

Since inorganic phosphates are not known to be phosphorylating agents, our results may be attributed to the partial thermal transformation of orthophosphates to condensed phosphates, the latter serving as the effective phosphoryl reagent ($X = OPO_3^{2-}$) [5,6]. Data in tables 1 and 2 and fig. 2 clearly establish that the conversion of orthophosphates to polyphosphates is possible at moderate temperatures. Most extensive polyphosphate formation occurred with H_3PO_4 and $Ca(H_2PO_4)_2.H_2O$. Significantly, no polyphosphates were detectable with dibasic and tribasic phosphates which previously gave very low yields of nucleotides (table 1).

Of great interest is the fact that the calcium salt is converted to pyrophosphate at as low a temperature as 92 °C after 18 hr. After two months some polyphosphates are formed even at 65 °C (table 2). These are the lowest temperatures at which a direct thermal conversion of an orthophosphate to pyrophosphate has been observed. Fig. 2 shows that the ease of polyphosphate formation is in the order of $Ca(H_2PO_4)_2 . H_2O > NaH_2PO_4 . H_2O > NH_4H_2PO_4$.

In order to assess the role of polyphosphates in the nucleotide formation, phosphorylation yields were determined with a variety of ortho- and poly-

Table 1
Conversion of orthophosphates to condensed phosphates at 160 °C after 2 hr.

Orthophosphate used [a]	% of total phosphorus [b] as		% Uridine converted to monophosphate
	Pyrophosphate [b]	Tripolyphosphate [b]	
H_3PO_4	30–50	1–5	8.3
$NaH_2PO_4 \cdot H_2O$	5–10	1–5 [c]	16.0
$Na_2HPO_4 \cdot 7H_2O$	–	–	0.6
$Na_3PO_4 \cdot 12H_2O$	–	–	0.6
$NH_4H_2PO_4$	10–30	5–10 [c]	5.9
$(NH_4)_2HPO_4$	10–30	5–10 [c]	13.4
$Ca(H_2PO_4)_2 \cdot H_2O$	30–50	10–30 [c]	10.5
$CaHPO_4$	–	–	
KH_2PO_4	– [d]	–	
$Ca_{10}(PO_4)_6(OH)_2$	– [e]	–	
$Ca_3(PO_4)_2$	–	–	0.1

[a] Orthophosphates were used as obtained commercially. No condensed phosphates were detectable prior to heating.
[b] Visual estimates made by comparing the sizes and color intensities of spots with those obtained with standard mixtures having known ortho- and polyphosphate compositions. Less than 1% indicated by –.
[c] Polyphosphates more highly condensed than tripolyphosphate were observed at or near the origin.
[d] Pyrophosphate appears after 6 hr at 160 °C.
[e] No condensed phosphates appeared after 7 days at 160 °C.

Table 2
Conversion of orthophosphates to polyphosphates at 90 °C after one week and at 65 °C after two months.

Orthophosphate [a] used	T (°C)	% of total phosphorus [b] as	
		Pyrophosphate	Triphosphate
$Ca(H_2PO_4)_2 \cdot H_2O$	90 (after 1 week)	2–5	2–5
$NaH_2PO_4 \cdot H_2O$		2–5	2–5
$NH_4H_2PO_4$		~1	~1
$Ca(H_2PO_4)_2 \cdot H_2O$	65 (after 2 months)	~1	–
$NaH_2PO_4 \cdot H_2O$		~1	–

[a,b] See table 1.

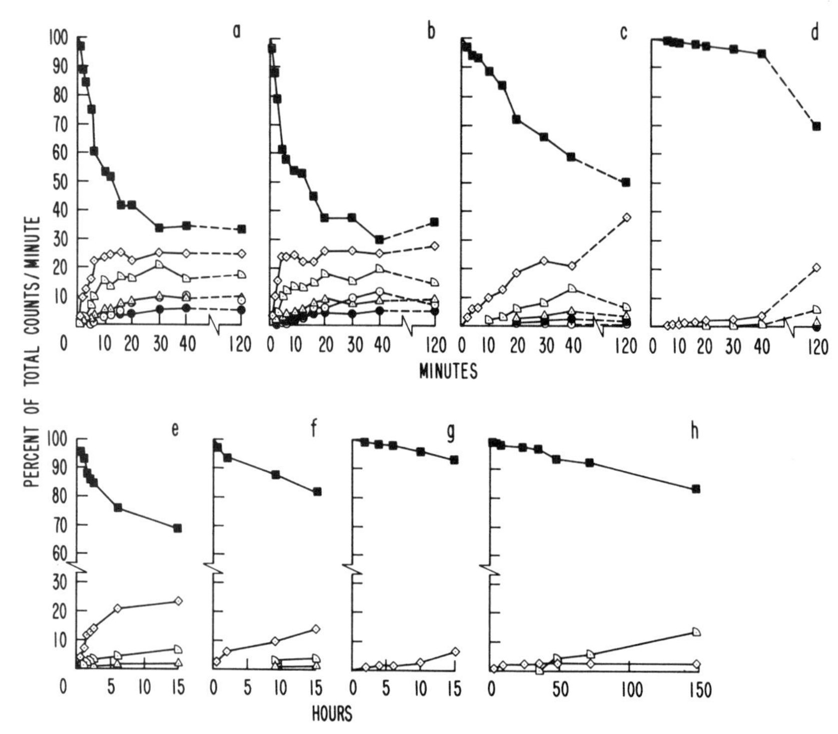

Fig. 2. Polyphosphate compositions produced after various times by heating (a) and (b) ^{32}P-$Ca(H_2PO_4)_2 \cdot H_2O$ at 162°C; (c) ^{32}P-$NaH_2PO_4 \cdot H_2O$ at 162°C; (d) ^{32}P-$NH_4H_2PO_4$ at 162 °C; (e) ^{32}P-$Ca(H_2PO_4)_2 \cdot H_2O$ at 126 °C; (f) ^{32}P-$NaH_2PO_4 \cdot H_2O$ at 126 °C; (g) ^{32}P-$NH_4H_2PO_4$ at 126 °C; and (h) ^{32}P-$Ca(H_2PO_4)_2 \cdot H_2O$ at 98 °C. Orthophosphate, ■——■, pyrophosphate, ◇——◇; triphosphate, ▷——▷; tetraphosphate, △——△; pentaphosphate, ●——●; hexa- and higher phosphates, ○——○ [7].

phosphates. In some cases the orthophosphates were preheated before their interaction with nucleosides. The results are summarized in table 3. In order to facilitate comparison between total phosphorylation yield and polyphosphate contents in parallel experiments, the latter are included as the last column in table 3.

Clearly, polyphosphates are more effective phosphorylating agents than orthophosphates. Reaction of uridine with polyphosphates formed during the period of heating could account for the small amounts of phosphorylation observed with $NaH_2PO_4.H_2O$, $NH_4H_2PO_4$, and the ortho and polyphosphate mixtures obtained by preheating $NH_4H_2PO_4$ and $NaH_2PO_4.H_2O$. Apparently heating uridine with $NaH_2PO_4.H_2O$ or $NH_4H_2PO_4$ results in rapid partial

Table 3

Yields of uridine phosphates obtained by heating uridine with various inorganic phosphates [7].

T (°C)	Time (hr)	Phosphate	Uracil (percent)	2′(3′)-UMP (percent)	5′-UMP (percent)	UDP (percent)	Total yield of uridine phosphates (percent)	Phosphorus [a] as poly-phosphates (percent)
162	0.1	$Na_2HPO_4 \cdot 7H_2O$	0.0	0.6	0.4	0.0	1.0	– [c]
162	0.1	$NaH_2PO_4 \cdot H_2O$	0.0	1.6 ± 0.2 [b]	0.2 ± 0.1	0.2 ± 0.1	2.0 ± 0.4	6
162	0.1	$NaH_2PO_4 \cdot H_2O$ (preheated)	0.4 ± 0.1	3.6 ± 0.3	1.2 ± 0.5	0.7 ± 0.2	5.5 ± 1.1	32 [d]
162	0.1	$Na_5P_3O_{10}$	0.0	7.5 ± 1.2	6.0 ± 2.0	1.9 ± 1.0	15.4 ± 4.2	
162	0.1	$Na_3P_3O_9$	0.7	3.9	1.7	1.4	7.0	
162	0.5	$NH_4H_2PO_4$	0.0	1.7 ± 0.2	0.2 ± 0.1	0.0	1.9 ± 0.3	4
162	0.5	$NH_4H_2PO_4$ (preheated)	0.0	3.7	2.4	0.0	6.1	64 [d]
162	2.0	KH_2PO_4	0.0	3.2 ± 0.4	1.7 ± 0.0	0.0	4.9 ± 0.4	–
126	6.0	$NaH_2PO_4 \cdot H_2O$	0.0	2.8	2.6	0.0	5.4	8
126	6.0	$Na_2HPO_4 \cdot 7H_2O$	0.0	0.5 ± 0.1	0.5 ± 0.1	0.0	1.0 ± 0.2	–
126	6.0	$Na_2H_2P_2O_7$	0.0	13.3	13.5	3.0	29.8	
126	6.0	$Na_5P_3O_{10}$	0.0	11.3	7.5	2.3	21.1	
126	6.0	$Na_3P_3O_9$	0.0	6.5	6.4	0.0	12.9	
126	6.0	Graham's salt	0.0	12.7	12.4	3.1	28.2	
126	6.0	$CaHPO_4$	0.0	0.0	0.0	0.0	0.0	–
126	0.1	$Ca(H_2PO_4)_2 \cdot H_2O$	0.8	2.6	3.8	3.3	9.7	1
98	48.0	$Ca(H_2PO_4)_2 \cdot H_2O$	2.7 ± 1.0	5.6 ± 0.1	16.1 ± 1.2	7.5 ± 0.8	29.2 ± 2.1	7
98	336.0	$Ca(H_2PO_4)_2 \cdot H_2O$	7.4 ± 1.8	3.8 ± 0.4	14.5 ± 1.0	8.5 ± 0.3	26.8 ± 1.7	

[a] Data taken from fig. 2 unless otherwise indicated.

[b] Mean ± average deviation of two independent experiments.

[c] Undetectable amounts, less than 1 percent indicated by –, taken from table 1.

[d] Content prior to heating with uridine, determined in separate experiments.

transformation of the orthophosphates to polyphosphates as well as in phosphorylation of the uridine. The resulting polyphosphates can phosphorylate nucleoside faster than the original orthophosphate. Therefore, strong circumstantial evidence exists for the intermediary role of polyphosphates in the phosphorylation of nucleosides with $NaH_2PO_4.H_2O$ and $NH_4H_2PO_4$. A mechanism involving orthophosphate cannot be eliminated, however.

Yields of uridine phosphates obtained by heating uridine with KH_2PO_4 at 162 °C and $Ca(H_2PO_4)_2.H_2O$ at 126 °C far exceed the amount of polyphosphate formed in parallelled experiments without uridine during the same periods of heating (table 3). It is improbable that nucleoside admixed with these orthophosphates catalyzed formation of polyphosphates to serve as phosphorylating agents. A more reasonable conclusion is that a direct phosphorylation mechanism was available involving orthophosphate and uridine. As 162 °C calcium polyphosphates are very readily formed, so that some phosphorylation at that temperature may involve polyphosphates.

A relationship has been suggested between weakening of the crystal lattice of orthophosphates due to loss of water of hydration, or loss of ammonia in the case of ammonium orthophosphates, and their tendency to form polyphosphates [3]. The effectiveness of orthophosphates in direct phosphorylation may be related to the same phenomenon. Possibly, diffusion of uridine molecules and H_2PO_2 ions together in crystalline phases occurs followed by thermal dehydration. Why such a process should be so favorable in the case of $Ca(H_2PO_4)_2.H_2O$ is not clear. Calcium ions are known to be catalysts for hydrolysis of phosphate esters; they may play a catalytic role in the phosphorylation reaction.

The results obtained in our laboratory point to several factors which may be significant for the study of chemical evolution. Orthophosphates can be readily converted into condensed phosphates which are effective phosphorylating agents. In some instances phosphorylation may take place directly via the orthophosphates. The calcium and sodium dihydrogen phosphates can be converted into condensed phosphates at temperatures as low as 65 °C.

The plausibility of prebiotic phosphorylation via thermal means either as linear phosphates or as orthophosphates may be questioned on the same grounds as any other presumed prebiotic phosphorylating agent. The concentration of phosphorus in the primitive ocean is believed to be extremely small. In the presence of calcium ions and in water whose pH is greater than 7, it has been suggested that all orthophosphate would be precipitated as insoluble hydroxyapatite [6]. It is possible that if localized geological phenomena permitted the existence of waters having pH values less than 7 there would be precipatation of various alkaline earth dihydrogen and hydrogen phosphates. The simplicity of the reaction would then make thermal polymerization of inorganic orthophosphates at moderate temperature very attractive as a general source of polyphosphates, providing efficient phosphorylating and condensing agents for primordial syntheses.

References

[1] Kirby A.J. and S.G. Warren, The Organic Chemistry of Phosphorus (Elsevier, New York, 1967).

[2] Miller S.L. and M. Parris, Nature 204 (1964) 1248.

[3] Newesley H., Monatsh. Chem. 98 (1967) 379.

[4] Ponnamperuma C. and R. Mack, Science 148 (1965) 1221.

[5] Rabinowitz J., S. Chang and C. Ponnamperuma, Nature 218 (1968) 442.

[6] Rabinowitz J., S. Chang and C. Ponnamperuma, in: The Proceedings of the Proteins and Nucleic Acid Symposium: Synthesis, Structure, and Evolution (Houston, Texas, 1968).

[7] Chang S., J.A. Williams, C. Ponnamperuma and J. Rabinowitz, Space Life Sci. (1970) in press.

Chemical Evolution and the Origin of Life, eds. R. Buvet and C. Ponnamperuma

MECHANISM OF THE CYANOGEN INDUCED PHOSPHORYLATION OF SUGARS IN AQUEOUS SOLUTION

Ch. DEGANI and M. HALMANN
Isotope Department, The Weizmann Institute of Science, Rehovot, Israel

Introduction

Sugar phosphates must have been formed in the earliest stages of the chemical evolution of life, because they are among the universal intermediates in all metabolic processes. Many attempts have therefore been made to provide laboratory models for prebiotic formation of sugar phosphates [3,21, 24,25]. In our work we have limited ourselves to phosphorylation of sugars in dilute aqueous, approximately neutral solutions. As was reported previously, aqueous solutions of D-ribose and orthophosphate in the presence of cyanogen or cyanamide undergo condensation to produce β-ribofuranose-1-phosphate as the only sugar phosphate formed [9]. We now wish to present further data on this reaction, which was applied also to other sugars, and whose mechanism was studied by ^{18}O-tracer experiments. We shall discuss the prebiotic plausibility of the system, the results so far obtained on the scope and the mechanism of the reaction, and the outlook for future studies.

Plausibility of the system

Hypotheses on the primitive earth atmosphere assume either a *completely reducing* atmosphere, containing hydrogen, methane, ammonia and water, as on the Jovian planets, or a *partly oxidized* atmosphere, of carbon dioxide (possibly carbon monoxide), nitrogen and water, as on our nearest neighbour planets (Mars and Venus) [12,28]. In S.L. Miller's classic experiments of passing an electric discharge through a mixture of hydrogen, methane, ammonia and water, one of the major intermediates formed is hydrogen cyanide

[18]. Already in 1863 it had been shown that cyanogen is formed readily by passing hydrogen cyanide through a red-hot-tube [2,4].

$$2HCN \rightarrow C_2H_2 + H_2$$

Cyanogen has also been obtained by passing a mixture of nitrogen and carbon dioxide through a carbon arc discharge, or by passing a mixture of "synthesis gas" ($CO + H_2 + N_2$) over ferrous oxide at 300–500 °C [2,20]. Thus cyanogen is produced readily, and by a variety of processes [2], in both completely reducing atmospheres and in partly oxidized atmospheres. Cyanamide is formed by ultraviolet light irradiation of aqueous ammonium cyanide [23] and also by sparking a gaseous mixture of carbon monoxide and ammonium [11].

Orthophosphate can be assumed as the most probable form of phosphorus in the primordial ocean and lakes, although reduced phosphorus oxyacids such as phosphite and hypophosphite cannot be excluded [8]. Sugars, from trioses to hexoses have been obtained readily in many studies on the formose condensation of formaldehyde [7,22]. Therefore, we may conclude that an aqueous solution of sugars, orthophosphate and cyanogen or cyanamide is a plausible starting system for prebiotic synthesis.

The scope of the reaction

Aqueous solutions of sodium or potassium orthophosphate and the sugar (initial concentration 0.01–0.10 M) in neutral or slightly alkaline solutions (pH 6.7 or 8.8) were evacuated and treated with cyanogen. As shown in table 1, glycerol was resistant to cyanogen induced phosphorylation, while most of the reducing sugars underwent phosphorylation. All the sugar phosphates produced were found to be alkali-stable (5 min in 0.2 N NaOH at 100 °C) and acid-labile (10 min in 1 M HCl at 100 °C). These properties are specific for glycosyl phosphates [15]. Thus, aldose sugars are converted into aldose-1-phosphates. The reaction with D-glucose was most carefully studied. The yield of sugar phosphate (after reaction of 0.1 M solutions of the reactants at pH 6.7 for 4 hr at 25 °C) was 15%, of which 4/5 consisted of α-D-glucopyranose-1-phosphate, identified by several paper chromatographic techniques, by the rate of acid-catalyzed hydrolysis and by a highly specific enzymatic assay (with phosphoglucomutase to glucose-6-phosphate and then with glucose-6-phosphate dehydrogenase to gluconate-6-phosphate). The only other sugar phosphate produced under our conditions is β-D-glucopyranose-1-phosphate, while no glucose-6-phosphate was formed. These results are dif-

Table 1
Cyanogen induced phosphorylation of sugars on addition of phosphorus compound (HPO_4^{2-}).

Sugar or sugar derivative	Qualitative result	Yield (%)	Product identity	Ref.
Glycerol	–	<1		30
Glyceraldehyde	+			19
D-Ribose	+	10–20	β-Ribofuranose-1-phosphate	9
Arabinose	+	{3–5 0.5–1.5	{Arabinopyranose-1-phosphate Arabinofuranose-1-phosphate	30
Xylose	+		Xylose-1-phosphate	30
2-Deoxyribose	–	<1		9
D-Glucose	+	10–25	{(4/5) α-Glucopyranose-1-phosphate (1/5) β-Glucopyranose-1-phospohate	30
D-Mannose	+		Mannose-1-phosphate	30
D-Galactose	+		Galactose phosphate	30
D-Fructose	+		Fructose phosphate	30
Glucosamine	+		Glucosamine phosphate	30
N-Acetyl-D-glucosamine	+		*N*-Acetylglucosamine phosphate	30

ferent from those reported for the phosphorylation of glucose with phosphoric acid and dicyanamide or cyanamide. Under these strongly acid conditions the sugar phosphate observed was glucose-6-phosphate [26,27]. It should however be noted that under such acid conditions any aldose-1-phosphates, if formed, would have been rapidly hydrolyzed [15].

The relatively abundant production of the natural isomer, α-D-glucose-1-phosphate (the Cori ester) seems significant in the context of chemical evolution, as this ester is a key intermediate in the biosynthesis and degradation of di- and polysaccharides. The cyanogen-induced phosphorylation was also successful with the reducing disaccharides maltose, melibiose, cellobiose and lactose, but failed with the non-reducing sugars sucrose and trehalose.

Mechanism of the reaction

It has been proposed that phosphorylation of alcohols with nitrilic reagents [16], by analogy with carbodiimide promoted phosphorylation [13], proceeds through an intermediate phosphate–nitrile adduct:

$$NC\,CN + HPO_4^{2-} \rightarrow NC\text{–}C(=NH)\text{–}O\text{–}PO_3^{2-} \quad (1)$$

which then reacts with the sugar (or with water) to produce a sugar phosphate (or again orthophosphate) and 1-cyanoformamide [29]:

$$\begin{aligned} &NC{-}C({=}NH){-}O{-}PO_3^{2-} + ROH \text{ (or } H_2O\text{)} \\ &\rightarrow ROPO_3^{2-} \text{ (or } HPO_4^{2-}\text{)} + NCCONH_2 \end{aligned} \tag{2}$$

Formation of a reactive phosphate adduct of this type was proven by studying the reaction of cyanogen with phosphate in ^{18}O-labelled water. We found that in the presence of cyanogen, the orthophosphate did become enriched in ^{18}O [6]. By successive addition of several equivalents of cyanogen to the reaction mixture the phosphate ion underwent a gradual enrichment in ^{18}O, successively approaching the isotopic composition of the water. Since it is known that orthophosphate under these conditions does not undergo appreciable exchange with water, these results show that the cyanogen–phosphate adduct reacts with water mainly with P–O bond fission, thus acting as a phosphorylating agent.

The preference for phosphorylation on the glycosidic hydroxyl among all the sugar hydroxyl groups, and the inactivity of glycerol towards phosphorylation, must be connected with the enhanced acidity of the glycosidic hydroxyl. The intermediate cyanoformimine–phosphate proposed above as a phosphorylating agent requires an activation process. Activation of the intermediate may be accomplished by protonation of an atom adjacent to the phosphorus atom, making it electron defficient, and thus causing increasing electron withdrawal from the phosphorus atom. Such activation of the intermediate by protonation can be achieved by the hemiacetalic hydroxyl group,

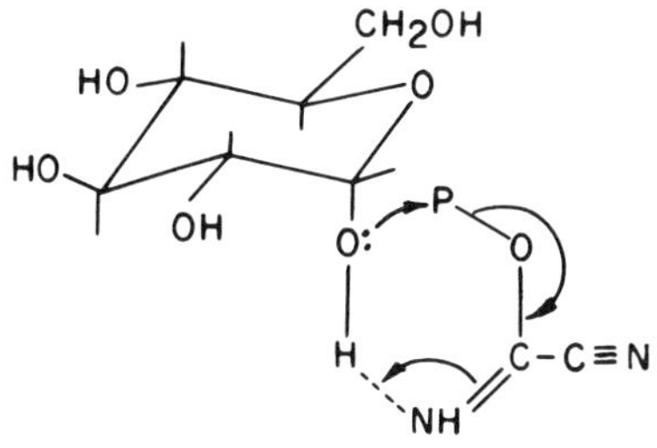

Fig. 1. Phosphorylation of α-D-glucose.

Table 2
Acid dissociation constants of sugars at 18 °C.

Sugar or sugar derivative	pK_a	Ref.
Fructose	12.1	17
D-Glucose	12.4	1
Galactose	12.3	17
Arabinose	12.4	17
Glycerol	14.1	17

which is relatively acidic, and the phosphorylation may proceed by a cyclic mechanism of the type shown in fig. 1.

The fact that the acidity of the hemiacetalic group is a predominant factor in the phosphorylation mechanism can explain the lack of phosphorylation in the case of glycerol. The acid dissociation constant of glycerol (pK_a = 14.1) is very much lower than that of the reducing sugars, for which pK_a is in the range 12.1–12.5 (see table 2) [1,17].

Results obtained in the phosphorylation of isotopically normal D-ribose in normal water with ^{18}O-enriched orthophosphate in the presence of cyanogen confirm this mechanism [10]. In the β-ribofuranose-1-phosphate produced, the oxygen atom of the glycosidic P–O–C group was found not to be appreciably enriched. This means that this oxygen atom must have been derived mainly from the ribose, which thus served as the nucleophile, while the cyanogen–phosphate adduct acted as a phosphorylating agent.

Conclusions

The cyanogen promoted phosphorylation of reducing sugars with aqueous orthophosphate to produce glycosyl phosphates seems plausible as a model for prebiotic synthesis of sugar phosphates. Further studies will be needed to demonstrate directly the formation of sugar phosphates from various presumed primitive earth atmospheres, possibly by achieving a reversal of the glycolytic pathway, i.e. by aldol condensation from triose phosphates to pentose and hexose phosphates. One model of such a reaction has been achieved for the condensation of triose phosphates into fructose diphosphate in the presence of ion-exchange resins [14], while a partial analogue of the anaerobic glycolysis of sugars was observed in the alkaline degradation of glucose-6-phosphate, via fructose-6-phosphate and glyceraldehyde-3-phosphate to lactate and orthophosphate [5]. Further work may be to apply the cyanogen promoted phosphorylation reaction to the synthesis of various high-energy phosphates.

Acknowledgement

We wish to thank the Israel Academy of Sciences and Humanities for partial support of this work.

References

[1] Albert A. and E.P. Serjeant, Ionization Constants of Acids and Bases (Methuen, London, 1962) p. 129.
[2] Brotherton T.K. and J.W. Lynn, Chem. Rev. 59 (1959) 841.
[3] Calvin M., Proc. Roy. Soc. London Ser. A 288. (1965) 441.
[4] Claire-Deville H. and L. Troost, Compt. Rend. 56 (1863) 897.
[5] Degani Ch. and M. Halmann, Nature 216 (1967) 1207.
[6] Degani Ch. and M. Halmann, J. Chem. Soc., in press.
[7] Gabel N.W. and C. Ponnamperuma, Nature 216 (1967) 453.
[8] Gulick A., Am. Scientist 43 (1955) 479.
[9] Halmann M., R.A. Sanchez and L.E. Orgel, J. Org. Chem. 34 (1969) 3702.
[10] Halmann M. and H.-L. Schmidt, J. Chem. Soc. (1970) 1191.
[11] Jackson H. and D. Northall-Laurie, J. Chem. Soc. 87 (1905) 433.
[12] Kenyon D.H. and G. Steinmann, Biochemical Predestination (McGraw-Hill, New York, 1969) chapter 3.
[13] Khorana H.G., Some Recent Developments in the Chemistry of Phosphate Esters of Biological Interest (John Wiley, New York, 1961).
[14] Kozlova N.Ya., I.V. Melnichenko and A.A. Yasnikov, Dopovidi Akad. Nauk Ukr. RSR B 8 (1967) 710.
[15] Leloir L.F. and C.E. Cardini, in: Comprehensive Biochemistry, eds. M. Florkin and E.H. Stotz, Vol. 5 (Elsevier, Amsterdam, 1963) p. 113.
[16] Lohrmann R. and L.E. Orgel, Science 160 (1968) 64.
[17] Michaelis L. and P. Rona, Biochem. Z. 49 (1913) 232.
[18] Miller S.L., J. Am. Chem. Soc. 77 (1955) 2351.
[19] Orgel L.E. and R.A. Sanchez, personal communication.
[20] Peters K., Naturwissenschaften 19 (1931) 402.
[21] Rabinowitz J., S. Chang and C. Ponnamperuma, Nature 218 (1968) 442.
[22] Reid C. and L.E. Orgel, Nature 216 (1967) 455.
[23] Schimpl A., R.M. Lemmon and M. Calvin, Science 147 (1965) 149.
[24] Schwartz A., and C. Ponnamperuma, Nature 218 (1968) 443.
[25] Schwartz A., Chem. Commun. (1969) 1393.

[26] Steinman G., R.M. Lemmon and M. Calvin, Proc. Natl. Acad. Sci. U.S. 52 (1964) 27.
[27] Steinman G., D.H. Kenyon and M. Calvin, Nature 206 (1965) 707.
[28] Urey H., The Planets (Yale University Press, New Haven, Conn., 1952).
[29] Welcher R.P., M.R. Castellion and V.P. Wystrach, J. Am. Chem. Soc. 81 (1959) 2541.
[30] This work.

Chemical Evolution and the Origin of Life, eds. R. Buvet and C. Ponnamperuma

THE ORIGIN OF PROTEINS: HETEROPOLYPEPTIDES FROM HYDROGEN CYANIDE AND WATER

Clifford N. MATTHEWS
Department of Chemistry, University of Illinois at Chicago Circle, Chicago, Illinois 60680, USA

The hypothesis that a long period of chemical evolution preceded the appearance of life on earth has led during the past two decades to considerable research on the origin of biological macromolecules, proteins in particular. Speculations of Oparin [23], Haldane [7], Bernal [2] and Urey [30] have stimulated many investigations which appear to support the view that the prebiotic formation of polypeptides occurred in two stages: (1) α-amino acid synthesis initiated by the action of natural high-energy sources on the components of a reducing atmosphere, followed by (2) polycondensation of the accumulated α-amino acids in the oceans or on land.

Results of this work have been summarized in several recent monographs and reviews, including those of Fox [6], Oparin [24], Pattee [27], Bernal [3], Keosian [10], Ponnamperuma and Gabel [28], Kenyon and Steinman [9] and Calvin [4].

A more critical examination of the experimental evidence, however, shows that the fundamental problem of overcoming the thermodynamic barrier to spontaneous α-amino acid polymerization has been solved only by invoking conditions – anhydrous locales, high-temperature milieu and acidic oceans or pools – that are not characteristic of a young, developing planet. We have therefore challenged the validity of this two-stage view of protein abiogenesis, proposing instead an atmospheric model according to which the original polypeptides on earth were synthesized spontaneously from hydrogen cyanide and water without the intervening formation of α-amino acids [11,13,14].

Our mechanistic studies suggest a pathway beginning with the dimerization of hydrogen cyanide to a reactive dipolar compound – aminocyanocarbene – which polymerizes to form chains that interact further with hydrogen cyanide to yield heteropolyamidines that are finally converted by water to heteropolypeptides containing up to fifteen of the twenty α-amino acid residues commonly found in proteins:

$$\mathrm{HCN} \xrightarrow{\mathrm{HCN}} \mathrm{H_2N}\overset{\delta+}{\cdots}\mathrm{C}\cdots\mathrm{C}\overset{\delta-}{\equiv}\mathrm{N} \rightarrow \left[\underset{\underset{\mathrm{NH}}{\|}}{\mathrm{C}}\mathrm{-CH{=}N}\right]_n \xrightarrow{\mathrm{HCN}} \left[\underset{\underset{\mathrm{HN}}{\|}}{\mathrm{C}}-\overset{\overset{\mathrm{H}}{|}}{\underset{\underset{\mathrm{CN}}{|}}{\mathrm{C}}}\mathrm{-NH}\right]_n$$

$$\xrightarrow{\mathrm{HCN}} \left[\underset{\underset{\mathrm{HN}}{\|}}{\mathrm{C}}-\overset{\overset{\mathrm{H}}{|}}{\underset{\underset{\mathrm{R'}}{|}}{\mathrm{C}}}\mathrm{-NH}\right]_n \xrightarrow{\mathrm{H_2O}} \left[\underset{\underset{\mathrm{O}}{\|}}{\mathrm{C}}-\overset{\overset{\mathrm{H}}{|}}{\underset{\underset{\mathrm{R''}}{|}}{\mathrm{C}}}\mathrm{-NH}\right]_n \xrightarrow{-\mathrm{CO_2}} \left[\underset{\underset{\mathrm{O}}{\|}}{\mathrm{C}}-\overset{\overset{\mathrm{H}}{|}}{\underset{\underset{\mathrm{R}}{|}}{\mathrm{C}}}\mathrm{-NH}\right]_n$$

Here R″ and R′ are precursors of the fifteen side groups R of the α-amino acids glycine, alanine, valine, leucine, isoleucine, serine, threonine, proline, aspartic acid, glutamic acid, asparagine, glutamine, lysine, arginine and histidine. It seems likely that the side chains of these fifteen α-amino acids originated from polyaminomalonitrile, an intermediate polymer of hydrogen cyanide with projecting nitrile groups capable of undergoing successive reactions with hydrogen cyanide before being hydrolyzed by water. The other side chains containing aromatic rings or sulfur radicals would have been formed through further interactions initially involving acetylene and hydrogen sulfide, two expected components of the primitive atmosphere.

Evidence consistent with this hypothesis has been obtained from several kinds of experiments:

(a) Peptide-like solids containing up to fourteen of the fifteen α-amino acid residues listed above are among the products resulting from base-catalyzed polymerization of anhydrous hydrogen cyanide followed by extraction with cold water [14]. Free α-amino acids were not found under the ambient conditions used, in contrast with earlier work carried out at higher temperatures [25,12]. We believe that in all these experiments peptide material was formed directly from hydrogen cyanide, but that in the heated reactions some hydrolysis to α-amino acids must also have taken place.

(b) Similar peptide-like solids can be extracted by water from the products obtained after applying electric discharges to mixtures of methane and ammonia, probable major components of earth's primitive atmosphere known to be high-yield sources of hydrogen cyanide [13].

(c) Similar peptide-like solids can be isolated after base-catalyzed hydrolysis of aminoacetonitriles (Strecker intermediates) [21], aminomalononitrile (HCN trimer) or diaminomaleonitrile (HCN tetramer) [18,19]. Results (b) and (c) suggest that the α-amino acids detected in various types of experiments by Miller [16], Abelson [1], Oró [25], Palm and Calvin [26], Harada and Fox [8], Ponnamperuma, Woeller and Flores [29] and others were not primary products, as had been assumed, but instead had been formed by alkaline hydrolysis of peptides arising directly from hydrogen cyanide that had been liberated and become a reaction intermediate. Peptides detected in some of these latter experiments most probably were formed directly in the same way from hydrogen cyanide rather than by condensation of α-amino acids.

(d) Spectroscopic studies (uv and esr) at low temperatures [20] favor a singlet structure for aminocyanocarbene, the dimer of hydrogen cyanide postulated as a key intermediate in polymerization reactions of HCN leading to

$$H_2N \overset{\delta+}{=\!=} C =\!= C \overset{\delta-}{\equiv} N$$

$$H_2N-\overset{\uparrow\downarrow}{C}-C{\equiv}N \leftrightarrow H_2\overset{+}{N}{=}\overset{-}{C}-C{\equiv}N \leftrightarrow H_2\overset{+}{N}{=}C{=}C{=}\overset{-}{N} \leftrightarrow H_2N-\overset{+}{C}{=}C{=}\overset{-}{N}$$

polypeptide formation [11]. Delocalization of charge in this manner occurs often with carbanions alpha to nitrile groups, as for example in solvolysis [5] or alkylation reactions [22] of hindered acetonitriles.

(e) Poly-α-cyanoglycine prepared by base-catalyzed polymerization of the *N*-carboxyanhydride of α-cyanoglycine [31] can be modified by hydrogen cyanide to polymers that on hydrolysis yield several α-amino acids in addition to glycine [17]. Rigorous studies are currently being carried out along these lines to establish the extent to which activated nitrile groups can be converted to the side chains of proteins by hydrogen cyanide attack followed by hydrolysis. Such reactions would be analogous to the modification of polyaminomalononitrile proposed in our stepwise mechanism.

Four aeons or so ago, then, prebiological synthesis of primitive proteins on earth probably proceeded in the following way:

Polypeptides originated spontaneously, not by condensation reactions of α-amino acids or related compounds such as aminoacetonitriles.

The main reactant was hydrogen cyanide formed by photolysis of methane and ammonia. Photolysis of water would not have occurred till most of the ammonia had been used up [15]. Hydrogen cyanide reactions yielded polymeric peptide precursors – heteropolyamidines – which interacted also with other atmospheric constituents related to acetylene and to hydrogen sulfide.

Primary polymerization reactions took place in the stratosphere at the time when it consisted largely of hydrogen cyanide lying above a cold trap of atmospheric water. As the resulting macromolecules gradually settled, contact with water yielded heteropolypeptides possessing essentially the twenty side groups found in today's proteins.

Earth's surface became covered with proteinoid material which became concentrated in the oceans together with other products of atmospheric chemistry. In this protein-dominated environment increasingly specific interactions of organic and inorganic molecules led to the emergence of life.

This detailed model of protein abiogenesis has great heuristic value both for chemical evolution studies and for research directed towards finding new methods for the controlled synthesis of heteropolypeptides. In our continuing program four main objectives are:

(i) to establish rigorously the validity of the overall reaction

$$HCN + H_2O \rightarrow \left[\begin{array}{c} \text{H} \\ | \\ -\underset{\underset{\text{O}}{\|}}{\text{C}}-\underset{\underset{\text{R}}{|}}{\text{C}}-\text{NH}- \end{array} \right]$$

by maximizing synthesis procedures, by characterizing unambiguously the nature of the backbone and side chain of the resulting polymers, and by searching for optical activity and any significant sequence data;

(ii) to elucidate the reaction mechanisms involved by synthesizing probable intermediates and demonstrating the plausibility of each proposed step leading from hydrogen cyanide to heteropolypeptides possessing all twenty side chains of today's proteins;

(iii) to uncover biochemical activity of the heteropolypeptides, for example, by looking in vitro for enzymatic behavior and in vivo for antibiotic and antitumor activity, as well as nutritional possibilities; and

(iv) to relate the mode of formation and the general characteristics of the heteropolypeptides to a universal theory of evolution.

Perhaps most exciting intellectually is the logical way in which such a model generates complexity from simplicity while accounting for the spontaneous origin of informational macromolecules from almost the simplest and most prevalent three-atom molecules in the universe. This first step towards understanding how hierarchies of order are created must surely lead to further insights into the origin of the genetic code and the phenomenon of life.

References

[1] Abelson P.H., Carnegie Inst. Wash. Yearbook 56 (1956) 171.
[2] Bernal J.D., The Physical Basis of Life (Routledge and Kegan Paul, London, 1951).
[3] Bernal J.D., The Origin of Life (World Publishing, New York, 1968).
[4] Calvin M., Chemical Evolution (Oxford University Press, New York, 1969).
[5] Cram D.J. and L. Gosser, J. Am. Chem. Soc. 86 (1964) 5457.
[6] Fox S.W., Ed. The Origins of Prebiological Systems (Academic Press, New York, 1965).
[7] Haldane J.B.S., in: The Origin of Life (New Biology No. 16, Penguin Books, London, 1954).
[8] Harada K. and S.W. Fox, Nature 201 (1964) 336.
[9] Kenyon D.H. and G. Steinman, Biochemical Predestination (McGraw-Hill, New York, 1969).
[10] Keosian J., The Origin of Life, 2nd ed. (Reinhold, New York, 1968).
[11] Kliss R.M. and C.N. Matthews, Proc. Natl. Acad. Sci. U.S. 48 (1962) 1300.
[12] Lowe C.U., M.W. Rees and R. Markham, Nature 199 (1963) 219.
[13] Matthews C.N. and R.E. Moser, Proc. Natl. Acad. Sci. U.S. 56 (1966) 1087.
[14] Matthews C.N. and R.E. Moser, Nature 215 (1967) 1230.
[15] McGovern W.E., Ph.D. Thesis under S.I. Rasool, New York University, Graduate Division of the School of Engineering and Science (1967).
[16] Miller S.L., J. Am. Chem. Soc. 77 (1955) 2351.
[17] Minard R.D. and C.N. Matthews (1970) in progress.
[18] Moser R.E., A.R. Claggett and C.N. Matthews, Tetradron Letters (1968) 1599.
[19] Moser R.E., A.R. Claggett and C.N. Matthews, Tetradron Letters (1968) 1605.
[20] Moser R.E., J.M. Fritsch, T.L. Westman, R.M. Kliss and C.N. Matthews, J. Am. Chem. Soc. 89 (1967) 5673.
[21] Moser R.E. and C.N. Matthews, Experientia 24 (1968) 658.
[22] Newman M.S., T. Fukunaga and T. Miwa, J. Am. Chem. Soc. 82 (1960) 873.
[23] Oparin A.I., The Origin of Life, 2nd ed. (Dover Publications, New York, 1953).
[24] Oparin A.I., Adv. Enzymol. 27 (1965) 347.
[25] Oró J., Ann. N.Y. Acad. Sci. 108 (1963) 464.
[26] Palm C. and M. Calvin, J. Am. Chem. Soc. 84 (1962) 2115.
[27] Pattee H.H., Adv. Enzymol. 27 (1965) 331.
[28] Ponnamperuma C. and N.W. Gabel, Space Life Sci. 1 (1968) 64.
[29] Ponnamperuma C., F. Woeller and J. Flores, Am. Chem. Soc. 153rd Natl. Meeting Proc. (1967) 142.
[30] Urey H.C., Proc. Natl. Acad. Sci. U.S. 38 (1952) 1147.
[31] Warren C.B. and C.N. Matthews (1970), to be published.

Chemical Evolution and the Origin of Life, eds. R. Buvet and C. Ponnamperuma

POLYMER FORMATION IN A SIMULATED JOVIAN ATMOSPHERE

Haruhiko NODA *, and Cyril PONNAMPERUMA
Exobiology Division, Ames Research Center,
Moffett Field, California, USA

Introduction

The atmosphere of Jupiter is believed to consist mainly of hydrogen, methane and ammonia. Sagan et al. [3,4] applied electric discharges to a gaseous mixture of hydrogen, methane and ammonia and obtained many kinds of compounds. The study [2,5] of compounds produced in a gaseous mixture of anhydrous methane and ammonia showed that a number of nitriles were formed in addition to a colored polymeric substance. The polymer appears to be a mixture of various compounds with various colors ranging from yellow to deep orange [5]. In the present experiment an attempt was made to fractionate the polymeric substances according to molecular weight, and to characterize them.

Material and methods

Reaction of ammonia and methane

A 1-liter glass flask was evacuated with a rotary pump to a vacuum of about 10 μm of Hg, and then flushed with pure methane three times, evacuating the flask after each flush. The flask was then filled with an equimolar mixture of methane and ammonia to half atmospheric pressure.

Three kinds of electrodes were used: brass wire, glass, and carbon. The glass electrode consisted of a piece of glass tubing of about 5 mm in outside diameter, filled with mercury, which was placed concentrically in the middle

* Present address: Department of Biophysics and Biochemistry, Faculty of Science, University of Tokyo, Tokyo, Japan.

of a glass tube of about 30 mm in diameter and was covered with aluminum foil on the outside (fig. 1).

The power supply was a luminous-tube transformer with a secondary winding of 15 kV, and the voltage supply to the primary was regulated by a variable-voltage autotransformer. When brass or carbon electrodes were used, carbon resistors of 5 to 10 MΩ were used in series in order to limit the current.

The experiment was continued for three to four days, and the material formed was recovered by washing the inside of the flask with methanol and stored in a glass stoppered bottle at room temperature.

Fractionation and characterization

Gel permeation chromatography with Sephadex LH–20 in a column of 22 cm in length and 25 mm in diameter was performed by using methanol as the solvent. Continuous measurement of refractive index increment was used for the purpose of detecting the eluted substance. When it was necessary, photometric measurements of fractions were made at a wave length of 500 nm. Each fraction was kept in a screw top test tube at room temperature.

Molecular weight determination

Equilibrium centrifugation in a Beckman analytical ultracentrifuge was performed in order to determine the average molecular weight for each fraction. Because the sample solutions were colored with a reddish tinge a Wratten No. 16 filter was used. For the determination of the concentration, synthetic boundary experiments were performed with a capillary type synthetic boundary cell.

Chemicals

All chemicals used were analytical grade. Methyl alcohol used as the solvent was spectrophotometric grade supplied by J.T. Baker Chemical Company.

Infrared spectrum

The infrared spectrum was observed as described in the following. Pure KCl was ground in an agate mortar and a few drops of methanol solution of sample were added. A small pellet of 1.5 mm in diameter was prepared, and observed in a Perkin-Elmer model 521 infrared spectrophotometer with a micro-beam attachment.

Amino acid analysis

The methanol solution of the sample was mixed with about ten times the volume of 6 N HCl, and the mixture was analysed by a Beckman amino acid analyser, both immediately after mixing and after 12 hr in an oil bath of 110 °C.

Results

Discharge reaction and products

After 3 to 4 days of discharge, the product with reddish color was collected from the inside of the reaction chamber by using solvents. Dimethyl sulfoxide, dimethylformamide were very good solvents, pyridine and methanol were fairly good, and tetrahydrofuran was rather poor. Methanol was used as the only solvent through the present investigation for the following reasons. Dimethylsulfoxide and dimethylformamide had the tendency to affect the chromatographic material used. Pyridine was suspected to have caused degradation of the polymeric substance because of its strong basicity, and in fact it was difficult to remove by evacuation when mixed with the sample.

When methanol solution of the discharge reaction product was kept on the bench at room temperature, precipitation of a black material was observed in many cases. It took place when metal electrodes and arc discharges were employed. When the semi-corona discharge between tungsten wire and aluminum foil wrapped on the outside of the reaction vessel was used, formation of precipitate was much less frequent. Therefore, a reaction vessel of the type shown in fig. 1 was made in which no metallic surface was exposed to the reaction mixture. Precipitation never took place when such a reaction vessel was used. The yield of the product in the reaction vessel of fig. 1 was about 200 mg out of 400 mg of starting material.

The black precipitate was some kind of a polymer, because it could be dissolved in dimethylsulfoxide. The cause for the formation of such precipitate is unknown, but probably the current density on the electrode is the most important factor. When a pure carbon electrode and arc discharge were used, a larger yield of polymeric substance was obtained and precipitation took place readily.

Chromatographic fractionation

The result of gel permeation chromatography of the product on Sephadex LH–20 is as shown in fig. 2. The distribution shown as the solid curve in

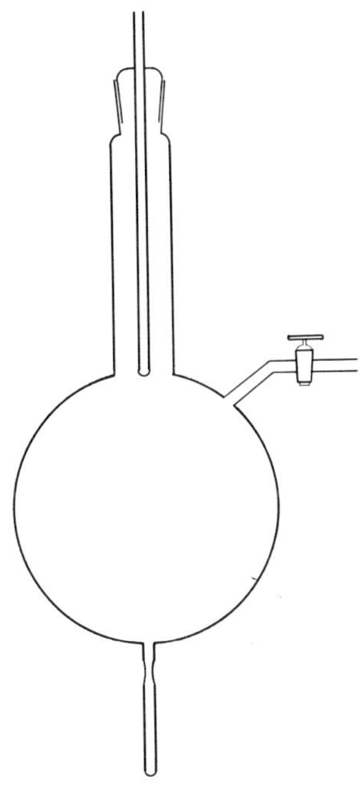

Fig. 1. The reaction vessel in which there is no direct contact between the reacting gas and metalic surface. The thin glass tubing extending down was immersed in a Dewar vessel with dry ice, avoiding direct contact of the tubing with the cooling mixture.

fig. 2 represents roughly the molecular weight distribution of various compounds included in the product. This was confirmed by rechromatography of the separate fractions.

The recovery of the total of all fractions was complete, and the same chromatographic column could be used over and over again.

As is seen in fig. 2, when the absorption at 500 nm is plotted against the elution volume, the peak of the curve does not coincide with that obtained by plotting the refractive index increment. This indicates that the fractions eluted earlier have stronger absorption. Various fractions were examined for their absorption spectra with a spectrophotometer, but the absorption curve was always a simple curve increasing towards the shorter wavelengths.

It was suspected that a large portion of the extinction was due to scattering, but no further analysis was made.

The value of partial specific volume was obtained by density determinations with glass picnometers at various concentrations of the test solution.

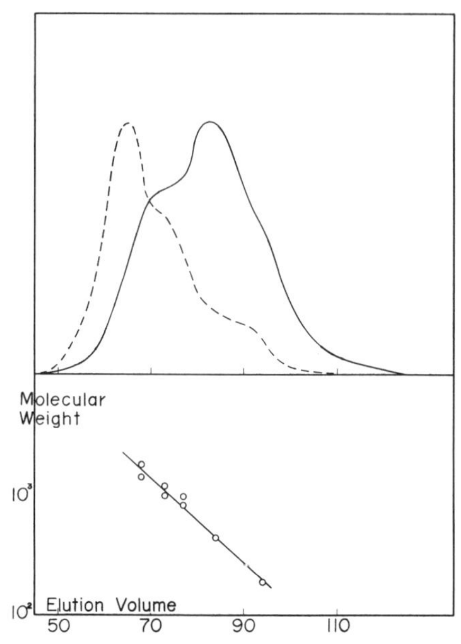

Fig. 2. Chromatographic fractionation of the reaction product by Sephadex LH–20. The abscissa is the elution volume of pure methanol. ——, the refractive increment for the fractions in arbitrary units; ---, the absorption at 500 nm. *Lower part:* logarithms of weight average molecular weights vs. elution volumes.

Table 1
Elution volume and average molecular weight of various fractions of gel permeation chromatography.

Fractions	Elution volume	Centrifugal field (*g*)	$\overline{MW}$	$\overline{MW}_a$	$\overline{MW}_b$
10-5	68	1.2×10^5	1.4×10^3	1.2×10^3	2.1×10^3
10-5	68	6.5×10^4	1.8×10^3	1.5×10^3	1.9×10^3
10-6	73	1.2×10^5	9.9×10^2	7.8×10^2	1.4×10^3
10-6	73	6.5×10^4	1.2×10^3	9.7×10^2	1.4×10^3
10-7	77	2.2×10^5	8.3×10^2	6.9×10^2	1.2×10^3
10-7	77	1.2×10^5	9.7×10^2	7.9×10^2	1.2×10^3
9-7	84	2.2×10^5	4.4×10^2	4.4×10^2	4.8×10^2
11-7	94	2.2×10^5	2.4×10^2	2.3×10^2	3.0×10^2

$\overline{MW}$: weight average molecular weight of the fraction
$\overline{MW}_a$: weight average molecular weight for the solution at the meniscus of the cell
$\overline{MW}_b$: weight average molecular weight of the solution at the bottom of the cell

Partial specific volume determination was made only for the starting material, which certainly was a mixture.

The molecular weight of various fractions was determined by the method of equilibrium ultracentrifugation, and the results are shown in table 1 and fig. 2. The values are for weight average molecular weight. The weight average molecular weight values at the bottom and meniscus of the cell are also shown in table 1, in order to indicate the degree of heterogeneity of the particular fraction used for the equilibrium ultracentrifugation. The fact that the logarithm of the molecular weight value lies on a straight line, as is shown in fig. 2 when they are plotted against the elution volume, is a good indication that LH–20 is functioning as a gel permeation medium and there is no specific absorption.

Color of the solution when illuminated sideways

Because the product obtained in the methane–ammonia reaction vessel had reddish color, it has been suggested that this could be the substance responsible for the red spot of Jupiter [3,5]. In order to find out the validity of the above statement, it is necessary for us to know the color of the substance when it is placed in a similar circumstance to that for the red spot of Jupiter. Therefore, the solutions of various fractions were illuminated with a Xenon arc lamp, and the light emitted to the right angle was examined with a spectrophotometer. The results are shown in figs. 3 and 4.

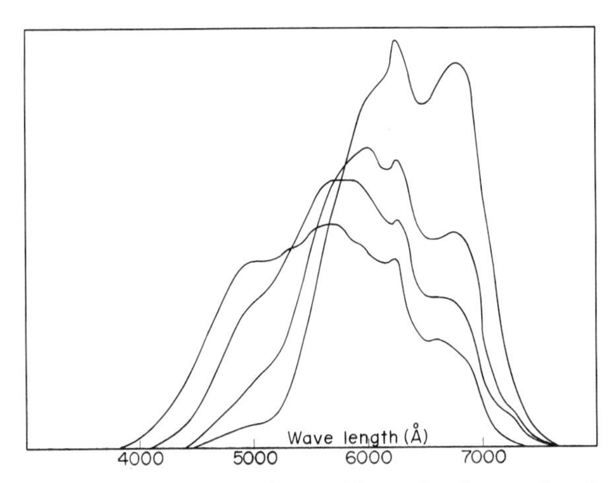

Fig. 3. The spectrum of light observed at a right angle when methanol solution of a fraction with a molecular weight of 2.4×10^3 was illuminated by a Xenon arc lamp. Relative concentration of the solute was 1, 1/2, 1/4 and 1/8 for curves from the top.

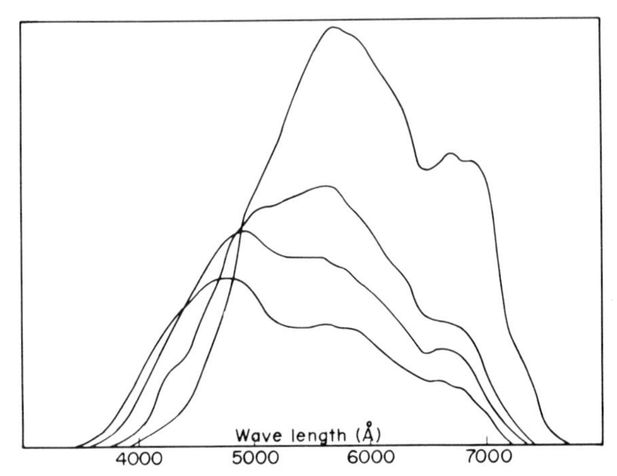

Fig. 4. The same kind of spectrum as for fig. 3 for a fraction with a molecular weight of 1.3×10^2.

The most interesting feature is that the position of the maximum of the spectrum changed when the concentration of the solution was varied. However, there are some specific features in the spectroscopic distribution, and if the spectroscopic analysis of the red spot of the Jupiter is available, it would probably be possible to determine whether this sort of substance is making the red color of Jupiter or not.

The spectrum observed in the above is the mixture of scattering and fluorescence. Therefore, effort was made to look for fluorescence characteristics by using monochromatic excitation. However, the emission spectrum varied when the excitation frequency was varied keeping an approximately constant difference in frequency, indicating that the solution contains a large number of compounds with varied spectrum.

Chemical composition

Because the starting material consisted of a mixture of methane and ammonia, it is natural to assume that the product will be compounds of C, H and N. When various fractions were analysed for their elementary composition, oxygen was found in every fraction. This suggest that hydrolysis took place when there was a slight amount of moisture. The result of elementary analysis is shown in table 2.

When analysed by the amino acid analyser, only traces of glycine were observed both in fractions of larger molecular weight and those of smaller molecular weight. However, after hydrolysis with 6 N HCl overnight, several amino acids appear to be jammed [1].

Table 2
Elementary analysis of various fractions of gel permeation chromatography.

Fraction	Elution volume	C (%)	H (%)	N (%)
2-2	60	47.8	6.2	20.9
2-6	72	50.4	6.7	25.3
2-15	100	50.5	8.5	25.3
2-23	124	39.0	6.0	33.4
2-31	148	39.4	5.6	28.0

When the infrared absorption spectrum was compared between the fraction for which the color absorption was maximum and the fraction for which refractive index increment was maximum, there was not much difference.

Discussion

As is seen in table 1 and fig. 2, the reddish material obtained by electric discharge in a gaseous mixture of methane and ammonia contained some polymerized material, the maximum molecular weight of which was about a few thousands. This maximum value probably was larger if the material was handled by strictly avoiding its contact with moisture. As is seen in table 2, the material always contained some oxygen, despite the fact that the starting material contained no oxygen and contamination of the reaction vessel with air was carefully avoided. It was found in a preliminary experiment in which reaction in methane, ammonia and water was compared with that in methane and ammonia only, that a large peak appeared upon gel permeation chromatography of the former at a place corresponding to a molecular weight of about 400, indicating the prevalence of hydrolysis reaction. Therefore, the possibility of hydrolysis due to atmospheric humidity cannot be overlooked.

As is shown in fig. 2 and mentioned in the section on results, the logarithms of the molecular weights of fractions of LH–20 chromatography lie close to a straight line when plotted against the elution volume. However, if examined closer, there are slight but obvious deviations. These are probably due to the difference in the values of specific volume for different fractions. As is shown in table 2, the atomic composition varies for different fractions. It is well known that the partial specific volume differs for compounds of different chemical structure. At the same time, it should be mentioned that, the atomic composition of the fraction with highest extinction at 500 nm and that for the fraction with highest value of refractive index increment were about the same, and larger differences took place for fractions eluted later. This seems to indicate that a large amount of material which is capable of polymerization is produced by electric discharge. Naturally, however, a number of other kinds are also produced, and these are fractionated according to the molecular size by chromatography, and eluted afterwards.

The formation of a precipitate which was purposely avoided in the present study is also a result of polymerization reaction and not carbonization. Such precipitates were dissolved well in dimethylsulfoxide, and to some extent in dimethylformamide. This confirms the notion that the current density in the discharge has a great influence in the mode of polymerization [5].

The substance obtained by discharge in methane and ammonia has often been assumed to be the substance responsible for the red color of Jupiter. The deposit formed on the reaction vessel really looks reddish. When the substance is fractionated according to the molecular weight, the color was darker the larger the molecular weight was. However, the spectrum of the color was, whether by absorption or by the combination of scattering and fluorescence, indistinguishable regardless of the molecular weight of the fraction. In addition, the apparent color changes when the concentration varies, as is seen in figs. 3 and 4. It would not be simple to identify the red substance in Jupiter.

References

[1] Chadha M.S., J. Lawless, J. Flores and C. Ponnamperuma, this volume, p. 143.
[2] Ponnamperuma C. and F. Woeller, Currents Mod. Biol. 1 (1967) 156.
[3] Sagan C.E., E.R. Lippincott, M.O. Dayhoff and R.E. Eck, Nature 213 (1967) 273.
[4] Sagan C.E. and S. Miller, Astronom. J. 65 (1960) 499.
[5] Woeller F. and C. Ponnamperuma, Icarus 10 (1969) 386.

Chemical Evolution and the Origin of Life, eds. R. Buvet and C. Ponnamperuma
© 1971, North-Holland Publishing Company

POLYMERIZATION OF AMINO ACID–PHOSPHATE ANHYDRIDES IN THE PRESENCE OF CLAY MINERALS

Mella PAECHT-HOROWITZ
Polymer Department, The Weizmann Institute of Science, Rehovot, Israel

In 1968, we reported our first experiments on polycondensation of amino acid phospho-anhydrides on clay surfaces [3]. At that time we had established that while in the absence of clay only peptides of low molecular weights are formed, in the presence of certain clays, higher degrees of polymerization are obtained displaying discrete spectra rather than a continuous distribution.

In our further studies, we have mainly investigated the mechanism of this polymerization. The monomer used was d,l-alanine adenylate (fig. 1). The anhydride bond between the carboxyl and the phosphate group is energy-rich; when the bond is split, enough energy is liberated to allow the formation of peptide bonds [1,4]. Also for polymerization to occur, the amino group of the amino acid or the peptide to be polymerized should be in the free and not in the ammonium form; therefore the experiments have to be carried out at slightly alkaline pH. The general plan of our experiments was as follows: to an aqueous solution, at constant pH (7.8–8.5) and at room temperature, 1 mmole of alanine adenylate was added, either at once or in small portions

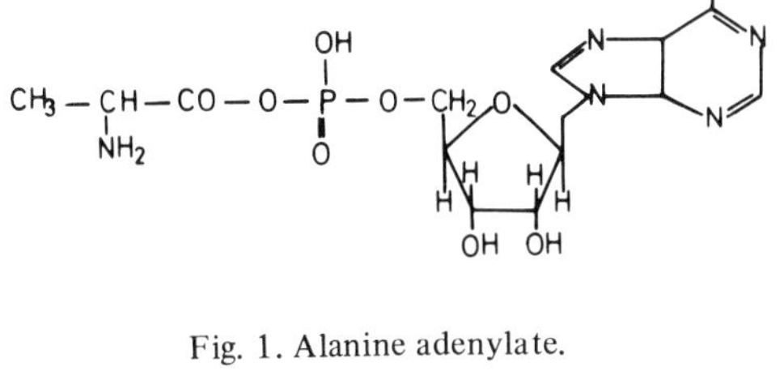

Fig. 1. Alanine adenylate.

to keep the concentration of the anhydrous bonds constant, thus obtaining essentially a steady state reaction. In the experiments with clay minerals, these were suspended in the aqueous solution to which the anhydride was added. At the end of the reaction, the clay was separated by centrifugation and the remaining solution concentrated and separated on a Sephadex G-25 column either with water or with 0.01 N HCl. The resulting fractions were freeze-dried, analyzed, and their molecular weights determined.

In the absence of clay, the polymerization mechanism is the following (fig. 2): the first step is hydrolysis; some alanine adenylate is split resulting in the formation of alanine and adenylic acid. The free alanine attacks the alanine adenylate giving di-alanine and free adenylic acid. The dipeptide now attacks another molecule of alanine adenylate to give tri-alanine and adenylic acid, etc. Only one molecule of amino acid is added at each stage and the reaction proceeds by interaction between one molecule of anhydride and one molecule of free amino acid or peptide. No interactions between two molecules of anhydride occur. As competition with hydrolysis is quite severe, the highest peptides obtained in this way, when the whole amount of alanine adenylate was introduced into solution in one portion, is 4–6-mers. Under steady state conditions, up to 10–12-mers can be obtained but only in very low yields. Here also a peptide is formed only when the peptide one unit smaller has reached a constant concentration. We then obtain a high amount of amino acid, less dipeptide, even less tripeptide, etc., and only traces of higher peptides [4] (fig. 3).

$$\mathrm{A} \sim \mathrm{P} \xrightarrow{h} \mathrm{A} + \mathrm{P}$$

$$\mathrm{A} + \mathrm{A} \sim \mathrm{P} \xrightarrow{k_1} \mathrm{A}_2 + \mathrm{P}$$

$$\mathrm{A}_2 + \mathrm{A} \sim \mathrm{P} \xrightarrow{k_2} \mathrm{A}_3 + \mathrm{P}$$

$$\mathrm{A}_3 + \mathrm{A} \sim \mathrm{P} \xrightarrow{k_3} \mathrm{A}_4 + \mathrm{P}$$

$$\vdots$$

$$A_n + \mathrm{A} \sim \mathrm{P} \xrightarrow{k_n} \mathrm{A}_{n+1} + \mathrm{P}$$

Fig. 2. Polymerization mechanism of alanine-adenylate, in the absence of clay.
A ~ P: alanine-adenylate
A : alanine; A_2-dialanine, A_3-trialanine
A_n : polyalanine of the *n*th degree of polymerization
P : adenylic acid

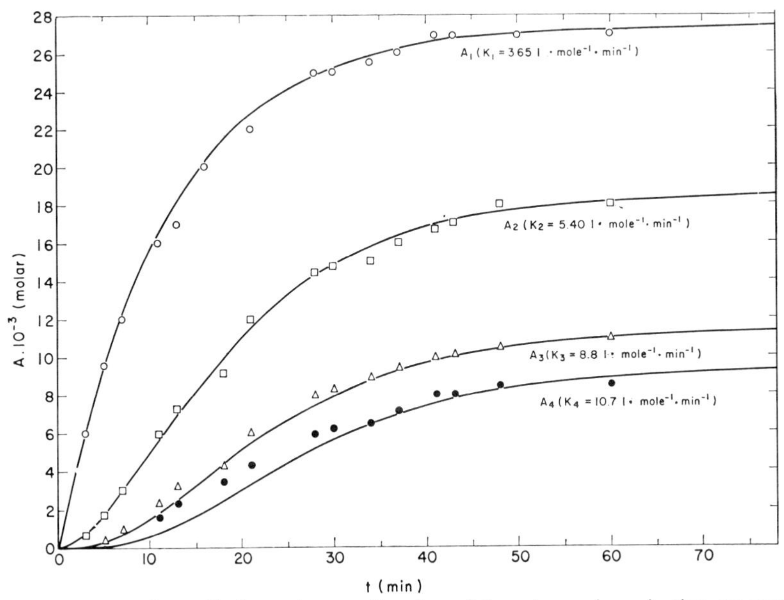

Fig. 3. Pattern of peptide formation as a function of time, in a polymerization process at constant phospho-anhydride concentration [4].

The results are entirely different when there is a suspension of a clay mineral having certain properties in the reaction mixture. Sodium montmorillonite is a clay mineral consisting of three layer sheets of SiO_2–Al_2O_3, $Al(OH)_3$–SiO_2 (fig. 4). It is able to swell and expand enormously, so that we may assume that at low concentrations we obtain monosheets with a very large surface area. We used in our experiments montmorillonite with a particle size of 700 Å. In the presence of this mineral, discrete groups of molecular weights were obtained rather than the continuous spectrum found in the experiments in its absence (fig. 5). The distribution of molecular weights is to a certain degree dependent on the concentration of the clay, but at all concentrations there are regions of molecular weights where the yields of peptides are very low and others where they are very high. Although the peptide values we obtain – 10–12-mers, 16–17-mers and 22–25-mers – are averages, when each fraction is analyzed by high voltage paper electrophoresis, it can be seen that most of the substance is indeed of the specified degree of polymerization and only traces of larger or shorter polypeptides are present.

With larger amounts of alanine adenylate, e.g. 2 mmoles instead of one,

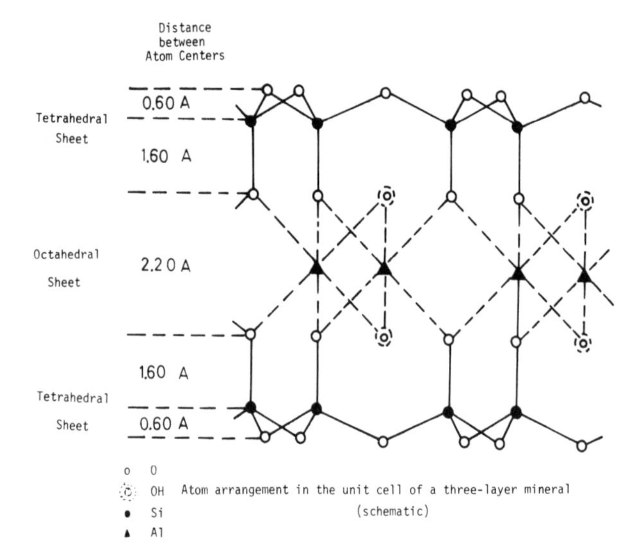

Fig. 4

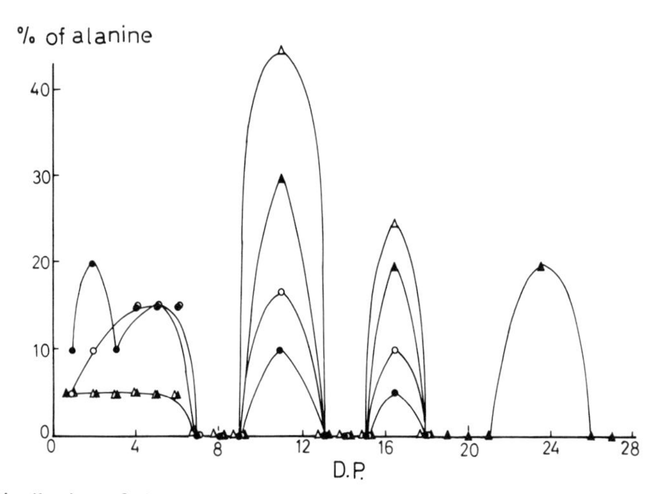

Fig. 5. Distribution of the molecular weights of polyalanines obtained when alanine–adenylate is introduced into an aqueous solution at pH 8.5, in the presence of various amounts of montmorillonite (mg/l): ● 20, ○ 80, △ 220, ▲ 480.

additional discrete groups of higher molecular weights are obtained, like 28–30- and 38–40-mers.

It is clear that in this case the mechanism of polymerization is no longer the addition of one unit to a previously formed peptide, but that some jumps occur from one group of molecular weights to another. A speculation was that we may have here a case of interactions between two molecules of anhydride, which would allow the formation of peptide anhydrides which could condensate further. In order to test this possibility, we added a large excess of free amino acid to the reaction mixture. When no clay is present, all the anhydride is immediately transformed into dipeptides by the free amino acid and the reaction stops. In the presence of clay, the free amino acid has no influence at all and the reaction proceeds according to the usual pattern. This becomes understandable when we take into account another set of experiments which were performed by Dr. Berger of our group. He measured the absorption of alanine, adenylic acid, and alanine adenylate by montmorillonite at pH 7, and found that at this pH, neither alanine nor adenylic acid are absorbed, while alanine adenylate is very strongly absorbed. Its components are liberated only very slowly from the montmorillonite after polymerization has taken place and the substance is no longer in its anhydride form. As we assume that our polymerization takes place on or between the layers of montmorillonite and as this clay absorbs only the anhydride, not the free amino acid, in the presence of montmorillonite the anhydride is practically isolated from the amino acid even if the latter is in very high concentration in the surrounding medium.

This seems a valid explanation of the fact that there is no interaction between anhydrides and free amino acids in the presence of montmorillonite. For polymerization purposes, anhydrides of peptides which are formed as a result of the interaction between two molecules of anhydride have two advantages. As anhydrides of peptides are much more stable than those of amino acids, competition with hydrolysis is very much reduced. Indeed, in experiments with montmorillonite, practically no amino acid is found at the end of the reaction, while it may be up to 50% of all substances present in experiments in the absence of clay. The second advantage is that two peptides of any length can combine and give directly a much higher peptide.

This reaction mechanism – interaction between anhydrides – might explain the much higher degrees of polymerization, but not the discrete groups of molecular weights obtained. We can only be speculative regarding this point. Initially, we believed that the catalytic properties of montmorillonite are due only to the large surface area it provides. If this is so, any large area of similar configuration should have the same properties. However, when we replaced montmorillonite by "Cabosil" (colloidal SiO_2) so well dispersed that a surface area of 700 m^2/g was obtained, no polymerization occurred. We repeated the experiment with very finely ground Al_2O_3 (the aluminium part of montmorillonite) and obtained the same polymerization pattern as when no clay was present. It seems therefore that the size of the surface area, although important, is not the only decisive factor. The configuration of the whole clay plays a role in the activation of the polymerization process. This idea is supported by some additional evidence. When we employed a montmorillonite in which the aluminium was exchanged by iron ("nontronite") we again obtained high polymers in discrete groups, but with a distribution different from that obtained with aluminium montmorillonite. This supports the notion that the polymerization is influenced by the whole configuration of the clay, and not by its individual components.

We also have to bear in mind that we have a superposition of reactions. In addition to polymerization and eventual hydrolysis, we have also a diffusion of monomer into the clay and one of the polymerization results out of it. As these diffusion processes, especially diffusion out of the clay, proceed at a much lower rate than polymerization, they have a very significant impact in regulating in some way the polymerization possibilities. Beside this blocking of the entrance of new monomer into the clay until the polymerization products of a previous portion have left, we have to consider also that at least half of the adenylic acid is liberated by polymerization, which produces a sharp local drop in pH. Yet we know that whatever the polymerization mechanism, the amino group of the amino acid or of the peptide, in order to be able to polymerize, has to be in its free form [2]; hence the pH of the reaction medium must be higher than 7. Therefore, until the liberated adenylic acid has diffused into the surrounding solution where it is neutralized by buffer or by added alkali, the polymerization will be stopped. It is very likely that the superposition of all these factors produce this oscillating form of polymerization and hence discrete groups of molecular weights.

We realize that all this is still very far from describing the mechanism of formation of living organisms, or even of natural peptides produced by living organisms. However, there is one additional point I would like to mention. About 50% of the peptides remain bound to one group of adenylic acid. At this stage, the adenylic acid undergoes a rearrangement, and is linked to the carboxyl group of the peptide no longer through its phosphate group by an anhydride bond, but through its sugar group by an esteric bond. As this is the bond by which amino acids are bound to nucleic acids in natural peptide formation, we hope that further investigation in this direction will be fruitful.

References

[1] Lewinsohn R., M. Paecht-Horowitz and A. Katchalsky, Biochim. Biophys. Acta 140 (1967) 24.
[2] Paecht M. and A. Katchalsky, Biochim. Biophys. Acta 90 (1964) 260.
[3] Paecht-Horowitz M., Abstracts of 5th International Symposium on the Chemistry of Natural Products, London July 8–13 (1968) p. 232.
[4] Paecht-Horowitz M. and A. Katchalsky, Biochim. Biophys. Acta 140 (1967) 14.

Chemical Evolution and the Origin of Life, eds. R. Buvet and C. Ponnamperuma

THE PRIMORDIAL SEQUENCE, RIBOSOMES, AND THE GENETIC CODE

Sidney W. FOX, Atsushi YUKI, Thomas V. WAEHNELDT and James C. LACEY, Jr.
Institute of Molecular Evolution, University of Miami, Coral Gables, Florida 33134, USA

Some of the key questions of the origin of life concern the primordial sequence of nucleic acid, protein, and cell. Which came first? Whichever postulate has been considered has seemed to leave an unresolved question. One principal dilemma has comprised the two reflexive questions of how nucleic acids might have arisen without enzymes to make them [5] or, alternatively, how enzymic protein might have arisen without nucleic acid to direct the sequence of monomers.

Experiments in our laboratory are consistent with one answer to the question of primordial sequence; this answer comprises two stages. In the first stage, by this view, prebiotic proteins were the first to appear [7,12,13]. This interpretation has come to be known as the proteins–first hypothesis [30]. Moody analyzes this and the gene–first hypothesis. The experimental results based on the proteins–first hypothesis have explained *in principle* how enzymes could have arisen in the absence of enzymes to make them, how cells could have arisen in the absence of cells to produce them [8,10,14], and a number of other chicken-egg questions have received a first answer. Through extensive characterization of the polymers, much ordering of the monomers [16,17] has been demonstrated to result from interactions of the amino acids with the growing peptide chain. This ordering need not have been as precise as * the "residue-by-residue" arrangements of the evolved contemporary genetic code (cf. [35]). No other kinds of molecule beside amino acids and peptide were involved. The resultant protein-like molecules, proteinoid, have been shown to be informational in the sense that they display selective effects with contemporary metabolic substrates (many laboratories; [37]), with diverse polynucleotides, and in phenomena of

* The possibility of rapid transpeptidation at elevated temperatures conceptually allows selection of a small number of most stable sequences.

assembly. As part of the total sequence such polyamino acids would have led into a second stage in which proteins and nucleic acids arose simultaneously in the *presence* of enzymes and cells to make them (fig. 1). As yet, this second stage is less thoroughly modelled by experiments than is the first. Some possibilities suggested by the experiments for the second more complicated, stage are reviewed in this paper.

Conceptually, the first stage would have been more primitive *, the second more contemporary. One should not assume that other answers to the questions of primordial sequence are necessarily ruled out. The essence of the proteins–first hypothesis, however, appears to be in accord with the sense of Darwinian processes [7,25] in that it permits nucleic acids to have selected effective protein molecules – protein molecules which necessarily functionally preceded those nucleic acids which selected them.

Several considerations which predated the first experiments in the laboratory included those of evolution of proteins, the thermodynamics of formation of peptide bonds, kinetic ordering of monomers, the organic

* In another way, the experiments of Lipmann [20] indicate the possibility of protein synthesis without nucleic acid template in early evolution (see also zymosequential specificity [24]).

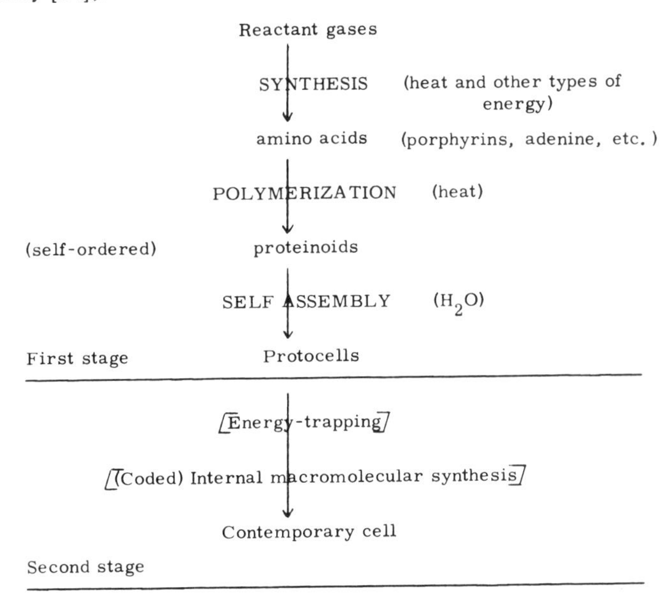

Fig. 1. Evolution of protocell from reactant gases, as modelled by experiments, followed by evolution of contemporary cell.

chemistry of decomposition, and the geological relevance of such studies [6]. An analysis of the thermodynamics of the formation of peptide bonds (fig. 2) has yielded equilibrium constants that state that only small yields of small peptides could be expected in aqueous solution unless the reactions were coupled to energy-yielding reactions [11]. In the biotic synthesis of peptide bonds phosphate energy is used, but contemporary systems also rely on ribosomes, so that the reaction is in fact not carried out in aqueous solution. We may therefore entertain the possibility that activation of amino acids by heat under hypohydrous terrestrial conditions evolved to activation by phosphate energy under the hypohydrous conditions of the ribosome. At least, the conditions are analogous.

The rates of the reactions are such that a sequence of gases → amino acids → proteins → microspheres ([13] for properties of microspheres) could easily occur in the nocturnal phase of a single diurnal cycle; the products would thus not have been exposed to decomposition from high-energy radiation during their production.

Heat was initially chosen as a form of energy for experimentation, since it is easily controllable and can be relatively gentle [9], especially for macromolecules. We believe that heat is most favorable for terrestrial organic chemistry, much as it is clearly preferred in the laboratory [21].

The route from monomers (now conceptually derivable below 200 °C from interstellar matter [40] to the first cells has been depicted (fig. 1) as proceeding through protein. This choice does not necessarily exclude the

$$H_2NCHRCOOH + H_2NCHR'COOH = H_2NCHRCONHCHR'COOH + H_2O$$

$$\Delta G^o_{298} = 1400 \text{ to } 3700 \text{ cal.}$$

energetics of formation of peptide bond

$$H_2NCHRCOOH + H_2NCHR'COOH \longrightarrow H_2NCHRCONHCHR'COOH + H_2O\uparrow$$

synthesis by elevation of temperature

$$nH_2NCHR^{18}COOH \xrightarrow[H_3PO_4]{100^o} (HNCHR^{20}CO)_n \quad \text{(thermal proteinoids)}$$

or P-O-P
or PPA

polymerization aided by phosphates

$$\text{AMP-anhydrides of 20 amino acids} \xrightarrow[25^o]{H_2O} \text{adenylate proteinoids}$$

polymerization *via* anhydrides with adenosine monophosphate, in aqueous solution

$$\text{amino acids} \xrightarrow[\text{ribosomes}]{} \text{protein}$$

Fig. 2. Formation of peptide bonds, prebiotically and biotically.

alternative route, proceeding first through functional nucleic acids [3,31]. It signifies rather that the route through prebiotic proteins first is consistent with a comprehensive set of experimental results. This route, also, is a Darwinian one [7,25], which is often preferred on logical grounds (cf. bibliography in [13,15]). Indeed, the ease of making protein-like polymers, as contrasted to nucleic acid-like polymers [39], has been cited as a reason for believing they appeared first on earth [27]. The argument against the appearance of functional nucleic acids before proteins or proteinaceous cells has been presented in various ways. Lederberg [28], for example, has stated, "... The point of faith is: make the polypeptide sequences at the right time and in the right amounts, and the organization will take care of itself. This is not far from suggesting that a cell will crystallize itself out of the soup when the right components are present." Jukes [25] has pointed out that the code had originally to be read backwards to resolve the "seemingly insuperable enigma of the initial appearance of a genetic nucleic acid molecule carrying the coded information for biologically active proteins that were previously non-existent."

The primitive cell, as represented by the proteinoid microsphere, had within it all that was necessary for a nearly heterotrophic [22,34,42] type of proliferation [16] that might have continued throughout eons. This process could have occurred repeatedly through polyamino acid preformed in the environment. In order to evolve to a *contemporary* cell, however, additional processes must have entered. At some stage, trapping of solar energy [45] and an internal synthesis of peptide bonds plus a contemporary type of genetic code must have originated through natural experiments. On logical grounds, this last development could have occurred through a nearly simultaneous appearance of proteins and nucleic acids [12].

An appropriate intermediate for such a development is the aminoacyl adenylate. This contains both the amino acid residue and the adenylic acid moiety. This compound is universally used in protein biosynthesis by contemporary organisms [33]. Some questions of the origin of adenylic acid have been answered in principle [26]; others have not. Experiments *in open systems* have emphasized that when one amino acid was synthesized [23,41], many others were simultaneously synthesized; accordingly, the adenylates of numerous amino acids were copolymerized in experiments [26,32]. Only single aminoacyl adenylates had been polymerized previously (e.g. Berg [1]; Paecht-Horowitz and Katchalsky [36]). The possibility of simultaneous synthesis of all proteinogenous aminoacyl adenylates was not promising, partly because Berg had indicated that the individual adenylic acid anhydrides of aspartic acid, glutamic acid, and histidine were difficult to study in the laboratory. Copolymerization, however, often succeeds where homopolymerization fails [19].

When all of the amino acids common to protein were simultaneously condensed as the adenylates, a proteinoid having a composition (table 1) remarkably like that of the average of contemporary proteins [43] was observed. This result occurred from an equimolar mixture of amino acids converted to AMP-anhydrides. The reaction required two steps: (a) amino acids DCCD aminoacyl adenylates and (b) aminoacyl adenylates → polyamino acids. We know now that kinetic differentiation occurs in step (a) and in step (b). We have no a priori reason to believe that protein biosynthesis proceeds from pools of equimolar mixtures of amino acids; accordingly, the results deserve further explanation. The polymerization does occur from aqueous solution, and appears not to proceed to large polymers in aqueous solution *.

* Lemmon [29] has incorrectly referred to this adenylate condensation as occurring in the presence of DCCD. DCCD was used in step (a), not step (b).

Table 1
Composition of hydrolyzate of proteinoid from amino acid adenylates alone compared with an average protein. (Calculated without ammonia)

Amino acid	Composition of polymer (mole %)	Ratios of amino acids in average protein (mole %)
Lysine	6.5	5.9
Histidine	2.4	1.8
Arginine	4.2	4.9
Aspartic acid	10.3	9.7
Threonine	4.9	4.8
Serine	4.2	6.0
Glutamic acid	9.7	12.7
Proline	5.1	6.2
Glycine	11.1	12.6
Alanine	14.3	9.6
Valine	7.3	5.9
Methionine	0.7	1.8
Isoleucine	4.5	6.0
Leucine	9.6	6.0
Tyrosine	0.1	2.3
Phenylalanine	4.5	3.7

Also found as a product of the polymerization is AppA; an energy-rich diphosphate bond has been produced. The evidence on the origin of this bond is equivocal [32] in that the possibility of its formation in the presence of DCCD, in the step (a), is not ruled out.

Regardless of whether a polymer of adenylic acid has been produced, the adenylate intermediate invites further investigation. Key questions are (a) those of how adenylate condensation might have been influenced by polynucleotides, of whatever cellular source, to yield code-related polymers *, and (b) how would ribosomal mechanisms enter into the evolutionary stream?

To model the origin of ribosomes, Waehneldt prepared a series of proteinoids having histone-like compositions [18]. When the proteinoids were sufficiently basic, they were found to react with nucleic acids to produce nucleoproteinoids which separated from solution. These were fibers or spherules [44], depending upon whether the polynucleotide was deoxyribo- or ribo- respectively (fig. 3). The production of spherules did not require a large polynucleotide nor contemporary nucleic acid. The oligomer of cytidylic acid, of 5.5 residue size, prepared thermally [39] also gave sperules with the same basic proteinoid. These various complexes are dissociated by adjustment of pH, concentration of salt, etc. in much the same manner as nucleoprotein complexes.

The reaction to form nucleoproteinoid particles is not merely one of neutralization of charge; it exhibits an array of specificities. The first of these to be reported involved differences between arginine-rich and lysine-rich proteinoids, as models of arginine-rich and lysine-rich histones [48]. The arginine-rich proteinoids were found to react preferentially with polypurines to yield microparticles (table 2) whereas the lysine-rich polymer reacted specifically, under the same conditions, with polypyrimidines [48].

The particles that resulted from homopolynucleotides and basic proteinoids showed preferences in further reactions with polynucleotides (table 3). This is partly analogous to the preferential behavior of ribosomes as suggested by experiments with other objectives [38,46].

Fig. 4 illustrates preferences in the formation of particles by various homopolyribonucleotides of proteinoids of different contents of lysine. These proteinoids contain ten types of amino acid. As the proportion of lysine varies, the proportion of each amino acid varies according to its individual pattern. The optimal interactions involve proteinoids containing 20% or 30% lysine.

* Preliminary evidence of code-related effects by added polynucleotides in condensation of adenylates has been noted [32].

Fig. 3. Fibrous complex of lysine-rich proteinoid and calf thymus DNA on left; complex of same proteinoid and yeast RNA on right, as viewed microscopically.

Table 2
Comparison of lysine-rich proteinoid with arginine-rich proteinoid in binding with polynucleotides.

Polyribo-nucleotide	Turbidity	
	Lysine-rich (arginine-free) proteinoid	Arginine-rich (lysine-free) proteinoid
Poly C	0.253	0.002
Poly U	0.050	0.058
Poly A	0.001	0.060
Poly G	0.003	0.218
Poly I	0.003	0.248

[a] Signifies <1 spherule/ml.

From Yuki and Fox [48]. Proteinoid concentration 1.0 mg/ml, polynucleotide concentration 0.1 μmoles/ml, 0.05 M tris buffer, pH 7.0, 25.0 °C.

Table 3
Binding of ^{14}C-polycytidylic acid and ^{14}C-polyadenylic acid by complexes of lysine-rich proteinoid with poly A, poly U, or poly I.

	^{14}C-Poly C (counts/5 min/filter)		^{14}C-Poly A (counts/5 min/filter)		C/A
A complex	8,043	(7.1%)	23,232	(12.6%)	0.346
U complex	4,403	(3.9%)	30,709	(16.7%)	0.143
I complex	7,974	(7.0%)	34,860	(18.9%)	0.299
Total used		(100.0%)	–	(100.0%)	0.614

Complexes from 700 μg of proteinoid and 50 μg of polynucleotide in 2.0 ml of 0.033 M sodium chloride. Separated on a millipore filter, through which was passed ^{14}C-poly A, in parallel experiments, ^{14}C-poly C.

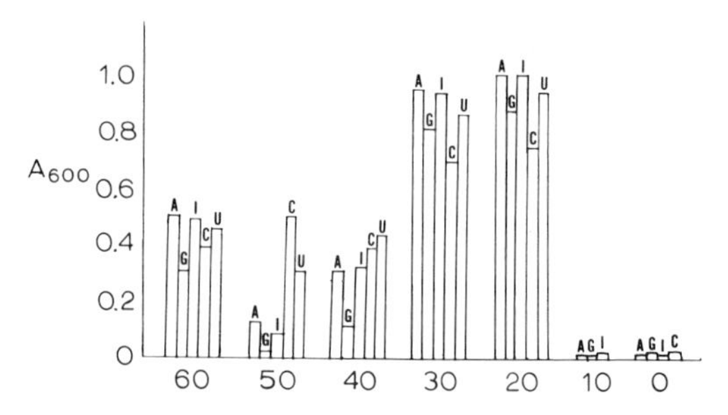

Fig. 4. Variation in interaction of each of five homopolynucleotides with lysine-rich proteinoids of various proportions of lysine, as measured by turbidity at 600 nm. The figures 60 to 0 correspond to mole percent of lysine in the reaction mixture yielding proteinoid. Conditions same as in table 2.

each amino acid varies according to its individual pattern. The optimal interactions involve proteinoids containing 20% or 30% lysine. At each of these levels, the turbidity generated by thermal proteinoids with polynucleotides exhibits a codonic relationship. At other levels, the turbidity displays other rank orders.

In these ways, and others, we observe that preferences for reaction with polynucleotides to form nucleoproteinoid microparticles are a function of a number of factors including content of specific amino acids. For some time after this study began, attempts to observe specificities in reaction of homopolynucleotides with amino acids or with aminoacyl adenylates were almost fruitless, comparable to what was reported from earlier studies [49]. What can now be observed is that strongly manifested specific reactions occur from polymer–polymer reaction. Whereas binding of single amino acids is weak, binding of polymers of several or many amino acid residues by other polymers is strong.

Also, the finding of many kinds of control of specificity suggests that evolution had a number of possibilities from which to select the mechanism and correspondences of the contemporary genetic code. The elaborate nature of the contemporary genetic code and apparatus can thus be understood as a means for imposing constraints in a device that is not yet fully evolved. We may also visualize in the overall evolution how contemporary residue-by-residue mechanisms replaced earlier ordering mechanisms. Again, the necessity for an intermediate primitive stage is rooted in the fact that, *so far*, this is the only way experimentally demonstrated to solve the problems of feedback mentioned earlier. Also, we visualize that the evolving cell adapted to progressively increasing independence of the environment from which the first ancestor emerged. An elaborate coding mechanism would have protected this biochemical separation from the chemical activities of the environment.

When viewed against the background of experiments on selective reaction of basic homopolyamino acids with mononucleotides [27,47], the experiments reported here help to establish a basis for understanding how information originally flowed from proteins to nucleic acids. The emergent picture suggests information from amino acids → proteinoids (prebiotic histones especially; see Dayhoff [4], Block [2]) → polynucleotides → contemporary proteins.

Acknowledgement

This work was aided by Grant NGR 10-007-008 of the National Aeronautics and Space Administration and by the General Foods Corporation. Contribution No. 154 of the Institute of Molecular Evolution. We thank Mr. C.R. Windsor for analyses of proteinoids.

References

[1] Berg P., J. Biol. Chem. 233 (1958) 608.
[2] Block R.J., J. Biol. Chem. 105 (1934) 663.
[3] Crick F.H.C., J. Mol. Biol. 38 (1968) 367.
[4] Dayhoff M.O., Atlas of Protein Sequence and Structure (National Biomedical Res. Foundation, Silver Springs, Madison, 1969) p. 42.
[5] Dixon M. and E.C. Webb, Enzymes (Academic Press, New York, 1958) p. 666.
[6] Fox S.W., Am. Scientist 108 (1956) 347.
[7] Fox S.W., Bull. Am. Inst. Biol. Sciences 9 (1959) 20.
[8] Fox S.W., Science 132 (1960) 200.
[9] Fox S.W., Bioscience 14 (1964) 13.
[10] Fox S.W., Nature 205 (1965) 328.
[11] Fox S.W., J. Sci. Ind. Res. (India) 27 (1968) 267.
[12] Fox S.W., Naturwissenschaften 56 (1969) 1.
[13] Fox S.W., in: Encyclopedia of Polymer Science and Technology 9 (1969) 284.
[14] Fox S.W., R.J. McCauley, P. O'B. Montgomery, T. Fukushima, K. Harada and C.R. Windsor, in: Physical Principles of Biological Membranes, eds. F. Snell, J. Wolken, G. Iverson and J. Lam (Gordon and Breach, Science Publishers, New York, 1970) p. 417.
[15] Fox S.W., R.J. McCauley and A. Wood, Comp. Biochem. Physiol. 20 (1967) 773.
[16] Fox S.W. and T. Nakashima, Biochim. Biophys. Acta 140 (1967) 155.

[17] Fox S.W., K. Harada and A. Vegotsky, Experientia 15 (1959) 81.
[18] Fox S.W. and T.V. Waehneldt, Biochim. Biophys. Acta 160 (1968) 246.
[19] Fox S.W., C.-T. Wang, T.V. Waehneldt, T. Nakashima, G. Krampitz, T. Hayakawa and K. Harada, in: Peptides: Chemistry and Biochemistry, eds. B. Weinstein and S. Lande (Marcel Dekker, New York, 1970) pp. 499–527.
[20] Gevers W., H. Kleinkauf and F. Lipmann, Proc. Natl. Acad. Sci. U.S. 63 (1969) 1335.
[21] Gilman H., Organic Synthesis, Coll. Vol. 1 (1932); A.H. Blatt, Coll. Vol. 2 (1943); E.C. Horning, Coll. Vol. 3 (1955); N. Rabjohn, Coll. Vol. 4 (1963) (John Wiley, New York).
[22] Haldane J.B.S., The origin of life, The Rationalist Annual (1929) 1969.
[23] Harada K. and S.W. Fox, Nature 201 (1964) 335.
[24] Haurowitz F., The Chemistry and Function of Proteins, 2nd ed. (Academic Press, New York, 1963) p. 424.
[25] Jukes T.H., Molecules and Evolution (Columbia University Press, New York, 1966) p. 187.
[26] Krampitz G. and S.W. Fox, Proc. Natl. Acad. Sci. U.S. 62 (1969) 399.
[27] Lacey J.C., Jr. and K.M. Pruitt, Nature 223 (1969) 799.
[28] Lederberg J., Current Topics Developmental Biol. 7 (1966) ix.
[29] Lemmon R.M., Chem. Rev. 70 (1970) 95.
[30] Moody P.A., Introduction to Evolution, 3rd ed. (Harper and Row, New York, 1970).
[31] Muller H.J., Perspectives Biol. Med. 5 (1961) 1.
[32] Nakashima T., J.C. Lacey, Jr., J. Jungck and S.W. Fox, Naturwissenschaften 57 (1970) 67.
[33] Novelli G.D., Ann. Rev. Biochem. 36 (1967) 449.
[34] Oparin A.I., The Origin of Life on Earth (Academic Press, New York, 1957).
[35] Orgel L., J. Mol. Biol. 38 (1968) 381.
[36] Paecht-Horowitz M. and A. Katchalsky, Biochim. Biophys. Acta 140 (1967) 14.
[37] Rohlfing D.L. and S.W. Fox, Advan. Catalysis 20 (1969) 373.
[38] Salas J. and F.J. Bollum, J. Biol. Chem. 244 (1969) 1152.
[39] Schwartz A.W. and S.W. Fox, Biochim. Biophys. Acta 134 (1967) 9.
[40] Snyder L.E., D. Buhl, B. Zuckerman and P. Palmer, Phys. Rev. Letters 22 (1969) 679.
[41] Taube M., S.Z. Zdrojewski, K. Samochocka and K. Jezierska, Angew. Chem. Intern Ed. Eng. 6 (1967) 247.
[42] Van Niel C.B., in: The Microbe's Contribution to Biology, eds. A.J. Kluyver and C.B. Van Niel (Academic Press, New York, 1956) p. 155.
[43] Vegotsky A. and S.W. Fox, in: Comparative Biochemistry, Vol. 4 (Academic Press, New York, 1962) p. 185.
[44] Waehneldt T.V. and S.W. Fox, Biochim. Biophys. Acta 160 (1968) 239.
[45] Weber A.L., Federation Proc. Abstr. 29 (1970) 939.
[46] Williamson A.R., Biochem. J. 111 (1969) 515.
[47] Woese C.R., Proc. Natl. Acad. Sci. U.S. 59 (1968) 110.
[48] Yuki A. and S.W. Fox, Biochem. Biophys. Res. Commun. 36 (1969) 657.
[49] Zubay G. and P. Doty, Biochim. Biophys. Acta 29 (1958) 47.

Chemical Evolution and the Origin of Life, eds. R. Buvet and C. Ponnamperuma

RECENT PROGRESS IN THE STUDY AND ABIOTIC PRODUCTION OF CATALYTICALLY ACTIVE POLYMERS OF α-AMINO ACIDS

Klaus DOSE and Laila ZAKI

Max-Planck-Institut für Biophysik
and
University of Frankfurt/M; Kennedy-Allee 70
6 Frankfurt/M, Germany

All catalytically active polymers which are discussed in this lecture are thermal polymers of amino acids. As has been suggested by Fox [5], these polymers are generally produced by heating dry amino acids above the boiling point of water, for an instance, for a few hours at 180 °C. The temperature may be lowered to 100 °C and even less. But then the reaction time has to be increased accordingly. Addition of polyphosphates, polyphosphoric acid and other dehydrating agents has favorable effects. This applies particularly for the use of lower temperatures. The polymers which result e.g. in the presence of polyphosphates [2] are extremely low in phosphate. The polymers which are obtained after thorough dialysis (3 days) and lyophilization resemble contemporary proteins in so many properties that in fact a clear distinction is problematical in numerous instances [7]. The name *proteinoids* [10] given to them reflects this relationship.

Earlier results

So far, the catalytic activities listed in table 1 have been reported.

Table 2 summarizes the reports on the hydrolytic activities of thermal polyamino acids. *p*-Nitrophenyl acetate and *p*-nitrophenyl phosphate, which are often used for hydrolytic tests, are not biological substrates however. But sufficient evidence has been accumulated to indicate that the polymers themselves are catalyzing the hydrolysis of these two substrates, whereas in the case of the ATPase activity the presence of Zn ions is required, the catalytic activity being largely ascribed to the effect of these metal ions.

Table 1
Catalytic activities so far ascribed to thermal polymers of α-amino acids.

A) Hydrolytic reactions
B) Decarboxylation reactions
C) Transfer of amino groups
  (a) Reductive amination of α-keto acids
  (b) Oxidative deamination of α-amino acids
D) Catalatic and peroxidatic reactions
E) Oxidase reactions
  glucose oxidase (?)

The decarboxylation reactions which have been found to be promoted by thermal polyamino acids are summarized, with comments, in table 3. The decarboxylation of glucose is obviously not truly catalytic, because glucose first is oxidized to glucuronic acid by the polymer. The promotion of both the decarboxylation of glucuronic and pyruvic acids appears to take place at an extremely slow rate. A high polymer/substrate ratio is required to produce measurable amounts of carbon dioxide. Also for this reason a strictly catalytic activity is questionable in both cases. The catalytic decarboxylation of oxalacetic acid, on the other hand, is well established by many data. The activity (0.3 μmole CO_2/mg/min) is comparable with that of biological decarboxylases (30–900 μmoles CO_2/mg/min). In some cases a photosensitized decarboxylation has been found to compete with the polymer promoted decarboxylation of various keto acids. A yellow pigment, which can be separated from the polymer, is the sensitizer. Decarboxylation reactions have been found to be up to 80 times faster in the light than in the dark.

The amination and deamination reactions which reportedly are promoted by proteinoids are summarized in table 4. In all cases the substrate must be reduced or oxidized by the polymer. In this respect the reactions are not strictly catalytic. So far the investigators have not presented sufficient evidence to indicate that the reactions reported can be compared with transaminations catalyzed by contemporary transaminases. The data reported by Krampitz and his associates indicate that the reductive amination of α-ketoglutaric acid is not the reverse reaction to the oxidative deamination of glutamic acid. Proteinoids with transaminase activities like those indigenous to contemporary glutamic acid–pyruvic acid transaminase have not been reported. Particularly the data on the reductive amination of pyruvic, phenylpyruvic, and oxalacetic acids and the oxidative deamination of glutamic acid are too incomplete to permit an appropriate evaluation.

Table 2
Hydrolytic reactions.

Substrates	Authors	Active AA in polymer	Activity (max) (μmoles/mg/min)	Other results and comments
p-Nitrophenyl acetate	Noguchi and Saito [21]			
	Rohlfing and Fox [26]	His, Lys, Asp-imide	(min–hr) [a]	1.3 to 10 times more active than free His. Inactivation by heat, Michaelis–Menten kinetics
	Usdin, Mitz, and Killos [29]	Glu, Asp, Lys, His, Ser, Thr	1.5×10^{-3} (min–hr)	(Chymotrypsin: 3×10^{-2} μmoles/mg/min). Inhibition by DFP (reversible)
p-Nitrophenyl phosphate	Oshima [22]	His, other basic and neutral AA cooperative Asp-imide	2×10^{-3} (hr–days)	Free AA inactive. Inhibition by arsenate, Cu^{2+} 10% degradation by pronase → 30% inhibition. Heat inhibition, Michaelis–Menten kinetics
ATP	Fox and Joseph [6]	content in Zn^{2+} essent.	(hr–days)	Zn active in ptd microspheres
	Durant and Fox [3]			Promotion of Zn effects in Lys-rich proteinoids
	Tetas and Löwenstein [28]	No polymer		Zn^{2+} as active as Zn-ptd microspheres

[a] (min–hr) or (hr–days) indicate duration of the particular experiment.

Table 3
Decarboxylations.

Substrate	Authors	Active AA in polymer	Activity	Other results and comments
Glucose or glucuronic acid	Fox and Krampitz [7]	Lys	Only counts of $^{14}CO_2$ (react. time: hr–day)	Anaerobic conditions. ATP stimulates two-fold. Glucose oxidation non-catalytic.
	Hardebeck and Fox [8]			Large polymer/substrate ratio. Strictly catalytic? No Michaelis–Menten kinetics.
Pyruvic acid	Krampitz and Hardebeck [18]	Glu, Thr	Only counts of $^{14}CO_2$ (react. time: hr–days)	Activity declines with reaction time. Strictly catalytic?
	Hardebeck, Krampitz and Wulf [9]	Ileu inhibitory		Heating reduces activity. Oxalacetic acid not decarboxylated.
$HO{-}\overset{O}{\overset{\parallel}{C}}{-}\overset{O}{\overset{\parallel}{C}}{-}CH_2{-}\overset{O}{\overset{\parallel}{C}}{-}OH$	Rohlfing [26]	Lys	0.3 μmole/mg/min (min–hr) [a]	Pseudo-first order kinetics. Thermal polylysine 10 times more active than free Lys. (Enzymes 30–900 μmoles/mg/min). No measurable decarboxylation of pyruvic, malic, malonic, α-ketoglutaric, glucuronic, oxalic and aspartic acid.
Pyruvic, Glyoxalic, Glucuronic acid	Weber, Wood, Hardebeck and Fox [30]	Yellow pigment	Photosensitized reaction (hr–days) [a]	Up to 80 times the dark reaction.

[a] Reaction time.

Table 4
Amination and deamination.

Substrate	Authors	Active AA in polymer	Activity (μmoles/mg/min)	Comments and other results
α-Ketoglutaric acid with urea and similar NH_2-donors	Krampitz, Diehl, and Nakashima [16]	Lys, and other basic AA;	2×10^{-3} (hr–day) [a]	Cu^{2+} required, reducing agent (?) 73% L and 27% D Glu (by Lys-rich ptd)
	Krampitz, Baars-Diehl, Haas, and Nakashima [15]	Neutral and acidic ptds not active		Polyanhydrolysine from Leuchs' anhydride inactive. Michaelis–Menten kinetics. Acylation inactivates. No heat inactivation
Pyruvic, phenyl-pyruvic and oxalacetic acid + NH_2-donor	Krampitz [14]		(hr) [a]	
Glutamic acid ↓ α-ketoglutaric acid	Krampitz, Haas, and Baars-Diehl [17]		(hr) [a]	Cu^{+} required. Oxidizing agent? Strictly catalytic? Part of transaminase reaction.

[a] Reaction time.

Recent results on peroxidatic and catalatic activities

We have added to the list of enzyme-like activities which are exhibited by proteinoids or thermal polymers of α-amino acids, the catalase- and peroxidase-like activities of hemoproteinoids. These proteinoids which carry the heme as a prosthetic group are most conveniently synthesized by heating an appropriate mixture of amino acids with 0.25–2% hemin (the chloride of hematin). When only 0.25% hemin is present in a mixture containing lysine (80%) and all the other protein amino acids, a largely homogenous polymer containing about 2.5% heme will be produced by heating the materials for 2 hr at 180 °C. Under these conditions the molecular weight of the polymer is about 20,000. With only 0.25% hemin among the reactants an average of one heme residue is attached to one polymeric molecule.

The possible prebiotic formation of porphyrins appears well established by the work of Hodgson and Baker [11], Hodgson and Ponnamperuma [12], Szutka [27], Krasnovskii and Umrikhina [19]. The amino acid composition of the polymer can be controlled by the amino acid composition of the reactant mixture (table 5). The resulting polymers show only a limited degree of heterogeneity. The lysine-rich hemoproteinoids (e.g. HP 83a) which exhibit the highest peroxidatic activity must be regarded as homogenous with respect to gel-filtration and gel-electrophoresis. Fig. 1 shows that HP 83a appears in a sharp single fraction after gel-electrophoresis at pH 8.3 (the isoelectric point of this fraction is about 8). The UV absorption spectrum in the range of the Soret band may be compared with that of peroxidase (see fig. 2). The shoulder in the absorption spectrum of the hemoproteinoids disappears in the case of hemoproteinoids which contain only one heme nucleus per molecule. The kind of interaction of the π-electrons of the Fe-porphyrin nucleus with the proteinoid part of the molecule may be related to the interaction found in peroxidase. In the case of the hemoproteinoids, however, the bonds between heme and proteinoid are very stable. An amide-type link between the carboxylic groups of the heme nucleus and the ϵ-amino groups of lysine appears very likely. Gel-filtration on Sephadex G-75 indicates that the molecular weight of the lysine-rich hemoproteinoids, such as HP 83a, is about 18,000. The molecular weights found for the other hemoproteinoids synthesized decrease with the content of lysine. Simultaneously their heterogeneity increases; e.g., hemoproteinoid 47a, which has an isoelectric point in the neutral pH range, consists of at least three different molecular species with molecular weights around 10,000.

The catalase- and peroxidase-like activities of these polymers are defined

Table 5
Amino acid analysis of various hemoproteinoids.

Amino acid	% (weight) 83a		87b		79a		47a		73a	
	reac [a]	prod [a]	reac	prod	reac	prod	reac	prod	reac	prod
Asp and Asn	1.0	1.5	1.0	1.2	1.0	3.0	15.4	17.5	41	40
Thr	0.9	0.3	0.9	tr [a]	0.9	tr	1.3	1.3	0.9	tr
Ser	0.8	0.3	0.8	tr	0.8	tr	8.1	1.7	0.8	tr
Glu and Gln	1.1	1.6	1.1	tr	1.1	7.8	8.6	13.1	41	15.7
Pro	0.9	tr	0.9	tr	0.9	tr	1.2	tr	0.9	tr
Gly	0.6	1.6	0.6	2.9	0.6	7.4	7.8	22.5	0.5	1.1
Ala	0.7	1.0	0.7	1.6	0.7	5.8	0.9	8.8	0.7	1.9
Val	0.9	tr	0.9	tr	0.9	4.6	1.2	tr	0.9	tr
1/2 $(Cys)_2$	1.9	tr	1.9	tr	1.9	3.7	21.6	2.1	1.8	1.8
Met	1.2	tr	1.2	tr	1.2	4.7	1.6	tr	1.2	13 ?
Ileu	1.0	tr	1.0	0.6	1.0	5.2	1.4	tr	1.0	1.2
Leu	1.0	0.7	1.0	1.4	1.0	5.5	1.4	2.1	1.0	2.0
Tyr	1.4	1.5	1.4	2.3	1.4	tr	1.9	tr	1.4	1.8
Phe	1.3	tr	1.3	3.0	1.3	7.5	1.7	3.9	1.3	2.6
Lys	81.0	89.5	1.1	15.8	1.1	123	8.6	23.0	1.1	19
His	1.6	0.8	1.2	5.8	81.0	26.8	15.7	6.4	1.2	1.0
Try	–	–	1.5	tr	1.5	tr	–	–	1.6	tr
Arg	1.3	tr	81.1	65.6	1.3	6.1	1.8	2.8	1.3	1.1

[a] reac = reactants; prod = final products; tr = traces.

by equations (1) and (2):

$$H_2O_2 + H_2O_2 \xrightarrow[\Delta \text{ catalase}]{} 2\,H_2O + O_2 \qquad (1)$$

$$H_2O_2 + H_2\text{-donor} \xrightarrow[\Delta \text{ peroxidase}]{} 2\,H_2O + \text{donor} \qquad (2)$$

The broad spectrum of substrate specificities as found for the hemoproteinoids may be compared with that of the biogenous heme peroxidases. Guaiacol is conveniently used as organic hydrogen-donor. But as shown in table 7, NADH is oxidized significantly faster than guaiacol. Typical reaction curves obtained by the guaiacol test are shown in fig. 3. In both the cases the same concentration of heme was present in the system. The heme bound to the proteinoid is up to 50 times more active than the free heme. The reaction is completed within a few minutes. It may be started again by the addition of

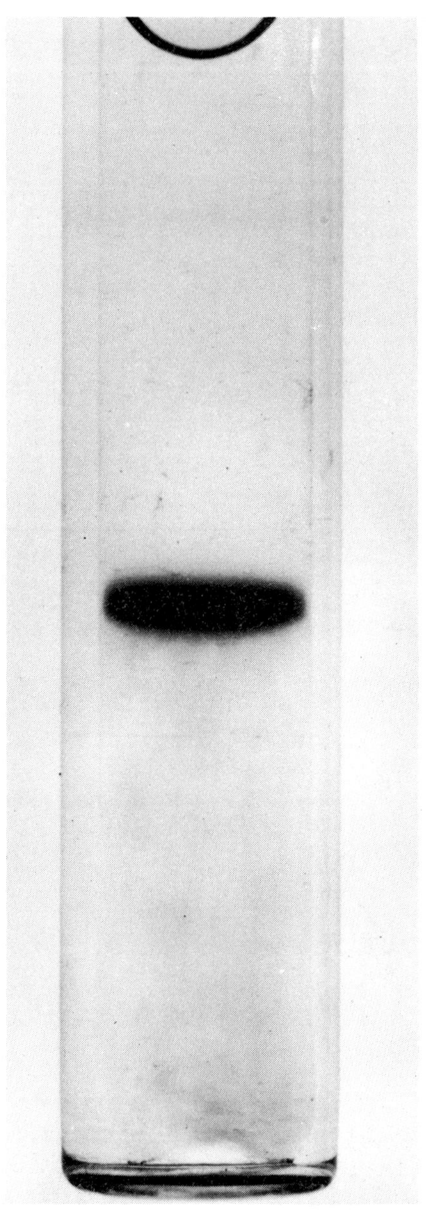

Fig. 1. Disc electrophoresis (polyacrylamide). Hemoproteinoid 83a; tris-gly buffer pH 8.3, 5 mA per tube (Shandon apparatus), staining with amido black 10B.

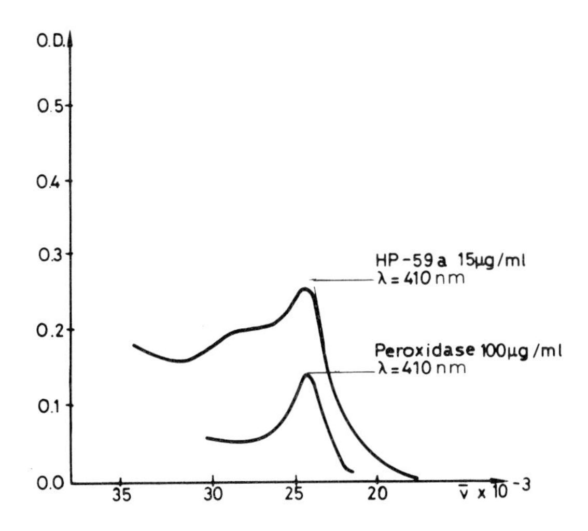

Fig. 2. Position of Soretband in hemoproteinoid HP 59a and horseradish peroxidase. 25°C; pH 6.8; 0.01 M phosphate buffer.

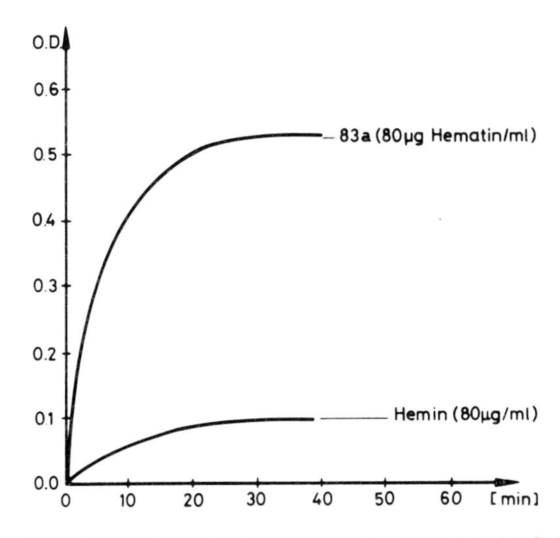

Fig. 3. Peroxidase activity of porphyrin proteinoid 83a and hemin. Substrates: H_2O_2 and guaiacol; 25 °C; pH 6.8; change of optical density measured at λ = 436 nm.

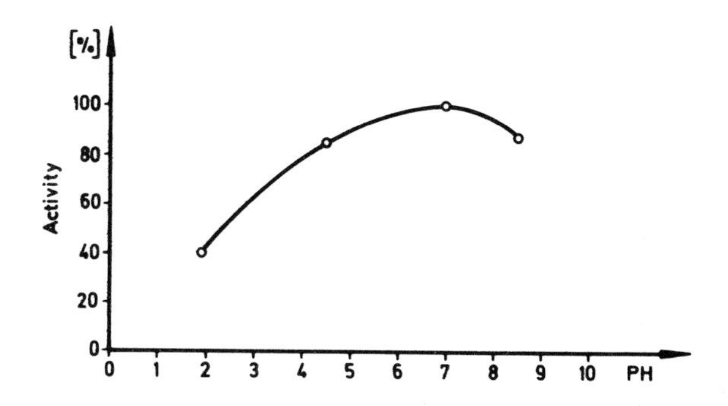

Fig. 4. pH-activity diagram for peroxidatic activity of hemoproteinoid 73b.

more H_2O_2. The reaction can proceed even at very low concentrations of H_2O_2 (10^{-6} M). Thus a high affinity of H_2O_2 to the hemoproteinoid is indicated. Fig. 4 shows the pH-activity diagram of the lysine-rich hemoproteinoid 73b. The isoelectric point of this proteinoid is close to pH 8.5; nevertheless, the activity optimum is in the neutral pH region as is the case for horseradish peroxidase. Related results have been obtained for other hemoproteinoids.

Table 6 shows a number of characteristic data on the peroxidase- and catalase-like properties of hemoproteinoids. A few additional comments may be made: The catalatic activity of the heme nucleus is always decreased when the heme is incorporated into the proteinoid. This increase is particularly striking in the case of polymer 68b which is relatively rich in phenylalanine. This property of phenylalanine has been observed earlier, though in different context [20]. The peroxidatic activity of hemoproteinoids can be increased by heating the polymer in aqueous solution for several minutes at 100 °C at neutral pH (as shown for HP 73b). This effect may be related to the hydrolytic splitting of some aspartoylimide bonds [24]. The results on the catalatic and peroxidatic activity of hemoproteinoids are summarized in table 7. More detailed data will be published elsewhere. The proteinoids lacking the heme group show no peroxidase activities (see e.g. no. 63, bottom of table 6). But if hematin is allowed to interact with such a polymer an increase in peroxidatic activity results.

Table 6
Catalatic and peroxidatic activity of various compounds containing the hematin nucleus.

Sample	Catalytic activity [a] (pH 6.8)		Composition of materials	
	Peroxidase (25°C)	Catalase (37°C)	Hematin % (weight)	Predominance of amino acids % (weight) of total amino acids
Hemin (Fe^{3+}) Cl, hematin	1.0 [b]	1.0 [c]	100	
83a	30 [b]	0.5 [c]	15	90 Lys, see table 5
83b	50	0.5	2	90 Lys, like 83a
73b	25	0.5	15	80 Lys
73b (after 60 min at 100°C)	50	0.5	15	80 Lys
87b	20	0.5	15	16 Lys, 6 His, 66 Arg, see table 5
79a	10	0.5	12	12 Lys, 27 His, 6 Arg, see table 5
79b	1	0.3	15	No basic amino acids
73a	10	0.2	20	66% Asp–Glu, 19 Lys, see table 5
47a	5	0.5	20	Neutral proteinoid, 23 Lys, see table 5
68b	15	0.05	15	Phe:Lys = 1:1
87a	30	0.5	10	95 Lys (thermal polylys)
91	15	0.5	6	Hemin heated with thermal poly-Lys
92	1	0.5	2.5	Hematin heated with Leuchs' poly-Lys
63	0	0	0	Neutral proteinoid
63 + 10% heme	4.4	1.0	10	Neutral proteinoid

a Relative to porphyrin content. Hematin = 1 unit of activity.
b Peroxidase activity (guaiacol test according to Chance [27]), 0.03 units/mg: purpurogallin number per mg approximately 1.6.
c Catalatic activity (turnover number; mole substrate per mole Fe) = 0.05 sec^{-1}.

Table 7
Summary of catalatic and peroxidatic properties of hemoproteinoids.

Substrate	Authors	Active AA in polymer	Activity (μmoles/mg/ml)	Other results and comments
H_2O_2 (catalatic react.)	Dose and Zaki [1]	Phe inhibitory	1 (sec–min)	50% less effective than free hematin. Liver catalase 10^5 times as effective. Pseudo-first order Michaelis–Menten kinetics. Inhibition by complexing agents.
H_2O_2 and H-donor	Dose and Zaki [1]	Lys and other basic AA cooperate with thermal by-product. Asp-imide inhibitory (?)	0.2 for guaiacol 2.2 for NADH (sec–min)	Up to 50 times more effective than hematin (guaiacol as substrate) Horseradish peroxidase 10^3 times as effective (NADH oxidation). Inhibition by complexing agents. 100% increase of activity after heating. Hematin heated with Leuchs' poly-Lys only as active as free hematin. Pseudo-first order Michaelis–Menten kinetics

Discussion

The decrease of the catalatic activity of hematin after incorporation into a proteinoid contrasts the strong increase of the peroxidatic activity particularly in the case of the lysine-rich proteinoids. In an anaerobic environment this effect would have been beneficial for any primitive biotic or prebiotic system which had a need for an oxidizing agent; a mere catalatic activity would have just destroyed hydrogen peroxide with the molecular oxygen formed being easily lost in the non-oxidizing atmosphere, whereas primitive peroxidative activity would have provided a way to direct use of hydrogen peroxide for a sequence of oxidation processes within the system. In this respect the oxidation of NADH or related compounds is of particular interest.

Energetically, the dehydrogenation of NADH in a peroxidatic reaction is a relatively inefficient reaction if it is compared with other redox reactions of contemporary biological systems [23]. This question of efficiency of energy conversion must be regarded under less critical aspects, however, in the case of primitive biotic systems.

The need for an oxidizing system in organisms which had to exist heterotrophically in an anaerobic environment is indicated by the kinds of substrates (food) available. These substrates probably originated abiotically from a non-oxidizing or reducing atmosphere. Both thermodynamic calculations [4] and experimental results [13] clearly show that the formation of the (oxygen-rich) carbohydrate-type aliphatic compound was considerably less favored than the formation of oxygen-poor aliphatic (amino) acids, other aliphatic hydrocarbons and aromatic compounds. These oxygen-poor compounds can only be metabolized efficiently if they are degraded oxidatively. In primitive times peroxidase systems would have permitted such an oxidation in the absence of free oxygen.

Acknowledgements

The authors are indebted to Dr. S.W. Fox for many discussions, to Dr. G. Horneck for her microbiological tests, to Miss C. Brand for carrying out the amino acid analyses, and to Mr. E. Friedrich for his help in synthesizing most of the hemoproteinoids. The sponsorship of Professor Dr. R. Schlögl, the acting director of the Max-Planck-Institut für Biophysik, and the Max-Planck-Gesellschaft with respect to providing funds and a fellowship to one of us (L.Z.) is also gratefully acknowledged.

References

[1] Dose K. and L. Zaki (1970), Z. Naturforsch., in press.
[2] Dose K. et al. (1970) in preparation.
[3] Durant D.H. and S.W. Fox, Federation Proc. 25 (1966) 342.
[4] Eck R.V., E.R. Lippincott, M.O. Dayhoff and Y.T. Pratt, Science 153 (1966) 628.
[5] Fox S.W., Encyclopedia Polymer Sci. Technol. 9 (1968) 284.
[6] Fox S.W. and D. Joseph, in: The Origins of Prebiological Systems, ed. S.W. Fox (Academic Press, New York, 1965) p. 371.
[7] Fox S.W. and G. Krampitz, Nature 203 (1964) 1362.
[8] Hardebeck H.G. and S.W. Fox, 3rd Annual Report Institute Mol. Evolution, Coral Gables, Forida, 1967).
[9] Hardebeck H.G., G. Krampitz and L. Wulf, Arch. Biochem. Biophys. 123 (1968) 72.
[10] Hayakawa T., C.R. Windsor and S.W. Fox, Arch. Biochem. Biophys. 118 (1967) 265.
[11] Hodgson G.W. and B.C. Baker, Nature 216 (1967) 29.
[12] Hodgson G.W. and C. Ponnamperuma, Proc. Natl. Acad. Sci. U.S. 59 (1968) 22.
[13] Keosian J., The Origin of Life, 2nd ed. (Reinhold, New York, 1968).
[14] Krampitz G., 156th Natl. Meeting Am. Chem. Soc., Atlantic City (1968), Abstr. Papers, Div. Biol. Chem. 087.
[15] Krampitz G., S. Baars-Diehl, W. Haas and T. Nakashima, Experientia 24 (1968) 140.
[16] Krampitz G., S. Diehl and T. Nakashima, Naturwissenschaften 54 (1967) 516.
[17] Krampitz G., W. Haas and S. Baars-Diehl, Naturwissenschaften 55 (1968) 345.
[18] Krampitz G. and H. Hardebeck, Naturwissenschaften 53 (1966) 81.
[19] Krasnovskii A.A. and A.V. Umrikhina, Dokl. Akad. Nauk. SSSR 155 (1964) 69.
[20] Lautsch W., W. Broser, E. Höfling, H. Gnichtel, E. Schröder, R. Krüger, J. Woldt, G. Schulz, R. Wiechert, W. Bandel, G. Kurth, H.-H. Kraege, W. Gehrmann, K. Prater, G. Parsiegla, R. Pasedag and W. Hunger, Kolloid-Z. 144 (1955) 82.
[21] Noguchi J. and T. Saito, in: Polyamino Acids, Polypeptides and Proteins, ed. M.A. Stahmann (Univ. of Wisconsin Press, Madison, Wisconsin, 1962) p. 313.
[22] Oshima T., Arch. Biochem. Biophys. 126 (1968) 478.
[23] Paul K.G., in: The Enzymes, eds. P.D. Boyer, H. Lardy and K. Myrbäck (Academic Press, New York, 1963) p. 263.
[24] Rohlfing D.L., Ph.D. dissertation, Florida State University (1964).
[25] Rohlfing D.L., Nature 216 (1967) 657.
[26] Rohlfing D.L. and S.W. Fox, Arch. Biochem. Biophys. 118 (1967) 122.
[27] Szutka A., Nature 202 (1964) 1231.
[28] Tetas M. and J.M. Lowenstein, Biochemistry 2 (1963) 350.
[29] Usdin V.R., M.A. Mitz and P.J. Killos, Arch. Biochem. Biophys. 122 (1968) 258.
[30] Weber A.L., A. Wood, H.G. Hardebeck and S.W. Fox, Federation Proc. 27 (1968) 830.

PART V

PHOTOCHEMICAL PROCESSES

Chemical Evolution and the Origin of Life, eds. R. Buvet and C. Ponnamperuma
© 1971, North-Holland Publishing Company

THE MODELS OF THE EVOLUTION OF PHOTOCHEMICAL ELECTRON TRANSFER

A.A. KRASNOVSKY
A.N. Bakh Institute of Biochemistry,
USSR Academy of Sciences, Moscow, USSR

The work of the photosynthetic electron transfer chain underlies the mode of light energy conversion in photosynthesis which sustains life on our planet.

The comparative biochemical approach facilitates the study of photosynthesis in contemporary photoautotrophic organisms, among them the most ancient forms – anaerobic photosynthetic bacteria.

All contemporary photosynthetic organisms possess an elaborate electron transfer chain with a pigment system and a set of biocatalysts arranged in the lammellar structure of chromatophores and chloroplasts. There are no longer any primitive photoautotrophic organisms on the earth and the imagination of the scientists is turned to other sources of information hidden in the fields of paleontology and exobiology, and to the study of models imitating the path of chemical evolution. By using photochemical models it is possible to present hypothetic stages of photochemical electron transfer evolution. The photoreceptor (photosensitizer) absorbing and transforming the light energy of the sun is an inevitable component of the system under study. At first we turn to inorganic constituents of the earth crust.

Inorganic photoreceptors

Some metal oxides possess photosensitizing activity, among them are white titanium and zinc oxides which are strongly absorbing in near ultraviolet (up to 400 nm), yellow wolframic acid, etc. It is known that titanium and zinc oxides are capable of photosensitizing the oxidation-reduction of some dyes. We revealed that using these photocatalysts it is possible to create an inorganic model of Hill reaction [2].

Under the action of the longwave ultraviolet or violet part of the spectrum the oxides listed above photosensitize the reduction of ferric ions, coupled

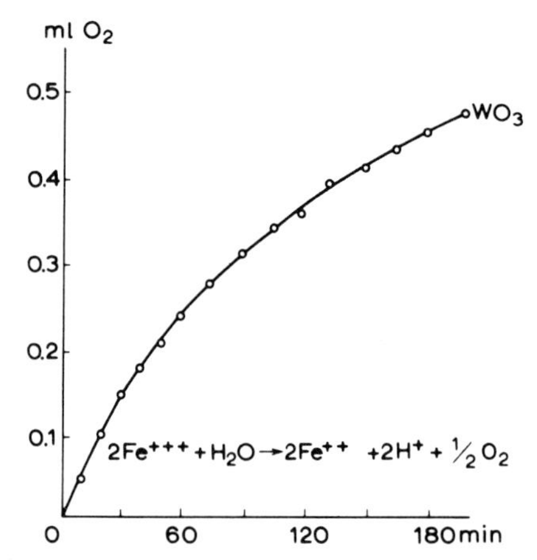

Fig. 1. Wolframic acid or titanium dioxyde sensitized oxygen evolution. 10 ml H_2O, 0.2 g WO_3 or TiO_2 0.05 g $FeNH_4(SO_4)_2$. Light intensity (365–436 nm) 2×10^5 erg/cm^{2}sec.

with the oxygen evolution (fig. 1):

$$2\,Fe^{3+} + H_2O \rightarrow 2\,Fe^{2+} + 2\,H^+ + \tfrac{1}{2}O_2$$

Such evolution of oxygen has a quantum yield near 1%. No organisms which live on our planet now use pure inorganic photoreceptors. But in the course of chemical evolution we cannot exclude such a possibility. For example, titanium is more abundant on the surface of the moon than on the earth.

Now we turn to the formation of abiogenic organic photosensitizers.

Abiogenic synthesis of porphyrins

There is some data on the possible formation of coloured substances in the mixture of gases imitating the primary reducing atmosphere; among them the porphyrins found in recent work in Ponnamperuma's laboratory [1]. Some years ago we studied the conditions of the Rothemund synthesis of porphin in the mixture of formaldehyde and pyrrole, the aid of fluorescence spectra measurements in reaction mixtures being a very sensitive method to see the traces of porphyrins formed [3], (fig. 2). In the absence of air the reaction proceeds slowly, and along with porphin more reduced pigments are formed – chlorin and probably bacteriochlorin. In the presence of air the porphin

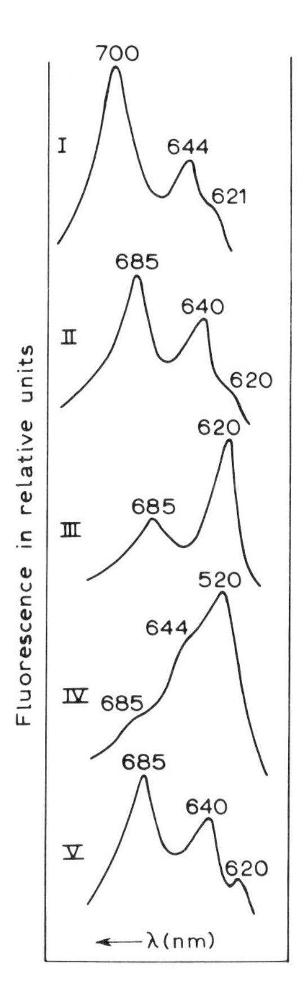

Fig. 2. Fluorescence spectra of porphyrins formed in the mixture of pyrrole and formaldehyde. (I) in methanol, vacuo; (II) in methanol, aerobic; (III) 50% aqueous methanol, O_2; (IV) as III + ZnO; (V) methanol, SiO_2, O_2.

formation proceeds more quickly. In the presence of silica, alumina or titanium dioxide, representing the main components of the earth's crust, the porphin fluorescence appears very quickly (table 1). (In the presence of zinc oxide the zinc complexes of porphin are probably formed.) In some cases the red fluorescence appears if the reacting mixture is left to stand at room temperature.

Porphin isolated from the reaction mixture is an active photosensitizer. We revealed that porphin sensitizes the electron transfer from ferrous ions to an electron acceptor – methyl red, for instance (fig. 3). So porphyrins are

Table 1
The porphyrin formation in the mixture of pyrrole and formaldehyde in 50% aqueous methanol.

Condition of experiment	Appearance of red fluorescence (min)
$-O_2$	200
$+O_2$	30
$+O_2 + SiO_2$	15

capable of using ferrous compounds, probably present in abundance in primary oceans, as electron donors.

In contemporary organisms no inorganic pigments nor free porphyrins are used as sensitizers. The complexes of porphyrins with metals – magnesium and iron – became the most efficient. The possible explanation of this choice in the long process of chemical and biological evolution may be presented in the following simple deduction [4].

The most abundant metals in the earth's crust are: Si, Al, Fe, Ca, K, Na, Mg, Ti. There are no Si, Al and Ti porphyrin complexes in organisms, probably due to the low solubility of Al, Ti and Si compounds. Ca, K and Na complexes of porphyrins easily hydrolyze in aqueous media. So only Fe and Mg complexes might win the competition. Fe and Mg atoms in the middle of the porphyrin ring possess coordinative vacancies perpendicular to the plane of the tetrapyrrolic ring, having additional bonding possibilities. The Fe com-

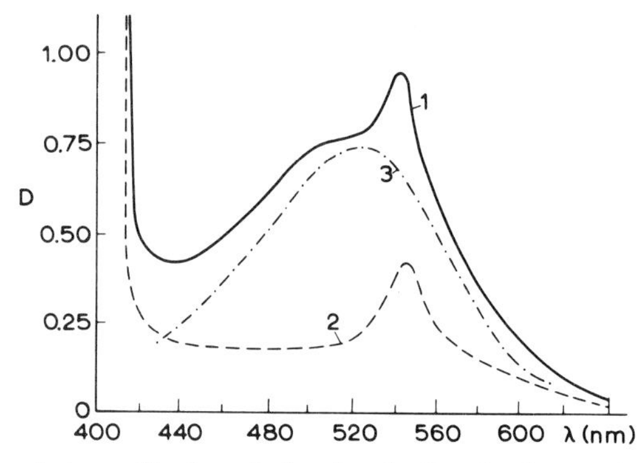

Fig. 3. Porphin photosensitized methyl red reduction by ferrous ions in acid water solution. 1) initial absorption spectra; 2) after illumination; 3) difference spectra (1–2) corresponding to methyl red absorbtion.

plexes are non-fluorescent, photochemically inactive, but very active catalysts capable of catalyzing "dark" electron transfer with a decrease of free energy of the reacting system. In contrast, the magnesium complexes are catalytically inactive, fluorescent, and very active photochemically. Excited by light quanta absorption, they are capable of electron transfer, with a storage of light energy in the reaction products. The excited molecules of chlorophyll and analogs are capable of reversible acceptance of electrons from donor molecules or of reversible donation of electrons to suitable acceptors. The compounds used in photochemical reactions are shown in table 2.

The ability of excited porphyrins and their magnesium complexes to oxido-reductive transformations underline the photosensitizing activity of the pigments in the photochemical electron transfer from electron donors to acceptors.

The primitive organisms could use both catalytic and photochemical modes of substrate activation, but if a chemically active substrate of abiogenic origin were abundant there was probably no need for photochemistry and the primary organisms were heterotrophic according to Oparin's hypothesis; so the Fe complexes were probably used as catalysts along with pyridine nucleotides, flavins etc.

The exhaustion of active substrates inevitably led to the use of photochemistry, and the Mg complexes were incorporated into the primitive catalytic chain. New stages of primary metabolism were created – the coupling of pigment-sensitizer which is active when excited with "non-excited" biocatalyst – the device widely distributed in all contemporary photoautotrophic organisms. We can get models of such primary systems using chlorophyll or its analogs as pigment sensitizer and cytochromes, flavins

Table 2
Electron donors and acceptors which interact with excited chlorophyll.

Reversible photoreduction by electron donors	Reversible photooxidation by electron acceptors
Ascorbic acid	Oxygen
Dienols	Benzoquinones
Cysteine	Ubi- and plastoquinones
Phenylhydrazine	Nitrocompounds
Hydroquinone	Dyes, viologens
$NADH_2$. $NADPH_2$	Ferric compounds
Ferrous compounds	
Tryptophan	

or pyridine nucleotides as catalysts – electron donors or acceptors. When chlorophyll is introduced into an aqueous solution of cytochrome *c* (more active system, in the presence of detergents) and the system is illuminated by red light which is absorbed by chlorophyll, we observe in anaerobic media the photosensitized reduction of cytochrome, and in the presence of oxygen – sensitized oxidation of reduced cytochrome [5,6] (figs. 4–6). When chlorophyll is introduced into a solution of pyridine nucleotides, riboflavin or safranine in the presence of ascorbic acid (reducing media) sensitized photoreduction is observed of the electron acceptors listed above. In the presence of oxygen, the reduced pyridine nucleotides are oxidized under the action of the red light absorbed by chlorophyll (fig. 7).

To summarize: in reducing media, chlorophyll and analogs catalyze photosensitized photoreduction of various electron acceptors and in oxidizing media, oxidation of electron donors (review [7]).

From these model blocks (pigment-biocatalyst) it is possible to reconstruct different types of electron transfer chains.

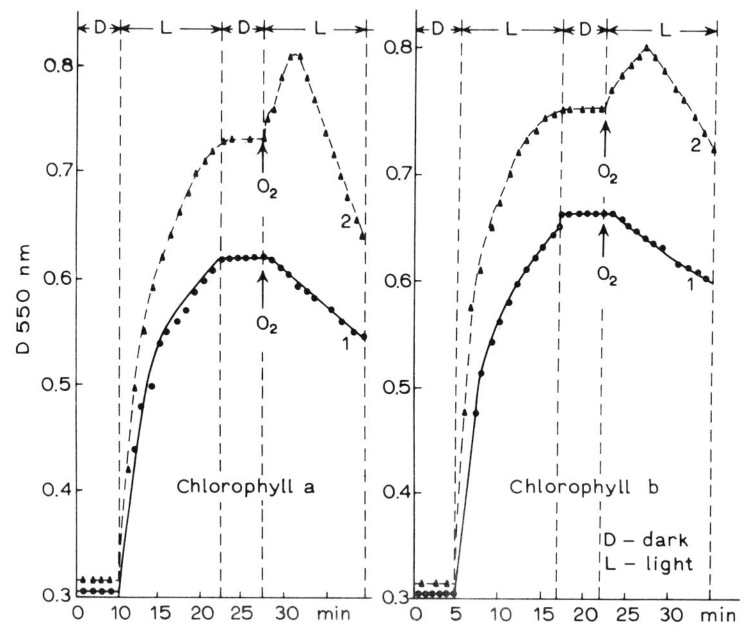

Fig. 4. Kinetics of chlorophyll photosensitized oxidoreduction of cytochrome *c* in 1% aqueous solution of Triton X-100.

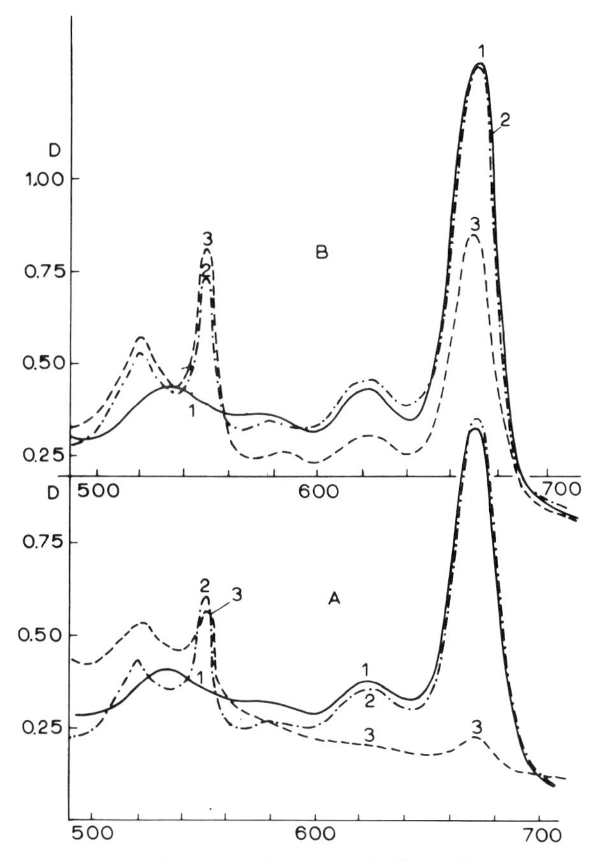

Fig. 5. Absorbtion spectra changes during chlorophyll sensitized oxido-reduction of cytochrome *c*. 1) initial spectrum; 2) after illumination in vacuo; 3) after illumination in air.

The orientation of pigments in primary lipoprotein membranes requires an improvement of pigment structure – the lipophylic phytol or farnesol tail gradually become a part of the pigment molecule.

In the course of evolution the pigment biosynthesis was greatly improved; to use the solar energy more efficiently a large quantity of light-absorbing pigment must be present. The accumulation of pigment in membranes led to the phenomena of pigment aggregation; in the contemporary photoautotrophs the bulk of pigment exists in highly aggregated quasi-crystalline forms. The phenomenon of light energy migration from the bulk of pigment to

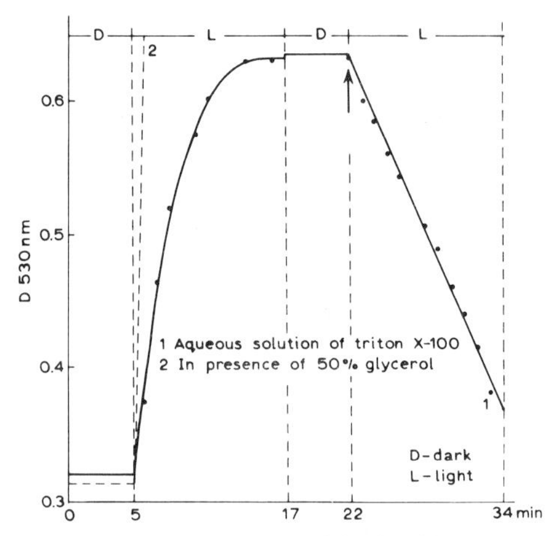

Fig. 6. Kinetics of hematoporphyrin-photosensitized oxido-reduction of cytochrome *c*. 1) in aqueous solution 1% Triton X-100; 2) +50% glycerol.

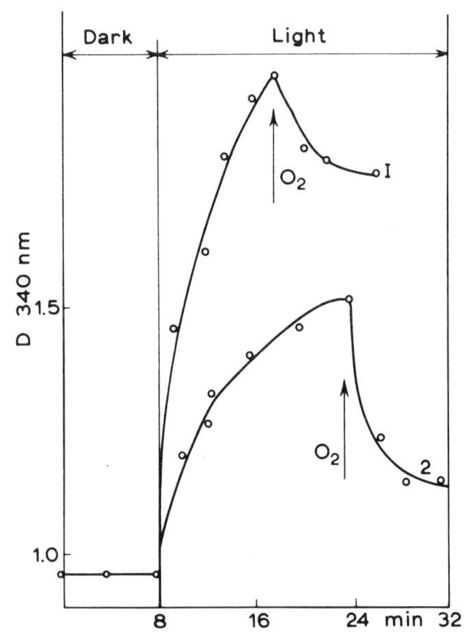

Fig. 7. Kinetics of photosensitized by chlorophyll photoreduction (by ascorbic acid) and photo-oxidation (oxygen) of pyridine nucleotides. 1) NAD; 2) NADP.

pigment molecules in a reactive center working directly in electron transfer was improved. So an elementary block of electron transfer chain consisted of pigment aggregate bound with biocatalytic moiety.
In primary anaerobic environment the most reduced porphyrins were active (bacteriochlorophyll), in the process of gradual aerobiosis the more oxidized porphyrin became abundant in organisms – chlorophylls *a* and *b*. The evolution of effective photosynthetic organisms proceeded by sequence of proper combination of these photocatalytic blocks in the elaborated photosynthetic electron transfer chain. We proposed in 1957 [4] the gradual evolution from "one quantum" to "two quantum" types of electron transfer. It is possible to propose the following primary types of electron transfer chains in anaerobic conditions: one quantum non-cyclic transfer from electron donors (for instance H_2S, organic substances) to electron acceptors with the formation of active substances such as reduced NADP or ferredoxin, used in the carbon dioxide cycle. One quantum cyclic electron transfer leading to formation of ATP from ADP and inorganic phosphate was possible. The combination of these types of "one quantum" reaction probably takes place in various photosynthetic bacteria.

The use of the water molecule as the ultimate electron donor required a more elaborated system of electron transfer where two photocatalytic blocks were involved in the electron transfer chain; this case is denoted in widely used schemes of photosynthesis as photosystems I and II.

References

[1] Hodgson G.W. and C. Ponnamperuma, Proc. Natl. Acad. Sci. U.S. 59 (1968) 22.
[2] Krasnovsky A.A. and G.P. Brin, Dokl. Acad. Nauk SSSR 147 (1962) 655; 168 (1966) 1100.
[3] Krasnovsky A.A. and A.V. Umrikhina, Dokl. Acad. Nauk SSSR 155 (1964) 691.
[4] Krasnovsky A.A., in: The Origin of Life on the Earth, eds. A.I. Oparin et al. (Pergamon Press, London, 1959) p. 606.
[5] Krasnovsky A.A. and K.K. Voinovskaya, Biofisika 1 (1956) 120.
[6] Krasnovsky A.A. and E.S. Mikhailova, Dokl. Acad. Nauk SSSR 185 (1969) 938.
[7] Krasnovsky A.A., A review in: Progress in Photosynthesis Research, Vol. II, Tübingen (1969).

Chemical Evolution and the Origin of Life, eds. R. Buvet and C. Ponnamperuma

POSSIBLE ROLE OF THE ACID–BASE EQUILIBRIUM IN THE EVOLUTION OF THE MECHANISM REGULATING PRIMARY PHOTOCHEMICAL PROCESSES OF PHOTOSYNTHESIS

V.B. EVSTIGNEEV
Institute of Photosynthesis, USSR Academy of Sciences, Moscow, USSR

It is beyond any doubt that the photochemical stage of photosynthesis, i.e. the processes directly involved in the primary transformations of chlorophyll-absorbed light energy, is the key stage whose progress determines the rate of photosynthesis as a whole. For this reason research on the evolution of the photosynthesis regulating mechanisms at the pigment level is of particular importance for a better understanding of the evolution of the photosynthetic apparatus in general.

Obviously, this problem can be solved only if one takes into consideration all the factors which are present in chloroplasts and chromatophores and which may affect the activity of photosynthetic pigments acting as sensitizers of redox reactions of photosynthetic electron transfer chain. Such factors are obviously many but their contribution to the evolution of the photosynthetic apparatus are not necessarily similar. Some of them may have developed at the later stages of the formation of this apparatus, and these apparently include diversity of the pigment aggregate states, structure of membranes surrounding the thylakoid reaction centre, possible feedback effects of subsequent reactions on the preceding ones, etc.

Other factors existed at the very beginning of the appearance of the autotrophic mode of life; they are of major interest for the understanding of the early evolution of the pigment activity regulation mechanism. In our opinion, one of these factors (probably the most significant) is the acid-base equilibrium in the medium, which is determined by the hydrogen ion concentration.

This assumption has been derived from recent experiments which have demonstrated that acidity within thylakoids of chloroplasts changes during photosynthesis and that the proton concentration gradient acts as a factor of energy storing utilized by photophosphorylation [9–14,16,17]. This is also confirmed by the results of many investigations conducted in our laboratory which have shown that the acid–base equilibrium plays a decisive role in the photochemical properties of chlorophyll in vitro [1–8]. Principal results of our recent studies are described below.

According to the hypothesis of Oparin [15] primary concentration of organic substances in coacervate-type formations that served as the initial material for the origin of life on the earth occurred in the aqueous medium of the primary world ocean. It should then be recognized that hydrogen ion concentration must have proceeded during the earliest and subsequent periods of life development under anaerobic and, later, under aerobic conditions. This factor, without doubt, played an important role in the processes that took place in those periods.

Hydrogen ion concentration may have played a decisive role in the very primary processes of the formation of organic matter that developed under the influence of different energy factors. The pH value of the aqueous medium that was favourable for these processes may have been brought about in different locations of the world's ocean by either the occurrence near by (on the bottom or on the shore) of specific rocks or gas discharges from the entrails of the earth which alkalified or acidified the water locally in comparison with the great bulk of the ocean water. These locations could have been the sites where the pH value thus obtained was optimal for both the primary formation of organic matter and the subsequent emergence of coacervate-type formations. The pH factor acquired an all the more important role for the maintenance of primitive metabolism in primary "organism-like" formations that had developed by that time. This was quite natural because a great number of metabolic processes are oxidation–reduction reactions associated with electron and proton transfer.

It appears that the "apparatus" regulating these processes through the control of hydrogen ion concentration emerged at the earliest stages of the appearance of life on the earth and improved concurrently with life development and complication. As long as the primary "living" formations remained extremely simple and to a great extent dependent on the environment, processes that developed in them were completely determined by the properties of the medium, and particularly by hydrogen ion concentration. The level of the concentration, determined as mentioned above by the environmental factors, may have been very important in maintaining the required activity of the necessary processes as well as in insuring the conditions under which the formed organic substances (biopolymers) remained in their native state. The regulation of vital activity processes by means of pH was, therefore, in that period entirely or almost entirely passive from the side of primary organisms.

When evolution resulted in the development of stronger membranes separating the internal medium from the environment, protecting organisms from adverse environmental influences and letting go only vitally required substances, the pH value remained a very important factor in the reaction regulation and maintenance of the proper condition of the most significant substances. Owing to the evolution of internal membranes, organisms may have developed an apparatus which actively controlled hydrogen ion concentration within their various parts, changing the membranes' permeability specifically for these or other ions. The effect of H^+ ions upon various processes may have been both direct and indirect.

The experimental data obtained on model systems with respect to the effects of the medium acid–base equilibrium on chlorophyll photochemical reactions and photosensitizing action suggest that, following the emergence of the autotrophic mode of life based on the utilization of solar energy, the pH value may well have become a factor regulating also processes directly involved in the light energy absorption and storage.

According to modern concepts, the photosensitizing effect of chlorophyll, the basic photosynthetic pigment of green plants, is essentially related to rapidly reversible oxidation reduction transformations of the pigment sensitizer itself. In view of this our hypothesis does not seem unlikely. At the same time the propagation of hydrogen ions as a universal constituent of the aqueous medium, their easy penetration through membrane pores (as they are the smallest ions) and some other properties made them a suitable vehicle for transmitting information or orders from one part of the plant organism to another.

Below is a brief review of the main results of our experiments with model systems: the idea that chlorophyll is a photocatalyst, which during its photosensitizing activity rapidly undergoes intermediate changes associated either with accepting or donating electrons, makes it necessary to ascertain first of all the effect of medium acidity on the capacity of the pigment for photooxidation or photoreduction.

Our numerous experiments have demonstrated that a lowering of the pH in the middle interval, situated not far from the neutral point, stimulates the chlorophyll's ability to serve as a photochemical electron donor, i.e. to undergo photooxidation, whereas a pH increase in this range contributes to its capacity to accept electrons, i.e. to undergo photoreduction. This is illustrated by figs. 1 and 2.

Fig. 1a shows the relationship between the medium pH and the photo-

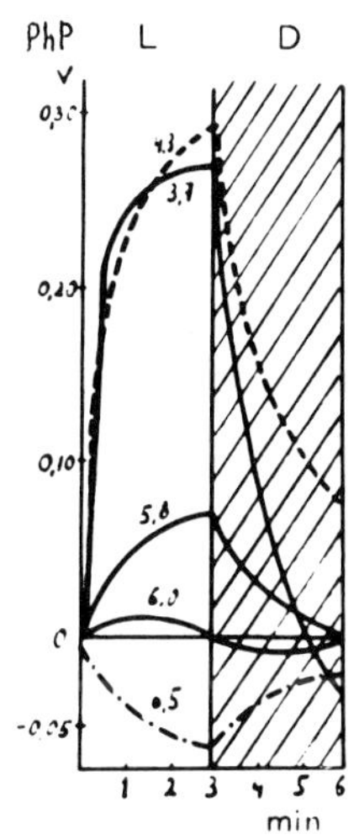

Fig. 1. pH-dependence of photopotential sign and value of chlorophyll *a* (10^{-5} M) solution in alcohol containing quinone (2×10^{-2} M) at room temperature. PhP: photopotential; L: light; D: darkness; figures on the curves: pH values.

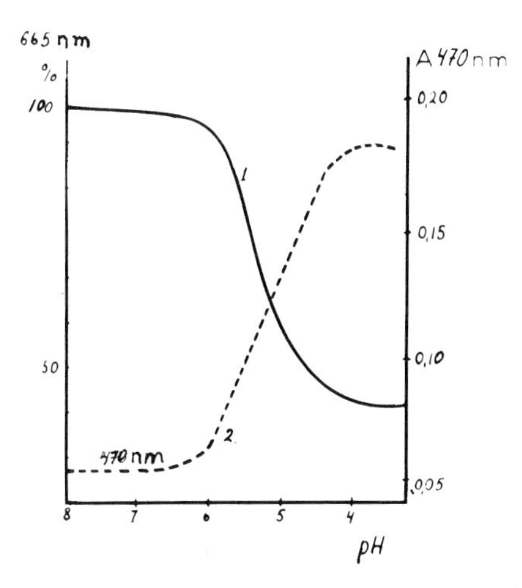

Fig. 1b. pH-dependence of photooxidation rate of chlorophyll *a* (10^{-5} M), by quinone (2×10^{-5} M) in ethanol at -70 °C. 1) The absorption value at 665 nm after 3 min illumination by red light. *Left ordinate*: A 665 nm in percentage of the initial values, 2) The corresponding absorption changes at 470 nm. *Right ordinate*: A 470 nm.

potential generated as chlorophyll a is photooxidized by parabenzoquinone in ethanol. The positive photopotential of several dozen millivolts at pH 4 indicating the occurrence in the solution of a labile electrode-active photooxidized form of the pigment decreases with a pH increase to disappear at pH 5.5–6.0.

Similar relationships between the pH value and the level of chlorophyll photooxidation can be demonstrated spectrophotometrically. Fig. 1b shows that spectral changes typical of pigment photooxidation begin at pH 6 and grow at pH lowering to 4.0.

Fig. 2 presents data concerning the pH effect on photoreduction with phenylhydrazine in ethanol of pheophytin, a chlorophyll analogue, which is easier to photoreduce than chlorophyll itself.

The negative photopotential which is indicative of the formation of a labile electrode-active reduced form of the pigment increases with an acid to mild alkaline shift of the pH value (fig. 2a). The intensity of reductive alterations observed by the absorption spectrum changes in relation to pH in a similar manner (fig. 2b).

Our findings as discussed from the point of view of the above scheme in which chlorophyll acts as a photocatalyst chemically involved in the reaction (this hypothesis seems the most plausible) suggest that acidity of the medium must necessarily influence the rate of photosensitized reactions involving both a donor and an acceptor.

Fig. 3 illustrates and proves this statement. The illustration shows the pH dependence of the rate of the pheophytin photosensitized methyl red reduction with phenylhydrazine in ethanol. The dependence curve has a minimum at pH 5.56 and two rising tails. This shape of the curve may be explained by the fact that in the weak acid pH range the primary photochemical reaction is the pigment-acceptor (methyl red) interaction, which is accelerated at lower pH, and in the neutral and weak alkaline medium it follows the "reduction" path starting as the pigment–donor (phenylhydrazine) interaction which is stimulated by a higher pH.

Of particular interest are the data regarding the pH effect on the rate of oxidation–reduction reactions photosensitized by chlorophyll-type pigments, that were obtained in the experiments carried out with solutions containing not one, but two electron acceptors which differed in their capacity for photochemical interaction with the pigment-sensitizer.

It has been shown experimentally that under certain conditions one or the other of the two acceptors may be subjected to predominant reduction by means of changes of the medium acidity.

Fig. 4 gives the data on the photoreduction by ascorbic acid of a mixture

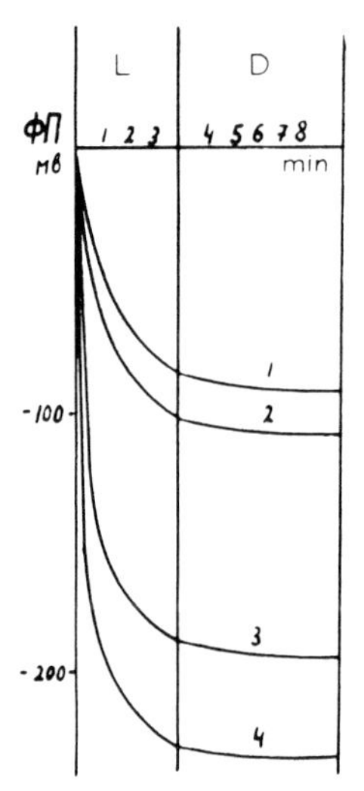

Fig. 2a. pH-dependence of negative photopotential value by photoreduction of pheophytin (*a* + *b*) with phenylhydrazine chloride in ethanol. Temperature −70 °C. pH: (1) 3.3; (2) 4.0; (3) 5.2; (4) 8.0.

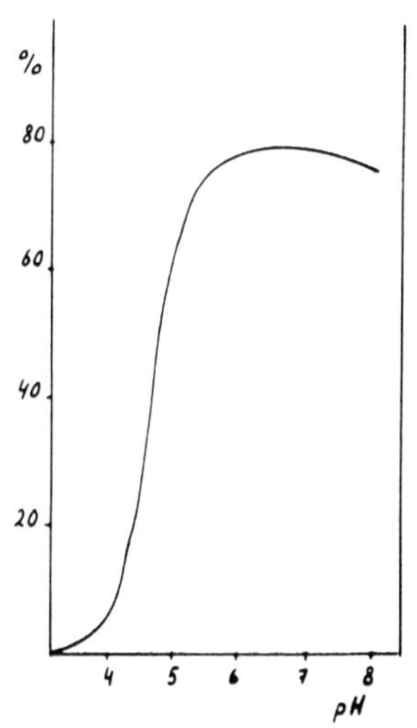

Fig. 2b. pH-dependence of lowering of pheophytin absorption maximum (665 nm) after 5 min illumination under the same conditions as at A. *Ordinate*: pH lowering at 665 nm in percentage of the initial value.

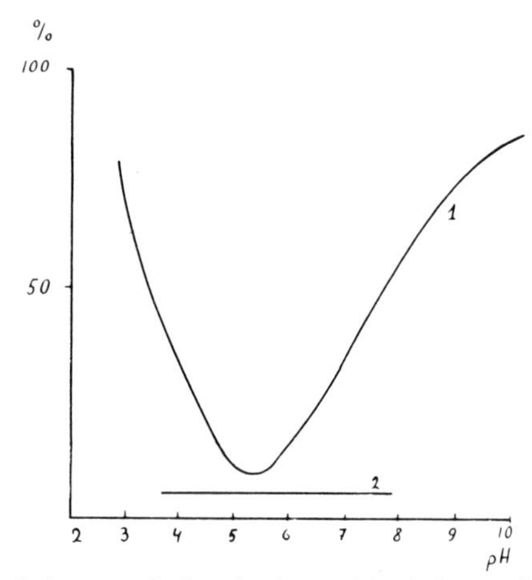

Fig. 3. Dependence of the rate of pheophytin-sensitized photoreduction of methyl red by phenylhydrazine (1). 2) Reaction course in the absence of the pigment. *Ordinate*: the amount of reduced methyl red in percentage of the initial one ($\sim 10^{-5}$ M), after 1 min of illumination.

of two acceptors – methyl red and triphenyltretrazolium chloride sensitized with magnesium phthalocyanine in ethanol. We have specifically selected the experimental results for magnesium phthalocyanine as, in contrast to chlorophyll, this analogue does not absorb in the short-wave spectrum where the above acceptors exhibit absorbance changes. In the case of chlorophyll itself the results are essentially the same.

Fig. 4a shows that at pH 4 illumination with phthalocyanine absorbed light brings about a rapid fall of methyl red absorbance; complete reduction of this acceptor being achieved, absorbance indicating triphenyltetrazolium reduction begins to increase.

On the contrary (fig. 4b), at pH higher than 7 triphenyltetrazolium reduction is the first to occur; once maximum absorbance of its reduced form at 550 nm is attained, methyl red absorbance starts to decrease.

The results of these and other experiments with two acceptors give direct evidence that under certain conditions acid–base equilibrium may act as a factor controlling chlorophyll photosensitized oxidation-reduction reactions in vitro.

As mentioned above, experimental data have been obtained testifying to the fact that acidity may play an important role in photosynthetic reactions in vivo.

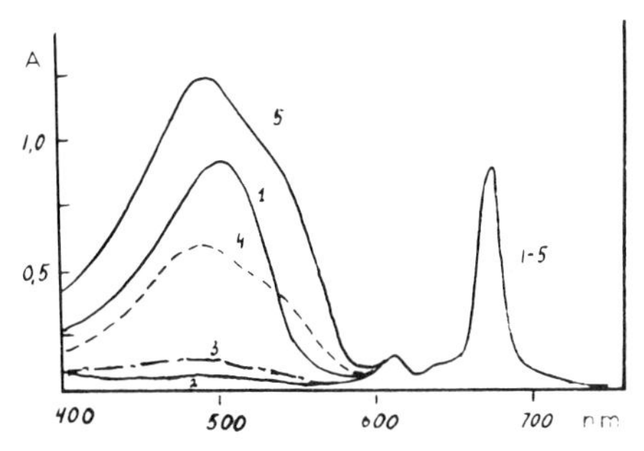

Fig. 4a. Magnesium phtalocyanine-sensitized photoreduction of methyl red and triphenyltetrazolium chloride with ascorbic acid (10^{-2} M) in ethanol; pH 3.9. The absorption spectra: (1) before illumination, (2) after 1.5 min illumination (the full bleaching of methyl red); (3) after 3.5 min illumination (the absorption of reduced triphenyltetrazolium begins to appear); (4) after 9.5 min illumination; (5) after 20 min illumination.

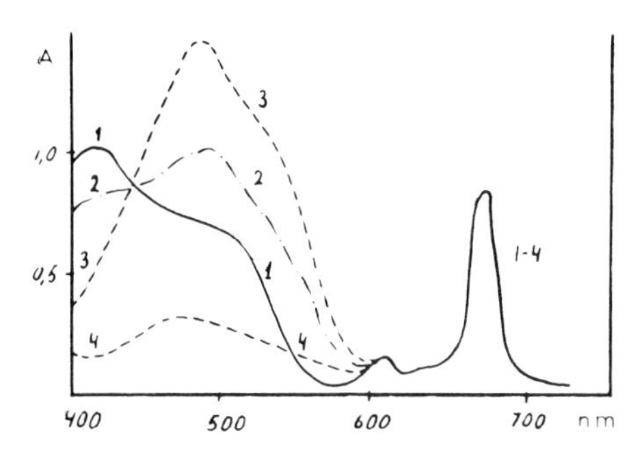

Fig. 4b. The same at pH 7.3. The absorption spectra: (1) before illumination; (2) after 2 min illumination; (3) after 9 min illumination; (4) after 16 min illumination.

I have no time to discuss all the pertinent information published in the literature. I would like to emphasize only that some of the investigators, e.g. Rumberg et al. (from the laboratory headed by Witt) [16,17] have experimentally demonstrated a direct relationship between acidity of the medium within chloroplast thylakoids and the rate of photosynthetic electron transfer. We are convinced that further studies will furnish more data in favour of the hypothesis that proton concentration exerts a controlling effect on primary photosynthetic reactions and, possibly, on relative intensity of the competing reactions: (1) direct electron transfer to NADP and (2) ATP formation.

Thus, the above facts give serious grounds to believe that the acid–base equilibrium acts as an important factor influencing and, consequently, regulating photosynthetic reactions at the pigment level.

The above scheme of the development of the regulation mechanism based on this factor is obviously very primitive. However, a comparative study from this point of view of the photochemical apparatus of photosynthetic organisms at different stages of evolution (bacteria, unicellular algae and green plants) holds great promise, as follows from the above consideration and experimental data.

We are convinced that a better understanding of this factor (and its importance for biochemical reactions has become practically a trivial statement) is indispensable if we are to learn more about the evolution of photosynthesis regulation mechanisms.

References

[1] Evstigneev V.B., Biofizika 8 (1963) 664.
[2] Evstigneev V.B., Photochem. Photobiol. 4 (1965) 171.
[3] Evstigneev V.B. and V.A. Gavrilova, Dokl. Akad. Nauk SSSR 165 (1965) 1435; 174 (1967) 476; Biofizika 11 (1966) 593; 14 (1969) 43.
[4] Evstigneev V.B., V.A. Gavrilova and O.D. Bekasova, Biofizika 11 (1966) 584.
[5] Evstigneev V.B., V.A. Gavrilova and N.A. Sadovnikova, Biochimia 31 (1966) 1229.
[6] Evstigneev V.B., V.A. Gavrilova and G.D. Olovianishnikova, Molekul. Biol. 1 (1967) 59.
[7] Evstigneev V.B., N.A. Sadovnikova and G.D. Olovianishnikova, Molekul. Biol. 2 (1968) 21; 3 (1969) 41; Dokl. Akad. Nauk SSSR 187 (1969) 1184.
[8] Evstigneev V.B., J. Chim. Phys. 65 (1968) 1447.
[9] Hager A., Planta 89 (1969) 224.
[10] Jagendorf A.T. and E. Uribe, Brookhaven Symp. Biol. 19 (1966) 215.
[11] Karlish S.J.D. and M. Avron, Biochim. Biophys. Acta 153 (1968) 878.
[12] Mitchell P., Nature 191 (1961) 144.
[13] Mitchell P., Biol. Rev. Cambridge Phil. Soc. 41 (1966) 445.
[14] Neumann J. and A.T. Jagendorf, Arch. Biochem. Biophys. 107 (1964) 109.
[15] Oparin A.I. ed. The Origin of Life on the Earth (Nauka, 1957) (in Russian).
[16] Rumberg R., E. Reinwald, H. Schroder and U. Siggel, Naturwissenschaften 55 (1968) 77.
[17] Rumberg B. and U. Siggel, Naturwissenschatten 56 (1969) 130.

Chemical Evolution and the Origin of Life, eds. R. Buvet and C. Ponnamperuma

POSSIBLE ROLE OF STRUCTURAL LIPIDS IN ACCUMULATING THE ENERGY OF LIGHT

K.B. SEREBROVSKAYA
A.N. Bach Institute of Biochemistry, Academy of Sciences of the USSR, Moscow, USSR

For a number of years, together with Lozovaya [13,14,20] we studied the transmission of electrons in lipoprotein coacervates exposed to light by using chlorophyll and irs analogues as sensitizers. Soaps and phosphatides were used as lipid components in lipoprotein systems. In such systems we could observe considerable activation of oxidation and oxido-reduction processes. This can be exclusively explained by the structural peculiarities of lipids, which are conducive to creating a more oriented flow of electrons.

Two years ago, together with Lutsik, Korneeva and Samsonova [9,15,16, 21–23], we succeeded in discovering a phenomenon testifying to the fact that lipids could possibly participate in oxidation and oxido-reduction processes. Indirect indications of possible participation of lipids in transmission of electrons are available in literature. However, these indications intermingle with ideas about the principal role of lipids as bearers of proper organization in biological structures.

Studying lipid–water systems containing chlorophyll in vacuum and in the absence of donors and acceptors of electrons we noticed changes in the conductance of the system at light–dark alterations. This creates an impression that under light a pigment molecule transmits an excited electron to a lipid molecule which returns it to the pigment molecule in the darkness (fig. 1). Such phenomena were noticed in the soap–water system, (curve 1) as well as in the phosphatide–water system (curve 2).

Further, we studied the influence of electron acceptors (O_2, quinone, cytochrome *c*) and donors (phenylhydrazine) on the photochemical activity of a lipid–pigment–water system. It was found (fig. 2) that the presence of acceptors sharply stimulate the process and increase the conductance of the system (curve 1). This can be explained by an increase of oxidized products (oxidized lipids). Addition of the donor, on the contrary, reduces the

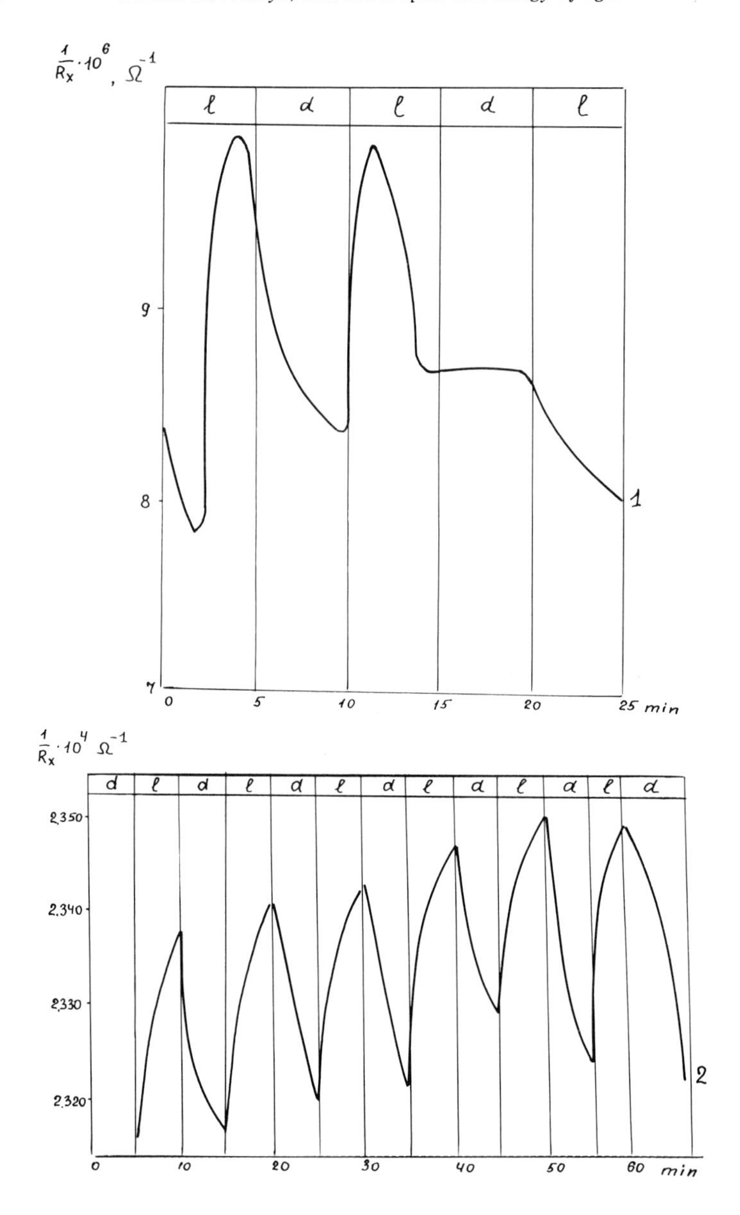

Fig. 1. Conductance of the lipid–chlorophyll–water system upon light–dark alterations: 1) lecithine + chlorophyll; 2) sodium oleate + chlorophyll.

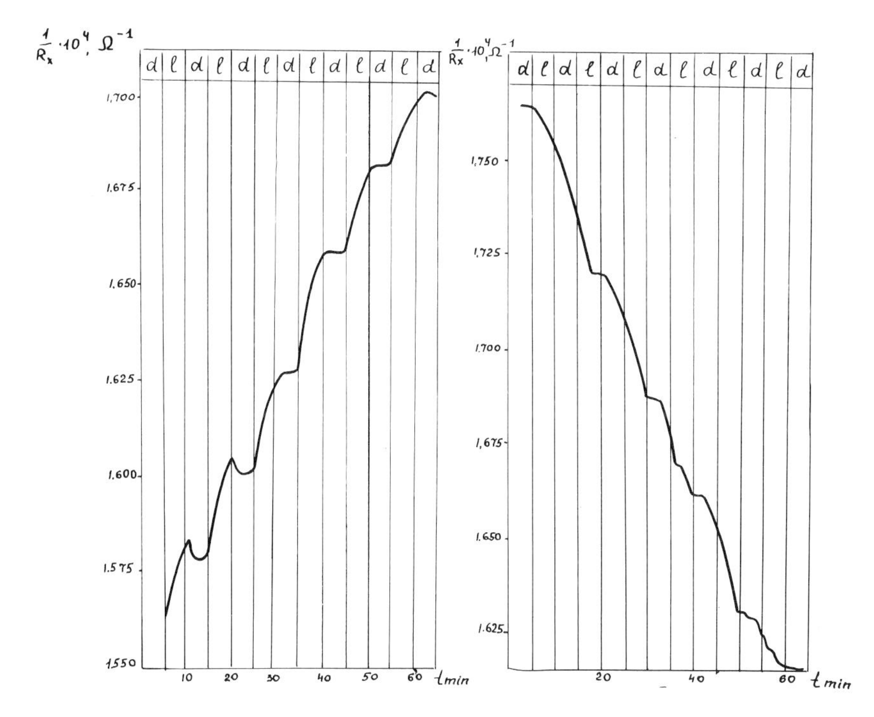

Fig. 2. Conductance of the sodium oleate–chlorophyll–water system in presence of: 1) acceptor (quinone); 2) donor (phenylhydrozin).

conductance (curve 2). At critical concentration of micella formation the system does not function as before in the presence neither of donor nor acceptor.

We suppose that in all three cases described the sequence of reactions was as in scheme 1.

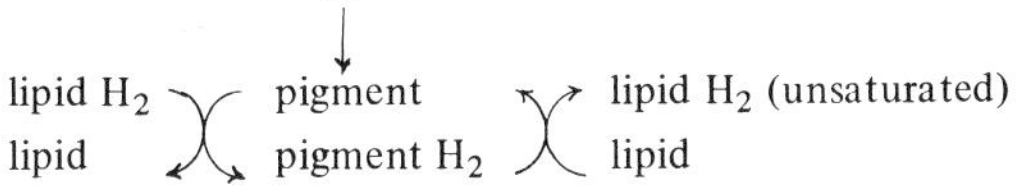

$$
\begin{array}{ccccc}
 & & h\nu & & \\
 & & \downarrow & & \\
\text{lipid } H_2 & \diagdown\!\!\diagup & \text{pigment} & \nwarrow\!\nearrow & \text{lipid } H_2 \text{ (unsaturated)} \\
\text{lipid} & \swarrow\!\searrow & \text{pigment } H_2 & \diagup\!\!\diagdown & \text{lipid}
\end{array}
$$

Scheme 1. Lipid–pigment system.

The lipid does not undergo any changes.

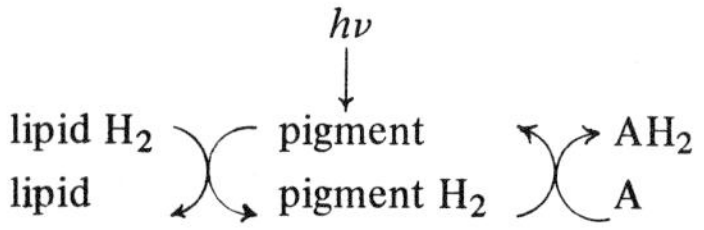

$$\begin{array}{ccccc} & & h\nu & & \\ & & \downarrow & & \\ \text{lipid } H_2 & \rightleftarrows & \text{pigment} & \rightleftarrows & AH_2 \\ \text{lipid} & & \text{pigment } H_2 & & A \end{array}$$

Scheme 2. Lipid–pigment–acceptor system.

The lipid is oxidized as in scheme 2.

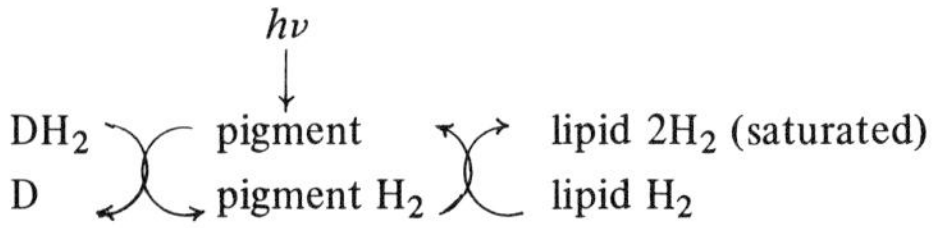

$$\begin{array}{ccccc} & & h\nu & & \\ & & \downarrow & & \\ DH_2 & \rightleftarrows & \text{pigment} & \rightleftarrows & \text{lipid } 2H_2 \text{ (saturated)} \\ D & & \text{pigment } H_2 & & \text{lipid } H_2 \end{array}$$

Scheme 3. Lipid–pigment–donor system.

The lipid is regenerated as in scheme 3.

Based on the above, when in the presence of a donor and acceptor the system should function as in scheme 4.

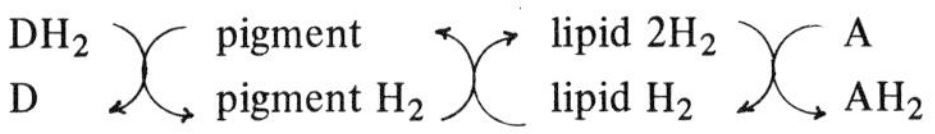

$$\begin{array}{ccccccc} DH_2 & \rightleftarrows & \text{pigment} & \rightleftarrows & \text{lipid } 2H_2 & \rightleftarrows & A \\ D & & \text{pigment } H_2 & & \text{lipid } H_2 & & AH_2 \end{array}$$

Scheme 4

The lipid and the pigment here do not change.

In all of these systems the energy of an electron excited by light will be wasted. However, this will not happen if there is some substance capable of accumulating the energy of a photon in some other form, perhaps in a conformative transition of a molecule.

In literature there are references to work by Engelhardt and Wenkstern [3,27] on photochemical oxidation of miozene in methylene blue. Exposed to light the miozene–methylene blue system, as mentioned by the above authors, is polymerized due to photooxidation of miozene in the presence of oxygen.

Methylene blue, as we know, regenerates irreversibly. That is why we used chlorophyll *a+b*, capable of reversible oxido–reduction. Above that we studied photoprocesses in miozene–pigment and miozene–pigment–lipid systems under the influence of light–dark alterations but without permanent exposure. The results can be seen in fig. 3. Curve 1 shows changes of viscosity in the miozene–pigment system, while curve 2 shows the same in the miozene–pigment–lipid system. Since the system returns to its original state, the oxidation-reduction processes are evidently reversible. We suppose that the process functions through photoreduction of pigment (scheme 5).

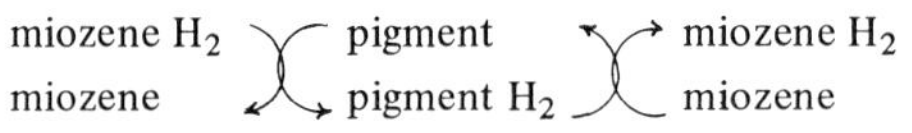

miozene H_2		pigment		miozene H_2
miozene		pigment H_2		miozene

Scheme 5

It should be noted that oxidation by oxygen does not take place as the pigment is again capable to release hydrogen.

In presence of a lipid in the system the process sharply activates. But in this case the system does not reverse in the darkness, as the lipid releases hydrogen to oxygen rather than to pigment.

The process goes as in scheme 6.

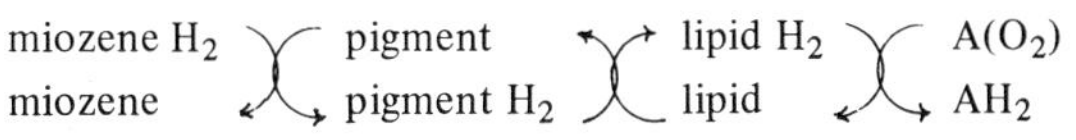

miozene H_2		pigment		lipid H_2		$A(O_2)$
miozene		pigment H_2		lipid		AH_2

Scheme 6

There we see the process of irreversible oxidation of miozene by oxygen. When the concentration of the lipid increases further (curve 3) the system gets rapidly polymerized. Reduction of viscosity does not take place when the light is switched off.

This means that the energy of a photon can accumulate directly in the system, causing polymerization of miozene.

A number of experiments carried out in red light show possible participation of lipids in oxido-reduction processes, sensitized by the pigment and exposed by light.

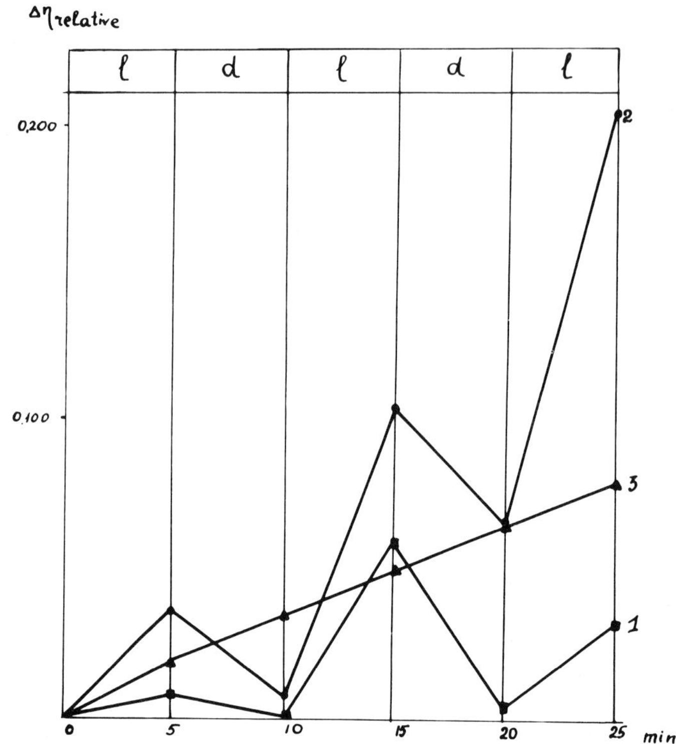

Fig. 3. Structural alterations in myosin pigment (curve 1) and myosin–pigment–lipid systems (curves 2 and 3).

Another group of experiments carried out in white light show that under these conditions direct dehydration of lipids is possible. A solution of lecithin in methanol exposed to white light increases the iodine number of the lipid in proportion to the time of exposure.

In the presence of spontaneously oxidizing phospholipid (lecithin) a number of ferments, such as ribonuclease, hexokinase and miozene not only undergo changes in their activity but even alter the direction of reactions, catalysed by them [17].

We believe that all these processes are connected with the changes in the lipid structures of the solution during its spontaneous dehydration.

In the biological role of lipids as "matrixes" on which the protein changes its conformation, what is required for its activity is determined by the properties most common to these substances – ability to create structural formations. The lipid molecules are diphilic, i.e. one part carries polarity, which determines their certain solubility in water, while the other does not carry any polarity. Due to the presence of sufficiently developed hydrocarbon chains the lipid molecules are capable of aggregation in the water medium. Hydrophobic interaction of the lipid molecules is the reason for their association in the water medium.

It has been proved experimentally [5,6] that the ability of lipids to associate is correlated with the degree of saturation of hydrocarbon radicals in lipid molecules. The higher the degree of saturation, the lower the critical concentration of micelle formation, i.e. formation of micelles starts at lower concentrations of surface-active agents. On the other hand, it is known [19] that with the increase in concentration of the surface-active agents and after reaching critical concentration of micella formation the form of associates also changes: spherical micelles change into lamellar. Hence, concentrations at which associates of unsaturated lipids will still represent spherical micella are sufficient for the associates of saturated ones to become lamellar. Micelles that contain lipid molecules with saturated hydrocarbon chains, being more hydrophilic than micelles composed of molecules with saturated radicals, appear to be looser formations containing a considerable amount of water [24].

This means that one of the factors determining the forms of associates composed of lipid molecules is the degree of oxidation (unsaturation) of hydrocarbon chains of lipid molecules. Thus, Green notes that phospholipids can be obtained in the form of micelles only in the case when hydrocarbon chains of lipids not contain less than one double bond [4].

According to the theory of a dynamic membrane, elaborated by Kavanau [8] on the basis of model experiments carried out by Luzzati and Stockenius [26], lipids in membrane structures of a cell can be found in micellar or lamellar forms. These forms are very labile and reversible transitions between them can be observed.

We may suppose that such reversible transitions of micellar and lamellar forms of lipids are regulated by constant oxidation and oxido-reduction processes that could take place in lipid molecules. Structural changes in lipid associates cause conformative changes in the proteins connected with them, and finally lead to changes in its biological activity.

Removal of hydrogen from a molecule of saturated lipid leads to transformation of a lamellar micelle into a spherical one. This is accompanied by an increase in the water content between the micelles. Reversible regeneration of the lipid well revert the structure to a lamellar one.

Based on the data presented regarding the dependence of a micelles form at a definite concentration of the surface-active agent, on the degree of saturation of hydrocarbon radicals, an attempt can be made to explain the mechanism of swelling and shrinking (contraction) of natural lipoprotein membranes, chloroplasts and mitochondria in case of an oriented flow of electrons.

Changes in the structure and the volume of mitochondria described by Lehninger [12] might proceed in the following way: Removal of hydrogen from structural lipids in case of their participation in the chain of electron transmission, leads to their hydrophylization, i.e. to the appearance of double bonds in the molecule. The excess molecules of fatty acids, of which the membrane was composed in a lamellar form, come into the surrounding medium, thus promoting transformation of membrane lipids into spheroid aggregates. Evidently, a swollen mitochondrion consists of loose watery micelles. When hydrogen is supplied by the donors, the hydrocarbon radical becomes hydrophobic. This leads to the transformation of spheroid micelles into more compact and less watery lamellar aggregates. The process is accompanied by accomodation of excess molecules of fatty acids into the membrane carcass. Such a state of lipid molecules corresponds to a shrunken (contracted) mitochondrion. Dehydration energy can easily convert itself into the energy of conformation transition of a protein molecule closely bound with the lipid.

Such conformation transition is the result of the changes that take place in the structure of associates of lipid molecules during a spontaneous dehydration of the latter (fig. 4). During such transition energy-rich bonds can appear, the energy of which could easily transform itself into the energy of phosphate bonds [1,2,7,25]. The above suggests a possible role of lipids in prebiological evolution.

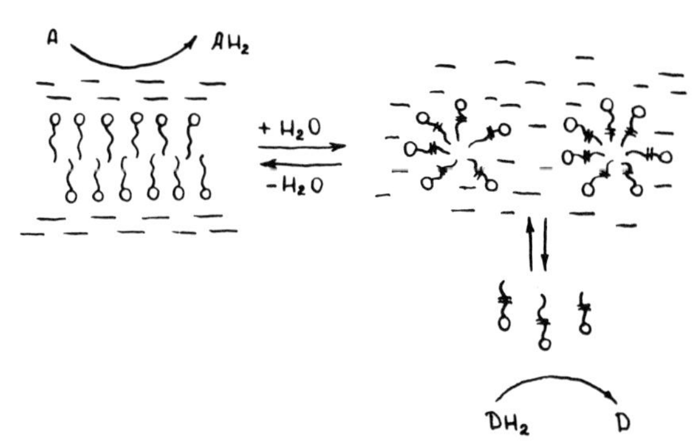

Fig. 4. Possible mechanism of a relationship of structure and function at the level of the lipid association.

Along with polymeric substances long-chain hydrocarbons were being created on the protoplanet. Abiogenic formation of paraffins in the regenerating atmosphere was the result of the vast tectonic activity on the forming planet [11]. After formation in the depth of the planet, hydrocarbons penetrated into the water reservoirs through the cracks in the earth's crust during volcanic eruptions. In the regenerating atmosphere paraffins could have the form of a deep layer of methane white oil. At high temperatures and pressures and in the presence of acid magma, relatively rich with oxygen and ions of metals which could act as catalyzers, some oxidation of paraffins to long-chain fatty acids could also occur.

Discovery of fatty acids with an even number of carbon atoms (C_{14}, C_{16}, C_{20}, C_{22}, C_{24} and $C_{26 \leftrightarrow 28}$) made by Mainstein in the meteorite named after Orgueil is being critically analyzed. However, the possibility of creating fatty acids abiogenically cannot be rejected, as there is data on the synthesis of alyphatic hydrocarbons and long-chain fatty acids in a pseudo-glow discharge [18].

Having appeared abiogenically, fatty acids could have reacted with alkaline and alkaline-earth metals forming various salts of fatty acids capable, at a critical concentration, of association in a water solution. The presence of protoproteins (which appeared through abiogenic synthesis) in the terrestrial oceans, contributed to the creation of various lipoprotein complexes (protomembranes). Such protomembranes, consisting of saturated soaps, were able to respond to ultraviolet radiation by dehydrating their composite molecules and accumulating the energy of light quanta by changing the conformation of protein bound to them as described above. Possible evolution of electron transmission could be described as follows.

Light, being a constant factor on the earth, changed its properties during the prebiological evolution. Along with the changes in its wavelength characteristics, alterations occurred in the mechanism of lipid participation in oxidation and oxido-reduction processes. Under hard ultraviolet radiation primary lipids could easily dehydrate. When passing to the period of long-wave radiation, the main role was acquired by unsaturated lipids. Furthermore, for more extensive dehydration of lipids in long-wave radiation, there appeared a need for sensitizers – the pigments [10]. Their formation through synthesis was contributed to by accumulation of oxygen in the atmosphere of the planet. At last its pressure reached the level sufficient to start oxidation and oxido-reduction processes in iron-porphyrins. Participation of cytochromes in these processes not only intensified transmission of electrons in the light, but facilitated the creation of systems no longer dependent on light, capable of oxidizing and regenerating lipids with the help of acceptors and donors.

All the above considerations regarding the possible role of the lipids in accumulating the energy of light are based on the analysis of data found in literature and on the experimental work quoted in this report.

References

[1] Barsiboim G.M., A.N. Domanski and K.K. Turoverov, Luminiscencia biopolimerov i kletok (Nauka, Moscow, 1966) p. 124.
[2] Boijer P.D. in: Molecularnaja Biologia (Nauka, Moscow, 1964) p. 227.
[3] Engelhardt W.A., N.S. Demjanovskaja and T.W. Wenkstern, Dokl. Acad. Nauk SSSR 72 (1950) 923.
[4] Green D.E. and S. Fleischer, Biochim. Biophys. Acta 70 (1963) 5, 554.
[5] Ivanova N.Ja. and A.I. Jurgenko, Kolloidn. Zh. 24 (1962) 278.
[6] Jurgenko A.I. and G.F. Storozh, Kolloidn. Zh. 22 (1962) 376.
[7] Zhuravlev A.I., Ju.N. Filippov and V.V. Simonov, Biophisica 10 (1965) 246.
[8] Kavanau I.L., Nature 198 (1963) 525.
[9] Korneeva G.A. and K.B. Serebrovskaja, Dokl. Acad. Nauk SSSR 187 (1969) 5.
[10] Krasnovskii A.A. and A.W. Umrikhina, Dokl. Acad. Nauk SSSR 155 (1964) 691.
[11] Kropotkin P. in: Vosniknovenie Jisni na Semle, Tr. Medjd. Simp. M. Isd. AN SSSR (1967) p. 88.
[12] Leninger A., The Mitochondrion (Benjamin, New York, 1964).
[13] Lozovaja G.I. and K.B. Serebrovskaja, Ukr. Biochem. J. 39 (1967) 78.
[14] Oparin A.I., K.B. Serebrovskaja and G.I. Lozovaja, Dokl. Acad. Nauk SSSR 179 (1968) 5.
[15] Oparin A.I., K.B. Serebrovskaja, H.W. Vasiljeva and W.M. Samsonova, Dokl. Acad. Nauk SSSR 179 (1968) 4.
[16] Oparin A.I., K.B. Serebrovskaja and G.A. Korneeva, Dokl. Acad. Nauk 183 (1968) 3.
[17] Oparin A.I. and K.B. Serebrovskaja, Dokl. Acad. Nauk SSSR 185 (1969) 3.
[18] Sb. Vozniknovenie organicheskogo vetchestva v solnetchnoi sistem, M. Mir (1969).
[19] Plaxin I.N. and W.I. Solnyshkin, Inphracrasnaja spectrokopia poverchnostnych sloev reagentov na mineralach (Nauka, Moscow, 1966) p. 182.
[20] Serebrovskaja K.B., G.I. Lozovaja and E.G. Sudjina, Naukova dumca, Kiev (1966).
[21] Serebrovskaja K.B., W.M. Samsonova and N.S. Minojedinova, Dokl. Acad. Nauk SSSR 180 (1968) 4.
[22] Serebrovskaja K.B. and G.A. Korneeva, Koll. J., in press.
[23] Serebrovskaja K.B., G.A. Korneeva, T.K. Lutzik and W.M. Samsonova, Biophisica, in press.
[24] Shinoda K. and T. Nakagawa, Colloidal Surfactants. Some Physicochemical Properties (Academic Press, New York, London, 1963).
[25] Skulachev V.P. in: Mechanismy Dychania, Photosintesa i Fixacii Azota (Nauka, Moscow, 1967) p. 7.
[26] Stockenius in: The Interpretation of Ultrastructure, ed. R.I.C. Harris (Academic Press, New York, 1962).
[27] Wenkstern T.W., Biokhimiya 18 (1953) 97.

Chemical Evolution and the Origin of Life, eds. R. Buvet and C. Ponnamperuma

EVOLUTION OF THE PIGMENT SYSTEM AND PRIMARY PROCESSES OF PHOTOSYNTHESIS

N.V. KARAPETYAN
Bakh Institute of Biochemistry, Academy of Sciences of the USSR, Moscow, USSR

In accordance with the evolutionary theory of Oparin, autotrophic assimilation emerged at first as photoautotrophy based upon a well developed heterotrophic assimilation. This seems to be true since the capacity of an organism to grow by assimilation of simple mineral compounds requires a more complicated cellular organization than that needed for assimilation of readily available organic compounds [12]. It is assumed that the pre-photosynthetic period witnessed heterotrophs which appeared to contain certain enzymic systems and systems synthesizing energy-rich compounds, and to be able to utilize strongly reduced organic compounds as electron donors. This heterotrophic activity gave rise to an accumulation on the earth's surface of more oxidized compounds, i.e. induced an oxidative shift of the biosphere potential. Therefore, at a certain stage, evolution demanded a mechanism allowing utilization of these more oxidized compounds as electron donors. The new mechanism is based on the pigment capacity for photosensitization which brings about a transition of an electron to a higher energy level; the transfer of this electron along the thermodynamic gradient of the chain finally results in the formation of NAD–H (or NADP–H) and ATP required to reduce assimilated CO_2.

It can be supposed that photosynthesis emerged approximately 2 billion years ago. At that time the earth's atmosphere was anaerobic and reducing, it contained substantial amounts of CO_2 formed as a result of decomposition of organic compounds by primary heterotrophs and O_2 traces in the upper layers which were caused by radiation and ultraviolet effects on water vapors. This emerged photosynthetic apparatus which allowed utilization of solar energy for synthetic processes was subjected to significant evolutionary changes. Photosynthetic bacteria that were more ancient from the evolutionary point of view, were capable of utilizing only reduced compounds whereas algae and plants were able to utilize oxidized compounds such as water. The pigment system, primary processes and structure of the photosynthetic apparatus underwent the most significant evolutionary changes. Some problems of the evolution of photosynthesis have been discussed by Krasnovsky [9], Gaffron [4,5] and Calvin [1]. The present communication is an attempt to show a possible pathway for the evolution of the pigment system and primary processes of photosynthesis on the basis of a comparative analysis of the photosynthetic organisms that have survived until today.

An analysis of the pigment system of the oldest representatives – photosynthetic bacteria – and of the most highly organized representatives – green algae and plants – gives evidence (fig. 1, table 1) that evolution of the pigment system proceeded as follows:

1) Photosynthetic organisms accumulated more oxidized chlorophyll types possibly brought about by an increase of the medium oxidation level. Chlorophyll *a* and *b* molecules contain one hydrated double bond (in the IV pyrrole ring) whereas bacteriochlorophyll molecules contain two hydrated double bonds (in the II and IV rings). However, some algae (e.g. blue–green) contain, in addition to chlorophyll, minor amounts of pigments of the bacteriochlorophyll type [6], these pigments are photochemically inactive and formed apparently during the synthesis of chlorophyll.

2) The pigment oxidative capacity increased: the double oxidized form of chlorophyll *a* – the pigment that emerged at a later stage – had a greater

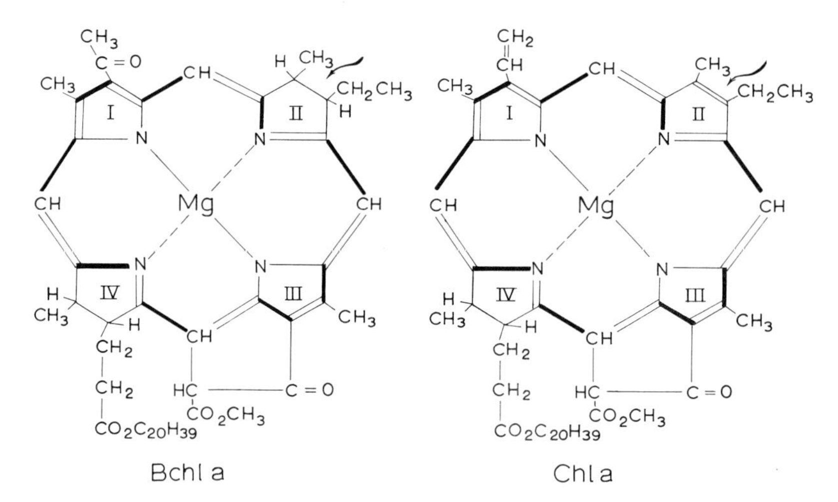

Fig. 1. Structures of bacteriochlorophyll and chlorophyll *a* molecules.

Table 1

	Bacteriochlorophyll		Bacterio-viridin	Chlorophyll	
	b	*a*		*a*	*b*
λ_{nm}	1050	900	750	680	650
eV	1.19	1.39	1.70	1.85	2.92
kcal	27	32	38	42	44
E^{2+} V	–	+0.60	+0.74	+0.90	+0.92
P/Chl	1:20 (1:40)		1:100	1:200 (1:400)	

oxidized potential (+0.90 V) than the double oxidized form of bacteriochlorophyll (+0.7 V). This was very important for participation in water photolysis [10].

3) The long wavelength absorption bands of chlorophyll shifted from the infrared part of the spectrum (900 nm in purple bacteria) to the far red region (750 nm in green bacteria) and red region (650–680 nm in plants and algae). This enabled effective absorption of light quanta of higher energy. Below are the values of energy of 1 Einstein of light quanta for different wavelengths which can be maximally absorbed by chlorophylls in vivo: E_{900} = 1.39 eV or 32 kcal (bacteriochlorophyll); E_{750} = 1.70 eV or 38 kcal (bacterioviridin); E_{680} = 1.85 eV or 42 kcal (chlorophyll *a*), E_{650} = 1.92 eV or 44 kcal (chlorophyll *b*).

4) The number of chlorophyll forms increased in vivo: from 3–4 forms of bacteriochlorophyll in purple and green bacteria to 6–8 (or more) forms of chlorophyll in plants and algae. In contrast to bacteriochlorophyll forms, chlorophyll forms in vivo are characterized by a stronger overlapping of absorption spectra which accelerates the efficiency of energy migration between various forms [11].

5) The amount of chlorophyll molecules increased compared to that of chlorophyll in the reaction centre, thus enhancing the efficiency of energy transfer to the reaction centre and of the function of the entire transfer chain. Thus, the ratio between bacteriochlorophyll molecules in the reaction centre and their bulk is 1:20 (up to 1:40) for purple bacteria; 1:100 (1:150) for green bacteria; and 1:200 (1:400) for plants and algae [3].

In the course of evolution the photosynthetic apparatus acquired the capacity for water decomposition to evolve O_2. The O_2 accumulation gave rise to significant changes not only in the photosynthetic apparatus but also in the environment on the earth's surface, thus causing the emergence of aerobic respiration and chemosynthesis and the determination of the subsequent evolution of living matter.

A more advanced type of photosynthesis involving O_2 evolution developed due to the following reasons:

1) Emergence of photoreaction II, functioning in a more oxidized range (beginning with +0.8 V) than photoreaction I (+0.43 V). While functioning, photoreaction II donates electrons to photoreaction I (fig. 2). A close structural relationship between the two photosystems determines a more effective performance of the photosynthetic apparatus.

2) Acceleration of electron transfer reactions from $10^{-2} - 10^{-1}$ sec in photosystem I to $10^{-3} - 10^{-5}$ sec in photosystem II [14].

3) Increase of the concentration of a metal with a high oxidative potential, e.g. Mn^{2+} in algae and plants. The formation and accumulation of Mn^{2+}-containing compounds with a high oxidative capability resulted in the functioning of photosystem II in a more oxidative range.

4) Development of a cooperative mechanism of energy accumulation of two quanta in a molecule of one compound which induced a higher oxidative potential of this compound [7].

From the evolutionary point of view, pigment system II emerged later since it can be found only in organisms that are capable of evolving oxygen in the light. This photosystem is more thermo- and photolabile whereas

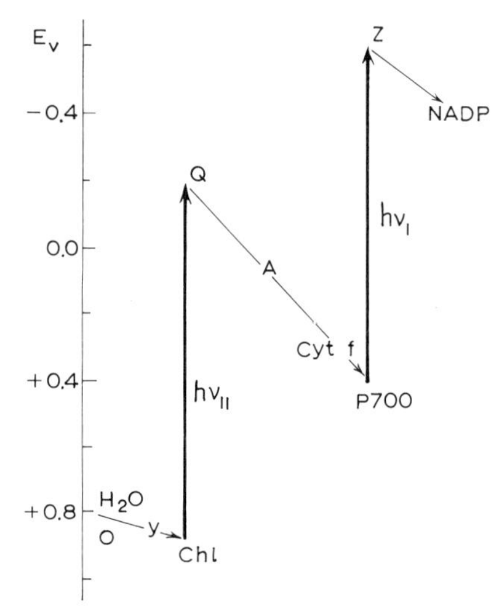

Fig. 2. Scheme of the photosynthetic electron transport chain of algae and plants [3].

photosystem I (as well as the pigment system of bacteria) is very stable to damaging effects. In general, plant photosystem I and the bacterial photosystem have many similar features. Ontogenetically, plants developed at first photosystem I and subsequently photosystem II. The concept that photosystem II developed at a later evolutionary stage is also suggested by the fact that during the adaptation of algae to H_2, photosystem II does not work, photosystem I is the only one that functions. On the contrary, the capacity of certain purple bacteria to grow under aerobic conditions (in the dark) has been developed later. Under these circumstances, the bacteria have no chromatophores which can be formed when the bacteria are transferred to light and kept under aerobic conditions. Despite the fact that purple bacteria may have two photosystems and two photoreactions [2,8,13], they are unable to use water as an electron donor. This complex reaction (which has so far been poorly studied) could have become possible not only because plants and algae developed photoreaction II which functions in a more oxidized range but also due to the development of a two-quanta mechanism of photolysis and formation of a compound with a high oxidative potential and participating in water decomposition.

The evolutionary process involved not only functional but also structural changes of the photosynthetic apparatus which resulted in its complication. As is well known, photosynthesis occurs in special organelles: in chromatophores of purple and green bacteria, and in chloroplasts of plants and green algae. However, chromatophores show only an orderly lamellar structure, whereas, chloroplasts display differentiation into lamellae of grana and lamellae of stroma. In this respect it seems interesting to note that blue–green algae which are similar to plants in the mechanism of primary processes (two photoreactions, Emerson effect) have organelles that are very similar to the bacterial photosynthetic apparatus, i.e. they have no lamellae of grana.

Thus, the evolution of the photosynthetic apparatus involved an isolation of structures which are the site of the pigment apparatus and light reactions of photosynthesis. Complication of the photosynthetic apparatus structure increased the energetic efficiency of photosynthesis providing closer conjugation of oxido-reductive reactions and their better coupling with formations of energy-rich compounds. A more complicated structure of chloroplasts makes photoreducing agents inaccessible to such a strong oxidant as oxygen. This accelerates its binding and removal from the synthesis sites in the chloroplast and assures spatial separation of photoproducts, eliminating a useless back reaction.

The evolution of the photosynthetic apparatus can be regarded as completed when it has acquired the capability of utilizing water and evolving oxygen. The amounts of the electron donor (water) are practically inexhaustible; therefore, the subsequent evolution of photosynthesis appears to be independent from the factor that determined the evolution of photosynthesis and other processes during billions of years.

References

[1] Calvin M., Horizons in Biochemistry (Academic Press, New York, London, 1962) p. 23.
[2] Cusanovich M.A., R.G. Bartsch and M.D. Kamen, Biochim. Biophys. Acta 153 (1968) 397.
[3] Duysens L.N.M., Progress in Biophysics, Vol. 14 (Pergamon Press, New York, London, Paris, 1964).
[4] Gaffron, H., Horizons in Biochemistry (Academic Press, New York, London, 1952) p. 59.
[5] Gaffron H., Origin of Pre-Biological Systems and of their Molecular Matrices (Academic Press, New York, London, 1965).
[6] Gassner E.B., Plant Physiol. 37 (1962) 637.
[7] Joliot P., G. Barieri and R. Chabaud, Photochem. Photobiol. 10 (1969) 309.
[8] Karapetyan N.V. and A.A. Krasnovsky, Dokl. Akad. Nauk SSSR 180 (1968) 989.
[9] Krasnovsky A.A., Origin of Life on the Earth, Symposium, Moscow (Nauka, Moscow, 1957) p. 355.
[10] Kutyurin V.M., Y.U. Artamkina and I.N. Anisimova, Dokl. Akad. Nauk SSSR 180 (1968) 1002.
[11] Litvin F.F., Biochemistry and Biophysics of Photosynthesis (Nauka, Moscow, 1965) p. 96.
[12] Oparin A.I., Origin of Life on the Earth (Nauka, Moscow, 1957) p. 362.
[13] Sybesma C., Progress in Photosynthesis Research, Vol. II (1969) p. 1091.
[14] Witt H.T., Nobel Symposium 5 (1968) 291.

Chemical Evolution and the Origin of Life, eds. R. Buvet and C. Ponnamperuma

PARTICIPATION OF FLAVINS IN PHOTOBIOLOGICAL PROCESSES IN CONTEMPORARY ORGANISMS

M.S. KRITSKY
A.N. Bakh Institute of Biochemistry, Academy of Sciences of the USSR, Moscow, USSR

It is believed that the photochemical processes in primordial living beings under the conditions of the reduced atmosphere were related to the consumption of energy-rich quanta of ultraviolet light. [8]. Due to the necessity to consume the quanta of ultraviolet or short-wave visible light, nicotinamide or flavin derivatives have been suggested to play the role of photoreceptor in such processes, because light may cause oxidative–reductive transformations of these compounds [5]. Further, when the oxidized atmosphere and ozone screen developed, the consumption of energy-rich quanta of ultraviolet became impossible, and the organisms had to adapt themselves for utilization of light with longer wavelengths and form perfect photosynthetic apparatus. We may suggest that, at a certain stage of the evolution of earth's atmosphere, photobiological processes – the reactions photocatalyzed by flavins with absorption maxima at the shorter range of visible light – could become very important.

The question arises whether all of the most ancient types of photobiological processes completely disappeared in the course of evolution, or do nicotinamide nucleotides or flavins, even now, serve as photoreceptors in contemporary metabolic systems?

Besides rather well studied metabolic processes forming the basis of important phenomena such as photosynthesis, vision, and also regulatory processes in plants connected with phytochrome, some relatively poorly studied light-dependent physiological reactions are present in certain groups of organisms. First, we would point out several light-controlled physiological processes in fungi. Among the best studied processes are phototrophic responses of *Phycomyces* sporangiophores [3], initiation and inhibition of certain morphogenetic processes by light [2,4,7], and light-induced formation of enzymes catalyzing carotenoid biosynthesis in *Neurospora* and *Fusarium* [10,17].

In this paper we are not going to discuss all metabolic events which are triggered by an absorption of light quantum by photoreceptors. The important point is that all these processes have one common feature: they are all induced only by shortwave visible light (blue range). There were some indications, however, that ultraviolet light is also active [3]. A comparison of the action spectra of various photobiological processes in fungi with the absorption spectra of various compounds revealed their close similarity to the spectra of flavin derivatives, especially of certain flavoproteins [10]. Experiments with inhibitors of flavin metabolism [9,16] and competitive suppression of light action by flavins [7] also support the idea of participation of flavins as photoreceptors.

Thus, we may suggest that in fungi metabolic processes exist in which flavins or probably flavoproteins serve as photoreceptors. The mechanisms of these photochemical reactions have been studied very poorly. We can only say that flavin to work as a photoreceptor must exist in the oxidized form. It has not yet been proved that illumination in fact causes reduction of flavins in vivo. We can point out, however, that the primary photochemical processes cause significant change in the balance of oxidized and reduced compounds in fungal cells. This was shown, for instance, for NAD/$NADH_2$ and NADP/$NADPH_2$ ratios for the light-induced morphogenesis of a mushroom *Lentinus tigrimus* [1,6].

Since certain physiological processes in which flavin acts as a photoreceptor have been found, not only in various groups of fungi, but also in *Mycobacteria* [11], some algae [13,15] and higher plants [12,14,18,19], we may believe that they are specific for a rather wide group of organisms. Such a wide distribution suggests their ancient origin. We may think that the photocatalytic reactions are to some extent similar to those that existed during the early stages of evolution of photobiological processes. In the course of evolution they were driven away from the role of energy-supplying mechanisms by much better contemporary photosynthetic processes with chlorophyll as a photosensitizer, and at present are connected only with different regulatory processes.

However, the facts obtained still do not allow us to ascribe definitely the role of a kind of metabolic rudiment or metabolic "fossil" to processes photocatalysed by flavins or flavoproteins.

We may hope, however, that the study of their mechanism will permit a closer understanding of the evolution of photobiological processes on the earth.

References

[1] Belozerskaya T.A. and M.S. Kritsky, 2nd All-Union Congress of Biochemistry, Abstracts of Section Communications, Section 18 (FAN, Tashkent, 1969) p. 74 (in Russian).
[2] Bjornsson I.P., J. Wash. Acad. Sci. 49 (1960) 317.
[3] Delbruck M. and W. Shropshire, Plant Physiol. 35 (1960) 194.
[4] Etzold H., Arch. Microbiol. 37 (1960) 226.
[5] Krasnovsky A.A., 2nd All-Union Congress of Biochemistry, Abstracts of Symposia Communications (FAN, Tashkent, 1969) p. 22 (in Russian).
[6] Kritsky M.S. and T.A. Belozerskaya, 6th Meeting of FEBS, Madrid (1969) Abstracts p. 3160.
[7] Lukens R.J., Am. J. Botany 50 (1963) 720.
[8] Oparin A.I., The Origin of Life, 2nd ed. (Dover Publications, New York, 1953).
[9] Page R.M., Mycologia 48 (1956) 206.
[10] Rau W., Planta (Berlin) 72 (1967) 14.
[11] Rilling H.C., Biochim. Biophys. Acta 79 (1964) 464.
[12] Schmid G.H., Z. Physiol. Chem. 350 (1969) 1035.
[13] Schmid G.H. and P. Schwarz, Z. Physiol. Chem. 350 (1969) 1513.
[14] Shropshire W. and R. Withrow, Plant Physiol. 33 (1960) 360.
[15] Tollin G. and M. Robinson, Photobiology 9 (1969) 411.
[16] Wooley D.W., Symp. Soc. Gen. Microbiol. 8 (1958) 139.
[17] Zalokar M., Arch. Biochem. Biophys. 56 (1955) 318.
[18] Zenk M.H., Z. Phlanzenphysiol. 56 (1956) 206.
[19] Zurzycki J., Acta Soc. Botan. Polon. 36 (1967) 133.

PART VI

ORIGIN OF BIOLOGICAL STRUCTURES

Chemical Evolution and the Origin of Life, eds. R. Buvet and C. Ponnamperuma
©1971, North-Holland Publishing Company

THE PROBLEM OF CHANCE IN FORMATION OF PROTOBIONTS BY RANDOM AGGREGATION OF MACROMOLECULES

R.W. KAPLAN
Institut fur Mikrobiologie, Universität
Frankfurt/Main, Germany

Reproduction and mutation can be considered as the key processes of life. Systems with these two abilities will evolve infinitely to a great variety of types by Darwin's process, as did the organisms on earth.

Mutations are, as we know now, "accidental" changes in the base pattern of nucleic acids. Several causes of spontaneous "errors" during replication are well known [6,15]. Some were certainly already active when the first organisms (protobionts) arose. Mutations are therefore no serious problem for the understanding of the origin of life.

The most difficult problem is certainly the origin of the apparatus of reproduction. Reproduction of present living cells is based on 2 groups of functions of macromolecules: (1) heterocatalytic activities within the cell as well as in the environment; (2) autocatalytic activities causing replication of the information which is used for the generation of those heterocatalytic activities. Since no macromolecules seem to exist which can perform both functions at the same time, both are allotted separately to 2 substances, proteins and nucleic acids, in contemporary organisms. As a consequence, the syntheses of both polymers must be coupled mutually, proteins being formed due to information from nucleic acids (translation), and nucleic acids are replicated by catalysis due to proteins (replication and transcription). * The main aim of the low molecular metabolism catalysed by a large number of proteins is to produce building stones of the polymers from nutrients.

The apparatus of reproduction of present life is constructed from proteins and nucleic acids of many different functions (replication, transcription, translation). The special function of a polymer is caused by a particular sequence of its monomers. In present day organisms these patterns are deter-

* In the case of an RNA genome – certain viruses and probably protobionts – both processes may not be differentiated.

mined by genetic information inherited from ancestors. At the time of the origin of life no such information was available. However, origin of life necessarily involves the origin of some apparatus of reproduction [13,14]. The apparatus must consist of a series of proteins as well as nucleic acids with the "right" sequences. The only source of these functional patterns could have been chance, i.e. abiotic production of polymers with a large variety of random sequences, and selection of the right ones from this mass of random polymers by trial and error, i.e. by Darwin's mechanism.

Random processes must have played a role in biogenesis at 2 levels: (1) in producing the right pattern of proteins and nucleic acids, (2) in constructing the apparatus of reproduction from a series of proteins and nucleic acids with the right patterns. Consideration and estimation of these chances are therefore essential for the understanding of the origin of life. Even if both polymers originated abiotically in large quantities on the early earth, life would scarcely or not have originated if these chances were too small for the generation of at least one protobiont. In the following paragraphs attempts are made to estimate the orders of magnitude of these chances [13,14,20].

The probability of functioning proteins

Proteins can originate abiotically by polymerisation of amino acids [9,10]. While the sequence of amino acids in such proteinoid is random with respect to biological function, e.g. as an enzyme, it is not exactly random with respect to the chance of incorporating certain amino acids into the polymer. Nevertheless, a large variety of sequences can be assumed to be present in such random proteins of the primitive earth. To simplify the estimation of the probability for random origin of an enzymatic activity in such a molecule the chances for all amino acids to be incorporated are assumed on the average to be equal. This simplification appears not too unrealistic when one takes into consideration that large quantities of abiogenic proteins probably have been produced on the primitive earth during a long period by different chemical reactions from different mixtures of amino acids.

Assuming this, the probability of finding one of the 20 types of amino acids at a certain position within the molecule is in this case 1:20, the probability for a certain type of amino acid in 2 positions is 20^{-2}, in 3 positions 20^{-3}, etc. The chain length of a modern enzyme is at least 100 amino acids. If a specific catalytic activity were to be equated with only one particular amino acid sequence, the probability of the enzyme would be $f = 20^{-100} = 1{:}10^{130}$. The number of atoms in the universe is assumed to be only in the order of 10^{80} [5]. Even if all atoms of the universe were used for making random protein, it would be almost impossible to obtain at least one molecule with the pattern of just this enzyme. One could conclude from this result that life could not have originated without a donor of information.

However, enzyme research has found that a particular enzyme function can be caused by many amino acid sequences. Only part of the positions within the chain contributes to the function, many are non-functional, e.g. lysozymes of several different organisms from man to phage are shown to have very different molecular structures and sizes but all of them attack specifically bacterial murein [3]. Work with mutants has shown that exchanges of amino acids often do not inactivate or even significantly lower enzyme function. This holds especially when the exchanged amino acids belong to the same group, e.g. the apolar group [7,11]. The 10 apolar amino acids make up about 30 to 50% of a molecule. They seem to be mainly responsible for the folding of the chain, their individuality appears to be of no great importance. Furthermore, it seems probable that only few amino acid positions have to be occupied by a particular type of amino acid, mainly those participating directly in the catalytic reaction [19]. In hen lysozyme, their number is apparently about 6, most of them being of a polar species [3]. Unfortunately, not enough is known for at least one enzyme investigated on the function and therefore on the exchangeability of all the positions.

In the following estimations it is assumed that to obtain a specific enzymatic activity of a modern protein, about 40 positions have to be occupied by one of the 10 apolar amino acids which is exchangeable within the group, and about 10 to 6 by only one species of amino acid which is not exchangeable, the other positions can be occupied by freely exchangeable amino acids. The probabilities f estimated for such enzymes are in the order of 10^{-25} to 10^{-20}. If the number of essential positions is smaller the probability becomes laeger (table 1). Assuming $f = 10^{-20}$ and a molecule of weight 2×10^{-20} g (100 amino acids), one molecule with the sequence functional for a given enzyme could be expected in 2 g random protein.

The sequences of modern proteins have developed to a functional optimum during a long evolution. Certainly less specific and lower activities can be obtained by sequences with smaller numbers of group-exchangeable as well as non-exchangeable amino acids. As table 1 shows, the chance f would be several orders of magnitude higher for such less effective enzymes. They could have arisen in abiotic proteins on the early earth. Experimental evidence for catalytic activities in proteinoids has been reported [9,12,16]. Probability f would be still higher (table 1) if the number of primordial amino acids was less than 20. Possibly only about 10 amino acids (e.g. Gly, Ala, Asp,

Table 1
Probabilities of functional proteins.

g	s	Probability f	
		$T = 20$	$T = 10$
40	10	10^{-25}	10^{-22}
40	6	10^{-20}	10^{-18}
30	5	10^{-15}	10^{-14}
20	4	10^{-11}	10^{-10}
10	4	10^{-8}	10^{-7}

T total number of types of amino acids. $T = 20$ for modern, 10 for primitive proteins;
n number of positions of amino acids being essential for function of the protein molecule;
F_i number of amino acid types one of which must occupy the given position i (= 1,2,3, ... or n) for function of the protein;

$$f = \left(\frac{F_1}{T}\right)\left(\frac{F_2}{T}\right)\left(\frac{F_3}{T}\right)\cdots\left(\frac{F_n}{T}\right)$$ *probability of a functional protein molecule.*

F_i = 1 for a non-exchangeable single type of amino acid;
s number of single-type positions within a molecule;
F_i = 10 for modern, 5 for primitive proteins, at positions where 10 or 5, resp., types are exchangeable within the group (e.g. apolar);
g number of group-type positions within a molecule;

$$f \approx \left(\frac{10}{20}\right)^g \left(\frac{1}{20}\right)^s$$ for proteins with g group-type and s single-type positions and with $T = 20$ amino acid types.

Glu, Thr, Ser, Lys, Leu, Ile, Val) were building blocks of the primitive proteins [2].

Although abiotic enzymes (protoenzymes) were much less effective than modern ones they could have functioned in protobionts constituting at least a slow life. In the following discussion the probability of a certain type of protoenzyme is assumed to be $f = 10^{-8}$ to 10^{-12}.

Probability of a nucleic acid to be a gene for a protein

Although abiotic formation of polynucleotides on the early earth seems not as clear as that of polyamino acids, a large variety of random sequences of bases can also be expected. Such base sequences would be informational for protein sequences if they are translated by an apparatus of reproduction, e.g. within a protobiont. According to information theory [8] the amount of information of a signal pattern is a simple function of the probability of this sequence. Therefore, the probability of a certain nucleic acid being a gene for a protein is the same as the probability of this protein. The probability of obtaining a gene for a certain protoenzyme by random polymerisation of nucleotides can be assumed $f = 10^{-8}$ to 10^{-12}, as for protoenzymes. Such a gene, formed abiotically, may be called a protogene.

Probabilities of formation of protobionts

The functional center of a cell is the apparatus of reproduction. It consists of nucleic acids and proteins performing replication and transcription of nucleic acids as well as translation of the base sequences into amino acid sequences. In this way the apparatus reproduces itself and the other cell components. If amino acids and nucleoside triphosphates are available as nutrients, a simple organism could consist solely of an apparatus of reproduction [13,14]. Since those nutrients were presumably present in the primordial soup of the early earth, a protobiont could arise if such an apparatus were formed from protoenzymes and protogenes.

However, reproduction, and therefore life, would result only if *all* necessary components were present at the same time and would remain within a small confined region i.e. within a protocell. Only such living individuals (protocells) are able to compete with one another and can evolve by Darwin's mechanism. Isolated components cannot participate in biological evolution because changes are not inherited. Therefore, a protobiont could arise only as an aggregate of abiotic protein and nucleic acid molecules, each having the correct sequence. This would be a multi-hit process. Here again chance plays a deciding role. Possible representatives of such aggregates are coacervates [16], microspheres [7], mineral (e.g. clay) particles covered with both types of polymers [18], or even pores in sand or mud filled with those substances. Such aggregates (precells) would have to be formed in large numbers (see below).

The number of components making up the reproduction apparatus of present life is large, at least about 80 special proteins and 100 genes coding for them as well as for tRNA's and rRNA's are necessary. One can assume that protobionts possessed a much simpler apparatus. It would be simple if amino acids could specifically bind colinearly to polynucleotides ("direct translation"), as supposed by some authors [17,21]; in this case the apparatus could perhaps consist of only 2+2 polymers: 1 replicase being also transcriptase if the genes are 1-stranded nucleic acid, further 1 protein polymerase (making peptide bonds between the adsorbed amino acids) plus the 2 genes coding for both proteins. If tRNA's are necessary as adaptors between gene and amino acids [4,19] as in contemporary organisms ("indirect translation") the apparatus would have been much more complicated. If in a primitive state only few tRNA's as well as aminoacyl synthetases with low specificity (e.g. one for apolar, one for acidic and one for basic amino acids) were available for 10 primitive amino acids, a minimum of about 5 proteins + 10 to 15 genes would be necessary.

In order to estimate the probability of expecting a protobiont among many aggregates, the number of necessary kinds of functional proteins as well as protogenes, respectively, is assumed to be 20 to 40. This seems sufficient for a primitive indirect translation. Direct translation would function with a lower number but since it is not yet clear whether it can exist or not, it is not taken into consideration for the estimations.

The above number B (= 20 to 40) is called a "basic set" of necessary types (sequences) of polymer molecules. In order to calculate the probability of obtaining an aggregate with at least one basic set of *protoenzymes* it is assumed for simplification that the probabilities of these functional molecules are all equal ($=f$). The mean number of functional molecules of *one* given pattern is $m = fA$ if aggregates have the size A (= number of random molecules per aggregate) (table 2, lower part). Since the probability of aggregates with at least one such molecule of a basic type is $1-e^{-m}$ the probability of aggregates containing at least one basic set of B functional proteins is $P = (1-e^{-m})^B$ [13,14,20].

The same reasoning is valid for *nucleic acid* molecules and leads to a P of the same formula. The probability of forming a protobiont among many aggregates consisting of random protein plus random nucleic acid is $P_{Pr} \times P_{Na}$. Since the probabilities of functional molecules of both polymers are about equal and since the basic sets of both are not very different in size, the approximation $P_{Pr} \times P_{Na} = P^2$ is made. This is the probability of forming one protobiont among P^{-2} aggregates of the size A.

One could assume that a protobiontic aggregate may consist simply of the two basic sets. In this case the mean number of a certain necessary molecule is $m = fB \ll 1$ (e.g. 20×10^{-8}). From this value $P^2 \approx (fB)^{2B} \approx (10^{-7})^{2\times 20} = 10^{-280}$ is obtained. Such a rare event would scarcely become real in this world.

However, "real" chances of protobionts can be obtained if the sizes of aggregates are much larger than the basic sets. A suitable size is $A = 1/f$ because in this case $m = 1$ and the probability of a protobiont is simply $P^2 = (1-e^{-1})^{2B} = 0.63^{2B}$.

Table 2
Probabilities of protobionts.

f	B	Size of aggregate			P^2	Mass of $1/P^2$ aggregates [c]
		$A = 1/f$	Weight [a] (g)	Diameter [b] (μm)		
10^{-8}	20	10^8	10^{-11}	2	10^{-8}	1 mg
10^{-8}	40	10^8	10^{-11}	2	10^{-16}	100 kg
10^{-10}	20	10^{10}	10^{-9}	10	10^{-8}	0.1 g
10^{-10}	30	10^{10}	10^{-9}	10	10^{-12}	1 kg
10^{-10}	40	10^{10}	10^{-9}	10	10^{-16}	10 tons
10^{-11}	20	10^{11}	10^{-8}	20	10^{-8}	1 g
10^{-11}	30	10^{11}	10^{-8}	20	10^{-12}	10 kg
10^{-11}	40	10^{11}	10^{-8}	20	10^{-16}	10^2 tons
10^{-12}	20	10^{12}	10^{-7}	50	10^{-8}	10 g
10^{-12}	40	10^{12}	10^{-7}	50	10^{-16}	10^3 tons
10^{-14}	20	10^{14}	10^{-5}	200	10^{-8}	1 kg
10^{-14}	40	10^{14}	10^{-5}	200	10^{-16}	10^5 tons

[a] Weight of 1 molecule = 10^{-19} g
[b] (Weight of 1 aggregate in $cm^3)^{1/3}$
[c] Weight of 1 aggregate × P^{-2}

f probability of a functional polymer pattern;
B number of types (patterns) of functional polymers (protein or nucleic acid, respectively) necessary for life (basic set),
A number of random polymer molecules within an aggregate (size of an aggregate);
$m = fA$ mean number of molecules of one necessary type per aggregate;
$P = (1-e^{-m})^B$, probability for an aggregate to contain at least one molecule of each of the B necessary types;
$P_{Pr} = P$ for proteins, $P_{Na} = P$ for nucleic acids.
$P_{Pr} \times P_{Na} \approx P^2 = (1-e^{-m})^{2B}$, *probability of a protobiont*;
$A = 1/f$, $m = 1$, $P^2 = (1-e^{-1})^{2B} = 0.63^{2B}$.

Table 2 shows a series of values obtained with this assumption. Using basic sets of $B = 20$ to 40 the protobiontic chances P^2 are 10^{-8} to 10^{-16}. Probabilities of functional polymers of $f = 10^{-8}$ to 10^{-14} and, accordingly, aggregate sizes $A = 10^8$ to 10^{14} molecules give weights of an aggregate of 10^{-11} to 10^{-5}g and diameters of 2 to 200 μm. These dimensions are in the order of present cells.

Another interesting value is the mass of the number of random aggregates $1/P^2$ which can be expected to contain one protobiont. As shown in table 2 these masses are of the order of grams to kilograms for basic sets of $B = 20$ to 30 and enzyme probability of $f = 10^{-8}$ to 10^{-14}. For larger sets they are several to many tons. These masses had to be produced on the early earth to give rise to life by chance.

Among the molecules of a protobiont some could have injurious activities. Therefore aggregates should not be large enough to include such molecules with a high chance. With sizes $A = 1/p$ the probability for exclusion of damaging proteins as well as genes is $P_d^2 = (e^{-1})^{2d}$ where d is the number of possible injurious activities. The P^2 given above would have to be multiplied by P_d^2 if such injurious molecules are to be considered. Assuming $d = 2$, e.g. for nuclease and protease, the probability of exclusion of P_d^2 is about 10^{-1}, for $d = 10$ about 10^{-4}. These chances do not seriously decrease the chances of protobionts given above.

All estimations are based on strong simplifications and are, therefore, very rough. They can give only an idea of orders of magnitude for the probabilities and masses of material involved in the origin of protobionts due to random aggregation of abiotic polymers. Nevertheless, one can conclude from them that biogenesis was possible on the primitive earth without pre-existing information on life-producing sequences of macromolecules. It also seems even probable that protobionts arose repeatedly independently within a geologically short period. If direct translation was possible needing basic sets of only $B = 2$ to 5, the number of such direct translating protobionts arising independently could be high. A multiple origin of protobionts (also of indirect translators of different types, e.g. of different genetic codes) would have led to many competing strains. One of these strains survived and grew to the phylogenetic tree which all present organisms seem to belong to.

Early evolution of protobionts

Protobionts of the type considered would be highly heterotrophic, requiring all amino acids as well as nucleosides or their triphosphates. These nutrients can be expected to have been available in the primordial soup. The growth was certainly very slow because of the low efficiency of proto-enzymes as well as of the probable low concentration of nucleoside-triphosphates.

The nucleic acids of a protobiontic aggregate are replicated by the primordial apparatus of reproduction but only proteins coded by the protogenes included are synthesized. The newly formed proteins are not the same as those included originally in the aggregate. However, they would have to contain a basic set of enzymes which were also not the same as the original. This is insured in our formula by P_{Na} for a basic set of nucleic acids with the right patterns. Aggregates with only a basic set of proteins would not be protobionts and they would be much more frequent (= P_{Pr}) than protobionts.

A protobiont will grow to a size where it becomes fragile and breaks into irregular pieces [13]. Only pieces receiving, by chance, at least one basic set of proteins as well as one of nucleic acids would be living progeny. During further multiplication of the clone, the original proteins are gradually replaced by the coded ones. Most of these proteins (and genes) are non-functional. However, they represent a large reserve for evolution of additional new functions due to mutation and selection. Because of the very low efficiency of protoenzymes, the selective advantage of even small improvements would be high. A much higher fraction of mutations would be positive than today. Such further evolved and improved progeny of protobionts may be called "eobionts" and are bridges to the procaryonts.

Important steps of the evolution of eobionts in the direction to familiar cells could be as follows [13]:

(1) If direct translators can exist they would have arisen with much higher frequency than indirect translators due to their greater simplicity. However, very probably their translation was rather inaccurate due to the binding of amino acids to nucleotide sequences with low specificity giving a high fraction of worthless proteins. Unfortunately, the accuracy could not be essentially improved by evolution since it is determined by these binding forces given. In contrast, indirect translators were able to improve their translation apparatus stepwise to a very high accuracy of function, as it is present now, by evolving the tRNA's, aminoacyl synthetases, ribosomes etc. Though indirect translators were much more rarely formed they would have overgrown and extinguished the direct translators due to the great selective advantage of more accurate translation. Nevertheless, the direct translators could have been of great value since they would have been a source of proteins as well as nucleic acids with many sequences. In particular, their reproduction of nucleic acids could have been an essential way of forming great masses of these polymers, whose abiotic formation seems at the moment to be scarcer than that of proteins. Nucleid acids are needed for the origin of every type of protobiont.

(2) The chance of an eobiontic strain breaking into living pieces could be increased if a solid pellicle evolved by mutation from non-functional protein. Such eocells would grow to a larger size and to a larger ploidy of genes and proteins; they would break into larger pieces receiving complete basic sets

more often than their parents. The further evolution of a fragile belt within the pellicle would cause frequent breakage into two approximately equal pieces.

(3) Coupling of genes, including the necessary ones, to a chain would further increase the chance for living pieces and therefore the rate of multiplication. A ligase could do this.

(4) Differentiation of two-stranded genes from one-stranded messengers and thus replication from transcription – would increase the stability of the genome against damage without impairing the usefulness of one-stranded nucleic acid for protein synthesis.

(5) Evolution of new enzyme systems producing rare building blocks, such as nucleoside triphosphates, from more frequent components of the primordial soup would increase the speed of growth and multiplication, giving great advantage over the more heterotrophic strains of eobionts.

In the evolution of proto- and eobionts the large reserve of non-functional polymers derived from the protobiontic aggregate would be of great advantage compared with protobionts consisting only of the basic sets of functional polymers. New functions could develop by mutations from the large variety of not yet used sequences without removing at first the "old" functions and without being confined to the few sequences of already functioning polymers.

References

[1] Bernal J.D., The Origin of Life (World Publishers, Cleveland, New York, 1967).
[2] Bloom B., Sci. Res. 3 (1967) 62.
[3] Chipman D.M. and N. Sharon, Science 165 (1969) 454.
[4] Crick F.H.C., J. Mol. Biol. 38 (1969) 367–379.
[5] Dicke R.H., in: Die Erde im Weltraum (Colloquium Verlag, Berlin, 1968) p. 220.
[6] Drake J.W., Ann. Rev. Genet. 3 (1969) 247.
[7] Epstein Ch. I., Nature 210 (1966) 25.
[8] Flechtner H.I., Grundbegriffe der Kybernetik (Wiss. Verl. Ges., Stuttgart, 1966).
[9] Fox S.F., Nature 205 (1965) 328.
[10] Fox S.F., Naturwissenschaften 56 (1969) 1.
[11] Goldberg A.L. and R.E. Wittes, Science 153 (1966) 420.
[12] Hardenbeeck H.G., G. Krampitz and L. Wulf, Arch. Biochem. Biophys. 123 (1968) 72.
[13] Kaplan R.W., in: Evolution d. Organismen, ed. G. Heberer (Fischer, Jena, 1967) p. 493.
[14] Kaplan R.W., Naturwissenschaften 55 (1968) 97.
[15] Kaplan R.W., in: Molekularbiologie, (Umschau-Verlag, Frankfurt, 1969) p. 73.

[16] Krampitz G., S. Baars-Diehl, W. Haas and T. Nakashima, Experientia 24 (1968) 140.
[17] Lacey J.C. and K.M. Pruit, Nature 223 (1969) 799.
[18] Oparin A.I., Genesis and Evolutionary Development of Life (Academic Press, New York, London, 1968).
[19] Orgel L.E., J. Mol. Biol. 35 (1969) 381.
[20] Quastler H. The Emergence of Biological Organization (Yale Univ. Press, 1964).
[21] Woese C.R., Proc. Natl. Acad. Sci. U.S. 59 (1968) 110.

Chemical Evolution and the Origin of Life, eds. R. Buvet and C. Ponnamperuma

A STUDY ON INTERRELATION OF MODEL STRUCTURES WITH BIOCHEMICAL PROCESSES OCCURRING IN THESE STRUCTURES

N.V. VASILYEVA
A.N. Bakh Institute of Biochemistry, Academy of Sciences of the USSR, Moscow, USSR

The basis of metabolism of living systems is the biochemical processes catalyzed by enzymes. Modelling of these processes in coacervated systems and studies of their regulation make it possible to investigate the effect of the system on the processes occurring therein, and vice versa [2]. Regulation of the simplest biochemical processes in such systems could be affected through the participation of components of the system itself and was perhaps the only type of regulation in primitive systems which had not yet developed complex regulatory mechanisms of living organisms.

This primitive regulation gave rise to the formation of self-preserving and growing systems which formed the basis of their stability against disintegration and lead to further evolution. The next evolutionary stage of pre-biological systems was the appearance of processes enabling growth either by synthesis of organic compounds or by using substances from the ambient medium. This principle could underlie primary selection of such systems. Their capacity for self-preservation and growth are conditions that make them able to resist destruction.

It is known that one of the fundamental prerequisites for protection of organic compounds against decomposition is the formation of complexes. Since formation of coacervates involves complex interactions between the components, the appearance of coacervated systems was in itself the first step in forming stability against decomposition of compounds therein.

But static coacervates could not play any significant role in evolution, even though they possessed a certain stability (continuity of existence). Their stability associated with immobility differs from that of living matter. Living matter is stable as long as a combination of processes known as metabolism continuously occurs. An increase in the resistance of primitive static systems to destruction in the course of evolution, their conversion into dynamic systems and formation on this basis of self-preserving systems and systems capable of growth, have played a key role in their conversion into the simplest living systems.

In our investigations we used nucleoprotein coacervates as models, and ribonuclease and polynucleotide phosphorylase as enzymes.

A model for obtaining a dynamically stable system was the coacervated system histone–RNA with incorporated enzyme ribonuclease. The rate of hydrolysis of RNA by ribonuclease in this system was only 15% of that in the absence of histone. This was shown by kinetic and polarographic methods to be due to the formation of a histone–RNA complex which is in agreement with published data on the protection of nucleic acids by basic proteins [1,5]. This factor is the main prerequisite for the formation of dynamically stable systems. An extemely high rate of hydrolysis would have caused rapid destruction of the coacervated system because of the loss of its RNA.

However, despite the fact that the destruction of histone–RNA is slow, the system still gradually disintegrates with time.

Our object was to obtain a system in which a decomposing substance would be replenished from the surrounding medium. Such a system, once formed, could exist in its structural integrity for an indefinitely long time, despite the disintegration.

To attain this object – formation of a dynamically stable drop functioning through continuous inflow of the substance from the outside – it was necessary to find out whether the substrate in the equilibrium liquid could be supplied to the coacervate drops, and what further conversions it would undergo. We therefore measured the inflow of RNA into the coacervate drops as a function of its concentration in the equilibrium liquid. We incubated the coacervates obtained with additional RNA at 37 °C in a water thermostat for 15 min, and then centrifuged the samples into different components. In the coacervate drops, we determined the amount of nucleic acid phosphorus according to Spirin's method with subsequent calculation of the result in terms of nucleic acid [4]. At the same time we measured the quantity of nucleotides that emerged from the drops during incubation (table 1).

The table shows that the addition of RNA to coacervate without an enzyme causes partial dissolution of the coacervate drops. If RNA is added to the coacervate containing an enzyme, the amount of RNA in the coacervate decreased at first, then increases somewhat, and then maintains constant level regardless of the concentration of RNA in the equilibrium liquid. The stability of the drops, nonetheless, is not a result of their static state, because,

Table 1
Amount of added RNA bound by coacervate drops and emergence of nucleotides from drops versus concentration of RNA in coacervate.

Amount of RNA in coacervate (mg)	Amount of RNA in coacervate drops 5 min after incubation (mg)		Accumulation of nucleotides in equilibrium liquid during incubation (mg)
	– enzyme	+ enzyme	
5	3.460	3.310	0.069
10	3.240	3.740	0.266
15	3.060	3.820	0.390
20	2.860	3.780	0.613

as is evident from the table, hydrolysis does take place in the drops, and the amount of nucleotides accumulating in the equilibrium liquid is proportional to the content of RNA in the reaction medium.

Another subject of our investigations was to find out whether in the same conditions, i.e. in the presence of an excessive substrate, the coacervate drops would retain their stability with time while hydrolysis was going on in the drops. For this purpose, we carried out a comparative study of the dynamics of the coacervate drops with time, with and without additional substrate. After incubation and subsequent centrifugation of the coacervates into individual components, we estimated the RNA content in the coacervate drops (fig. 1). Coacervate drops without an enzyme were used as controls (fig. 1, curves b and c).

As can be observed from the figure, in the coacervate drops with enzyme but no additional RNA (curve a), the quantity of the nucleic acid component decreases with incubation time as a result of hydrolysis of RNA in the drops. In coacervate drops without enzyme in the presence of additional RNA (curve b), the amount of the nucleic acid component is constant with time. The absolute quantity is reduced through dissolution of the drops in the excess of one of the components. In the coacervate drops without enzyme and in the absence of additional RNA (curve c), the RNA content also remains unchanged with time. While in the coacervate drops with the enzyme and additional RNA (curve d), the amount of RNA at first increases and then remains constant. It should be remembered, however, that as shown above, hydrolysis in the drops continues, since in the course of coacervate incubation, nucleotides are accumulated in the equilibrium liquid.

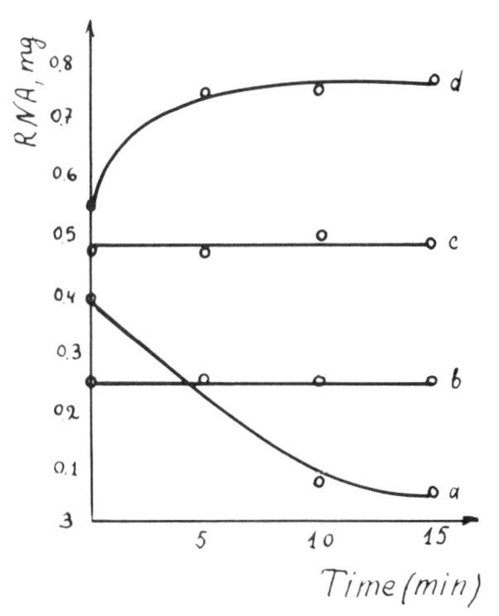

Fig. 1. The content of RNA in coacervate drops in the course of incubation. a) with enzyme, without additional RNA; b) without enzyme, with additional RNA; c) without enzyme, without additional RNA; d) coacervate drops with enzyme, with additional RNA.

Fig. 2 gives data on the emergence of nucleotides from the coacervate drops containing additional RNA, in the course of incubation.

Summing up, the coacervate drops containing the enzyme are radically different from those without the enzyme. The latter are typically static systems. Drops containing the enzyme in the absence of additional RNA are broken down during incubation. When RNA is supplied from the outside, the drops become dynamically stable and their composition remains constant with time.

The main feature of the system obtained is that it is stable not only functionally, but also structurally on account of continuous exchange between a component in the system and outside.

The model of a system able to grow was in our experiments: histone–RNA containing the polynucleotide phosphorylase. A study of the rate of poly A synthesis with time demonstrated that the rate of synthesis is much higher than that of a homogeneous solution. The results of the study are given in fig. 3. Comparison of initial rates of the processes occurring in a homogeneous solution and in a coacervate reveals that in the latter the rate of synthesis of poly A is 17 times higher than that in the homogeneous solution. In earlier experiments in our laboratory it was shown that the synthesis of poly A catalyzed by polynucleotide phosphorylase in the presence of histone

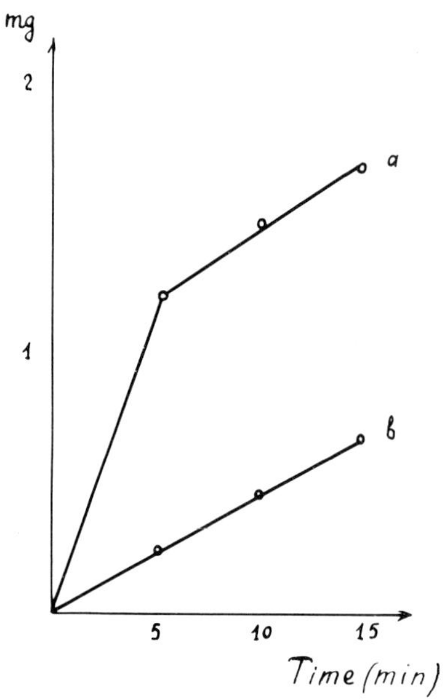

Fig. 2. Digestion of added RNA by coacervate drops as a function of time: a) accumulation of nucleotides in the coacervates; b) emergence of nucleotides from the coacervate drops.

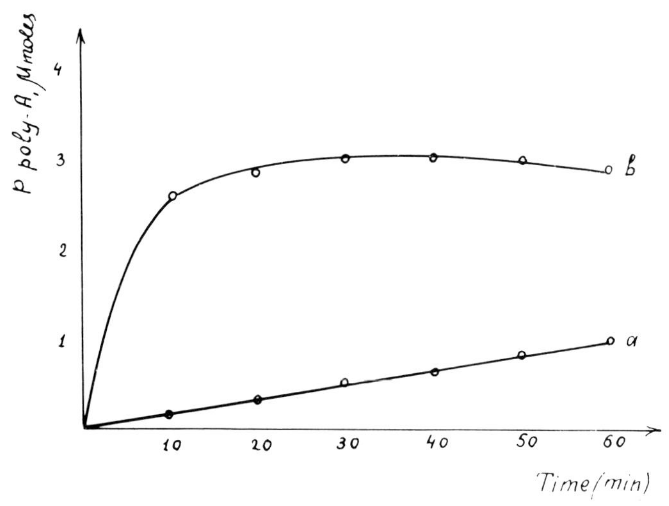

Fig. 3. Synthesis of poly A in coacervate drops: a) homogeneous solution; b) coacervate system.

proceeded much more rapidly [3]. The other component of the coacervate, RNA, did not affect the synthesis of poly A.

Thus, it can be assumed that polynucleotide phosphorylase is activated as a result of binding of synthesized polymer with the histone of the equilibrium liquid which causes a shift of the equilibrium towards synthesis of poly A. This leads to growth of the coacervate drops not only through an increase in the amount of the polymer, but also at the expense of the histone bound by the synthesized polymer in the equilibrium liquid. The total increase in the mass of the coacervate deposit is 14%. However, a certain period after sharp activation of synthesis, equally abrupt inhibition is observed, irrespective of whether there is sufficient substrate in the ambient medium.

As is known, any system can grow only to a certain limit, because disproportion between the volume of the system and its surface-area will interfere with the exchange of pre-biological systems with the ambient medium and cause their destruction. But if a system could somehow divide, such a system could again grow to a certain limit on condition that there is extensive exchange with the medium.

In preliminary experiments we have shown that partial disintegration of coacervate drops under the influence of ribonuclease causes these drops to start functioning again. However, an increase in the amount of poly A in the coacervate drops treated with ribonuclease is approximately equal to that of hydrolyzed RNA. In other words, we have obtained a dynamic system in which hydrolysis and synthesis are mutually compensated. No increase of growth is observed. We believe that to achieve this increase, deeper disintegration of he coacervate drops is required – down to their complete breakage into parts.

We have thus analyzed two model systems in which the hydrolysis and synthesis lead to the formation of new properties which can be regarded as specific functions of these systems.

Such models have profound significance, since the formation of self-preserving systems and systems capable of growth at a certain stage of evolution certainly played an important role in the conversion of these pre-biological systems into the simplest living systems.

References

[1] Liebl B., Mezhd. Biokhem. Kongress, M. 1 (1961) 269.
[2] Oparin A.I., Life, its Nature, Origin and Development (Nauka, Moscow, 1968) p. 70.
[3] Oparin A.I. and K.B. Serebrovskaya, Dokl. Acad. Nauk SSSR 148 (1963) 243.
[4] Spirin A.S., Biokhimiya 23 (1958) 656.
[5] Tashiro J., J. Biochem. (Tokyo) 45 (1958) 803.

Chemical Evolution and the Origin of Life, eds. R. Buvet and C. Ponnamperuma

OXIDOREDUCTASES AND THE STABILITY OF COACERVATE DROPS

T.N. EVREINOVA, T.W. MAMONTOVA, W.N. KARNAUCHOV
and A.N. DUDAEV

Moscow University, Lomonosov Biological Faculty,
Department of Plant Biochemistry and Biophysical
Institute, Academy of Science of the USSR, Moscow, USSR

Coacervate systems are one stage of association of molecules in the ancient ocean (primary soup) according to Oparin's theory of the Origin of Life. The primary soup consists of an aqueous solution of different and complex organic substances, monomers and polymers [5–7]. The hydrophilic coacervate systems consist of drops and equilibrium liquid. The drops have a size

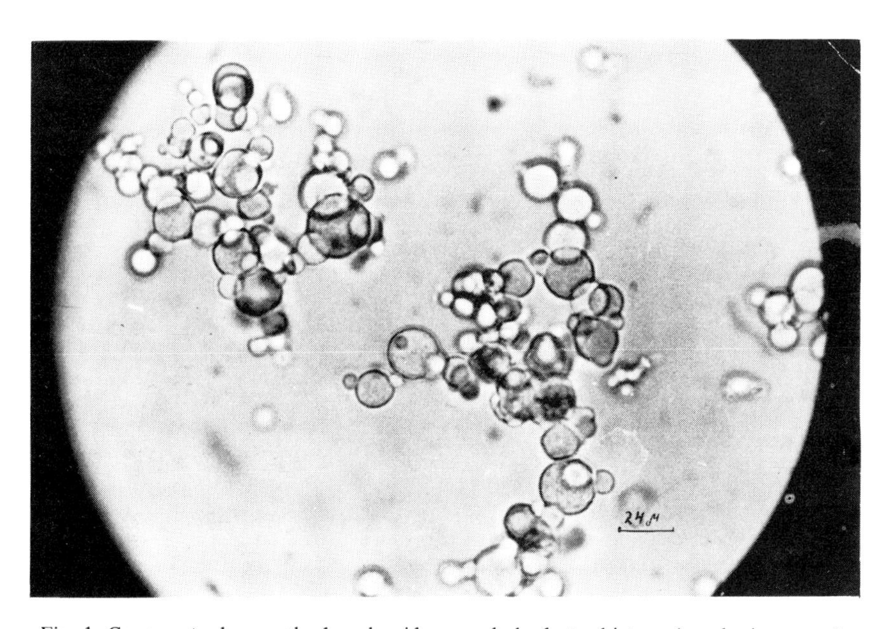

Fig. 1. Coacervate drops polyphenol oxidase–carbohydrate–histone (usual microscope).

range 0.5–640 μm in diameter. They have many new properties, for instance: 1) the association of molecules in drops; 2) the formation of surfaces between drops and equilibrium liquid, as well as surfaces forming the drops.

The coacervate systems contain drops with different distributions of polymer molecules (polypeptides, polynucleotides, proteins, nucleic acids, carbohydrates, etc.). The distribution may be homogenous (diffused more or less evenly) or heterogenous. In the latter case there is very wide variety. Drops can be found in which the molecules form clumps with zones in which polymer is present in negligible amounts, for example in vacuoles. That is why there are the specific conditions for the investigation of the behavior of enzymes in coacervate systems [1]. The evolution of drops was possible if protoenzymes could act in coacervate systems during prebiological stages. More than 200 differing hydrophilic coacervate systems are known. Recently hydrolytic synthetic and oxido-reduced enzymatic reactions have been investigated in coacervate systems. The drops were shown to settle down to the bottom of the vessel and disappear. They were converted into a thin layer.

The aim of this report is the demonstration of stable drops. They also settle down but they do not disappear. They have remained intact for over two years.

The composition of coacervate systems

The coacervate systems were prepared as follows:
(1) *Carbohydrate–histone.* 0.8 ml H_2O (distilled); 0.2 ml 0.5 M acetic buffer (pH = 6.0); 0.5 ml arabinat Na 0.67%; 0.5 ml substrate; 0.1 ml enzyme; 0.4 ml histone 1%.
(2) *DNA–histone.* 0.2 ml H_2O (distilled); 0.1–0.2 ml 0.5 M acetic buffer (pH = 6.0); 0.5 ml histone 1%; 0.5 ml substrate; 0.1 ml enzyme; 0.4 ml DNA 0.5%.

The enzymes were polyphenol oxidase (EC 1.1031) and peroxidase (EC 1.11.1.7). The substrates of polyphenol oxidases were tyrosine and pyrocatechol, and the substrates of peroxidase used pyrocallol, H_2O_2, *o*-dianisidin. The behavior of the enzymes was investigated by varying the concentration of enzymes and substrates and the reaction times. When the reaction was completed, the coacervate drops were dissolved in 2.5 ml 1 M

* The activity of enzymes was equal: 500 units polyphenoloxidase and 572 units peroxidase [8].

acetic buffer. Optimal concentration per 1 ml was equal (in %): enzymes 0.04, pyrocatechol 0.04, tyrosine 0.07, pyrogallol 0.4, *o*-dianisidin 0.4 and H_2O_2 0.4. The reactions continued for 15–30 min. The drops became yellow–brown. One such drop was very stable. In this case the substrates were oxidized by enzymes to different quinones. That is why the distribution

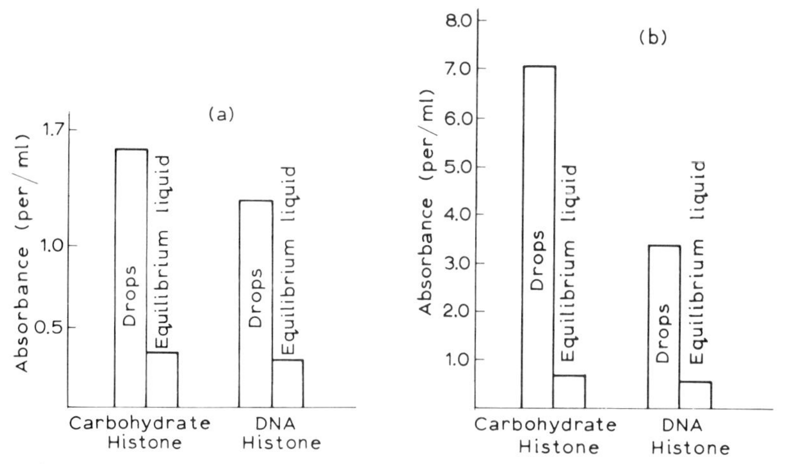

Fig. 2. The distribution of quinones in coacervate systems: a) polyphenol oxidase; b) peroxidase.

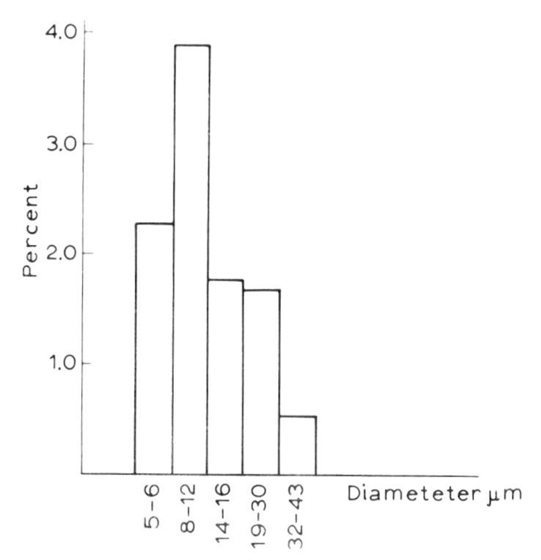

Fig. 3. The distribution of coacervate drops, polyphenol oxidase–carbohydrate–histone–quinones.

A

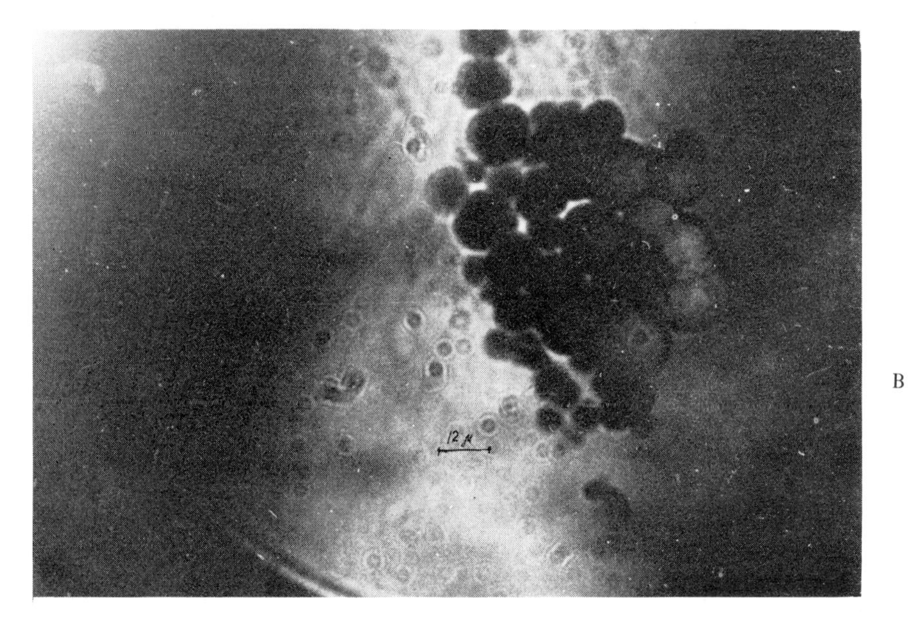

Fig. 4. A,B Coacervate drops: polyphenoloxidase–carbohydrate–histone (interference microscope).

Table 1
Individual coacervate drops
(polyphenol oxidase–protein–carbohydrate)

No.	Diameter (10^{-4} cm)	Volume (10^{-12} cm^3)	Weight (10^{-12} g)	Concentration (C) (%)	C drop / C solution
1	0.53	0.08	0.05	52.4	105
2	2.22	5.73	2.03	35.4	71
3	4.0	33.44	7.34	21.94	44
4	7.33	204.94	23.85	11.64	23
5	8.44	314.13	31.73	10.10	20

of quinones in coacervate systems was measured and calculated in units/per 1 ml absorbance at 395–400 nm (polyphenol oxidase systems) and 420 nm (peroxidase systems). The results are shown in fig. 2 A–B. More quinones are in the drops than in the equilibrium liquid [4]. The sizes and the dry mass of individual stable drops are shown in fig. 3–4 and in table 1. The sizes were calculated by AB-analysator of biological particles (Laboratory Dr. G.R. Ivanizcov of Biophysical Institute) and dry mass of molecules was measured by interference microscope with Dr. A. Bailey (Laboratory Prof. H. Davis, King's College, London University).

The limit of measurement of dry mass was 10^{-14} g with an error of 2–5 percent [2,3].

The coacervate drops under an interference microscope are shown in fig. 4. There every drop is separate.

The content of quinones in individual drops was measured by a cytospectrophotometer MUV-5 with automatic recording (Laboratory, Academician G.M. Frank, Biophysical Institute). The limit was $10^{-14}-10^{-15}$ g of quinones with an error of 2–7 percent. The quantity of quinones was calculated by special formulas [1]. The data are shown in fig. 5 and in table 2. The quinones were found (in practice) only in the drops. The stability of drops is connected with very small quantity of quinones. Thus the size and dry mass of individual stable coacervate drops are the same as in other coacervate systems. The size, dry mass and number of polymer molecules in living cells and in coacervate drops are shown in tables 3 and 4. We often compare coacervate drops and living things. This does not mean that drops and cells are the same. We wish only to emphasize that both cells and drops may have similar properties. It seems that these phenomena are the result of properties of polymer molecules which are both in cytoplasma and

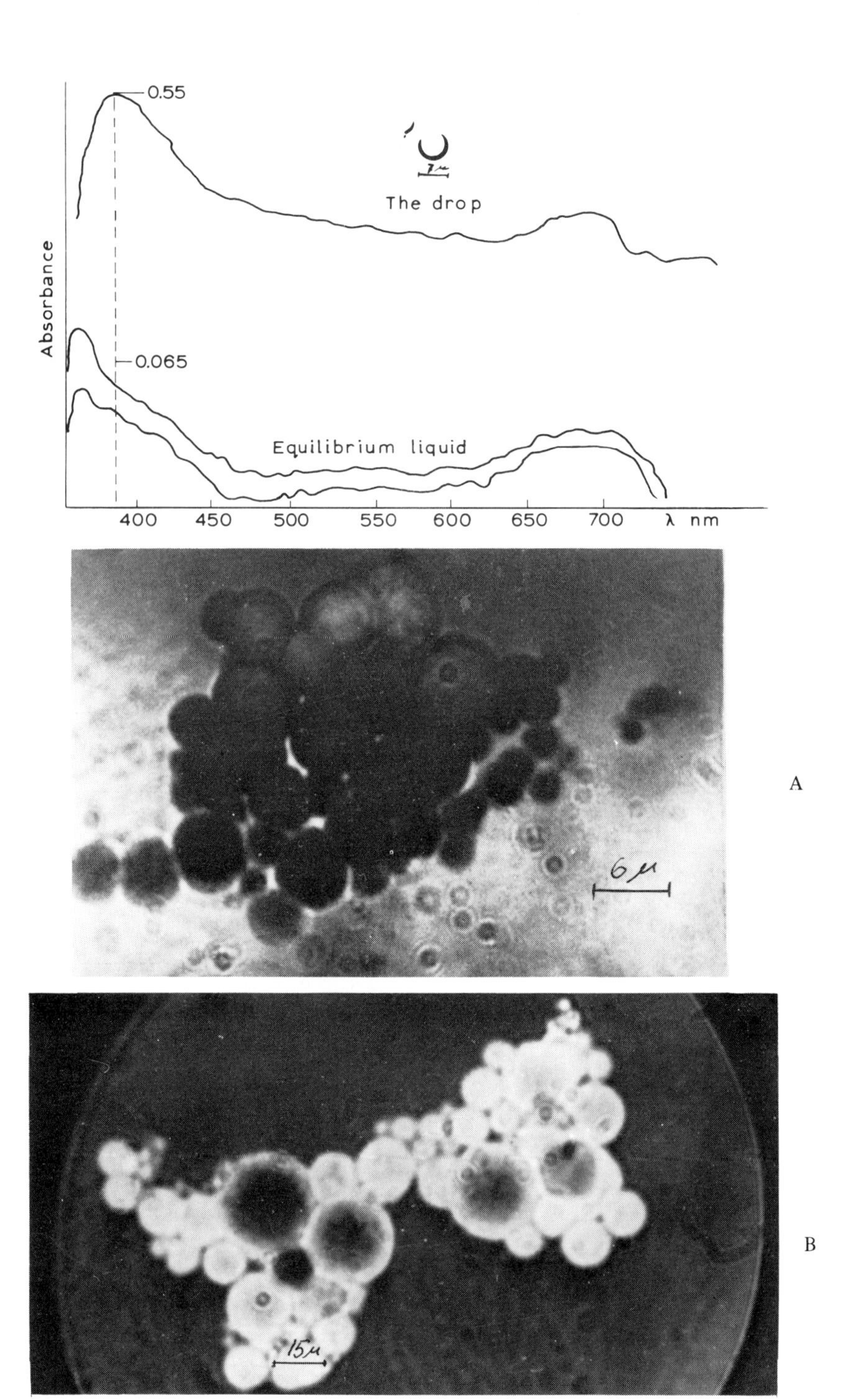

Fig. 5. Polyphenol oxidase–carbohydrate–histone–quinone.

Table 2
The content of quinones in individual drops

No.	Diameter (10^{-4} cm)	Volume (10^{-12} cm^3)	Weight (10^{-12} g)	Concentration (%)
Peroxidase–DNA–histone				
1	2.00	4.2	0.06	1.4
2	4.4	44.6	0.09	0.2
3	6.3	130.8	0.17	0.1
Peroxidase–carbohydrate–histone				
4	5.6	98.85	0.07	0.07
5	24	7229.95	0.19	0.003

Table 3
Weight and concentration of dry mass in various objects

Objects	Diameter (cm)	Volume (cm^3)	Weight (g)	Concentration (%)
Bacteria	2.5×10^{-4}	8.5×10^{-12}	2.5×10^{-12}	30–25
Mammalian cells	2.5×10^{-3}	8.5×10^{-9}	2.1×10^{-12}	25–10
Amoeba	1.0×10^{-2}	5.2×10^{-7}	7.8×10^{-8}	15–10
Coacervate drops	2.42×10^{-4} -1.01×10^{-2}	7.4×10^{-12} -5.3×10^{-7}	2.5×10^{-12} -3.5×10^{-8}	34–7

Table 4
The occupied space by polymer molecules in various objects

Molecules	Objects	
	Number	Volume (cm^3)
Bacteria	4×10^{7}	4.4×10^{-12}
Mammalian cells	3.6×10^{10}	4×10^{-9}
Amoeba	1.2×10^{12}	1.3×10^{-7}
Coacervate drops	5×10^{7} -7.2×10^{11}	5.5×10^{-12} -7.9×10^{-8}

coacervate drops. Perhaps protomolecules (protoproteins, protonucleic acid etc.) were present in primitive prebiological systems (coacervate systems and other ones) many billions of years ago. We feel that the differing stability of coacervate drops is of interest from the point of view of selection of drops best suited to evolution.

Acknowledgments

We wish to thank Academician A.I. Oparin, Prof. John Bernal, Prof. Harry Carlisle, Prof. Horward Davis, Dr. Anita Bailey, Prof. G.M. Frank and Dr. G.R. Ivanizkii.

References

[1] Evreinova T.N. in: Concentration of Substances and the Action of Enzymes in Coacervate Systems, ed. A.I. Oparin (Nauka, Moscow, 1966).
[2] Evreinova T.N. and A. Bailey, Dokl. Acad. Sci. USSR, 179 (1968) 723.
[3] Evreinova T.N., A. Bailey, T.W. Mamontova and M.M. Garber, J. Evolutionary Biochem. SSSR (1970) in press.
[4] Mamontova T.W. and M.M. Garber, The action of oxidoreductases in coacervate systems, in: Symposium of Science (Moscow University Publication, 1968) p. 3.
[5] Oparin A.I., Life, its Nature and Origin (Academy of Science USSR, Moscow, 1960).
[6] Oparin, A.I., K.B. Serebrovskaja, Dokl. Acad. Sci. SSSR 154 (1964) 407.
[7] Pavlovskaja T.E., A.G. Pasinskii, W.S. Sedorov and A.J. Ladiginskaja, in: Abiogenes and the Initial Stages of Evolution of Life (Nauka, Moscow, 1968) p. 41.
[8] Worthington Catalog of Enzymes (New-Yersey, 1968).

Chemical Evolution and the Origin of Life, eds. R. Buvet and C. Ponnamperuma

POSSIBLE PARTICIPATION OF PIGMENTS IN FORMATION OF SIMPLEST STRUCTURES

G.A. KORNEEVA
A.N. Bakh Institute of Biochemistry, Academy of Sciences of the USSR, Moscow, USSR

It is now generally accepted that single membranes are made up from layers of proteins and lipids; a peculiar feature of chloroplasts is that they also contain layers of chlorophyll and carotinoids. It is assumed that the formation of membrane structures in cells depends on the ability of associated colloids (soaps, detergents, phosphatides) to form micelles in a solution. Studies of proteins which form the structures of protoplasm organelles lead to the conclusion that the components of mitochondria and submitochondrial particles are linked by hydrophobic bonds [6], thus determining their high affinity to lipids. Therefore, a complex formed by structural protein, lipids and enzymes results from the action of attractive forces between the hydrophobic groups.

The above investigations have been conducted at the molecular level, while the colloid nature of these complexes remains unstudied.

We believe that modelling is a promising experimental approach that can provide an insight into the colloidal nature of natural lipoproteides. X-ray diffraction of lipid–water systems as a function of concentration and temperature performed by V. Luzzati and coworkers and others reveals that polymorphous changes in such systems are not very different from those in natural lipoprotein cell structures [9]. In this connection structures formed in aqueous solutions of surface-active agents (SAA) seem to be highly suitable for modelling lamellar systems of a cell. Such structures incorporating biologically active agents can be used as models of primary systems.

Our purpose was to study lipid systems both before and during coacervation. The lipids used were soaps (oleates) and phospholipid (a lecithin fraction obtained from bull's brain). Phospholipids are of particular interest as a group of associated colloids of a biological origin. Chlorophyll and protochlorophyll served as pigments.

In earlier experiments [12], we investigated the effect of biologically active proteins with different hydrophobic properties on coacervation in a soap–water–KCl system, proceeding from D.E. Green's concept of the key role of hydrophobic interactions in the formation of membrane structures and using the technique developed by H.G. Bungenberg de Jong and H.L. Booij [1]. The experimental results have demonstrated that the structure-forming ability of the protein increases with increasing hydrophobic capacity.

The mechanism of the structure-forming action of globular protein may be conceived as follows (fig. 1). The underlying assumption of our scheme of interaction between protein and the micelles of SAA is that the principal effect of most ionogenic surface-active agents consists in rearrangement and straightening of the peptide chains of a protein molecule.

The interaction of globular protein with a lamellar-type lipid micelle must involve the implantation of he hydrophobic part of the protein molecule between the hydrophobic groups of the SAA. The hydrophilic part of the molecule is still able to bind with the membrane through Coulomb forces.

In conformity with this scheme, attempts to break down the membrane do not revert the structure to the original conformation and the membrane desintegrates into individual fragments.

Using Bungenberg de Jong's technique of coacervation of oleate systems under the influence of KCl, we studied the effect of chlorophyll-group pigments with different extents of hydrophobic capacity on the volume of the coacervate layer. The data we obtained are given in table 1. Since the pigments were added as solutions in ethanol, the effect of additions of ethanol had been previously investigated. From refraction data, we determined graphically the beginning of the division of Na-oleate water solution into a coacervate layer and equilibrium liquid under the influence of KCl in the presence of added ethanol, protochlorophyll and chlorophyll. The addition of chlorophyll causes earliest separation of the system (table 1). The

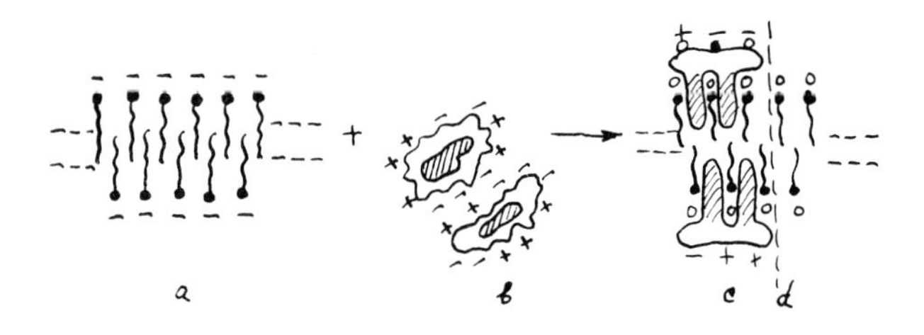

Fig. 1. Scheme of protein interection with surface-action agent micelle: a) the part of the flat micelle; b) the globular protein molecules in water solution; c) lipoprotein micelle; α-line boundary of the sector.

Table 1
Effect of the pigment addition on the coacervate formation in the system:
Na oleate + KCl + H_2O.

Additions	Beginning of the exfoliation in KCl (M)	Coacervate volume (%) by 1.52 M KCl
Ethanol	1.78	100
Protochlorophyll	1.48	38
Chlorophyll	1.34	32

table also gives the volume of the coacervate layer in per cent of the total volume of the system at the same coacervation of KCl. In the presence of the pigments the volume of the coacervate layer is much smaller than in their absence. It can, therefore, be supposed that these pigments are structure-forming substances, and chlorophyll possesses a much stronger structure-forming capacity than protochlorophyll.

Further, we studied the structure-forming action of the pigments in diluted aqueous solutions of surface-active agents.

Modern theory of the structure of soap hydrosols developed by Rebinder, Markina and others makes an assumption that a large amount of free energy stored at the interface between the hydrocarbon chains of polar lipid molecules and water provides thermodynamic advantages to the formation of micelles in a solution [10]. The SAA solutions are true solutions only up to a certain limit – until the so-called critical concentration of micelle formation (CCM) is reached. Above this concentration they become micellar solutions, with an extremely wide range of types and forms of micelles.

The effect of various additions on the structure of SAA can be assessed from variations in CCM values, because this concentration is constant for each particular surface-active agent. The criterion for the structure-forming action of the pigment which we used was a shift of the CCM of a surface-active agent in the presence of the pigments. A shift of the CCM towards lower concentration testifies to the structure-forming role of the addition, while a shift towards higher values indicates the destructive effect of the pigment added.

Table 2 shows the effect of chlorophyll-group pigments on the CCM of the water solutions of potassium oleate and lecithin. CCM was measured conductometrically at 37°C. Chlorophyll-group pigments belong to chlorophyllide and phytol esters [13] and are diphilic compounds able to implant

Table 2
Effect pigment additions on the critical concentration of micelle formation by surface-active agents.

	Critical concentration of micelle formation (M/L)			
	No additions	Ethanol (1×10^{-2} M)	Protochlorophyll (1.4×10^{-6} M)	Chlorophyll (1.4×10^{-6} M)
Potassium oleate	1.25×10^{-3}	1.54×10^{-3}	1.35×10^{-3}	1.10×10^{-3}
Lecithin	0.5×10^{-5}	0.75×10^{-5}	0.56×10^{-5}	0.30×10^{-5}

into a lipid micelle. Ethanol raises CCM. It can be supposed that, as a low-molecular compound, ethanol has a destructive effect on the oleate or lecithin micelles when implanted into the latter. Addition of protochlorophyll reverts CCM to its initial value; in other words, a diphilic pigment assists formation of micelles in SAA solutions.

Ethiolated seedlings are known to contain protochlorophyll [5]; however, this pigment fails to promote formation of lamellar structures to any substantial degree. As soon as the plant is exposed to light, synthesis of chlorophyll starts in its leaves. The synthesis of the pigment is accompanied by an active structure-formation process.

Addition of chlorophyll into our model shift CCM towards lower values; so chlorophyll is a stronger structure-forming agent than protochlorophyll. Perhaps because of its stronger hydrophobic capacity (as is known, a protochlorophyll molecule has an unsaturated bond in the fourth ring between C_7 and C_8 which renders protochlorophyll more polar).

The structure-forming capacity of pigments is important, since it can be assumed that in the course of its implantation into the natural lecithin-based membrane of a diphilic pigment, on the one hand it facilitates structure formation and, on the other, acquiring regular orientation on the membrane structure, enhances its specific activity as a sensitizer.

The above has given rise to the two following problems that were of interest for our research: 1) whether chlorophyll exhibits photochemical activity in SAA systems; and 2) what is the effect of the type of structure of a surface-active agent on the photochemical activity of the pigment.

The photochemical activity of chlorophyll was assessed by a conductometric technique developed by Evstigneev et al. [3]. SAA solutions with chlorophyll additions were exposed to red light after which changes in the conductance of the solutions were studied. In the presence of the pigment, the conductance of the system increased in light and returned to its initial value in dark.

The results of our study of the effect of the structure of the water solution of Na oleate on the photoactivity of chlorophyll (fig. 2) show that in the region of true solution (up to CCM – critical concentration of micelle formation), the nature of the process is close to that in solar solvents in the presence of electron acceptors, namely, the conductance of the system increases in light and reverts to its initial value in dark.

In the CCM region the pigment displays no photoactivity. Studies on the temperature dependence of the structure of aqueous SAA solutions conducted by Markina et al. [11] have confirmed that spherical micelles possess maximum thermodynamic stability. Lamellar micelles, under the influence of temperatures, become spherical micelles, but are not reduced into a molecular state. The thermodynamic stability of spherical micelles, formed at the

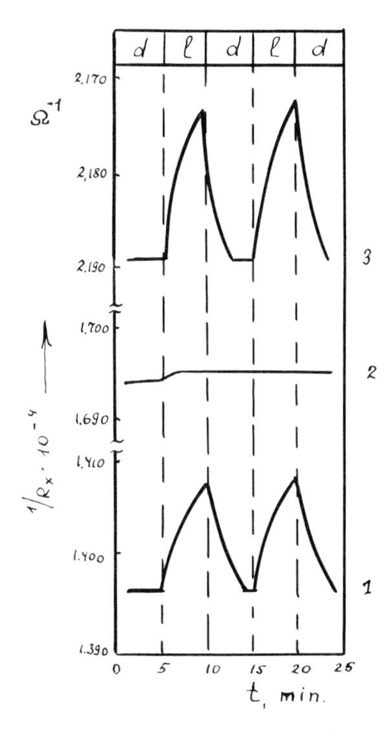

Fig. 2. Effect of the structure of water solution of sodium oleate on the chlorophyll photoactivity: 1) true solution; 2) region CCM; 3) micellar solution.

moment of the transition of true solution to micellar solution, accounts for an unchanged conductance in the CCM region. In the micellar region, conductance sharply increases in light and drops in dark.

Experiments with aqueous lecithin solutions have also revealed photochemical activity of chlorophyll and the dependence of this activity on the structure of the solution [7].

The lecithin (phosphatidylcholine) molecule contains an acid radical (phosphor group) and a quaternary ammonium base dissociating as an alkali. This provides conditions for amphoteric behaviour of lecithin at various pH. Therefore, we studied solutions at pH 6.5 and 10.9. In the alkaline region the photoactivity of chlorophyll is more pronounced and its changes are the reverse to those described above: in light the conductance of the system decreased while in dark it increased or remained at the same level as at the end of the illumination.

The dual nature of the process observed in aqueous lecithin solutions can be explained by the amphoteric behaviour of lecithin at various pH of the solution. Besides, recent discoveries indicate that depending on the surrounding conditions, chlorophyll itself can be either an electron donor or an electron acceptor, thus determining the *oxidation* or *reduction* pathway of photosensitization [4,8].

The pigment in vivo seems to play an important role in formation of chloroplast lamellae, because in light active structure formation goes side by side with chlorophyll synthesis. The interpretation of the mechanism of the structure-forming action of pigments is important both for studying the photosynthetic apparatus of contemporaneous organisms, and for investigating the phylogeny of this apparatus down to the level of chemical evolution in formation of the simplest structures for primary photosynthesis in the conditions of the primeval earth. According to Calvin [2], "rudimentary beginnings of the chlorophyll type of photosynthesis may have occurred in the precellular period in which one had already seen the occurrence of lipid-type coacervates and in which disk-like molecules of chlorophyll could assume a pseudo-crystalline arrangement facilitating the energy transport and conversion".

References

[1] Booij H.L., J.C. Lycklama and C.J. Vogelsang, Proc. Acad. Sci. Amsterdam 52 (1949) 9.

[2] Calvin M., Structure and Function Photosynthetic Apparatus (I.L. Public House, Moscow, 1962) p. 163.

[3] Evstigneev V.B., Elementary Photoprocess in Molecules (Nauka, Moscow, 1966) p. 243.
[4] Evstigneev V.B., Bioenergetics and Biological Spectrophotometry (Nauka, Moscow, 1965) p. 66.
[5] Godnev G.N., Chlorophyll, its Structure and Formation in Plants (Minsk, 1963).
[6] Green D.E., Structure and Function of Subcellular Particles (V. Intern. Biochem. Congr., Moscow, 1962).
[7] Korneeva G.A. and K.B. Serebrovskaya, Dokl. Akad. Nauk SSSR 183 (1968) 712.
[8] Krasnovsky A.A., Elementary Photoprocess in Molecules (Nauka, Moscow, 1966) p. 213.
[9] Luzzati V., F. Reiss-Husson and P. Saludjian, Symposium on Principles of Biomolecular Organizations (1966).
[10] Markina Z.N., A.V. Chinnikova and P.A. Rebinder, Physical and Chemical Mechanics (Nauka, Moscow, 1966) p. 53.
[11] Markina Z.N., A.V. Chinnikova and P.A. Rebinder, Dokl. Akad. Nauk. SSSR 174 (1967) 131.
[12] Oparin A.I., K.B. Serebrovskaya and G.A. Korneeva, Dokl. Akad. Nauk SSSR 183 (1968) 719.
[13] Rabinovich, Photosynthesis (I.L. Publ. House, Moscow, 1959).

Chemical Evolution and the Origin of Life, eds. R. Buvet and C. Ponnamperuma

A POSSIBLE PATHWAY OF BIOLOGICAL MEMBRANE EVOLUTION

D.N. OSTROVSKY

A.N. Bakh Institute of Biochemistry, Academy of Sciences of the USSR, Moscow, USSR

Molecular organization of biological membranes is the most challenging problem of contemporary biochemistry. There are half a dozen hypothetical schemes – models of a biomembrane – but none are accepted by all the investigators [1]. Certainly it is very difficult at the moment to escape a subjective approach while attempting to analyze a possible sequence of events in biomembrane evolution. Let us try to deduce some implications based on bacterial membrane biochemistry which we are engaged in, taking only selected facts from other fields of membranology.

Let us admit, as suggested by the theory of the origin of life [2] that hydrocarbons and their derivatives, e.g. fatty acids, fatty alcohols, etc, were among the first "inhabitants" of the earth. Then we are bound to assume that the primary biomembrane – a film which subdivided the ocean between protoplasma and medium – looked very much like today's lipid bilayer, its polar groups being oriented toward the water-lipid interface.

The appearance of polysaccharides and protein bodies in the course of evolution gave a strong impetus to the structure forming processes and made the problem of the molecular organization of biomembranes much more difficult for investigation.

Two types of protein–lipid interactions are considered to be most probable: polar (mainly ionic) and nonpolar (hydrophobic). In accordance with this two models of biological membranes are mainly discussed: a model of Davson and Danielli [3] (bimolecular leaflet of lipid squeezed between two protein layers), and the other one proposed by Green and Perdue [4]. The latter postulates and partly proves the existence of "structural protein" which forms a basic structural unit of the membrane when complexing with lipids and enzymic proteins.

In our investigation we made an attempt to find out whether the membranes of a bacterium, *Micrococcus lysodeikticus*, preserved some features of the hypothetic protomembrane. In other words, "Is the Davson–Danielli model adequate to the structure of a bacterial (protocariotic) membrane? Two approaches were used:

1) Synthetic nitroxile free radicals are known. to give a characteristic signal–isotropic triplet when dissolved in nonviscous solutions. The shape of the signal drastically changes on freezing the solution.

We inserted a free radical of this type with lipophylic side groups (caprylic ether of 2,2,6,6-tetramethyl-4-oxy-piperydine-1-oxyl) into the bacterial membrane. Its behavior corresponded to that of the radical in a very viscous fluid [5]. This may be interpreted according to McConnell [6] and Koltover et al. [7] as a proof of lipid present in the membrane as a liquid phase. This assumption is further supported by the differential thermal analysis data obtained by Steim and his colleagues on the membranes of *Mycoplasma* [8]. They demonstrated a phase transition in *Mycoplasma* membranes depending on the lipid composition.

2) The second approach was based on the other side of the problem. We tried to detect a quantitatively dominating structural protein in the membranes of *M.lysodeikticus*. Protein fractionation was performed by electrophoresis in polyacrylamide gels in a phenol acetic acid urea system after Takayama et al. [9] and molecular weights were determined as a function of retardation coefficient [10]. In addition the molecular weights of some membrane enzymes were determined by the radiation inactivation method. Proteins were labeled in vivo with ^{14}C-amino acids.

The molecular weights of the bacterial membrane proteins ranged from 12,000 to 160,000 of the two dozen proteins identified, none exceeded 10% of the total amount [11]. Therefore, the idea of one structural protein as a major component of membranes is not experimentally confirmed in case of bacterial membranes, although the methods used for protein separation are not free of criticism.

We may think that at least a part of the membrane is occupied by a continuous layer(s) of lipid, but certainly this lipid is tightly bound to the protein components of the membrane, since it can be separated from the protein only by very drastic treatment.

Is there any difference in molecular structure of biomembranes derived from different organisms? Electron microscopy, optical rotatory dispersion, infrared spectroscopy and other methods show a striking similarity among these organelles from *Mycoplasma* to human brain. In this respect we can see an illustration of "the rule of Prof. J. Monod": "That which is true for an elephant is true for a bacterium". However, a bacterium is a bacterium and the basis for the chemotherapy of infectious diseases. Bacterial membranes are more multifunctional than those of higher organisms and, as we conjecture, their enzymatic systems do not assemble with constant ratios of components, as shown, for example, for mitochondria.

Our implication is mainly based on the ability of the bacterial membranes under various treatments to fall into heterogeneous mixtures of particles, the smallest being of 8–12 nm Stokes radius, many containing certain respiratory enzymes [12]. This phenomenon probably means that those respiratory enzymes are in contact with various other components – different in different parts of the membrane. Besides, enzymatic composition of bacterial membranes is apt to change easily at the change of the cultivation conditions, as shown for bacterial cytochromes [13].

So a biological membrane is thought to develop from a bimolecular lipid film to an assembly of lipoprotein particles with a partial "survival" of the lipid as a phase. Later on evolution, driven by the necessity of higher specialization, proceeded from a multifunctional membrane having no constant boundaries between lipoprotein associations to a membrane built of standardized functional and structural subunits.

References

[1] Rothfield L. and A. Finkelstein, Ann. Rev. Biochem. 39 (1968) 675.
[2] Oparin A.I., Life, its Origin and Development, Moscow (Nauka, Moscow, 1960).
[3] Davson H.A. and J.F. Danielli, J. Cell. Comp. Physiol. 5 (1935) 495.
[4] Green D.E. and J.F. Perdue, Proc. Natl. Acad. Sci. U.S. 55 (1966) 1295.
[5] Goldfeld M.H., D.N. Ostrovsky and E.H. Rozantzev, Dokl. Akad. Nauk SSSR 191 (1970) 702.
[6] McConnel H., J. General Physiol. 54 (1969) 247.
[7] Koltover V., M. Goldfield, L. Hendel and E.G. Rozantzev, Biochem. Biophys. Res. Commun. 32 (1968) 421.
[8] Steim J.M., M.E. Tourtellote, J.C. Reinert, R. McElhaney and R. Rader, Proc. Natl. Acad. Sci. U.S. 63 (1969) 104.
[9] Takayama K., D. Mac Lennan, A. Tzagoloff and C. Stoner, Arch. Biochem. Biophys. 114 (1966) 223.
[10] Thorun W. and E. Mehl, Biochim. Biophys. Acta 160 (1968) 132.
[11] Ostrovsky D.N., I.M. Tzfasman and N.C. Helman, Biokhimiya 34 (1969) 993.
[12] Ostrovsky D.N., N.A. Pereversev, I.H. Jukova, S.M. Trutko and N.S. Helman, Biokhimia 33 (1968) 319.
[13] White D. and L. Smith, J. Biol. Chem. 239 (1964) 2956.

Chemical Evolution and the Origin of Life, eds. R. Buvet and C. Ponnamperuma

THE PROPERTIES OF AN ION SELECTIVE ENZYMATIC ASYMMETRIC SYNTHETIC MEMBRANE

M.A. MITZ

Office of Space Science and Applications, National Aeronautics and Space Administration, Washington, D.C., USA

Primitive cells may have had a simple membrane system. Sometime ago I conceived of a simple model membrane system which may be related to the primitive system. The model is based on materials which may be present on a primitive earth.

One of my long-term interests is the study of enzymes and how they function in a cell [1]. My present interest in the model membrane system is to develop a practical means of studying the functions of enzymes in membranes. It is my hope to derive some understanding about the origin and evolution of the biomembrane from these studies.

Since I conceived of this model system a few years ago, we have been able to demonstrate in principle, at least, that membranes of this type can be synthesized from readily available materials in the laboratory. Most of the cases that I use to demonstrate the potential of the model system are projected from fundamental principles.

The model is based on a synthetic membrane with the ability to sort or concentrate materials. This model is not intended to account for all the processes performed by the biomembrane. However, it is surprising how many different membrane functions can be visualized based on this model. The objective of this paper is to present an overview of my model membrane system in terms of how it is constructed, what are its properties, and what to expect in performance characteristics.

A primitive membrane needs only a few simple elements to exhibit a selective pump-like action associated with the cell. I believe that this model is achievable by an ion selective enzymatic asymmetric membrane. What I mean can best be shown diagrammatically in fig. 1. There are three general types of ion selective enzymatic asymmetric membrane in this figure. The dotted line represents the base membrane which is semi-permeable and each membrane

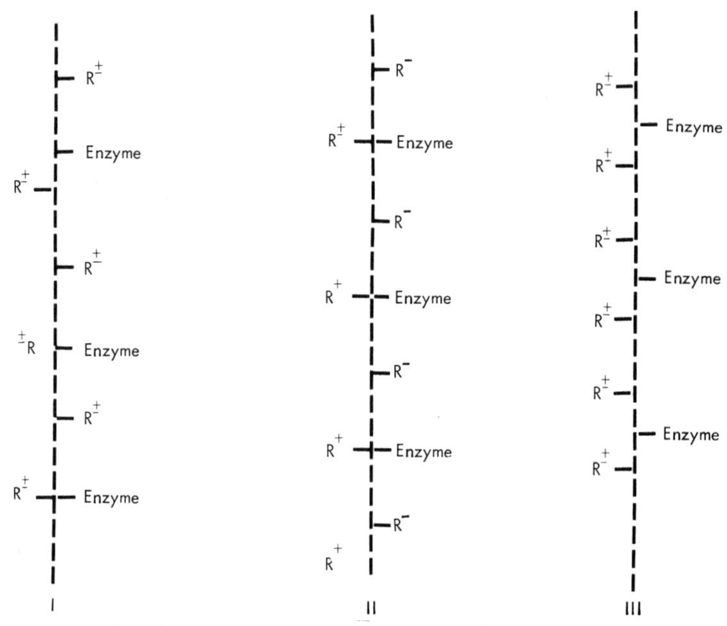

Fig. 1. Ion-selective enzymatic asymmetric membranes.

has a right and left hand side. One could also speak in terms of an inside and outside. When I think in these terms, the left side is usually outside and the right side the inside. The first type of membrane represents a uniform distribution of functional groups with the enzyme on only one side. The second type has a positive charge on one side and a negative charge on the other with the enzyme attached to one side. The third has only a positive or negative group on one side and the enzyme on the other.

Type I represents what I loosely call the "active transport type". Types II and III, I believe, have applications as electron transport and electrochemical membranes. Type I represents the simplest of the series and is the only one which I will discuss in this paper.

How does one synthesize this type of membrane? First, I have selected for my model system a base membrane of reconstituted cellulose, such as the cellulose dialysis membrane which has known properties of size limitation for the transfer of ions and molecules. Other materials such as synthetic polypeptides, collagen-type membranes and even porous glass membranes are also under consideration but the cellulose membrane will serve to illustrate the model. The second step is to graft charged groups onto the backbone to make a cellulose ion exchange membrane. The third step is to link by covalent bonds appropriate enzymes to one surface of the membrane.

The ion exchange membrane can be synthesized as follows: Cellulose in the form of sheets or casing is treated with 25% sodium hydroxide at 0 °C. The sodium salt is then treated with an alkyl halide at room temperature. The membrane is washed with salt, alkali and finally with water. For long term storage it is placed in a 40% glycerin solution and kept in the deep freezer. Some of the membranes which have been prepared are represented in table 1. The membranes are listed in order of increasing strength as a base or an acid. These membranes were prepared by adapting existing procedures [4] for powdered cellulose to cellulose membranes. This series of membranes represent a spectrum of ion exchangers which extend from the weakly basic to the strongly acidic. These membranes should have useful application on their own.

The physical properties of the product are similar to the starting membrane provided care is taken during the preparation to avoid physical damage to the membrane. Uniformity of the substitution is determined by visual observation of sorbed ionic dyes and direct titrations of samples taken from different parts of the sheet. Visual dye binding shows up gross effects such as streaks, creases, and unreacted segments. The titration is more sensitive as a measure of the semi micro distribution.

The properties of charged cellulose membranes are exactly as expected. The size of the molecules which passes freely through is slightly reduced by the introduction of the functional groups but basically all the cellulose membranes prepared in this manner retard large molecules (above 5000 M.W.), and most of the smaller oppositely charged ions. It is sensitive to

Table 1
Charged membranes which have been prepared.

Basic
Cellulose——$O-CH_2-\phi-NH_2$
Cellulose——$O-CH_2CH_2N(C_2H_5)_2$
Cellulose——$O-CH_2CH_2N^+(C_2H_5)_2$
Acidic
Cellulose——$O-CH_2\phi OH$
Cellulose——$O-CH_2COOH$
Cellulose——$O-PO_3H_2$
Cellulose——$O-C_2H_5SO_3H_2$

pH. One of the more dramatic types of experiments with this membrane is to mix an anionic and catonic dye in solution on one side and place a water or buffer solution on the other. The dark mixture of dyes on one side gives way to the bright colored solution of the dye on the other within a short period of time.

These membranes are similar to the charged cellulose powders used in column chromotography of proteins [5]. The present membrane structure combines the properties of the semi-permeable base membrane and an ion exchange membrane. One of the most interesting characteristics of this type of ion exchanger is the availability of the functional groups to macromolecules [2]. It appears as if the groups are on the surface of the fibers. Another fact to consider is that even though the number of charged groups is small, usually less than one milliequivalent per gram, the concentration may be relatively high if the function groups are located only on the exposed or accessible portions of the membrane. This kind of structural relationship may account for the observed rapid response of the ion exchanger to environmental changes such as pH and ion concentrations.

The next subject is the bonding of the enzyme to the cellulose backbone. Our approach is to activate part of the functional groups on the membrane to react with the protein to form a covalent link. This has been worked out on cellulose fibers as shown in fig. 2. Three of the better methods of bonding the protein to the surface of the membrane are shown here: the azide, the diazo, and the active alpha acyl halogen. I have studied a number of enzymes coupled in this manner to cellulose powder [3]. The coupled proteins have the following general properties: The insoluble enzyme derivatives can be dried and stored without loss in activity. They can be used and reused without appreciable loss in activity. Leaching of protein or enzyme activity has not been detectable in a large number of cases examined.

Some of the properties of cellulose enzyme are shown in table 2 [3].

$$\text{I Cellulose} - O - CH_2CON_3 + H_2N - \text{Enzyme} \longrightarrow \text{Cellulose} - O - CH_2CONH - \text{Enzyme}$$

$$\text{II Cellulose} - O - CH_2\Phi \overset{+}{N_3} + \underset{OH}{\Phi} - \text{Enzyme} \longrightarrow \text{Cellulose} - O - CH_2\Phi - N{=}N - \underset{OH}{\Phi} - \text{Enzyme}$$

$$\text{III Cellulose} - O - \overset{O}{\overset{\|}{C}} - CH_2X + NH_2 - \text{Enzyme} \longrightarrow \text{Cellulose} - O - \overset{O}{\overset{\|}{C}} - CH_2 - O - NH - \text{Enzyme}$$

Fig. 2. Cellulose enzyme derivatives.

Table 2
Properties of cellulose enzymes.

1) Rate of action on small substrates the same or greater than free enzyme (sometimes less active on macromolecular substrates).
2) Specificity unchanged for substrates (specificity unchanged for inhibitors except some macromolecular inhibitors).
3) All enzyme molecules appear to be equally active (equally available).
4) The pH optimum is unchanged.
5) Stability is often greatly increased.

Although these properties were derived from a study of the cellulose enzymes in powder form, my recent studies on representative cellulose enzymes in membrane forms show the same properties.

The next question is how do I propose to apply these selective enzymatic asymmetric membranes to specific cases? Table 3 reviews the expected properties. I wanted the enzyme on one side to change the size or the charge of the solute, the membrane to restrict the transfer based on the size of the molecule and a charge to pass ions selectively. The first case to demonstrate how this combination can be applied and the results expected are shown in case 1 in which the accelerated transfer of esters is presented. It is expected that the neutral ester will pass from left to right through the membrane against a maximum gradient because once it is through it will be converted. The esterase will catalyze its conversion to an acid which becomes charged at the alkaline pH. The net effect should be an apparent accelerated transfer of esters.

It should also be possible to reverse this first case. The esterase may be used to help transfer the acid to an ester under the proper conditions. We

Table 3
Overall expected properties of model membrane

Selectivity and pump-like action

Specific expected properties of model membrane

1) Semi-permeable – to limit size
2) Charged – to limit ionic species
3) Enzyme bound – to catalyze a change in charge and/or size of solute

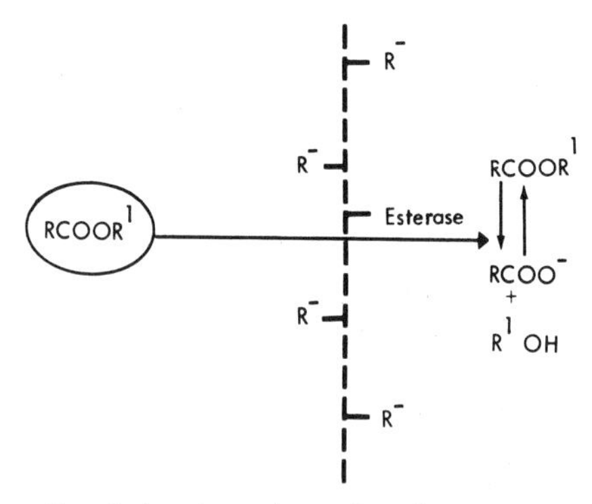

Case I. Accelerated transfer of esters.

know that at lower pH's the synthesis of the ester is favored with some esterases. It will be interesting to see if by regulating conditions and the availability of the alcohol, the system can shift from the transfer of ester in one direction to the transfer of acid in the opposite direction.

Another case is the transfer of amino acids and peptides from proteins. Our earlier studies indicated that the amino acid zwiterion will not be retarded by the weakly basic or acidic cellulose. As the peptidase chops off one amino acid after another, they will be able to migrate through the membrane as shown in case II. In this case amino acids and some peptides will transfer and others will not depending on the overall charge. We have made a fairly systematic study of the retention of peptides by cellulose ion ex-

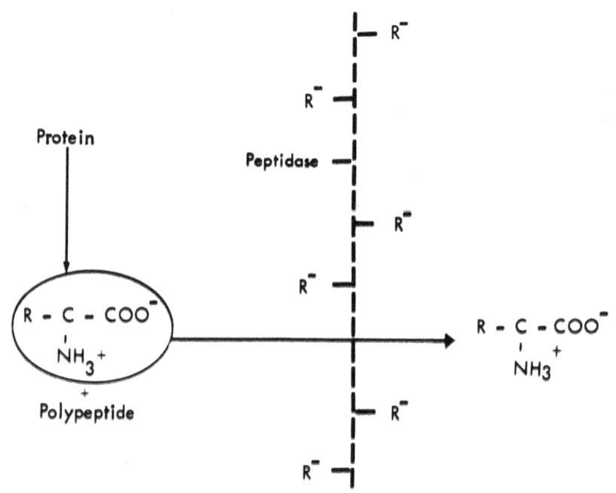

Case II. Transfer of amino acids.

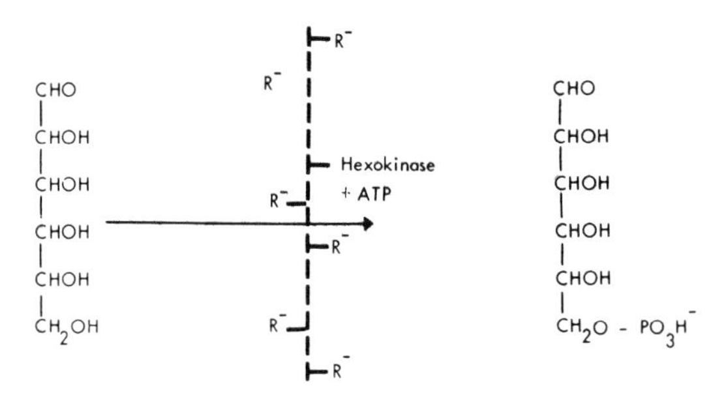

Case III. Accelerated transfer of sugars.

changers and found there is a correlation of retention time to the strength of the amino group as well as size of the peptide.

In case III is represented a mechanism for the accelerated transfer of sugars. Phosphorylation is employed for the accelerated transfer of sugars. This case should demonstrate the selectivity of the membrane based on the specificity of the enzyme. Optical specificity is also demonstrated by case IV in which acylase acts on acetylated dl-amino acids to selectively change one optical form and not another. The acetylated amino acid is too rapid to pass freely through the membrane but the deacylation of the l-form of the amino acid permits it to pass through leaving the d-form of the acetylated amino acidds.

Finally, I have taken a look at how one might produce hydrochloric acid from carbon dioxide using carbonic anhydrase as the enzyme. In the simple case shown in case V, the proton produced by the ionization of the carbonic acid is free to exchange with the sodium ion producing a more acid

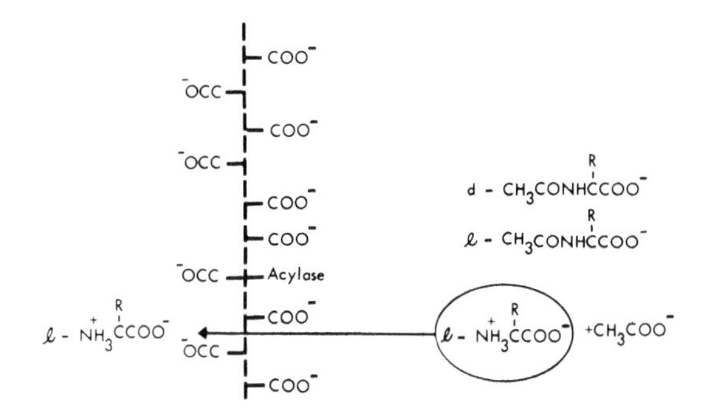

Case IV. Separation of optical isomers.

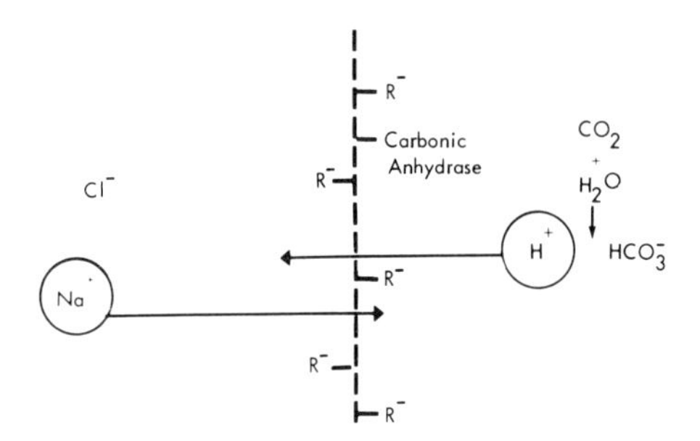

Case V. Production of acid.

condition on the left. As the partial pressure of carbon dioxide is increased on the inside, the concentration of carbon dioxide in solution increases and there is an increase in the production of hydrogen and bicarbonate ions. The hydrogen ion moves in one direction and the sodium ion in the other. The sodium bicarbonate buildup in the cell may also cause the cell to attract water from without which may help to concentrate the acid solution outside the cell.

In conclusion, I think we may have a useful model system to study the selective and in some cases accelerated transfer of nutrients and metabolites. It is based on a semi-permeable membrane, appropriately charged and containing attached enzymes which changes the size and/or the charge on the solute. The membrane may be constructed from cellulose, or other polymeric materials such as acrylamide, or collagen. Enzymes may even be attached to porous inorganic charged materials. The structure of certain minerals may have acted as the first membranes especially if the enzymes were synthesized inside the porous structure. Finally, I hope that the study of these model systems may also shed some light on the mechanisms of natural membranes and cell metabolism.

References

[1] Mitz M.A., Science 123 (1956) 1076.
[2] Mitz M.A. and R.J. Schueter, J. Am. Chem. Soc. 81 (1959) 4024.
[3] Mitz M.A. and L.J. Summaris, Nature 576 (1961) 576.
[4] Peterson E.A. and H.A. Sober, J. Am. Chem. Soc. 78 (1956) 751.
[5] Sober H.A., F.J. Cutler, M.M. Wycohoff and E.A. Peterson, J. Am. Chem. Soc. 78 (1956) 756.
[6] Yanari S.S., M. Volini and M.A. Mitz, Biochim. Biophys. Acta 46 (1960) 595.

Chemical Evolution and the Origin of Life, eds. R. Buvet and C. Ponnamperuma

A MODEL OF SELECTIVE ACCUMULATION OF CARBOHYDRATES DIFFUSING THROUGH ARTIFICIAL POLYMER MEMBRANES

L.N. MOISEEVA
A.N. Bakh Institute of Biochemistry, Academy of Science of the USSR, Moscow, USSR

The evolutionary scheme of the origin of life on earth involves two stages: chemical and biological evolution. Chemical evolution resulted in prebiological systems with certain properties such as material exchange with the surrounding medium, diffusion processes proceeding through interfacial surfaces, incorporation into their composition of the first primitive catalysts, such as inorganic salts, then more complex ones and finally, proenzymes and enzymes.

These properties relate the prebiological systems to open catalytic microsystems. An open catalytic system as a principal evolutionizing unit is characteristic of all stages of evolution. Without exception, such organization is intrinsic to all living cells and organisms at a higher stage of biological evolution as well. Thus, the theory of open systems is of great interest for understanding the process of life generation since the earliest stages of its development.

The theory of open systems has been described in a number of studies [1,4–6]. Open systems which exchange their substances and energy with the surrounding medium show dynamic and kinetic specificity. Instead of the equilibrium state that can be established in closed systems, the stationary state is established in open systems. In time the properties remain constant both in the equilibrium and stationary state. But the difference is that in the equilibrium state there are no changes in free energy ($\Delta F = 0$) whereas, in the steady state such changes take place during all processes and proceed at a constant rate ($\Delta F = \text{const.}$). Under equilibrium entropy reaches its maximum, whereas in the steady state it is maintained constant but different from maximum. Kinetic characteristics of open systems are most clearly seen on the basis of the study of the properties of stationary states, transition between them and conjugation between diffusion through the boundary of the system with chemical reactions proceeding inside the system.

The formation of mechanisms of release, accumulation and utilization of the energy of the proceeding reactions could take place even within chemical evolution. That was a prototype of metabolic processes [3].

Active accumulation of substances in prebiological systems and living cells can be accounted for in terms of the theory of open systems.

The localization of catalysts on or near the interfacial surface in the internal volume of an open system results in the disarrangement of a simple equation between concentrations of the penetrating substance of an external and internal medium. In this case the penetrating substance in the internal volume of an open system should undergo a chemical conversion to change its molecular state. For example, the diffused substance can undergo phosphorylation, amination, polymerization and so on. But the prime interest for this matter is the conversion of electroneutral organic molecules (carbohydrates, amino acids etc.) into electrically charged ions, for instance as a result of phosphorylation. If any of these conversions result in a substance, the penetration of which is hindered through a plasmatic shell, it would actively accumulate in the cell. Perhaps one of the membrane functions in the active transport and accumulation of substances is an efficient limiting of the reverse diffusion of the ionic reaction products due to the electrostatic interaction with the charges of the membrane surface. Apparently these considerations should be taken into account in studying the biological effect of the cell membrane charges.

Thus, active accumulation of substances may be due to the transition of the penetrating substances into another molecular state for which the interfacial surface of an open system is less permeable.

To support this point of view we used in our studies a model of the transfer and accumulation of carbohydrates (glucose, mannose, galactose) through artificial polymer charged membranes, under the conditions of an open system.

This model makes use of two factors:

1) Phosphorylation reaction conjugated with carbohydrate transfer from the external medium with an ATP–hexokinase system to yield ionic reaction products (hexosomonophosphates).

2) Selective permeability of membranes with respect to neutral substances and ionic products of reaction due to its electrical surface charges.

Ionite membranes obtained by radiation injection of various polymers to polyolefinic base have been used in the experiments. The permeability of the membranes at low concentrations of the substances (of the order of one hundredth normal) obeys the Fick diffusion equation. The permeability constant of various substances has been determined from a modified Fick equation (as applied to the diffusion through a membrane).

The density of the surface charges of the membrane has been determined by electroosmosis. The correlation between permeability of the membrane for neutral and ionized substances (P cm/min) and density of the surface charges of the membranes ($\sigma\ \bar{e}/cm^2$) has been found [2].

In fig. 1 it can be seen that the permeability of glucose-6-phosphate and fructose-1,6-diphosphate decreases as the surface charge of the membrane increases. With fructose-1,6-diphosphate which has two phosphate groups, this decrease is more pronounced.

It has been found that selective permeability of the membrane (the permeability constants ratio) for hexose and hexose-phosphates is 8–30, that is, the addition of one phosphate group to the molecule considerably decreases the permeability.

On the other hand the accumulation of substances in this model of an open system resulted from their conversion into a phosphorylated state. In the absence of phosphorylation when the experiment with galactose and ATP–hexokinase system was carried out, accumulation effect was not observed. The enzyme is not specific for the substrate and no conversion into phosphorylated form occurs. In the case of glucose and mannose which undergo phosphorylation by means of an ATP–hexokinase system the effect of accumulation of phosphorylated carbohydrates is observed.

The conjugation process of diffusion and enzymatic reaction of phosphorylation carbohydrates such as glucose under the conditions of an open system can be presented by the following scheme:

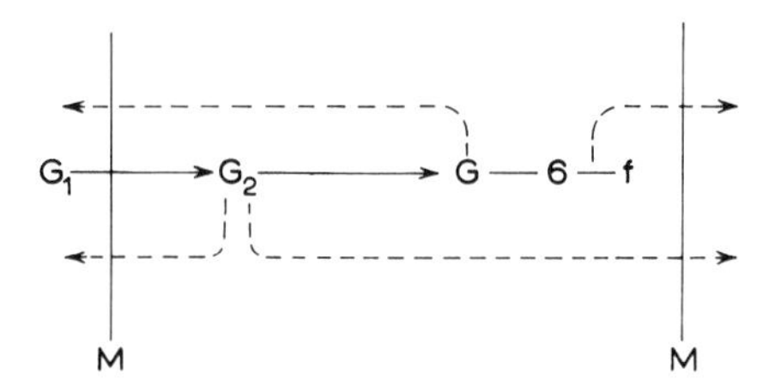

where M is membranes,

G_1 is the starting external concentration of glucose,

G_2 is the concentration of glucose diffusing into internal volume through membrane,

G-6-f is the concentration of the resulting reaction product of glucose-6-phosphate.

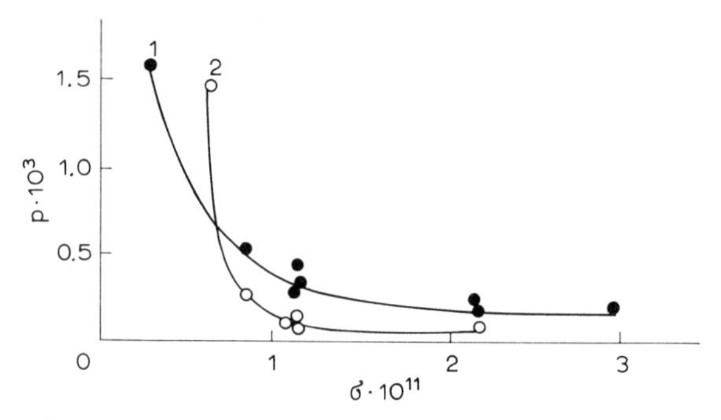

Fig. 1. Membrane permeability constant for ionized substances as a function of density of membrane surface charges. 1) glucose-6-phosphate, 2) fructose-1,6-diphosphate.

The dashed line shows the reverse diffusion of the substrate and reaction product from the internal volume into the external medium.

The enzymatic reaction rate in the open system and the establishment of the steady state are dependent equally on the kinetic parameters of the reaction itself and the constants of substrate transfer from the external medium. Here the conjugation between substrate diffusion and its chemical conversion in the open system manifests itself. Such conjugation is assumed as a simultaneous growth of the concentration gradient of the substrate with its continuous chemical conversion as a result of the reaction.

The ratio of the rate of kinetic processes of diffusion and reaction is of great importance from a point of view of the regulation of the enzymatic action in an open system.

The reaction conditions found in an open system for various concentration ratios of substrates enable two extreme regions to be selected. It is the kinetic one in which the rate of the process is determined only by kinetic parameters of the reaction and the diffusion one in which the reaction rate is determined only by the diffusion parameters of the system.

One of the characteristics of chemical open systems is the effect of the operating active concentration of a catalyst, not only on the reaction rate but also on the steady concentration of substances.

The operating active concentration of a catalyst in an open system can be changed under the action of different factors. They are various molar concentrations of an enzyme, temperature, pH, the effect of different activating and inhibiting agents including the inhibition of the catalytic reaction by its products.

Under the conditions when the rate of enzymatic phosphorylation of

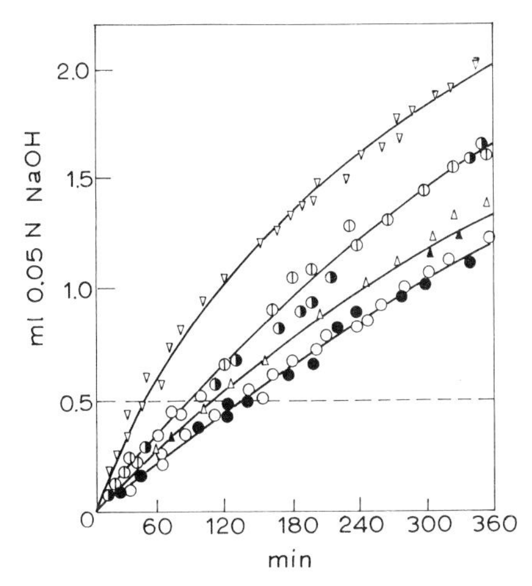

Fig. 2. Accumulation of glucose-6-phosphate and mannose-6-phosphate under the conditions of an open system, with various negatively charged polymer membranes used (curves 1–4), external concentration of carbohydrates is 10^{-3}, internal concentration of ATP is 5×10^{-3}–7×10^{-3} and that of hexokinase is 0.28 E/ml (pH 7.6; 37 °C).

glucose and mannose prevailed over that of the diffusion of these substrates from the external medium and when the reverse diffusion of phosphorylated products from the internal volume into the external medium was considerably hindered by membranes, the concentration of the accumulated phosphorylated products exceeded 2–4 times that of the substrate in the external medium (fig. 2). Fig. 2 illustrates the accumulation of glycose-6-phosphate (light dots) and mannose-6-phosphate (black dots) in time on the membranes of various density of the surface charges. The dashed line shows the substrate level in the external medium. The amount of the products is given in relative units. As seen from fig. 2 accumulation exceeds by several times the external concentration of the substrate.

Thus, the model under study enables the accumulation of substances during their conversion into ionic form to be explained by their electrostatic interaction with the membrane surface charges.

References

[1] Denbigh K., M. Hicks and F. Page, Trans. Faraday Soc. 44 (1948) 479.
[2] Moiseeva L.N. and A.G. Passynsky, Biokhimiya 31 (1966) vyp. 6, 1159.
[3] Oparin A.I., V Intern. Biochem. Congress, Symposium III, M. (1961).
[4] Passynsky A.G., Usp. Sovrem. Biol. 43 (1957) 263.
[5] Sugita M.I., Theoret. Biol. 1 (1961) 415.
[6] Walter C., Enzyme Kinetics, Open and Closed Systems (Ronald Press, New York, 1966).

Chemical Evolution and the Origin of Life, eds. R. Buvet and C. Ponnamperuma

EXCITABILITY, POLYPHOSPHATES AND PRECELLULAR ORGANIZATION

Norman W. GABEL
Research Department, Illinois State Psychiatric Institute, Chicago, Illinois 60612, USA

The occurrence and distribution of polyphosphate material in vertebrate tissue has been under investigation for slightly less than two years. This investigation was initiated in an effort to obtain substantive evidence for a hypothesis which related the excitability phenomenon of neural tissues to the concept of precellular organization [3,12]. It was recognized that many of the properties of a viable neuronal membrane would also be present in a macromolecular polyphosphate coordination complex in which alkaline earth metals would have a cross-linking function and in which alkali metals would serve as counterions. This type of metastable coordination complex is excitable by virtue of its structure and chemical properties, and it may have served as the evolutionary template for biological membranes.

Several different structural forms of polyphosphates can be prepared synthetically depending upon the conditions of condensation to which the monomeric orthophosphates are subjected. Polyphosphates can be linear, cyclic, branched, sheet-like, or adamantane-like in structure. The sheet-like

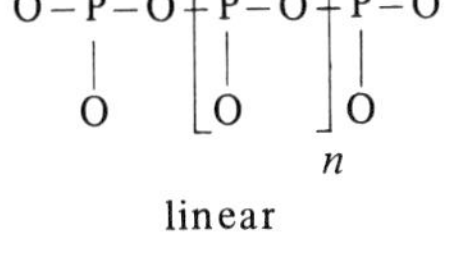

```
    O   ┌O  ┐O
    |   | | | |
O - P - O-P-O-P - O
    |   | | | |
    O   └O  ┘O
             n
```

linear

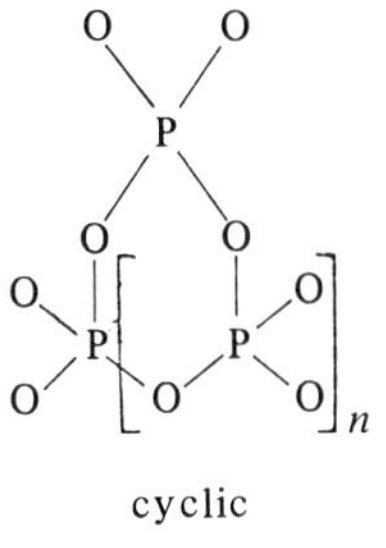

```
        O     O
         \   /
           P
         /   \
        O     O
   O   |┌     |   O┐
    \  P|  P /     |
   O  └O    O      O┘n
```

cyclic

```
     O   ┌O   ┐O
     |   |     |
 O - P - O-P-O-P - O
     |   |     |
     O   └O   ┘O
          |    n
      O - P - O
          |
          O
```

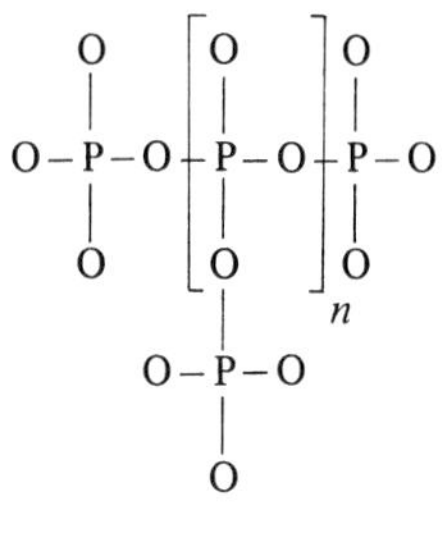

branched

and adamantane-like forms are three-dimensional polymers which are composed of cyclic trimetaphosphate rings (n=1) and are comparable to the graphite and diamond structures of elemental carbon.

It is a characteristic property of inorganic polyphosphates to form coordination complexes with metal cations in aqueous solution. Although the addition of a polyphosphate salt to water which contains Ca^{2+} or Mg^{2+} ions leads first to the precipitation of a high molecular weight metal polyphosphate, an excess of the polyphosphate, above that necessary to balance the valence charge of the alkaline earth cations, results in sequestration and resolublization of the precipitate [10]. It is because of this action that sodium polyphosphates have found commercial use as water softening agents.

Polarographic data on the complexing of alkaline earth cations with polyphosphates indicates that more than one polyphosphate chain is attached to an individual alkaline earth cation. The implication of these data is that alkaline earth cations (Mg^{2+}, Ca^{2+} and Ba^{2+}) can serve in a cross-linking capacity to bind a large number of polyphosphate chains together into a metastable, macromolecular coordination complex. Van Wazer and Campanella [15] have estimated the molecular weight of the complex formed from barium ions and polyphosphate chains to be between 10^3 and 10^7.

Conversely, alkali metal cations (Li^+, Na^+, and K^+) form very unstable complexes with polyphosphates the bond types of which are primarily electrostatic. Furthermore, alkali metals have a coordination number of two in these complexes and are bound to only one phosphate chain at a time thus precluding any participation in the cross-linking of polydentate phosphate chains. If inorganic polyphosphates were present in primordial waters, it can be expected that in the presence of dissolved alkali (Na^+ and K^+) and alkaline earth (Ca^{2+} and Mg^{2+}) cations, a metastable, macromolecular coordination complex would have formed due to the cross-linking capacity of the alkaline earth ions. Since Na^+ and K^+ do not participate in this cross-linking of the polyphosphate chains, their positions as counterions would have been determined by the ion exchange characteristics of this macromolecular coordination complex. As complexification (cross-linking) takes place, the cation with the least solvated volume, (K^+ rather than Na^+), is adsorbed. Thus, the structural organization of macromolecular, polyphosphate coordination complexes are, in themselves, responsible for and will determine the distribution of metal ions in their environs. These macromolecular complexes have an electrochemical gradient (resting potential) from their interior to their exterior by virtue of this selective distribution of metal ions within and about their structure. It can further be expected that organic molecules in the immediate environment which could be electrostatically attracted or could coordinate with exterior metal cations would form a film or envelope around a macromolecular polyphosphate coordination complex.

The formation of this metastable structure would proceed with a decrease in free energy of the system and, concomitantly, with an increase in entropy of the system due to the release (to the system) of solvent molecules previously bound to the individual parts of the macrostructure prior to their organization.

From an operational point of view excitability is the property of an organized group of particles to transfer environmental information along its parts resulting in the maintenance of its structural integrity. If this information were not transferred and utilized, the structural organization would disintegrate because it is the information, in itself, which is the disruptive reaction. On a creature level it would be called response to stimuli. Since life is a metastable, dynamic equilibrium state, it is profoundly affected by changes in its solvent environment. A chemical change which would have been disruptive to the organization of protolife very likely would have involved the solvent or atoms (groups of atoms) of the solvent. The chemical similarity of the excitability of biological materials to the excitability of polyphosphate complexes induced by exterior variations in the solvent has been discussed in detail [3].

The occurrence of polyphosphates in a large number of microorganisms is well documented [7]. However, their functionality has remained obscure and has led many investigators to regard them as metabolic fossils. The first report of polyphosphates in mammalian tissue was made by Griffin, Davidian and Penniall [6]. Under the conditions of their experiments rat liver nuclei formed polyphosphate chains of high molecular weight.

Polyphosphates having apparent chain lengths ranging from less than ten to over 5,000 orthophosphate units have been isolated from the adult rat brain and every other tissue which has been examined (table 1 and 2) [5]. The identity of the polyphosphate material was inferred from its isolation procedure (fig. 1), its chromatographic behavior and color development, and its metachromatic effect. The presence in sheep brains of phosphate units covalently linked by phosphoric anhydride bonds was unequivocally demonstrated through the use of ^{31}P nuclear magnetic resonance spectroscopy (fig. 2). The distribution of chain lengths is similar in rat brain, beef brain, chicken brain, frog brain, and shark brain. Approximately 40% of the polyphosphate material has an apparent chain-length of less than ten and 30–40% a chain length of over 5,000. The remainder lies between these two ranges. There appeared at first to be a quantitative variation but these differences were actually due to techniques by which the tissues were

Table 1
Rat brain polyphosphate

	Polyphosphate phosphorus (μg/g)	Chain length (% of recovered polyphosphate)			
		<10	11–117	117–5000	>5000
Fresh [a]	1.7	59.4	4.7	10.0	25.9
Frozen [b]	12.8	42.8	5.4	10.7	41.1
Pronase digested [c]	3.3	1.5	53.7	0	46.2

Chain lengths were estimated by a procedure adapted from Ohashi and Van Wazer [11]. The tissue homogenates were processed according to the procedure outlined in fig. 1.

[a] The rat brains were maintained in a "fresh" condition in a beaker immersed in an ice bath for approximately 30 min after excision before being homogenized in 1 N KOH at 0 °C.

[b] The rat brains were forcibly flattened between two slabs of dry ice within 30 sec after sacrificing the animal. The frozen tissues were homogenized directly in 1 N KOH at 0 °C.

[c] The frozen brains were homogenized at 2–4° in 2:1 chloroform:methanol, washed once with one volume of methanol, washed twice with equal volumes of water, and digested with pronase after the method of Cook and Eylar [1].

Table 2
Polyphosphate chain-length distribution in bovine tissue.

	Total protein (mg/g)	Total phos-phorus (mg/g)	Total poly-phosphate phosphorus (μg/g)	Chain length (% of recovered polyphosphate)			
				<10	11–117	117–5000	>5000
Brain	99	1.16	2.95	19.7	35.3	0	50.2
Liver	150	1.18	2.24	0	16.5	35.2	48.2
Pancreas	186	1.19	2.10	0	47.6	23.8	28.6
Kidney cortex	174	1.18	1.89	0	7.9	29.1	62.9
Spleen	148	1.35	1.98	1.5	29.4	20.0	55.1
Erythrocytes [a]	135	0.74	7.48	9.4	44.6	9.6	35.0
Eryth. membrane [a]	48	0.17	2.51	0	6.8	22.3	70.9

[a] Rabbit erythrocytes were used. The calculations were based on cm^3 of packed cells instead of g of fresh tissue.

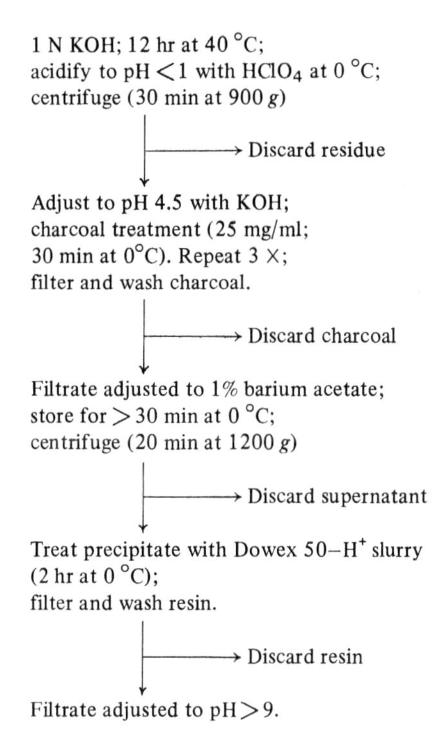

Fig. 1. Isolation of polyphosphates adapted from Griffin, Davidian and Penniall [6].

obtained. Neural polyphosphate is very rapidly degraded after the animal has been sacrificed and it would appear that the same difficulties of establishing the actual in vivo levels of neural ATP and phosphocreatine are also operative in these experiments [9]. The highest minimum value which has been obtained thus far is approx. 15 μg of phosphorus as polyphosphate per gram of wet tissue.

From the polyphosphate–protein ratios of the rat brain subcellular fractions obtained according to the Whittaker technique it would appear that the polyphosphate material is concentrated in subfractions P_2A and P_2B (table 3). However, caution must be exercised in interpreting these data since considerable degradation of the polyphosphate material occurred during the fractionation procedure. (Total recovered polyphosphate ranged from less than one to three μg of phosphorus per gram of fresh tissue.) Furthermore, leakage of water soluble materials is known to occur during fractionations. If the polyphosphates redistributed themselves during homogenization, they

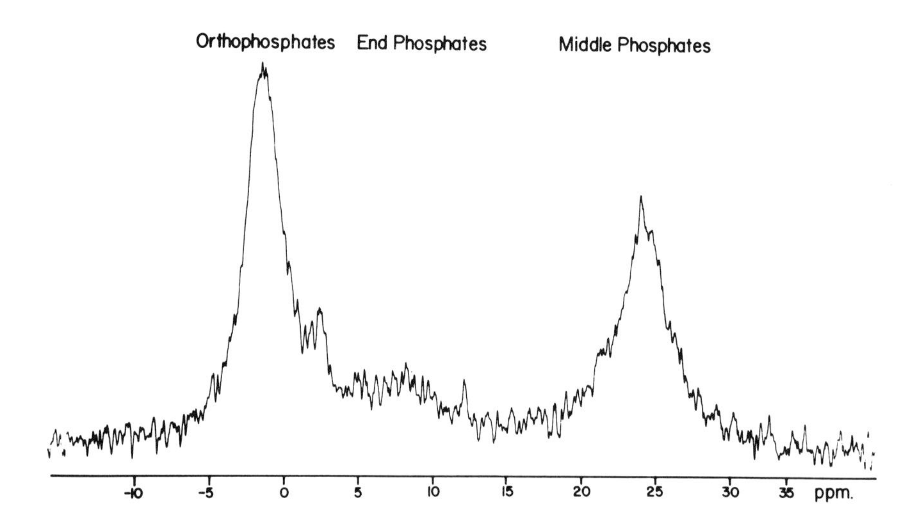

Fig. 2. The ^{31}P NMR spectrum of polyphosphates and orthophosphates which are present in an extract prepared from 4 kg of sheep brains according to the procedure outlined schematically in fig. 1. Orthophosphates were not removed by chromatography in order to illustrate the resolution and the difference between the chemical shifts of the three regions. The spectrum is divided into 3 regions: orthophosphates, −5 to +5 ppm, 45% of total phosphorus; end phosphate groups of polyphosphate chains, +5 to +13 ppm, 13%; middle phosphate groups of phosphate polymers, +20 to +30 ppm, 42%. A Bruker HFX−5 NMR spectrometer was employed operating at 90 MHz for ^{1}H and 36.4 MHz for ^{31}P. The high field homogeneity which is intrinsic in this spectrometer enables one to obtain high-resolution spectra while using large diameter spinning sample tubes and thereby significantly increases the inherent sensitivity. A resolution of 0.4 Hz was obtained using the 13 mm tube employed in these experiments. A heteronuclear ^{1}H lock was used for field-frequency stabilization with the water signal from the sample being used as the reference. All frequencies used were referenced to a single oven-controlled crystal of a Brucker Model ND30M−B frequency synthesizer. Field homogeneity was maintained by an automatic shimming device. The drift of the above system was measured over a period of 72 hr and was found to be less than 0.1 Hz. The signal-averager employed was a Fabri-Tek Model 1064; 400 scans were averaged requiring a total experimental time of ca. 18 hr (sweep speed, 1.5 Hz/sec.; time constant, approx. 1 sec); pH of the sample, 7.8. Under similar experimental conditions a ^{31}P signal arising from a polyphosphate of average chain length 100.8 units, 0.1 mM in phosphorus, was readily detected. The resonance of the middle phosphate groupings of this reference sample occurred at 21.3 ppm. The spectrum and its interpretation were kindly supplied through the courtesy of Drs. T. Glonek and T.C. Myers as part of a program in the study of the occurrence and structure of abiotic and biotic polyphosphates (Dept. Biol. Chem., U. of Ill., Coll. of Med., Chicago, Ill.).

Table 3
Polyphosphate/protein ratio in rat brain subcellular fractions
(μg of phosphorus per mg of protein)

Fractionation	Debris	P_1	P_2A	P_2B	P_2C	P_3	Soluble
I	0.024	0.028	1.12	1.25	0.035	0.049	
II	0.011	0.030	0.449	0.482	0.015	0.056	0.099
III	0.004	0.005	0.200	0.228	0.012	0.034	
IV	0.005	0.017	0.063	0.333	0.009	0.017	
V	0.022	0.051	8.20	2.42	0.231	1.30	0.232

P_1 = nuclei, P_2A = myelinated axons and myelin, P_2B = axons and nerve endings, P_2C = mitochondria, P_3 = microsomes, soluble = supernatant remaining after sedimentation of P_3.

might be expected to be concentrated in the subfractions containing the most calcium and magnesium. Metal analysis shows these subfractions to be P_2A and P_2B.

Three acellular fluids were assayed for polyphosphates. Rabbit blood plasma was found to contain 4.9 μg of polyphosphate phosphorus, 140 μg of total phosphorus, and 81.5 mg of protein per ml. Virtually all of the polyphosphate material had an apparent chain length of 10–5,000 orthophosphate units. Rabbit aqueous humor contained 190 μg of total phosphorus and 2.5 mg of protein per ml; polyphosphate was not detectable. Freshly voided human urine contained questionable trace amounts of polyphosphate.

Every tissue which has been examined contains polyphosphate material with apparent chain lengths ranging to over 5,000 orthophosphate units. There is approximately three times as much polyphosphate material in the adult rat brain as there is in the liver. Neural polyphosphate is very rapidly degraded after death but the chain length distribution seems to remain unchanged. Liver polyphosphate is much more catabolically stable and the chain lengths appear to follow a different distribution pattern. Although liver nuclei synthesize polyphosphates [6], it can not be ascertained at this time whether the polyphosphates are concentrated in the nucleus. The data obtained from the rabbit erythrocytes at least indicate that the presence of a cell nucleus is not necessary for the maintenance of a relatively high polyphosphate content. The results obtained from the three acellular fluids would seem to denote that the polyphosphate material is primarily a cellular constituent. The conspicuous absence of measurable quantities of polyphosphate material in aqueous humor and urine should at least obviate the remote possibility that the digestion and separation procedures produced polyphosphates as an artifact.

A brief study of the in vivo incorporation of ^{32}P-orthophosphate into the rat brain polyphosphate fractions was designed to elicit information relating to the metabolic activity of these arbitrarily delimited fractions (fig. 3). The specific activity of the total polyphosphate and the large molecular weight polyphosphate remained higher than the specific activity of the total homogenate throughout the experiment. The specific activity of low molecular weight polyphosphate reached a maximum value between 10 and 24 hr and it would appear that the low molecular weight material is the precursor of the high molecular weight material.

Two questions are immediately self-evident: (1) Why has a material of such an apparently large molecular weight thus far gone unnoticed? and (2) What is the function of this inorganic polymer? Within the last two decades many investigations have been directed toward the elucidation of the structure of acid-labile phosphoproteins which were characterized by a rapid

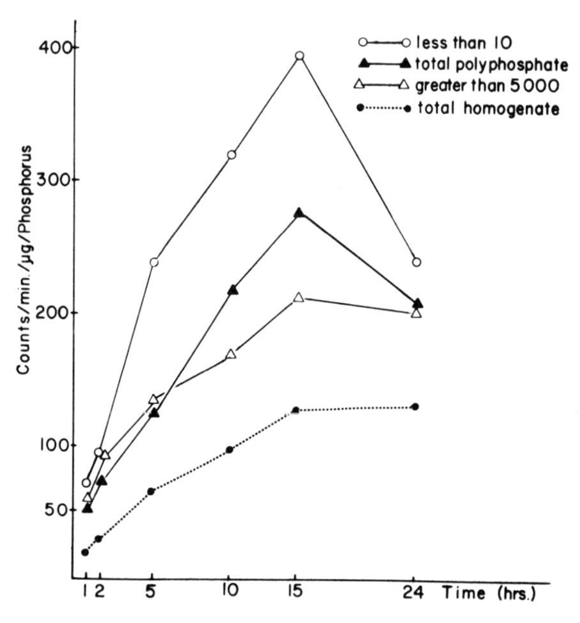

Fig. 3. Incorporation of ^{32}P-orthophosphate into rat brain polyphosphate. *Less than 10* and *greater than 5000* refers to apparent chain length.

rate of metabolic turnover. The phosphorus of these phosphoproteins was usually liberated from the protein component by digestion in concentrated aqueous base [8]. (This is exactly what is done in step 1 of fig. 1). The phosphorus was then determined quantitatively after strong acid hydrolysis at 100 °C. In the present investigation polyphosphates survive hydrolytic cleavage during the initial strong basic tissue digestion because their high negative charge density impedes the nucleophilic attack of water or OH^- ion at the P–O bond [14]. Strong acid hydrolysis at elevated temperatures rapidly converts polyphosphate into orthophosphate.

The difference in catabolic stability between neural and liver polyphosphates, as well as the variations in chain length ranges, may signify a functional polymodality for polyphosphates. There has been demonstrated a membrane phenomena associated function in yeast [2] and a nuclear synthesis of polyphosphate in liver [6]. It is entirely conceivable that all of the uncharacterized protein phosphate acceptors (including phosphoprotein ATPases) have as part of their structures polyphosphates which could be bound covalently or ionically.

It has been reported recently in experiments simulating primordial conditions that polyphosphates could have been present before the emergence of recognizable life-forms [4,13]. Ancient igneous phosphate deposits [16] would have separated from the cooling magma as polyphosphate salts. It now appears that polyphosphates are ubiquitous to living matter. Perhaps polyphosphates were ubiquitous before recognizable life-forms emerged.

References

[1] Cook G.M.W. and E.H. Eylar, Biochim. Biophys. Acta 101 (1965) 57.
[2] Deierkauf F.A. and H.L. Booij, Biochim. Biophys. Acta 150 (1968) 214.
[3] Gabel N.W., Life Sci. 4 (1965) 2085.
[4] Gabel N.W., Nature 218 (1968) 354.
[5] Gabel N.W. and V. Thomas, J. Neurochem. (1971) in press.
[6] Griffin J.B., N.M. Davidian and R. Penniall, J. Biol. Chem. 240 (1965) 4427.
[7] Harold F.M., Bacteriol. Rev. 30 (1966) 772.
[8] Heald P.J., Biochem. J. 66 (1957) 659.
[9] Heald P.J., Phosphorus Metabolism of Brain (Pergamon Press, New York, 1960) p. 7.
[10] Johnson R.D. and C.F. Callis, in: The Chemistry of the Coordination Compounds, ed. J.C. Bailer, Jr. (Reinhold, New York, 1956) Chap. 23.
[11] Ohashi S. and J.R. van Wazer, Anal. Chem. 35 (1963) 1984.
[12] Ponnamperuma C. and N.W. Gabel, Space Life Sci. 1 (1968) 64.
[13] Rabinowitz J., S. Chang and C. Ponnamperuma, Nature 218 (1968) 442.

[14] Thilo E., Angew. Chem. Intern. Edit. 4 (1965) 1061.
[15] Van Wazer J.R. and D.A. Campanella, J. Am. Chem. Soc. 72 (1950) 655.
[16] Van Wazer J.R., Phosphorus and Its Compounds (Interscience, New York, 1961) pp. 961, 450.

PART VII

PRIMITIVE BIOCHEMISTRY AND BIOLOGY

Chemical Evolution and the Origin of Life, eds. R. Buvet and C. Ponnamperuma

GRAMICIDIN S AND TYROCIDINE BIOSYNTHESIS: A PRIMITIVE PROCESS OF SEQUENTIAL ADDITION OF AMINO ACIDS ON POLYENZYMES

F. LIPMANN
The Rockefeller University, New York, New York, USA

At my first encounter with the problems of mapping prebiological events, expecting stepwise evolution, I hoped to recognize evolutionary paths by finding surviving fossils carried over into present-day metabolism. For example, I reviewed present-day metabolism for known instances where pyrophosphate functions as energy donor and carrier in parallel with ATP [8]. This problem is being more comprehensively discussed here in the section on phosphorylations.

Turning now to the evolution of enzymatic processes of the type I am going to discuss here, I would like to begin with a few remarks on the primitive polypeptide structure of the hydrogenase ferredoxin. Taking the *Clostridium butyricum* ferredoxin as example, it is made up of an incomplete set of only 13 rather simple amino acids: alanine, isoleucine, valine, glycine, phenylalanine, proline; serine, threonine, cysteine; aspartic acid, asparagine, glutamic acid, glycine, in a sequence of 55 [1] (fig. 1a); the more complex amino acids – tryptophan, tyrosine, methionine, histidine and also arginine, leucine and lysine – are missing. It contains eight essential cysteines that participate in the complexing of $Fe^{II}S$, which is the reactive center. Sequence determination clearly indicates that, as shown in fig. 1b, it contains two nearly equal sequences of 25 amino acids with four cysteines each. These may well have been derived from condensation of still smaller units [2]. I find some relevance in the prominent function of cysteine-SH in this quite primitive catalyst. On the other hand, SH groups of coenzymes [7] and enzyme proteins [9] are well known reagents in the polymerization of carboxyl derivatives. We will present here a polyenzyme system that catalyzes a polypeptide synthesis from thioester-linked amino acids on a protein template.

Our interest in exploring this mechanism was aroused by the hope of

Ala	Ser
Ile	Thr
Val	*Cys*
Gly	Asp
Phe	Asn
Pro	Glu
	Gln

Fig. 1a. Amino acids in *Chlost. butyricum* ferredoxin.

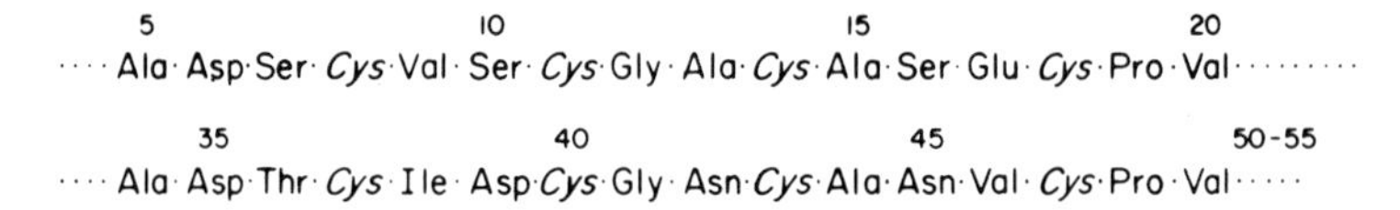

Fig. 1b. 16 amino acid repeats in ferredoxin, *Clost. pasteurianum*. The numbers indicate the position of the amino acids in the complete peptide.

meeting a much simpler process than the eventually developed ribosomal protein synthesis on nucleic acid templates. The carrier function of thioesters was first found in long-chain fatty acid biosynthesis. Here, equal subunits, the activated two-carbon fragments $-CH_2-CO\sim$, are condensed to β-keto-acyl-thioesters, and are reduced to fatty acyl thioesters. During elongation and reduction, these compounds remain thioester-linked to a polyenzyme system [9], and yield free fatty acid after condensation of eight to nine two-carbon units. The mechanism of the $-CH_2-CO\sim$ polymerization has long impressed me by its formal analogy to $H_2NCHR-CO\sim$ condensation into polypeptides, with the difference that in the latter case a repeating backbone structure with *variable* side chains is constructed. Here, the function of the product depends uniquely on the sequence in which the different side chains follow each other. This is essentially true for the relatively few amino acids strung together in an antibiotic polypeptide, as it is for the 100 or more amino acids linked in an enzyme protein.

One can look at these two polymerization processes, of fatty acid synthesis and the polypeptide synthesis described here, as stages in the evolution of a condensation of chains of homologous units to the more sophisticated condensation of heterologous units into predetermined sequences. The latter process evolved further into replicative protein synthesis through transfer of genetic information. This is encoded there in the base sequences of nucleic acids, which are expressed as amino acid sequences in the complex ribosomal process of protein synthesis. In the phase of amino acid activation as well as in the process of chain elongation the ribosomal process still shows a clear relationship to fatty acid and antibiotic synthesis. The analogy between the two elongation processes is depicted in fig. 2.

Our interest in antibiotic biosynthesis as a more primitive process was aroused when it became clear that we were dealing here with polypeptide synthesis without nucleic acids [2,12,13]. I felt that the proposition of initial enmeshing of nucleic acid coding and amino acid sequencing would have presupposed an understanding of the "genius" of the protein. This made me curious to elaborate the more primitive mechanism to see if it might give hints about a "try-out" synthesis of short amino acid sequences such as appear in ferrodoxin, for example. These, then, might have brought to light the eminent suitability of proteins as carriers of catalytic functions as well as the essentiality of their replication, by a chemical information transfer, for building living organisms and evolving organisms of increasing complexity.

We chose for our study the biosynthesis of gramicidin S (GS). As mentioned, many laboratories agreed and we have confirmed [4,5] that this reaction occurs in the supernatant fraction of bacterial homogenates and in the presence of pancreatic ribonuclease. This indicated a sequence determination by a protein structure. GS is a cyclic decapeptide; it contains a repeating chain of only five different amino acids. Its structure is presented in fig. 3. The repeating pentapeptide contains one D-amino acid, phenylalanine, and ornithine, which is not found in proteins. Extracts of the GS-producing strain of *Bacillus brevis* on filtration through Sephadex G-200, yield two complementary fractions [4,12] that synthesize GS when supplied with ATP, Mg^{2+},

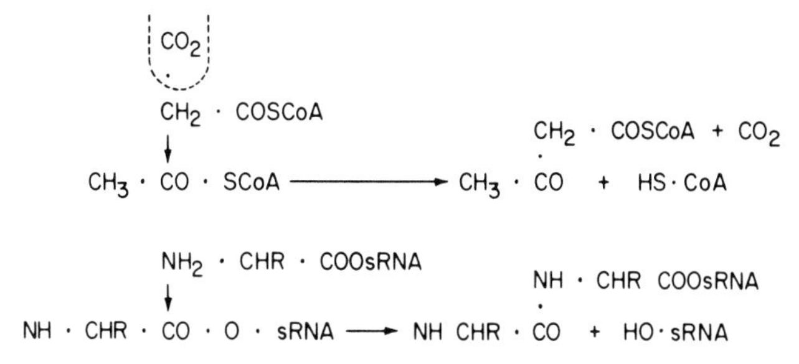

Fig. 2. Comparison of fatty acid (upper equation) with peptide synthesis (lower equation).

L-Leu
D-Phe
L-Pro
L-Val
L-Orn
L-Leu
D-Phe
L-Pro
L-Val
L-Orn

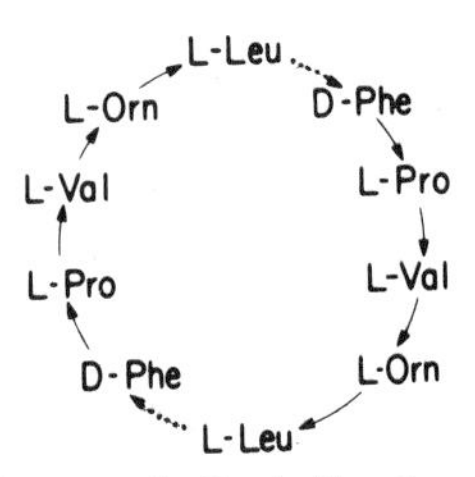

Fig. 3. Gramicidin S. The dotted arrows indicate the places and direction of cyclization.

and the corresponding amino acids. The molecular weight of the heavier fraction I is 280,000 and that of fraction II, 100,000; the lighter fraction activates and racemizes D- or L-phenylalanine. The heavy fraction activates the four L-amino acids. This activation is measured by ATP-$^{32}PP_i$ exchange.

The first hints of protein-bound intermediaries came through observations indicating a transfer of the activated amino acids from aminoacyl-adenylates to the enzyme proteins. This was obtained when, in addition to ATP-PP_i exchange, we observed an ATP–AMP exchange [4] which indicated a transfer from initial aminoacyl~AMP to a secondary acceptor which turned out to be an enzyme-bound SH (fig. 4). The activated amino acids are protein-bound, and the complexes can be isolated by Sephadex filtration [4]. We find that the protein fractions charged with an amino acid contain the aminoacyl-adenylate and an equivalent amount of amino acid covalently bound to protein, as indicated in the last line of fig. 4. After precipitation with trichloroacetic acid (TCA), the aminoacyl-AMP is discharged, but a stable protein-bound amino acid remains with the precipitate. This shows the linkage characteristics of a thioester (table 1), i.e. it is stable to acid, decomposed by alkali and mercury salts, and reacts with hydroxylamine at low pH (table 2). Furthermore, *N*-ethylmaleimide (NEM), when applied in

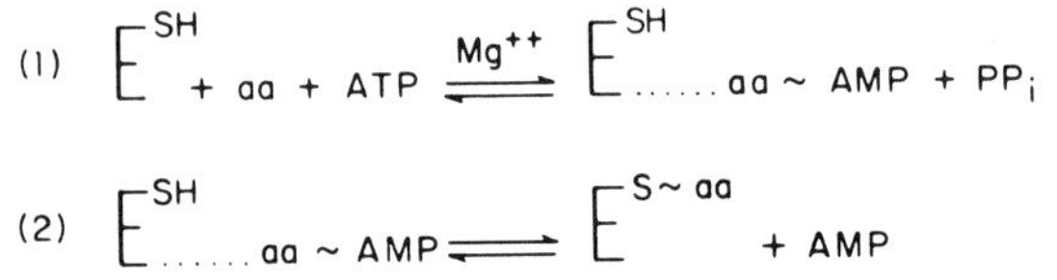

(1) E–SH + aa + ATP ⇌(Mg++) E–SH ...... aa ~ AMP + PPi
(2) E–SH ...... aa ~ AMP ⇌ E–S~aa + AMP

Fig. 4. Amino acid activation and binding to enzymes. Initially aminoacyl-AMP binds, reaction I, and then transfers the aminoacyl to an –SH on the enzyme, reaction 2. From ES~aa the amino acids enter into polymerization; after treatment with low NEM (cf. fig. 5), reaction 2 is blocked and polymerization is inhibited.

Table 1
Isolation of protein-bound amino acids free of aminoacyl-adenylates.

Materials	^{14}C-amino acid (pmoles)	^{3}H-AMP (pmoles)
A) Sephadex eluate	35.0	18.1
TCA-precipitable material, dissolved in dilute alkali	16.2	0.0
B) Sephadex eluate	34.0	19.0
TCA-precipitable material, dissolved in dilute alkali	17.0	0.0

Samples of A) fraction I (25 μg) and B) fraction II (10 μg), were incubated separately for 5 min at 37 °C with ^{14}C-L-leucine (2 μM, 0.5 μCi/tube) and ^{14}C-L-phenylalanine (1.3 μM, 0.5 μCi/tube), respectively. The reaction mixtures also contained, in 0.2 ml: 10 μmoles triethanolamine, pH 7.8; 5 μmoles magnesium acetate; 0.5 μmole dithiothreitol; 80 μg inorganic phosphatase; 12.8 nmoles of ^{3}H-ATP (2.5 μCi/tube). The mixtures were separately applied to Sephadex G-50 columns, eluted with buffer A, and the excluded protein collected. For further details see [4].

Table 2
Properties of protein-bound amino acids (fractions I and II).

1) Stable to acid (pH 6); labile to alkali; TCA-precipitable
2) Cleaved by mercuric acetate
3) Readily form hydroxamates at pH 6
4) Reductive cleavage with borohydride

Enzymes require dithiothreitol, are inhibited by NEM and *p*-chloromercuriphenylsulfate.

low concentration to block active SH-groups, completely inhibits polypeptide synthesis but not amino acid activation when measured by ATP-^{32}PP$_i$ exchange (fig. 5). After NEM treatment, the Sephadex filtrates contain only half as much bound amino acid as before, and this is in the form of aminoacyl-adenylate; practically no acid-stable amino acid is found to remain with the protein after TCA precipitation (table 3). This makes us assume that the first reaction product, aminoacyl-adenylate, binds to the polyenzyme and from it the aminoacyl is transferred to an active SH, cf. fig. 4. It is, then, the thioester-linked amino acid that donates its aminoacyl~group in polymerization.

Both light and heavy enzyme fractions, when separated, may carry charged amino acids without polymerization (table 4). For peptide synthesis, the two

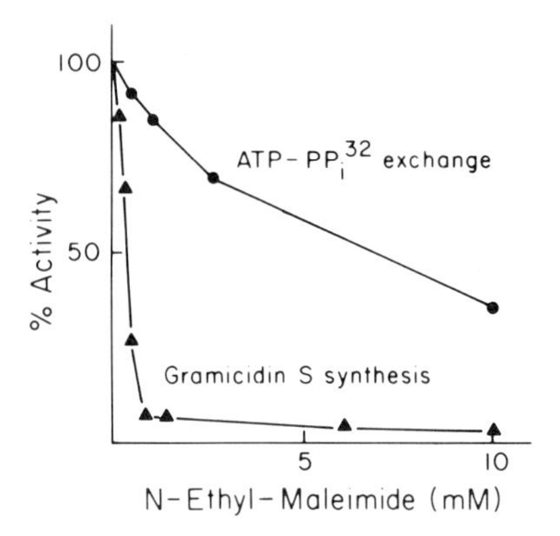

Fig. 5. Differential effects of NEM on activation and polymerization. L-ornithine-dependent ATP-^{32}PP$_i$ exchange catalyzed by fraction I, and GS formation carried out by the recombined fractions I and II, were measured as described [4].

Table 3
NEM effect on leucine binding (fraction I).

	L-Leucine bound (pmoles)		
	Total	TCA-precipitable (Leu~S-Enzyme)	Non-TCA-precipitable (Enzyme ... Leu~AMP)
Control	11.5	5.85	5.65
+ NEM (1 mM)	4.80	0.61	4.19

Table 4
Formation of GS from preformed amino acid complexes with complementary fractions.

No.	Additions	Gramicidin S	
		^{3}H	^{14}C
1	Fraction I: L-Pro, L-Val, ^{3}H-L-Orn, L-Leu complex (4500 cpm)	0	0
2	Fraction II: ^{14}C-Phe-complex (6000 cpm)	0	0
3	Both complexes together (10,500 cpm)	1300	1850

fractions have to be combined. The lighter protein, carrying D-phenylalanine, initiates on the heavier protein the sequential addition of proline, valine, ornithine, and leucine to form enzyme-bound polypeptide thioesters (table 5). Enzyme-bound Phe–Pro, Phe–Pro–Val, Phe–Pro–Val–Orn–Leu, can be isolated by TCA precipitation, and then identified electrophoretically, after addition of hydroxylamine, as the corresponding hydroxamates (fig. 6). If proline, the second amino acid in the pentapeptide, is omitted, polypeptide synthesis is completely inhibited (table 5). This indicates that chain elongation is permitted only when the residues are lined up in the right order on the protein of fraction II.

After leucine, the last amino acid in the repeating pentapeptide sequence, is added, the product must be promptly treated with TCA in order to be able to find the pentapeptidyl hydroxamate corresponding to the enzyme bound thioester, because addition of leucine causes a fast release of decapeptide (cf. table 5). As shown in fig. 6 and table 5, no peptides longer than pentapeptides can be found as protein-linked peptides. This we interpret to mean that after completion of a sequence of five, the chain does not grow further and the decapeptide forms by cyclization from two enzyme-bound pentapeptides, as is schematically shown in fig. 7.

Recent work on tyrocidine synthesis in extracts of a different strain of *B. brevis* [11] indicates this cyclic decapeptide to be synthesized by an analogous mechanism. Tyrocidine contains the ten amino acid sequences shown in fig. 8. The tyrocidine sequence includes the five amino acids (italicized) that constitute the half cycle of GS. Tyrocidine-synthesizing extracts [11] yield three fractions: the lighter, molecular weight 100,000, activates and racemizes phenylalanine; an intermediate fraction, molecular

Table 5
Formation of protein-bound chains with increasing numbers of amino acids.

Amino acids added	Radioactivity (cpm)	
	Thioesters	Gramicidin S
Phe-^{14}C	1,549	265
Phe-^{14}C, Pro	6,001	375
Phe-^{14}C, Pro, Val	10,005	1,531
Phe-^{14}C, Pro, Val, Orn	14,325	1,005
Phe-^{14}C, Pro, Val, Orn, Leu	2,029	25,409
Phe-^{14}C, Val, Orn	1,610	376

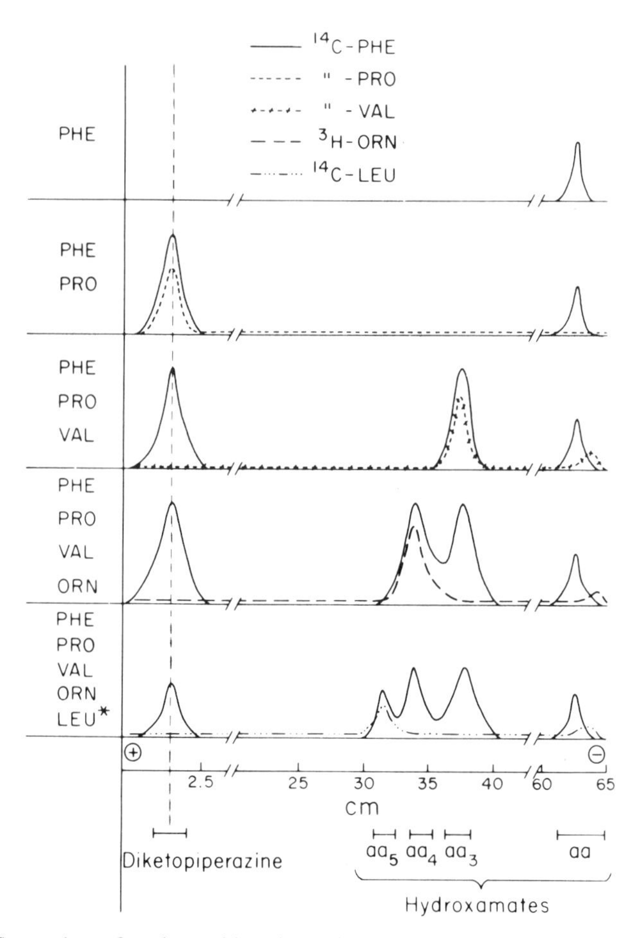

Fig. 6. Separation of amino acid and peptidyl hydroxamates by high-voltage paper electrophoresis. For details, see [5].

weight 230,000, activates proline; a third fraction, molecular weight 460,000, activates the other amino acids in the sequence: phenylalanine, asparagine, glutamine, tyrosine, valine, ornithine, and leucine. The latter fraction also racemizes a second phenylalanine. The enzyme-amino acid complexes formed after reaction with ATP can be isolated, and contain the protein-bound thioesters. For tyrocidine synthesis, all three fractions have to be combined; synthesis is initiated by the light fraction carrying phenylalanine in combination with the proline-activating fraction. The decapeptide again forms by

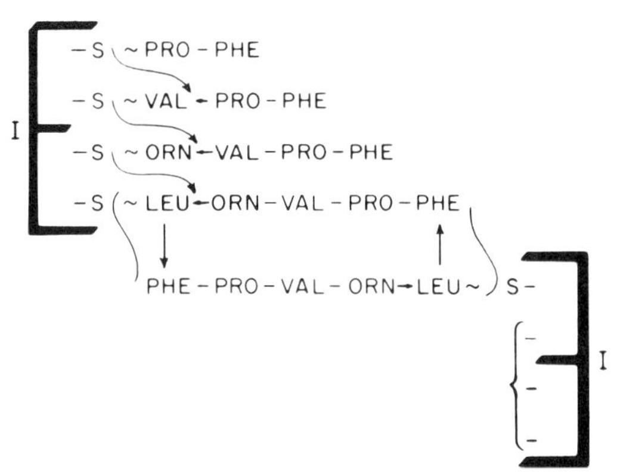

Fig. 7. Elongation and termination by cyclization between two pentapeptidyl-S–Enzymes (280,000 molecular weight) in GS biosynthesis. Phenylalanine is activated, bound, and racemized on the smaller enzyme, molecular weight 100,000. The reaction of this phenylalanine-carrying enzyme with the heavier one carrying the other four amino acids initiates polymerization (not shown in this scheme).

sequential addition of the amino acids in the direction and order indicated in fig. 8, with ring closure between the amino group of phenylalanine and the thioester-bound leucine carboxyl on fraction III.

To summarize, we find this amino acid polymerization to proceed on polyenzymes from protein-bound aminoacyl thioesters. The amino acids are activated by ATP on specific sites by forming aminoacyl-adenylate enzyme complexes from which the amino acids transfer to enzyme-linked SH groups with a displacement of AMP. The reaction proceeds from an amino terminal through one-by-one transpeptidation of the thio-esterified carboxyl to the

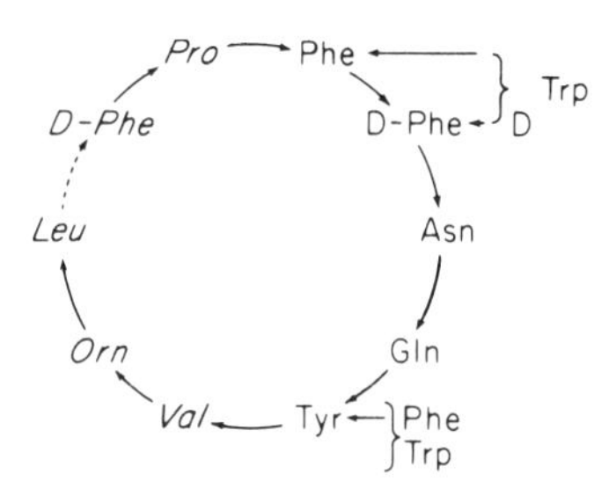

Fig. 8. Amino acid sequence in tyrocidine. The chain grows from Phe.... to Leu. The dotted arrows indicate the places of cyclization.

amino group of the next aminoacyl thioester. The lengthening peptide chain thus remains thioester-bound to the protein until GS and tyrocidine are liberated by cyclization (cf. fig. 8).

The specificity here is not as accurate as in the ribosome-linked process. Particularly in tyrocidine, replacements of amino acids, indicated in fig. 8, were noted first with living cells [10], e.g. phenylalanine may be replaced by tryptophan depending on its relative concentration in the mixture. Such replacements have also been observed in our in vitro synthesis; the antibiotic activity is not greatly altered. Furthermore, the length of the polypeptide is limited; the longest antibiotics of this group seem to be the gramicidins A, B, and C, originally isolated by Dubos and Hotchkiss using the same organism that makes the tyrocidines. These are straight chains of 15 amino acids, formylated at the amino terminal and blocked with ethanolamine at the carboxyl end. Partial reactions, similar to those described, have been found by us in tyrocidine-synthesizing extracts of the organism that also makes the gramicidins (unpublished).

Obviously the enzyme proteins described here that are isolated from present-day organisms are made on messenger RNA by ribosome-linked synthesis. In terms of process evolution, the antibiotic synthesis is closely related to the multi-enzyme fatty acid synthesis, where the growing fatty acid chain remains enzyme-linked through thioester bonds. A next step in the evolution toward protein synthesis is antibiotic polypeptide synthesis, a much

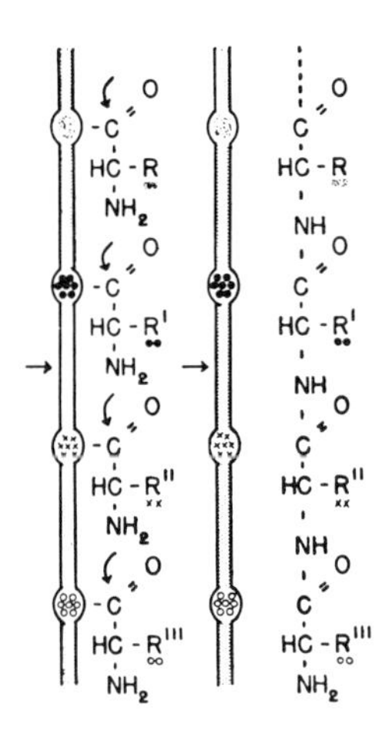

Fig. 9. The left side scheme depicts a number of different amino acids charged to an enzyme. The right side indicates their interaction and release to form the polypeptide chain starting from the amino terimnal.

more compact procedure than ribosome-linked protein synthesis with only enzymes, ATP, and amino acids as ingredients. And thus it gives us a primitive model for a process of sequential addition of amino acids to form functionally defined polypeptides.

In the introduction, ferredoxin, the short polypeptide containing an incomplete set of only 13 amino acids, was discussed as a model of an early enzyme. It is heavily loaded with cysteine: eight cysteines in a total of 55 amino acids. Its structure indicates that it may have arisen by gradual apposition of smaller peptides. In a similar manner, polypeptide structures may have been formed where, with polyphosphate as primary condensing agent, amino acids could be thioester-linked to specific cysteine residues to be zippered off in sequence. This is shown in fig. 9, which is taken from an early review of mine dating from before we understood the ribosomal process [6, cf. fig. 2, p. 602].

References

[1] Benson A.M., H.F. Mower and K.T. Yasunobu, Arch. Biochem. Biophys. 121 (1967) 563.
[2] Berg T.L., L.O. Froholm and S.L. Laland, Biochem. J. 96 (1965) 43.
[3] Eck R.V. and M.O. Dayhoff, Science 152 (1966) 366.
[4] Gevers W., H. Kleinkauf and F. Lipmann, Proc. Natl. Acad. Sci. U.S. 60 (1968) 269; 62 (1969) 226; 63 (1969) 1335.
[5] Kleinkauf H. and W. Gevers, Cold Spring Harbor Symp. Quant. Biol. 34 (1970) 805.
[6] Lipmann F., in: Mechanism of Enzyme Action, ed. W.D. McElroy and B. Glass (John Hopkins Press, Baltimore, 1954) p. 599.
[7] Lipmann F., Science 120 (1954) 855.
[8] Lipmann F., in: Origins of Prebiological Systems, ed. S.W. Fox (Academic Press, New York, 1964) p. 259.
[9] Lynen F., D. Oesterhelt, E. Schweizer and K. Willecke, in: Cellular Compartmentalization and Control of Fatty Acid Metabolism (Universitetsforlaget, Oslo, 1968) p. 1.
[10] Mach B. and E.L. Tatum, Proc. Natl. Acad. Sci. U.S. 52 (1964) 876.
[11] Roskoski R. Jr., H. Kleinkauf, W. Gevers and F. Lipmann, Federation Proc. 29 (1970) 468.
[12] Tomino S., M. Yamada, H. Itoh and K. Kurahashi, Biochemistry 6 (1967) 2552.
[13] Yukioka M., Y. Tsukamoto, Y. Saito, T. Tsuji, S. Otani, and S. Otani, Biochem. Biophys. Res. Commun. 19 (1965) 204.

Chemical Evolution and the Origin of Life, eds. R. Buvet and C. Ponnamperuma

EVOLUTION OF PROTEINS

M.O. DAYHOFF

National Biomedical Research Foundation,
11200 Lockwood Drive, Silver Spring, Md. 20901, USA

"Relics" of ancient organisms can be found in the biochemical systems of their living descendents. The exceedingly conservative nature of the evolutionary process has preserved such relics in all living species. Many basic reaction pathways and even many features of complicated polymer structures are derived from extremely remote ancestors, living long before the ordinary fossil record was formed. We shall see that the biochemical evidence in the protein and nucleic acid structures gives us a wealth of information from which to construct a phylogenetic tree of all life. Further, it gives us glimpses of the ancient molecules in ancestral organisms, and of the nature of the evolutionary process, and finally it provides a time scale for measuring the relative antiquity of divergences. Untimately a great deal will be learned from living organisms about protein and nucleic acid structures in a form which existed more than 3 billion years ago – the most recent ancestor of all presently living species and even in the single line which preceded it.

The interweaving of the inferences from this biochemical evidence with the absolute time of selected events derived from fossil evidence promises to fulfill a long-standing hope, that of working out the complete, detailed, quantitative phylogenetic tree – the history of the origin of all living species back to the very beginning of life.

In this chapter, I shall deal mainly with an important portion of the biochemical evidence, the amino acid sequences of proteins from living organisms. Potentially a great quantity of such information can be elicited; there are several thousand different kinds of proteins in the bacterium *Escherichia coli* and possibly a million kinds in a human being. There are a small number of generally applicable, theoretically understandable laws that have governed the evolution of proteins; therefore it is possible to apply logical and statistical methods quantitatively to the data.

The first complete sequence of a protein was published in 1953 by Sanger and his coworkers [1]. Since then the sequence information has grown exponentially; well over 300 complete sequences [2] have already been established by hundreds of workers in many fields. Their varied interests include chemical structure and function, catalysis, polymer chemistry, chemical synthesis, drug manufacture, medicine, genetics, and taxonomy. Due to this broad interest base and to recent major improvements in laboratory techniques, sequence information promises to accumulate rapidly for some time to come.

Because of our own research interest in the theoretical aspects of protein chemistry, my laboratory has long maintained a collection of known sequences. In order to integrate the information gathered by diverse groups, to focus attention on this important new area of study, and to establish a foundation of readily available, correct data, we decided in 1965 to publish all the known sequences annually in an *Atlas of Protein Sequence and Structure*, four volumes of which have already appeared. To aid workers in the correlation and understanding of the protein structures, we have included in the *Atlas* theoretical inferences and the results of computer-aided analyses which are necessary to illuminate such inferences. Many of the ideas that I will discuss are taken from the *Atlases* [2,3] to which many people in my

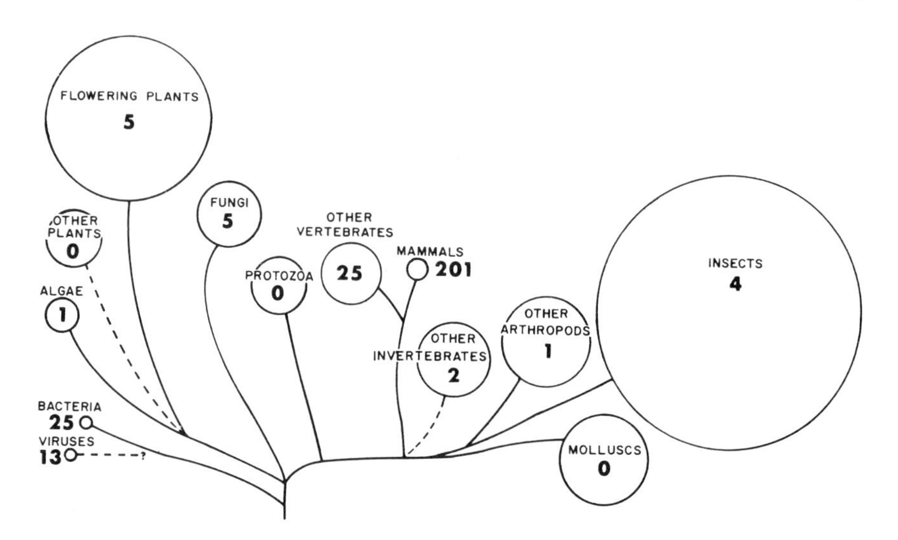

Fig. 1. Distribution of sequence data among biological groups [3]. The area of each circle is proportional to the number of species which have been described for each group (some groups are very incompletely known). The numbers indicate protein sequences more than 30 links long which are presented in the *Atlas of Protein Sequence and Structure 1969* [2].

group have contributed, including C.M. Park, L.T. Hunt, P.J. McLaughlin, W.C. Barker, R.V. Eck, and M.R. Sochard. A detailed bibliography on the original sequence work and related ideas of more than 1,400 authors is given in the 1969 *Atlas* [2]. The distribution of the sequenced proteins among the biological groups is shown in fig. 1. The area of each circle is proportional to the total number of species which have been described for that group. However, some groups are very incompletely known, and there is no generally accepted definition of "species" for bacteria or for viruses. The integer in each circle represents the number of sequenced proteins. It can readily be seen that the emphasis so far has centered on the mammals. Only rarely have proteins from species of widely different ancestry been considered. Yet a comprehensive approach is important because of the constraining interrelatedness of all living organisms through their common descent; every biological structure, function, and system is largely determined by its evolutionary history. Knowledge concerning the structure of a cellular component, such as a protein, in a selection of diverse organisms, together with knowledge of the evolutionary tree of biological species, and of the rules of evolution permits a kind of biological interpolation, the prediction, within ascertainable probability limits, of the nature of the structure in all other living organisms and even in ancestral organisms.

Structure of Proteins

Proteins are polymers composed of twenty different kinds of amino acids. Each protein is synthesized in the cell according to a corresponding message in the genetic material, the desoxyribonucleic acid (DNA) polymer. Each successive triplet of monomer nucleotides ultimately results in the incorporation of a particular amino acid into the succession of amino acids of the protein chain. The chemistry of this process is very complex, involving many nucleic acid and protein catalysts. As each amino acid is added to the chain, one molecule of water is given off. The growing chain begins to coil and adhere weakly to itself to form a three-dimensional structure of characteristic shape, flexibility, and chemical properties [4].

Many different kinds of proteins are required for the functioning of a living organism; each protein is delicately adapted to perform important structural, chemical, or control functions in the cell. Proteins are very important components of the complex feedback network of cell chemistry; they are the main determinants of the physical and chemical characteristics of a species. It is the very slight differences in the structures and quantities of some of the very many proteins in the human complement which lead to individuality.

It has become clear that the basic metabolic processes of all living cells are very similar. A number of identical structures, reaction pathways, and small compounds are found in all living species so far observed. Identical proteins are seldom found; however, a number of proteins have identifiable counterparts, known as homologues, in most living things. Biochemical reactions which involve cooperation between catalytic proteins from widely different species have been carried out in the laboratory [5]. During the past 15 years, homologous proteins have been shown to have very similar amino acid sequences as well as closely similar three-dimensional structures [2,6].

The variation in protein structure from one biological group to another is illustrated in fig. 2 by the first halves of the sequences of cytochrome *c* [7], a protein of fundamental importance which functions in mitochondrial electron transfer. A clearly, though very distantly, related protein has also been analysed from a simple photosynthetic bacterium, *Rhodospirillum rubrum*, in which the protein is used in photosynthetic electron transfer.

All the sequences, even though from very diverse organisms, contain so many identities that there is no question about the correct alignment. About 60% of the total number of amino acids at corresponding positions are identical in wheat and human chains while 30% are identical in human and *R. rubrum*. Certain positions contain the same amino acid in all the sequences; others are filled by any one of several amino acids which are usually of similar shape or chemical properties.

The different positions in a sequence may be treated as separate traits from which a phylogeny of the species can be deduced in much the same way as from biological traits. The mathematical techniques used are based on the premise that when identical changes in a trait are observed in two species, the most likely explanation is that the change occurred in a single ancestral organism which was common to both. Fig. 3 shows the phylogenetic tree which we have derived from the cytochrome *c* data. This tree, within the limitations of the small quantity of data now available, has the same topology (order of branching without regard for the lengths of branches or for the angles between them) as trees derived from morphological or other biological considerations. Each point on the tree represents a definite time, a particular species, and a definite protein structure within the majority of individuals of this species. There is a "point of earliest time" on any such tree. Radiating from this point, time increases on all branches. Protein sequences from living organisms all lie at the ends of branches which represent the present time. The pattern of the network of lines in the tree then indicates the relative order in which the proteins became distinct from one another. The branch

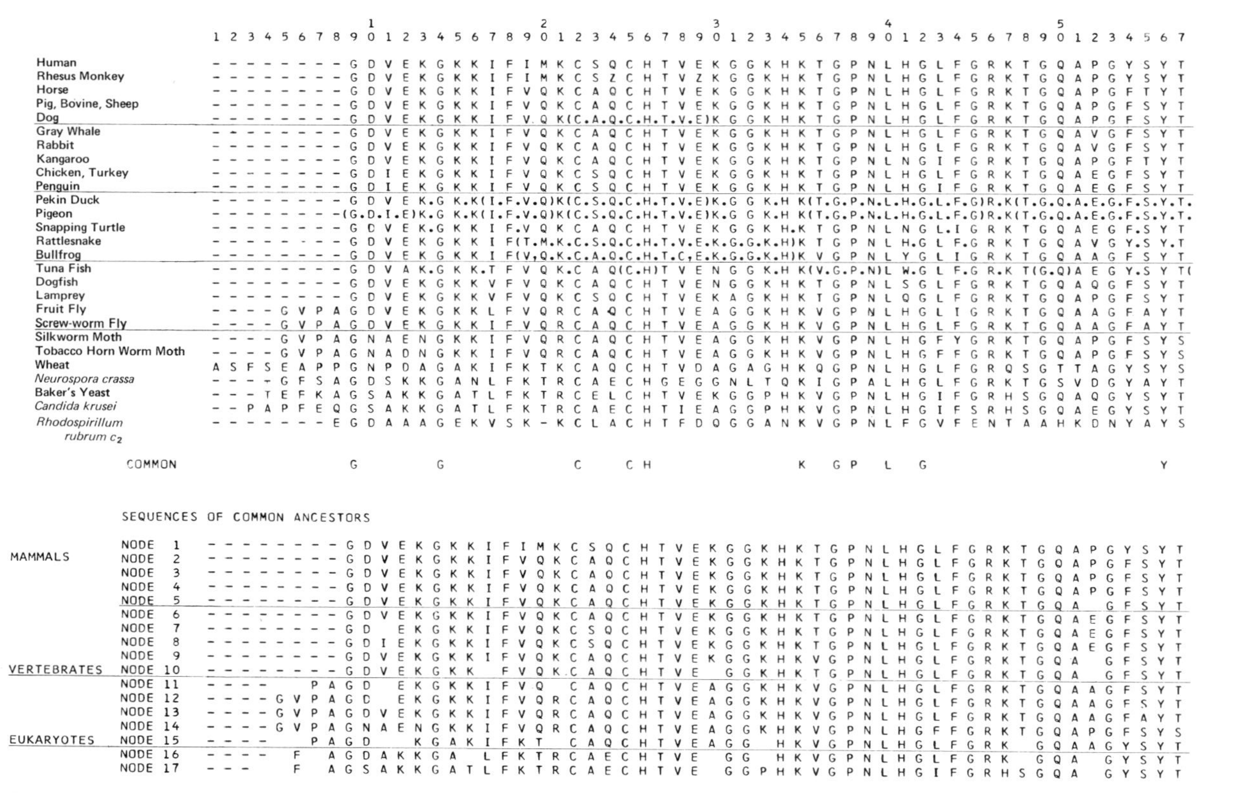

Fig. 2. The first halves of the cytochrome *c* proteins from 29 species [3]. Each amino acid residue is represented by a single letter. Sequences of closely related groups are similar. Even though the proteins come from widely different biological groups, only a single amino acid is found in 11 of the 57 positions shown. Very few changes have been accepted in the other positions. Dashes have been inserted to maintain the alignment where insertions or deletions of genetic material must have occurred. The sequences of the ancestors at the nodes of fig. 3, which were derived by the computer methods discussed are shown in the lower part of the figure. In the few places that are blank, two or more amino acids are equally likely to have been present. The abbreviations for the amino acids are: A,alanine; B,aspartic acid or asparagine; D,aspartic acid; E,glutamic acid; F,phenylalanine; G,glycine; H,histidine; I,isoleucine; K,lysine; L,leucine; M,methionine; N,asparagine; P,proline; Q,glutamine; R,arginine; S,serine; T,threonine; V,valine; W,tryptophan; Y,tyrosine; Z,glutamic acid or glutamine.

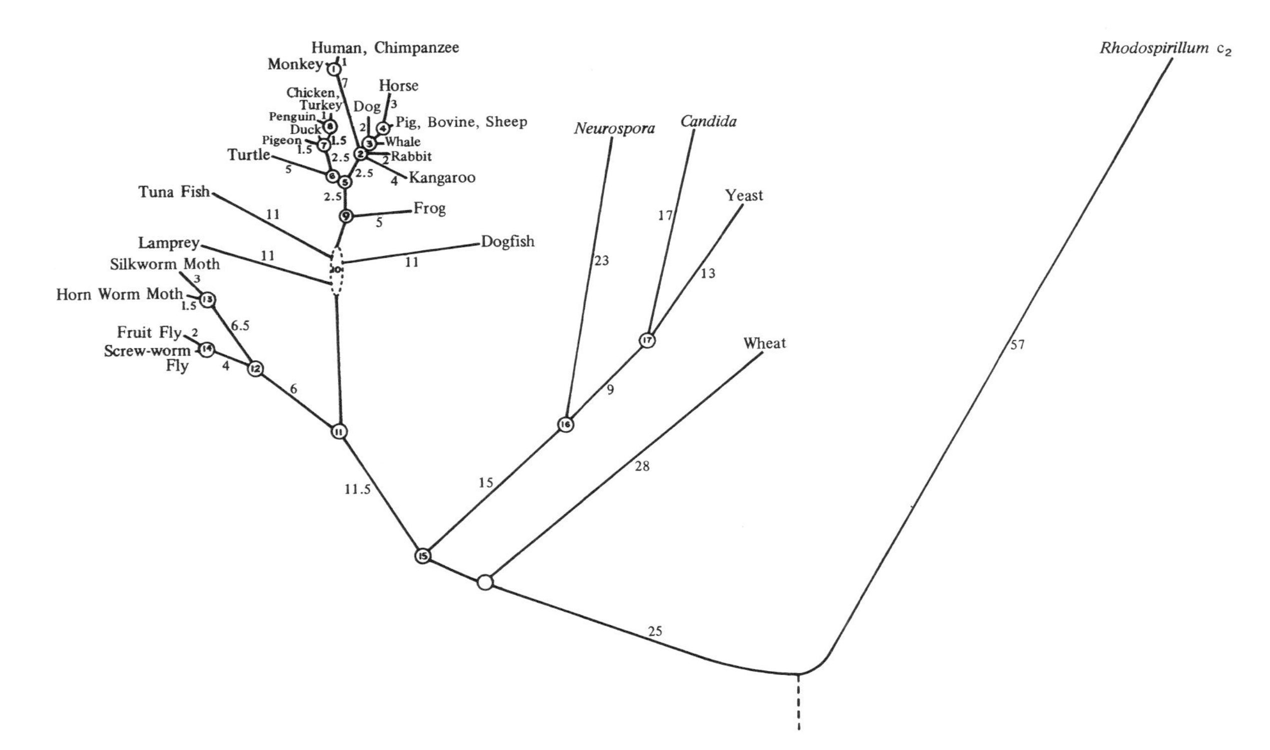

Fig. 3. Phylogenetic tree of cytochrome *c* [3]. The topology has been derived from the sequences as explained in the text. The numbers of inferred amino acid changes per 100 links are shown on the branches of the tree. The point of earliest time cannot be determined directly from the sequences; we have placed it by assuming that, on the average, the cytochrome sequences in different species change at the same rate.

lengths are proportional to the number of changes which we infer occurred in each interval. The location of the point of earliest time, that is, the connection of the trunk to the branching structure, cannot be inferred directly from the sequences, but must be estimated from other considerations. It seems likely that sequences from all organisms have changed with time, even though the morphology of some organisms may appear primitive. In the absence of any evidence to the contrary, we have used a point about equally distant from all of the observed sequences as the point of earliest time.

Computer methods

To derive the topology of the tree we have used a computer method based on the requirement that the minimum possible number of amino acid changes has occurred in the ancestral organisms. Since cytochrome *c* changes so slowly and since there are very few identical mutations occurring independently in different species, this tree reflects quite closely the true historical course of events.

In principle, the computer tries every pattern of connected branches and all possible sequences for the ancestral structures at the branch points, or "nodes". For each combination, the minimum number of mutations which must have occurred during the history of the tree can be calculated; it is derived from a count of the total number of changes between adjacent sequences at the nodes and at the ends of the branches. In practice, there are many shortcuts to finding this best tree. The detailed procedure is described in Chapter 2 of the *Atlas of Protein Sequence and Structure 1969* [2].

The ancestral sequences generated by the computer are shown at the bottom of fig. 2. If one amino acid is clearly most likely to have been ancestral, it is shown. Where two or more amino acids are nearly equal contenders, the position is left blank. One can see that as a result of the great conservation of proteins most positions of an ancestral structure can be inferred. With the advent of methods to synthesize proteins in the laboratory [8], these ancient structures take on added significance; they may one day be produced and their properties measured. A great deal about the chemical capabilities of ancient organisms may thus eventually be known.

The cytochrome *c* phylogenetic tree

Fig. 3 shows the best tree for the cytochromes obtained by these procedures. The branch lengths are expressed in PAMs (Accepted Point Mutations per 100 links). This unit includes a correction for the superimposed mutations estimated to have occurred [2]. The major groups fall clearly into the topology shown. However, the more recent divergences cannot always be resolved because either there have been no mutations in cytochrome *c* or the evidence of these mutations has been obliterated by subsequent changes.

One aspect of the topology is particularly interesting, the attachment of the bacterial sequence. The evidence found in positions 13, 29, and 57 indicates a connection to the wheat branch. Position 55 alone gives weak conflicting evidence for a connection to the fungus branch. If the preponderance of evidence is true, then the fungi and animals must have diverged from each other after their divergence from the line to green plants. In the resulting tree, the bacterial sequence is at the end of a very long branch.

Geological time scale

We are interested in knowing the relative geological times associated with the events of the phylogenetic tree of cytochrome *c*. We estimate these times from the number of mutations on the branches. For each family of proteins, the rate of accumulation of mutations is different. However, the risk of mutation within a family such as cytochrome *c* is evidently rather constant, at least for nucleated organisms, over the interval during which the species under consideration diverged. The variation of branch lengths which correspond to the same time interval is about what would be expected for a random process of mutation. The proportionality constant of time per mutation must be fixed from considerations such as the divergence point of the fish and mammalian lines. Paleontological evidence [9] places this particular divergence at about 400 million years ago. The cytochrome *c* tree places it at 11.5 PAMs, on the average. Therefore, 400 million years correspond to 11.5 PAMs. Using the constant rate assumption, let us consider the implications for the time of divergence of the major groups.

The information in fig. 3 is redrawn in fig. 4 with the meaning of the branch lengths altered. Now the vertical coordinate is proportional to time, with the present day at the top. The positions of the nodes are derived by averaging the number of mutations on branches which represent the same time interval. The connection of the branches to the trunk was presumed to be equidistant from the bacterial and the other sequences. From this one family of clearly related proteins, estimates of the time relationships of an extensive phylogenetic tree of life are derived.

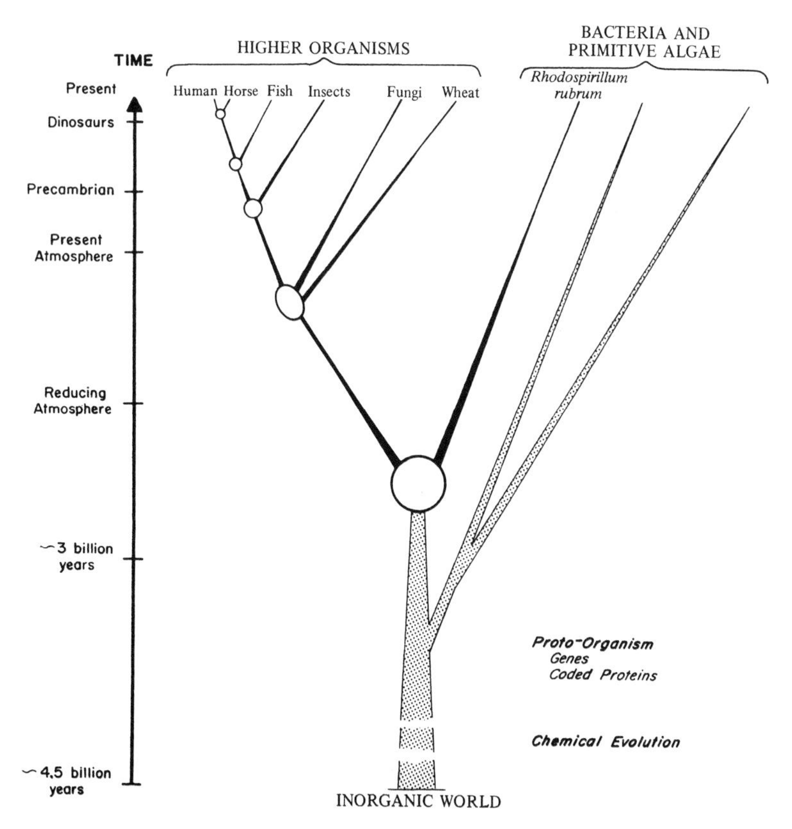

Fig. 4. Temporal aspects of the cytochrome *c* phylogenetic tree [3]. Details of the upper part of the tree, shown in black, are derived from information in the cytochrome *c* sequences. The size of any node reflects the uncertainty in its position. The stippled parts of the tree are inferred from the details of biochemical and biological structures and from geological and fossil evidence. Fossils resembling present-day bacteria and blue–green algae have been found in chert more than three billion years old [10]. All living phyla seem to be derived from a proto-organism which possessed many metabolic capacities, nucleic acid genes, and protein enzymes (produced by the genetic code). Preceding the time of the proto-organism, the genetic code and the metabolism of nucleic acids and small molecules evolved [11]. The level of the atmosphere is inferred from the oxidation states of ancient sediments [12].

The evolutionary process

The cell routinely forms a copy of its genetic nucleic acids before cell division, passing on one set to each daughter cell. The rare errors which occur in this duplication make possible the evolutionary process.

Several kinds of mistakes occur repeatedly [13]. First there are gross errors in chromosome structure. A duplicated portion may be incorporated in a chromosome, resulting in the synthesis of two identical sets of proteins coded from the original and the duplicated genes: or a section of a chromosome may be deleted altogether, in which case the corresponding proteins would no longer be produced at all. Occasionally sections of chromosomes are moved so that the order of the genes on the chromosome is altered. In the higher organisms, sections of genetic material can be exchanged between chromosomes; or an extra chromosome may be passed to the offspring, as in the case of human mongolism. Even an entire duplicate genetic complement is sometimes found in plants.

There are also errors which occur within single genes. Of these, the changes most commonly accepted by natural selection are changes of one single nucleotide within a gene to one of the other three nucleotides. The resulting protein may then have one amino acid changed. Occasionally an amino acid is inserted or deleted. Most such changes are deleterious, diminishing or destroying the effectiveness of the protein produced. Many examples of the effects of these errors, so-called "point mutations", can be seen in the cytochrome sequences of fig. 2.

Mutant proteins

Mutant proteins are found in some individuals of a species. An exhaustive search for such forms has been made in the case of the human hemoglobin proteins. Hemoglobin is the principal oxygen-transporting substance in vertebrate blood. In higher vertebrates, hemoglobin typically has a tetrameric structure consisting of two alpha-type and two beta-type protein chains; a heme group is attached to each chain. Two atoms of oxygen can be transported by each heme group. Fig. 5 shows the beginning of the amino acid sequences of the alpha and beta proteins of human adults. The abnormal amino acid replacements which have been reported are shown below these. In many cases it has been possible to correlate deficiencies in function of the red blood cells with the chemical defects in these chains [2,13]. The hemoglobin molecule participates in a number of chemical equilibria which can be

```
                              10                  20                  30
Alpha     V - L S P A D K T N V K A A W G K V G A H A G E Y G A E A L E R M F L S
                    D             D     D E           D Q             Q
                                                        V
                                                        K

                            10                      20                  30
Beta      V H L T P E E K S A V T A L W G K V N - - V D E V G G E A L G R L L V V
            Y       V G   C         R   D               A - R R K   P   S
                    K K                 R               K
                    -
```

Fig. 5. Mutant human adult hemoglobin chains [3]. The first parts of the sequences of normal alpha- and beta-chains are shown. Twenty-seven mutant chains are indicated by the bold letters beneath. Each mutant chain had one amino acid change which could have been the result of a single nucleotide change in the gene. Deleted amino acids, indicated by dashes, are the result of the deletion of three nucleotides. Surveys of healthy individuals of European origin indicate that fewer than 1% of the population produce abnormal chains.

drastically disturbed by the substitution of a single amino acid; a constituent normally present in low concentration, which has suboptimal functional ability, can become a major component. For example, the tetramer can dissociate into two alpha–beta dimers and finally into alpha- and beta-monomers. The solubility of the separate chains may be altered so that the hemoglobin precipitates and damages the physiological functioning of the cell. The normal folding of the protein may be impossible or the heme group may dissociate from the protein chains. The ferrous iron may be oxidized to the nonfunctional ferric state. The binding of oxygen may be altered in either direction: if the bond is strengthened, the oxygen will not be given up to the tissues readily enough; if it is weakened, not enough oxygen will be picked up in the lungs. The tetramers can form crystals which destroy the normal flexibility of the cell and impair its flow through the capillaries. Further, one mutation has recently been shown to have an effect at the nucleic acid level by decreasing the stability of the messenger RNA which codes for hemoglobin [14].

Because of the many interactions involved, almost any change in the structure of a protein, even though it might be of particular advantage for one function, would coincidentally disturb so many other interactions that it would be extremely disadvantageous to the organism. So severe are these constraints that an identical sequence of each protein is found in most individuals of a species. A given sequence often predominates within a species for several million years. A minor variant, usually differing by only one amino acid, may eventually become preferable because of changes either in the other cell constituents or in the environment. Only infrequently does a more profound structural change prove beneficial. The net effect of this extreme conservatism is the great similarity of protein structures in various species, as seen in fig. 2.

Of the almost infinite number of possible combinations of chromosomal aberrations and mutations that might have occurred, only a very insignificant fraction have actually occurred in living organisms. It seems very likely that many which could have occurred, but in fact did not, might have proved beneficial and have been accepted. The evolutionary paths that were actually followed were based on a succession of rare random events, the acceptable mistakes. At any given time, the future of biochemical evolution is largely indeterminate because of these many possible alternatives. Each diverging species is relatively free to accept a unique series of changes. The preservation of identical changes in different species today usually marks them as sharing common ancestors in which the mutations occurred.

Globins

Extensive sequence work has been done on the hemoglobins and on myoglobin, a related protein found in muscle. These studies reveal an intermixture of species divergences and duplications of genetic material in the last billion years [15]. The main events in the globin family are shown in fig. 6. Genes coding for alpha- and beta-type chains, presently located on different chromosomes, were derived originally from a single ancestral gene by an ancient duplication of genetic material. More recently, another duplication gave rise to separate genes coding for the human beta chain, which is synthesized after birth, and to the human gamma chain, a beta-type chain which is synthesized before birth. This separation in time of the production of the chains indicates that both the gene for the original protein structure and the mechanism which controlled protein synthesis were duplicated in each case. Some recent duplications have given rise to genes which lie close to each other on the same chromosome, for example, the human delta and beta chain genes.

Myoglobin is evidently the result of an ancient duplication. Extrapolating from the estimated geological times of divergence of some species on the globin tree (400 million years for mammals and bony fish), we could expect that the divergence of hemoglobin and myoglobin took place in the Precambrian, perhaps a billion years ago. Other related proteins which diverged from the ancestral globin stock a long time ago have also been examined, one from an annelid worm [6], one from an insect larva [6,16], and one from a plant [16].

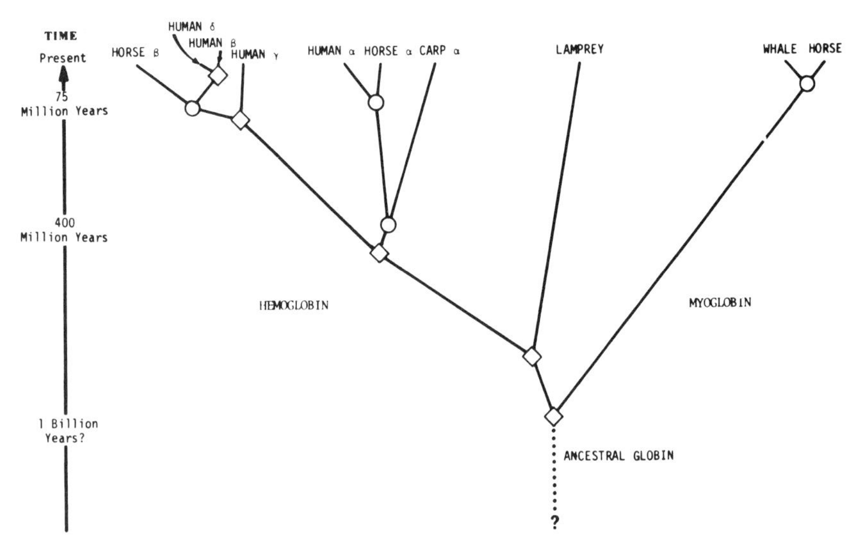

Fig. 6. Evolutionary tree of the globins [3]. The lengths of the branches are proportional to the estimated number of amino acid replacements per 100 links. The scale at the left shows two geological time markers. The time of the ancestral globin is estimated by extrapolation. Circles indicate species divergence, and diamonds, gene duplications. Note that, as would be expected, the horse alpha- and the horse beta-chains diverged from the human at about the same time. In general, it is impossible to construct a tree that shows both the number of changes on each branch and time levels. In this simple case it happens to be possible and is instructive. The insect sequence diverged very early, possibly even before the myoglobin–gemoglobin duplication.

Composite phylogenetic tree

Eventually the information from many proteins will be combined to yield a more precise tree than that derived from any single one, the precision increasing as the square root of the total length of the sequences used. Fig. 7 shows the tree derived from cytochrome *c* and the alpha- and beta-hemoglobin sequences. Even with this amount of information, no difference can be seen between the human and chimpanzee sequences.

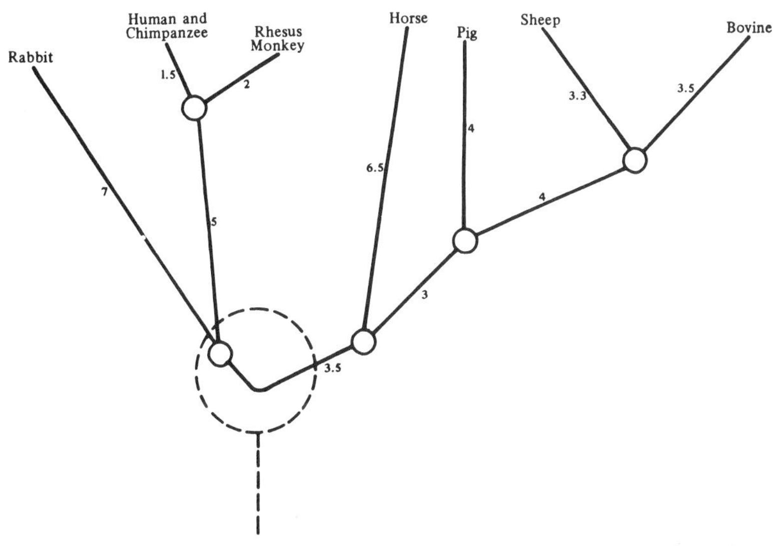

Fig. 7. Composite phylogenetic tree [3]. The proteins used to make this tree, cytochrome *c* and the α- and β-chains of hemoglobin, are combined to produce a total of nearly 400 residues for each organism. This total gives the tree a precision twice as great as one derived from a sequence of 100 residues. The tree should become larger and more meaningful as additional sequences accumulate. That chimpanzees have sequences identical to humans is surprising. On this basis it is very likely that human and chimpanzee diverged from each other within the last 10 million years, a considerably shorter interval than the 20 to 30 million years previously suggested by several authorities. If these species had diverged more than 10 million years ago, there would be less than a 1% probability of finding no mutations in these 400 links, on the basis of a theory of uniform risk of mutation over time.

Other protein families.

There is presently sufficient sequence information to draw interesting trees for a few other families [3]. Fig. 8 shows the tree inferred from immunoglobulin proteins, a tremendously complex family derived from hundreds of genes which have proliferated in the vertebrates [17]. There have been numerous duplications of genetic material, which include portions of genes, whole genes and even whole chromosomes, as well as point mutations. In this family there is even a splicing mechanism, as yet little understood which attaches two separate genes so that one single protein is produced. Knowledge

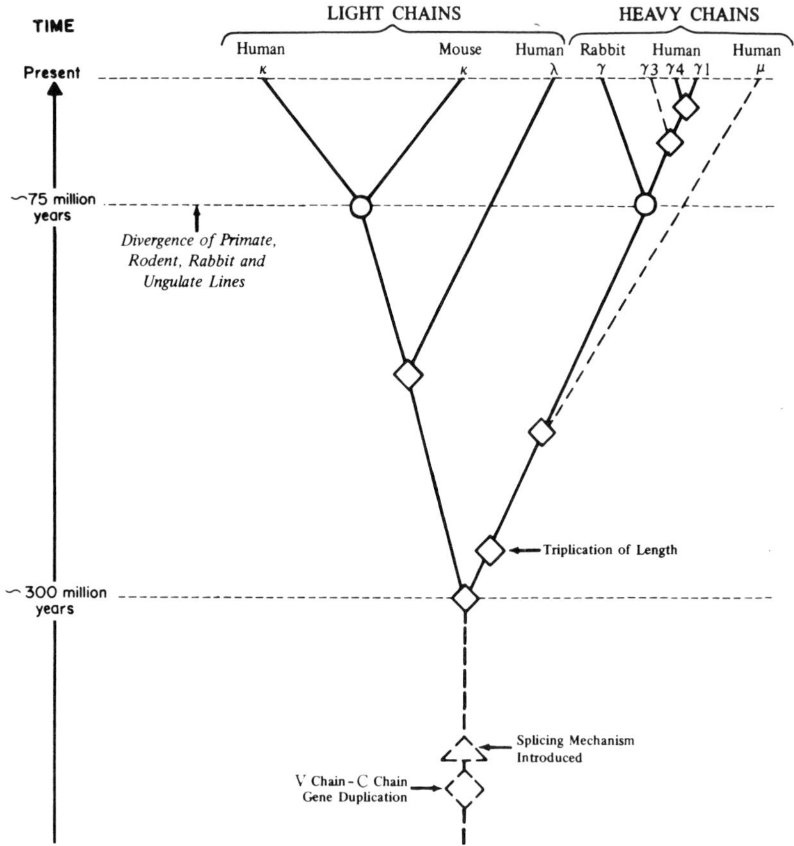

Fig. 8. Summary of the evolutionary course of events in the C genes of immunoglobulins [3]. The tree is derived from inferences made from amino acid sequence data presently available and from fossil evidence which gives the times of species divergence. A chromosomal duplication took place roughly 300 million years ago, giving rise to the ancestral genes for heavy and light chains. The light chain gene duplicated again about 170 million years ago, giving rise to the ancestral gene for the kappa- and lambda-chains. The heavy chain gene underwent triplication in length soon after the gene duplication which led to the light chain lines. Following this triplication, a duplication occurred which led to the ancestral genes for a μ-chain and for a γ-chain. In the human, a separate gene duplication, which took place after the species divergence of primate, rodent, rabbit, and ungulate lines, led to genes for $\gamma 1$, $\gamma 2$, $\gamma 3$, and $\gamma 4$. The special splicing mechanism must have originated *before* the light-heavy divergence, but could not have preceded the duplication giving rise to distinct V and C genes. These developments are indicated on the trunk of the tree leading from the ancestral gene. The species divergence between the mammal and bird lines most probably occurred after the heavy-light chain divergence. The heavy chains of the two lines would then be homologous, as would the light chains. The kappa- and lambda-light chain divergence may have occurred only in the mammalian line.

of the sequences has contributed to the basic understanding of the structure and evolution of this very complex, medically important group.

Extensive work has been done on trypsin and related proteolytic enzymes [18]. The evolutionary relationships are shown in fig. 9. There has been a proliferation of different types of vertebrate enzymes, related in structure, but each type having a slightly different catalytic specificity and physiological purpose. Further, fragments of a bacterial sequence have recently been reported which are clearly homologous to the vertebrate chains. It is apparent that the digestion of protein food is of very ancient origin.

Even the viruses, whose origin is little understood, can be organized on the basis of the evolution of their proteins. Fig. 10 shows the relationship of the coat proteins of a number of tobacco mosaic viruses. The protein sequence information may eventually answer the question whether viruses are primitive or degenerate forms.

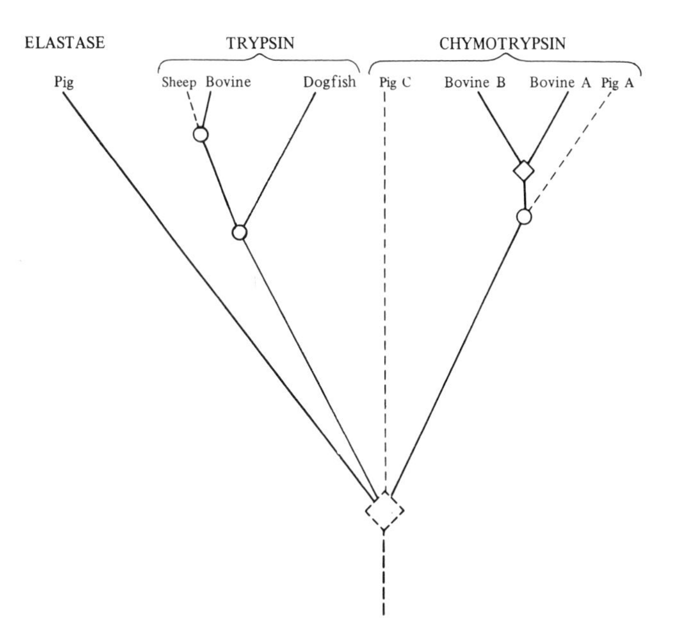

Fig. 9. Evolutionary tree of the proteolytic enzymes [3]. The sequences demonstrate that these different enzymes are the related products of gene duplications. The relationships of enzymes which are almost entirely sequenced are shown by solid lines, whereas the connections of partial sequences are traced by dashed lines.

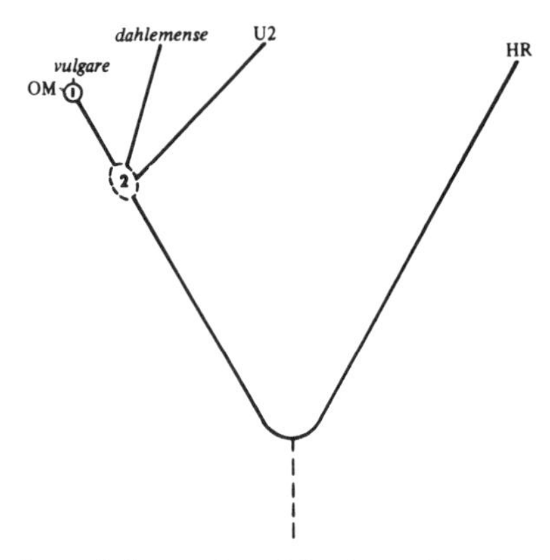

Fig. 10. Evolutionary tree of the coat proteins in strains of tobacco mosaic virus [3]. The HR, or Holmes ribgrass, strain is very distantly related to the other, more closely related strains.

Ferredoxin

The evolutionary tree derived from the protein of ferredoxin is particularly interesting in connection with the study of the early evolution of life. Ferredoxin is found in anaerobic and photosynthetic bacteria, in blue–green and green algae, and in the higher plants. The protein is bound to iron and inorganic sulfur through some of its cysteine residues. Ferredoxin is the most electronegative metabolic enzyme known, with a potential close to that of molecular hydrogen. It participates in a wide variety of biochemical processes fundamental to life, including photosynthesis, nitrogen fixation, sulfate reduction, and other oxidation-reduction reactions. It may have achieved these functions very early in the differentiation of the biological kingdoms, and its structure been conserved strongly ever since. The evidence from the plastids of higher plants indicates a rate of change of the protein comparable to that of cytochrome *c*.

The overall topology of the evolutionary tree of ferredoxin, shown in fig. 11, is clearly established by the sequences. Within the groups of plants and the group of nonphotosynthetic bacteria, the relative branch lengths have been estimated from the point mutations.

The sequences of ferredoxin from plant plastids are almost twice as long as those from anaerobic bacteria, whereas the one from *Chromatium*, a photosynthetic bacterium, is intermediate. A quantitative estimate of the

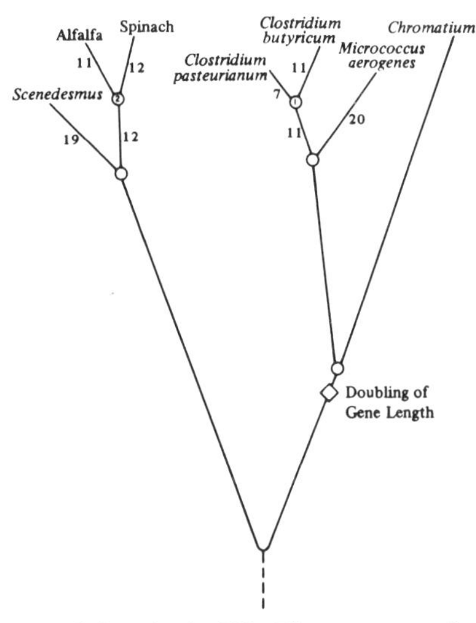

Fig. 11. Evolutionary tree of ferredoxin [3]. The sequences from bacteria and from plants are clearly related, but very distantly so. The doubling of gene length in the bacterial line probably occurred as shown here. The details of duplications in the plant line are obscured by the many subsequent changes.

evolutionary distance between the plant plastids and the bacteria in terms of point mutations is impossible from these sequences because of the many changes in length. It must be a very long distance. There has clearly been a duplication of genetic material in the bacterial line, and the two halves of the molecule are closely similar [19]. The plant line also appears to have incorporated duplications of genetic material, possibly independently of the event occurring in bacteria.

The biological meaning of the first divergence on the tree is not clear. It is likely that the plastids are symbionts acquired by a primitive higher organism about the time of the divergence of the plant and animal lines. The gene for plant ferredoxin may have derived either from the nuclear line or the plastid line. If it is a plastid gene, then the earliest branching point on the tree refers to the plastid – bacterium divergence rather than to the plant nucleus – bacterium divergence.

Evolution of transfer RNA

Of even more ancient origin than the coded proteins of the cell are the nucleic acids. One family of these, the transfer ribonucleic acids (tRNA's), has been subjected to extensive study [2,11]. The tRNA molecules are the translation catalysts in the synthesis of proteins, each being responsible for physically positioning one particular amino acid for incorporation into protein according to the sequence of nucleic acids in the genetic message. These structures are in part responsible for the specificity of the genetic code. The directions for their synthesis are contained in the nucleic acids. Their sequences have been shaped by natural selection acting on mutants of the genetic nucleic acids, just as in the case of proteins [2]. The sequences of over 14 tRNA's from bacteria and higher organisms, with specificity for eight of the amino acids, are now known. They are so closely similar that they can be aligned, just as in the case of related proteins. The nucleotide "traits" can even be used to form a phylogenetic tree reflecting the evolution of the amino acid specificities of the genetic code. Unfortunately, the quantity of data is still too small to make a convincing case for the exact order of branching.

All of the tRNA sequences studied are about the same length. Some positions always contain the same nucleotide and certain homologous regions contain complementary bases which evidently form hydrogen bonds; the common structure of all the tRNA's is shown in fig. 12.

The close similarity of all the tRNA's must be due to their origin from a common ancestral molecule, proto-tRNA, in a primeval organism. The probability is extremely small that a cell could independently synthesize even two structures which would function smoothly in the coding process. However, it is relatively easy for nature to create such similar structures by the production of redundant genetic material (by doubling or other chromosomal aberrations) followed by the accumulation of independent mutational changes in the separate genes. The development of the specificity of the genetic code seems to represent such a succession of doublings and minor changes both in the genes for the tRNA molecules and in those for the activating enzymes.

Since the very complicated genetic code is shared by all living species, it almost certainly arose in the common ancestral line before the divergence of biological species. The tRNA gene duplications and the evolution of amino acid specificity then took place in this primordial ancestral line. An extremely simple organism without the ability to manufacture coded proteins, routinely synthesized a molecule with the basic tRNA structure similar to that shown

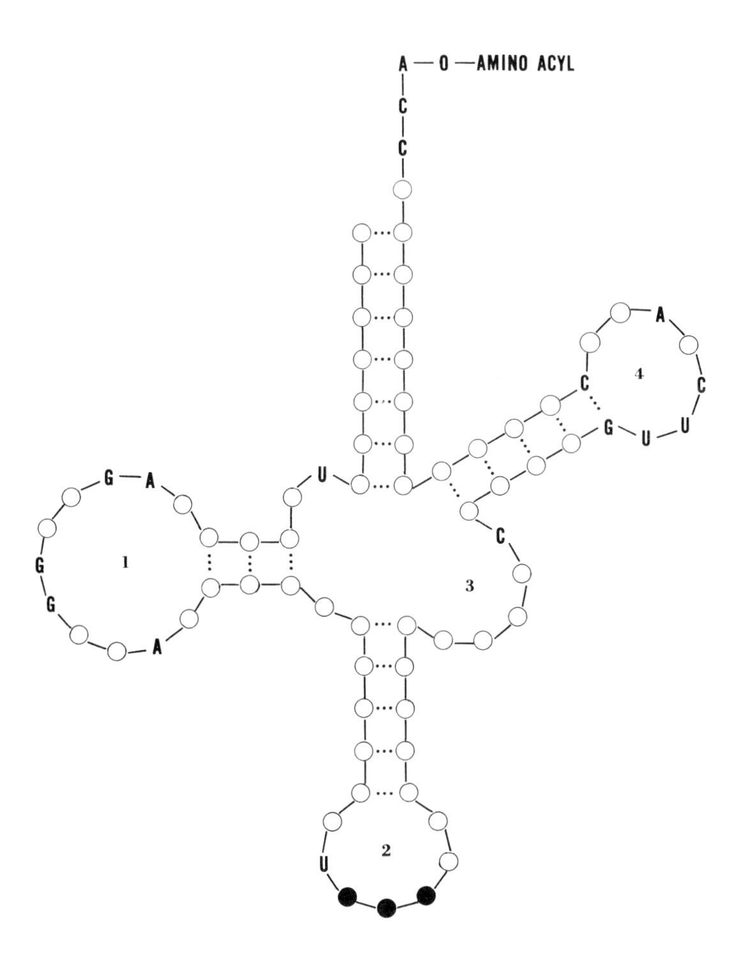

Fig. 12. Primitive transfer RNA[20]. This molecule must have been synthesized prior to the development of the genetic code and the synthesis of modern proteins. The structure shown incorporates the features which all known tRNA's have in common. The letters indicate positions always filled by the same base. The circles indicate positions which are filled by more than one base. The dotted lines show hydrogen bonding; complementary bases are usually found in the positions connected. The solid circles show the position of the anticodon triplet which recognizes the genetic message. The number of positions in loops 1 and 3 is variable, other regions are of constant size. There must be further interactions of the arms but the details have not been determined yet. The abbreviations for the bases are as follows: A, adenine; C, cytosine; G, guanine; and U, uracil.

in fig. 12. This simple form had evolved the ability to conserve and to propagate the structure of nucleic acids. The primitive tRNA is the oldest genetically produced molecule whose structure we can presently approximate. It must have imparted some selective advantage to the primeval organism possessing it, so that natural selection preserved them both, presumably because the proteins thus produced were beneficial. The primitive tRNA most probably acted as a nonspecific catalyst, polymerizing amino acids by a mechanism similar to the one still used today. This polymerizing mechsnism could hardly have arisen at a later time. Once the divergence of the specific tRNA's had occurred, it would have been almost impossible for nature to introduce a fundamental new feature; the simultaneous adaptation to this new feature by all the divergent tRNA's would be extremely unlikely.

General outline of biological evolution

In considering the general outlines of the origin of living things, three events stand out: the development of the genetic code, the divergence of bacteria from nucleated organisms and the divergence of nucleated plants and animals [20]. By a careful estimate of the number of genetic changes in cytochrome *c* and tRNA, McLaughlin and Dayhoff [21] have shown that the divergence between the nucleated organisms (eukaryotes) and bacteria (prokaryotes) was 2.6 times more remote in evolution than the divergences of the nucleated organisms into plants, animals and fungi. The divergence of these major groups of living organisms were related to an even earlier group of events in a very primitive organism – the development of the genetic code through the proliferation and distinction of genes for different tRNA molecules capable of transporting different amino acids. Using only the tRNA data, comparisons were made between the tRNA's which carry different amino acids. The ratio of the evolutionary distance between tRNA types to the evolutionary distance between prokaryotes and eukaryotes is 1.2. In fig. 13 this differentiation is placed at a distance proportional to the scale of the other divergences. By a comparison of the tRNA sequences of different amino acid specificity, it is possible to show that the overall rate of evolution of the tRNA sequences is the same in prokaryotes as in eukaryotes. If one assumes that the rate was constant through geological time, then the evolutionary distance is proportional to time. If the plants and animals diverged 1.1 billion years ago, then bacteria and nucleated organisms diverged some 2.9 billion years ago and the genetic code evolved approximately 3.4 billion years ago.

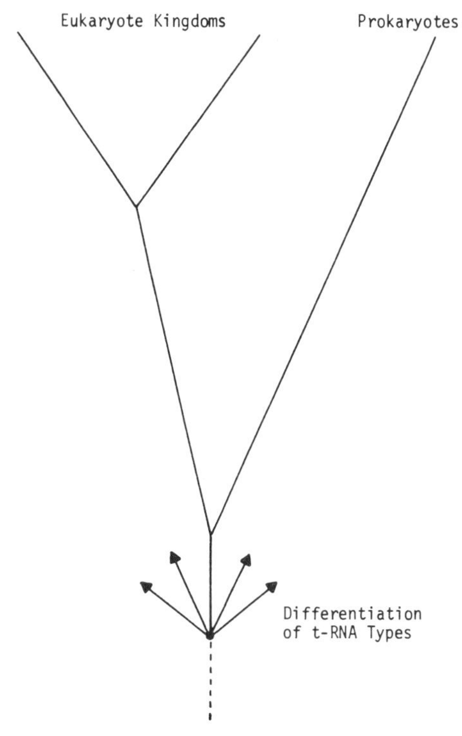

Fig. 13. The evolutionary distances between eukaryote kingdoms, between eukaryotes and prokaryotes, and between tRNA types. Not shown are the further differentiations of each type of tRNA with the divergences of groups of organisms [21].

Information potentially available

The total number of proteins which will eventually prove useful in establishing the relationships among the phyla and the nature of the earliest living things can be judged from the recent rate of change of proteins. Where the function of a protein is well established and where there is sufficient information, the rate of change has proven to be surprisingly constant over long periods of time. Table 1 shows the rates of change of all of the proteins so far examined in more than one higher species. A wide range is displayed for this heterogeneous collection. Presumably the other proteins would also fall into such a range. On the basis of constant rates, we would predict that the sequences of the bacterial homologues of proteins which change as slowly

Table 1
The rates of mutation of selected proteins [2].

Proteins	Mutations per 100 million years
Fibrinopeptides	90
Growth hormones	60
Immunoglobulins	34
Kappa C region	40
Kappa V region	34
Heavy C region	28
Ribonucleases	30
Hemoglobins	12
Beta	13
Alpha	11
Myoglobins	9
Gastrins	9
Adenohypophyseal hormones	9
Encephalitogenic proteins	7
Insulins	4
Cytochromes *c*	3
Glyceraldehyde-3-PO_4 dehydrogenases	2
Histones	0.06

The figures are given in PAMs, or Accepted Point Mutations per 100 residues, which we have estimated would occur in 100 million years of evolution. For most of the proteins, the rate was based on the time of divergence of the mammalian orders at 75 million years ago.

as cytochrome *c*, such as glyceraldehyde-3-phosphate dehydrogenase (GPDH), would also be recognizably related to the sequences in higher organisms. This prediction is supported by the fragmentary active site sequences of GPDH, which are almost identical. It would not be surprising if the hundreds of enzymes of the basic metabolic processes common to all living things proved to be changing just as slowly as cytochrome *c* or GPDH. It is likely that this sequence information will unravel the obscure phylogenetic history of the bacteria.

From table 1 we see that the histones, which function in the control of the transcription of the genome prerequisite to protein synthesis in the higher organisms, are much more rigidly conserved than cytochrome *c*. If bacteria are found to contain homologous compounds, this type of protein should provide valuable insight into early phylogeny.

Proteins which change as rapidly as the hemoglobins are useful in studying the relationships among the vertebrates, while the more rapidly changing proteins, such as ribonuclease or growth hormone, will be helpful in distinguishing closely spaced phyletic events among the mammalian groups.

It will also prove possible to learn about the evolution of the genome in the primordial single cell line which led to the most immediate common ancestor of all living species. Just as mammals today have several related genes coding for the various types of globins, so the proto-organism already had families of related genes which are still preserved. The tRNA molecules constitute such a family which must have arisen through gene duplication and subsequent adaptation to the specialized recognition of a codon and reaction with a specific amino acid [2,11]. The protein enzymes which catalyze the attachment of the amino acids to the tRNA's may also constitute such a system. The enzymes which control the replication of DNA and the transcription of RNA may be related. No sequences are presently known for these ancient proteins. It is difficult to even estimate the amount of sequence information still preserved in the metabolically functional nucleic acid sequences of the cell, such as the ribosomes and the control sections of DNA. These may very likely tell us something about the biochemistry of life before the genetic code.

Eventually it should be possible to extrapolate far back into this primeval era using the knowledge of macromolecular structures and the principles of the evolutionary process. It has long been the dream of biologists to work out the history of all the living species back to the very beginning. The many traits obtainable from biochemistry promise to make this a reality.

It is our hope that, when the sequences of the ribosomes, more tRNA's, the DNA, and the associated proteins have been worked out, it will be possible to smoothly connect the details of the evolution and structure of the early living organisms with the processes and products of chemical evolution.

Acknowledgements

This paper was supported by Contract 21–003–002 from the National Aeronautics and Space Administration and Grant GM-08710 from the National Institutes of Health.

References *

[1] Sanger F. and E.O.P. Thompson, Biochem. J. 53 (1953) 366–374 (Sequence of bovine insulin A).
Sanger F. and H. Tuppy, Biochem. J. 49 (1951) 481–490 (Sequence of bovine insulin B).

[2] Dayhoff M.O., ed., *Atlas of Protein Sequence and Structure 1969* (National Biomedical Research Foundation, Silver Spring, Md., 1969). Extensive references and results are given on all protein and nucleic acid sequences. Material from the following chapters has been used here:
Dayhoff M.O. and R.V. Eck, Chapter 1, pp. 1–6, Inferences from protein sequence studies.
Dayhoff M.O. and C.M. Park, Chapter 2, pp. 7–16, Cytochrome *c*: building a phylogenetic tree.
Dayhoff M.O. and R.V. Eck, Chapter 3, pp. 17–24, Evolution of the globins.
Dayhoff M.O., M.R. Sochard and P.J. McLaughlin, Chapter 4, pp. 25–32, The immunoglobulins: relationships and evolution.
Dayhoff M.O. and P.J. McLaughlin, Chapter 5, pp. 33–38, Evolution of other protein families.
McLaughlin P.J. and M.O. Dayhoff, Chapter 6, pp. 39–46, Evolution of species and proteins: a time scale.
Dayhoff M.O., L.T. Hunt, J.K. Hardman and W.C. Barker, Chapter 7, pp. 47–59, Enzyme structure and function.
Sochard M.R. and M.O. Dayhoff, Chapter 8, pp. 61–73, Abnormal human hemoglobins.
Dayhoff M.O., R.V. Eck and C.M. Park, Chapter 9, pp. 75–83, A model of evolutionary change in proteins.
Dayhoff M.O. and P.J. McLaughlin, Chapter 11, pp. 89–94, Transfer RNA.
Hunt L.T., W.C. Barker, P.J. McLaughlin and M.O. Dayhoff, Data Section, pp. D-1 to D-228. A complete, currently correct, critically reviewed collection of protein and nucleic acid sequence data with references and comments is presented here.

[3] The following figures and tables have been reproduced or adapted from the *Atlas of Protein Sequence and Structure 1969*, ed. M.O. Dayhoff (National Biomedical Research Foundation, Silver Spring, Md., 1969): and are used here with the permission of the publisher:
fig. 1, p. xxiv; fig. 2, p. D-213; fig. 3, p. 9; fig. 4, p. 15; fig. 5, p. D-221; fig. 7, p. 41; fig. 8, p. 28; fig. 9, p. 34; fig. 10, p. 33; fig. 11, p. 37; table 1, p. 42.
The following figure appeared originally in the *Atlas of Protein Sequence and Structure, 1967–68*, eds. M.O. Dayhoff and R.V. Eck (National Biomedical Research Foundation, Silver Spring, Md., 1968) and is reproduced here with the permission of the publisher:
fig. 6, p. 19.

* On account of the nature and extent of this bibliography, its format is different from that of the others.

[4] Epstein C.J., R.F. Goldberger and C.B. Anfinsen, Cold Spring Harbor Symp. Quant. Biol. 28 (1963) 439–449 (Refolding of denatured proteins).
Koshland D.E. Jr. and K.E. Neet, Ann. Rev. Biochem. 37 (1968) 359–410 (Enzyme structure and function).
Dickerson R.E. and I. Geis, The Structure and Action of Proteins (Harper and Row, New York, 1969) pp. 128.

[5] Sallach H.J. and R.W. McGilvery, Chart of Intermediary Metabolism (Gilson Medical Electronics, Middleton, Wis., 1963).
Margoliash E. and E.L. Smith, in: Evolving Genes and Proteins, ed. V. Bryson and H.J. Vogel (Academic Press, New York, 1965) pp. 221–242 (Evolution of cytochrome *c*).

[6] Kendrew J.C., R.E. Dickerson, B.E. Strandberg, R.G. Hart, D.R. Davies, D.C. Phillips and V.C. Shore, Nature 185 (1960) 422–427 (Structure of myoglobin at 2 Å resolution).
Perutz M.F., H. Muirhead, J.M. Cox and L.C.G. Goaman, Nature 219 (1968) 131–139 (Horse hemoglobin structure at 2.8 Å resolution).
Dickerson R.E., in: The Proteins, Vol. 2, ed. H. Neurath (Academic Press, New York, 1964) (X-ray crystallography with regard to hemoglobin and myoglobin structure) pp. 603–778.
Huber R., H. Formanek and O. Epp, Naturwissenschaften 2 (1968) 75–77 (Structure of insect hemoglobin at 5.5 Å resolution).
Padlan E.A. and W.E. Love, Nature 220 (1968) 376–378 (Structure of annelid worm hemoglobin at 5.5 Å resolution).

[7] The following describe the experimental work leading to the eludication of cytochrome *c* sequences:
Matsubara H. and E.L. Smith, J. Biol. Chem. 237 (1962) PC3575–3576 (Human).
Rothfus J.A. and E.L. Smith, J. Biol. Chem. 240 (1965) 4277–4283 (Rhesus Monkey).
Stewart J.W. and E. Margoliash, Can. J. Biochem. 43 (1965) 1187–1206 (Pig).
Nakashima T., H. Higa, H. Matsubara, A. Benson and K.T. Yasunobu, J. Biol. Chem. 241 (1966) 1166–1177 (Bovine).
Smith E.L. and E. Margoliash, Federation Proc. 23 (1964) 1243–1247 (Sheep).
Margoliash E., E.L. Smith, G. Kreil and H. Tuppy, Nature 192 (1961) 1125–1127 (Horse).
McDowall M.A. and E.L. Smith, J. Biol. Chem. 240 (1965) 4635–4647 (Dog).
Needleman S.B. and E. Margoliash, J. Biol. Chem. 241 (1966) 853–863 (Rabbit).
Goldstone A. and E.L. Smith, J. Biol. Chem. 241 (1966) 4480–4486 (Whale).
Nolan C. and E. Margoliash, J. Biol. Chem. 241 (1966) 1049–1059 (Kangaroo).
Chan S.K. and E. Margoliash, J. Biol. Chem. 241 (1966) 507–515 (Chicken).
Bahl O.P. and E.L. Smith, J. Biol. Chem. 240 (1965) 3585–3593 (Rattlesnake).
Chan S.K., I. Tulloss and E. Margoliash, Biochemistry 5 (1966) 2586–2597 (Snapping turtle).
Goldstone A. and E.L. Smith, J. Biol. Chem. 242 (1967) 4702–4710 (Dogfish).
Kreil G., Z. Physiol. Chemie 334 (1963) 154–166 (Tuna fish).
Chan S.K. and E. Margoliash, J. Biol. Chem. 241 (1966) 335–348 (Silkworm).
Stevens F.C., A.N. Glazer and E.L. Smith, J. Biol. Chem. 242 (1967) 2764–2779 (Wheat).
Yadi Y., K. Titani and K. Narita, J. Biochem. 59 (1966) 247–256 (Baker's yeast).

Narita K. and K. Titani, J. Biochem. 63 (1968) 226–241 (*Candida krusei*).
Heller J. and E.L. Smith, J. Biol. Chem. 241 (1966) 3158–3180 (*Neurospora crassa*).
Dus K., K. Sletten and M.D. Kamen, J. Biol. Chem. 243 (1968) 5507–5518 (*Rhodospirillum rubrum*).
The following unpublished results have been cited by R. Wojciech and E. Margoliash, in: Handbook of Biochemistry, ed. H.A. Sober (Chemical Rubber Co., Cleveland, 1968) pp. C158–C161.
Chan S.K., I. Tulloss and E. Margoliash (Pigeon).
Chan S.K. (Tobacco horn worm moth).
Nolan C., L.J. Weiss, J.J. Adams and E. Margoliash (Fruit fly).
The following results have been published in the *Atlas of Protein Sequence and Structure 1969*, ed. M.O. Dayhoff (National Biomedical Research Foundation, Silver Spring, Md., 1969).
Chan S.K., I. Tulloss and E. Margoliash (Penguin).
Chan S.K., I. Tulloss and E. Margoliash (Turkey).
Chan S.K., I. Tulloss and E. Margoliash (Pekin duck).
Chan S.K., O.F. Walasek, G.F. Barlow and E. Margoliash (Bullfrog).
Chan S.K., I. Tulloss and E. Margoliash (Screw worm fly).
Uzzell T., C. Nolan, W.M. Fitch and E. Margoliash (Lamprey).
[8] Hirschmann R., R.F. Nutt, D.F. Veber, R.A. Vitali, S.L. Varga, T.A. Jacob, F.W. Holly and R.G. Denkewalter, J. Am. Chem. Soc. 91 (1969) 507–508 (Synthesis of ribonuclease S).
Gutte B. and R.B. Merrifield, J. Am. Chem. Soc. 91 (1969) 501–502 (Synthesis of ribonuclease).
[9] Young J.Z., The Life of Vertebrates, 2nd ed. (Oxford Univ. Press, New York, 1962) pp. 820.
Romer A.S., Vertebrate Paleontology, 3rd ed. (Univ. Chicago Press, Chicago, 1966) pp. 468.
Mayr E., Animal Species and Evolution (Harvard Univ. Press, Cambridge, 1966) pp. 797.
[10] Barghoorn E.S. and J.W. Schopf, Science 152 (1966) 758–763 (Microorganisms three billion years old from the Precambrian of South Africa).
Barghoorn E.S. and S.A. Tyler, Science 147 (1965) 563–577 (Microorganisms from the Gunflint Chert).
[11] Jukes T.H., Biochem. Biophys. Res. Commun. 24 (1966) 744–749 (Evolutionary origin for tRNA's).
Eck R.V. and M.O. Dayhoff, *Atlas of Protein Sequence and Structure* 1966, (National Biomedical Research Foundation, Silver Spring, Md., 1966) pp. 141–147.
Crick F.H.C., J. Mol. Biol. 38 (1968) 367–379 (The origin of the genetic code).
Orgel L.E., J. Mol. Biol. 38 (1968) 381–393 (Evolution of the genetic apparatus).
[12] Rutten M.G., Space Life Sci. 1 (1970) 1–13 (History of atmospheric oxygen).
Rutten M.G., The Geological Aspects of the Origin of Life on Earth (Elsevier, New York, 1962) pp. 146.
Calvin M., Chemical Evolution (Oxford University Press, New York, 1969) pp. 278.
[13] McKusick V.A., Mendelian Inheritance in Man, 2nd ed. (Johns Hopkins Press, Baltimore, 1968) pp. 521 (Genetics of human traits).

Perutz M.F. and H. Lehmann, Nature 219 (1968) 902–909 (Review of molecular pathology of hemoglobins).
Schroeder W.A. and R.T. Jones, Progr. Chem. Org. Nat. Prod. 23 (1965) 113–194 (Review of abnormal hemoglobins; general information on hemoglobins).
Ohno S., U. Wolfe and N.B. Atkin, Hereditas 59 (1968) 169–187 (Chromosomal duplications).
Atkin N.B. and S. Ohno, Chromosoma 23 (1967) 10–13 (Chromosomal duplications).
[14] Schwartz E., Science 167 (1970) 1513–1514 (Unstable hemoglobin messenger RNA).
[15] Zuckerkandl E. and L. Pauling, in: Evolving Genes and Proteins, eds. V. Bryson and H.J. Vogel (Academic Press, New York, 1965) pp. 97–166 (Hemoglobin evolution).
Hill R.L. and J. Buettner-Janusch, Federation Proc. 23 (1964) 1236–1242 (Hemoglobin evolution).
Braunitzer G., K. Hilse, V. Rudloff and N. Hilschmann, Advan. Protein Chem. 19 (1964) 1–17 (Review article on hemoglobins).
[16] Buse G., S. Braig and G. Braunitzer, Z. Physiol. Chemie 350 (1969) 1686–1690 (Insect hemoglobin sequence).
Ellfolk N. and G. Sieces, Acta Chem. Scand. 23 (1969) 2994–3002 (Partial sequences of plant leghemoglobin).
[17] Cold Spring Harbor Symp. Quant. Biol. 32 (1967) pp. 619 (Antibodies).
Gamma Globulins, Nobel Symposium 3, eds. J. Killander (Almqvist and Wiksell, Stockholm, 1967) pp. 643 (Immunoglobulin structure).
Wikler M., H. Kohler, T. Shinoda and F. Putnam, Science 163 (1969) 75–78 (Evolution of light and heavy chains).
Dreyer W.J. and W.R. Gray, in: Nucleic Acids in Immunology, eds. O.J. Plescia and W. Braun (Springer, New York, 1968) pp. 614–643 (Evolutionary origin of immunoglobulins).
[18] Hartley B.S., J.R. Brown, D.L. Kauffman and L.B. Smillie, Nature 207 (1965) 1157–1159 (Evolution of trypsins).
Neurath H., K.A. Walsh and W.P. Winter, Science 158 (1967) 1638–1644 (Evolution of trypsins).
Shotton D.M. and B.S. Hartley, Nature 225 (1970) 802–806 (Sequence of porcine elastase).
Jurasek L., D. Fackre and L.B. Smillie, Biochem. Biophys. Res. Commun. 37 (1969) 99–105 (Partial sequence of a bacterial trypsin).
Brandshaw R.A., H. Neurath, R.W. Tye, K.A. Walsh and W.P. Winter, Nature 226 (1969) 237–239 (Dogfish trypsin sequence).
[19] Matsubara H., T.H. Jukes and C.R. Cantor, Structure, Function and Evolution in Proteins, Brookhaven Symposium in Biology 21 (1968) (Brookhaven National Laboratory, Upton, New York, 1969) pp. 201–216.
Eck R.V. and M.O. Dayhoff, Science 152 (1966) 363–366.
Jukes T.H., Molecules and Evolution (Columbia Univ. Press, New York, 1966) p. 229.
[20] Sagan L., J. Theoret. Biol. 14 (1967) 225–274 (Evolution of phyla).
Margulis L., Science 161 (1968) 1029–1022 (Evolution of phyla).
[21] McLaughlin P.J. and M.O. Dayhoff, Science 168 (1970) 1320.

Chemical Evolution and the Origin of Life, eds. R. Buvet and C. Ponnamperuma

DNA: ORIGIN, EVOLUTION AND VARIABILITY

A.S. ANTONOV
Moscow State University, Moscow, USSR

The present communication deals with three problems:

1) Are we in a position to estimate the "age" of DNA as genetic material based on the available data on the degree of variability of the DNA primary structures in the species belonging to various taxons?
2) What was the evolution pattern of the primary DNA structures?
3) May one apply the data on the variability degree of the DNA primary structures to determine the formal scale of taxons in the existing systems of animals, plants and microorganisms?

The degree of variability of the DNA primary structures is largely different in various defined groups of organisms. Protokaryotic organisms display maximum variability. For example, the content of GC pairs in bacterial DNA ranges from 20 to 80 mole%. The variability of the nucleotide composition in DNA of viruses and blue–green algae is almost the same. Amongst the eukaryotic organisms maximum variability is observed in protozoa and primitive multicellular organisms (sponges) (ΔGC = 45 and 25 mole%, respectively). The more recent the "evolutional age" of a given type of animals, the smaller the range of variability of the DNA composition of a species of that type (Antonov, 1964, 1965). Similar conclusions were made by Mazin and Vanyushin (1969) who studied the frequencies of isopliths in DNA. The chordates were shown to have the minimum variability of these two parameters (percentage of GC and degree of stacking). Within this type, animals of various classes display different degrees of DNA variability: maximum ΔGC = 10% in fishes; minimum Δ GC = 5 mole% in mammals.

Comparison of the degree of variability of DNA primary structures in plants has revealed a similar picture: in the angiosperms it varies in a narrower range than in lower plants. It should be noted, however, that only a few groups of organisms have been studied thoroughly enough in this respect. I confine myself to the data on three groups only: chordates, protozoa and microorganisms. I would like to emphasize, though, that the very fact of different degrees of variability in the DNA primary structures is now beyond doubt.

The existence of such differences has led to the conclusion (Antonov, 1964, 1965, 1968) about the proportional relationship between the degree of variability of the DNA primary structure in the species of a taxon and its evolutional age. Similar conclusions were made by other investigators from the analysis of the degree of DNA nucleotide sequence homology among various vertebrate animals (Hoyer et al., 1964) and the degree of clustering of pyrimidine nucleotides in animal DNA's (Mazin and Vanyushin, 1969).

All those engaged in the study of this problem have come to the conclusion that the divergence of the primary DNA structures is the result of fixed mutations. Naturally, when comparing contemporary forms, the longer the process of divergence of the DNA primary structures within a group of phylogenetically related organisms, the more differences in the DNA primary structures were revealed.

As we know well enough the variability ranges of DNA primary structures for many important groups of animals and plants and their true "evolutional age" from the paleontological evidence, there is a temptation to date back the taxons whose paleontological "chronicle" is either absent or incomplete. Logical as this approach is, the data it furnishes are certainly approximate. In fact, we do not know if the DNA primary structures have been changing at a constant rate in the course of evolution. Some indirect data indicate that it has not been constant and the difference may be quite noticeable from group to group of organisms.

The relationship between the degree of variability of the DNA nucleotide composition and evolutionary age of the two classes of chordates is illustrated in table 2. The time of origin of the first specimens of each class is known rather accurately. Comparing the degree of variability of DNA composition in fishes with that in mammals and assuming that the primitive fishes originated about 450–500 million years ago, we may date the mammals back to 200–250 million years ago. This figure agrees with the conclusions of paleontologists. It is appropriate to note that chordates and echinodermates have rather similar variability degrees which are in accordance with the conception of the common origin of these two types of *Deuterostomia*.

Let us attempt to use the degree of variability of DNA composition in protozoa to determine the time of origin of primitive eukaryotes. The data on variability degree of the DNA composition of the chordates and their evolutional age show that the composition of this DNA changes by 1 mole% within 45–50 million years. As the DNA of protozoa varies by 45–50 mole% this group of organisms could have existed approximately 2.0–2.5 billion years ago from the beginning of the photerozoyan era. Based on the data on the bacterial DNA composition variability degree (Δ GC = 60 mole%) and indulging in still bolder comparisons and extrapolations, we may deduce that 3 billion years ago bacteria like DNA-containing unicellular organisms existed on the earth.

Nevertheless, the main conclusion we can draw from the above speculations is that DNA in its present make-up was functioning in the earliest, most primitive forms of life.

Thus, we have no unambiguous solution to the problem of the origin of DNA and its conversion into a hereditary substrate. Subsequent stages of evolution of these polymer primary structures may be discussed at a much higher level of reliability.

Quite deliberately I refrain from discussing the question: was pre-DNA constituted of nucleotides of two types (A and I, according to Crick, 1968) or did it contain the four known types of nucleotides? I am going to consider hypothetical evolutionary patterns of the primary structures whose composition is identical to the contemporary forms, as sooner or later, such polymers must have appeared in the cells of primitive organisms. They were a valuable evolutionary gain, as all genetic material other than double stranded DNA has been almost completely discarded by selection. The very few exceptions which have narrowly escaped may be considered as a peculiar rudiment (storage of information in RNA and single-stranded DNA).

It is logical to suggest that by the time DNA became the heredity substrate, the general level of organization of organisms must have been sufficiently high. It is likely that there already existed an enzymatic mechanism of DNA replication, the information transfer being carried out with high accuracy comparable with that in unicellular organisms today. The total amount of genetic information was not very high as the primary organisms probably had a lesser degree of complexity than the contemporary forms.

Analyzing the data on the principles of organization of DNA primary structures in contemporary forms of organisms (Antonov, 1969), we make a sound conclusion that the early trends in the DNA evolution were to increase the genetic potential of organisms, to obtain increasing amounts of DNA in cells. The growing genetic potency of the organisms contributed to their metabolic potential by introducing new types of enzymes and also at the expense of improved engineering of metabolic reactions, transport of substances and energy. The increase in the amount of cellular DNA might have taken pathways similar to those observed in contemporary forms of prokaryotes. The mechanisms postulated by geneticists pre-supposed the slow rate of this process.

The growth of the genetic potential is not infinite in the sense that a unicellular organism may contain only a limited amount of DNA (as is the case in the present day prokaryotes). Hence, such an organism has a finite genetic potential. The evolution of genomes and the very organisms themselves would have taken the only way – i.e. improving a certain amount of the genetic material at their disposal and its very economical enrichment. In the course of evolution of DNA the most varied primary structures appeared.

Evolution of genetic material and the cell has had its landmarks; one of them is the formation of the system responsible for saltatory replication of a gene (Britten and Kohne, 1966). This really might have been the turning point of evolution. In our opinion the population of organisms which gave birth to this mechanism and mastered it, paved the way to the eukaryotes with their high amount of DNA per genome. The mechanism of saltatory replication facilitated rapid accumulation of genetic material.

This mechanism is still at work in contemporary eukaryotes. Repetitive DNA sequences are inherent in all genomes of the eukaryotes studied. The saltatory gene replication has been proved to be of vital importance in the processes of ovogenesis, cell differentiations, and formation of a polychromosome.

The existence of repetitive sequences ensures higher reliability of storage and transport of hereditary information. Having acquired the "families" of the initially identical genes, the organism could afford to produce clusters of enzymes of close specificity and hence provide a large amount of necessary compounds within short, actually critical, moments of ontogenesis. Gradual alterations of such groups of sequences in the course of evolution have finally resulted in the higher genetic potential and greater information pool of the organisms. Finally, as a result of parallel but different patterns of development we have two basically different groups of organisms, prokaryotes and eukaryotes.

The third problem I would like to touch upon is taxonomic rather than the one pertaining to evolutionary biochemistry, but it uses the data furnished by the latter. As I have said, the variability degree of DNA primary structures in specimens within single-scale taxons may be largely different. The data of tables 1 and 2 show that often single-scale (according to the accepted systems) taxons: types, classes, families, cover groups of organisms with an unequal degree of genetic relationship which is clear from largely different degrees of variability of their DNA's. Hence, when determining formal scales of taxons the DNA variability degree of its species should not be overlooked. The use of this criterion should add objectivity to the estimation of the taxon's scale.

I would like to conclude by reproaching biologists for their timidity in the application of the biochemical and molecular biology data on the evolu-

Table 1
Limits of variability of DNA base composition.

Group	GC (mole%)
1) Microorganisms	80–20 = 60
2) Blue–green algae	70–35 = 35
3) Viruses	75–30 = 45
4) Protozoa	65–20 = 45
5) Spongia	60–35 = 25
6) Chordata	45–35 = 10

Table 2
Limits of variability of DNA base composition and the evolutionary age of some animals.

Group	Age (years $\times 10^{-6}$)	GC (mole%)
Fishes	450–500	70
Mammals	200–225	4

tionary pattern of DNA primary structures and variability degree of its primary structures within groups of organisms. Those engaged in the solution of the phylogenetic and taxonomic problems may benefit from what is furnished by evolutionary biochemists. We are hoping that the evolutionary biochemistry of nucleic acids will soon find its way into taxonomy and phylogenetics.

References

Antonov A.S., Ph.D. Thesis 1964, Moscow State University, Moscow, USSR.
Antonov A.S., Usp. Sovrem. Biol. 60 (1965) 161–177.
Antonov A.S., in: Problemy Evolucii, Vol. 1 (Nauka, Moscow, 1968) p. 34–47.
Antonov A.S., Usp. Sovrem. Biol. 68 (1969) 299–317.
Britten R.J. and D.E. Kohne, in: Carnegie Institute Year Book (McGraw Hill, New York, 1965–1966) pp. 78–106.
Crick F., J. Mol. Biol. 38 (1968) 367–379.
Hoyer B.H., B.J. McCarthy and E.T. Bolton, Science 144 (1964) 959–967.
Mazin A.L. and B.F. Vanyushin, Molekul. Biol. 3 (1969) 846–855.

Chemical Evolution and the Origin of Life, eds. R. Buvet and C. Ponnamperuma

THE ORIGIN OF RIBOSOMES AND THE EVOLUTION OF rRNA

B.M. MEDNIKOV
Laboratory of Bioorganic Chemistry,
Moscow State University, Moscow, USSR

Not all the stages of biogenesis have been studied with equal thoroughness. At present there are grounds to suggest several experimentally different ways of emergence of low molecular predecessors of life (amino acids and nucleotides). Less is known about the second stage, i.e. that from the "primary broth" to coacervate or the microsphere. As to the third stage, from the coacervate to the protocell, we can offer little more than speculations. The systems of the third stage were too complex to be studied as experimental models by present day science and too simple to have survived till now. Therefore, if the first part of my report may seem somewhat speculative to you, please, do not forget that it is the specificity of the problem itself.

The main problem of the third stage of the origin of life is that of realization of the hereditary information, the question of development of the protein biosynthesis machinery.

At present the cells possess ribosomes which are complex systems of three RNA molecules and several dozen proteins. Ribosomes have a highly ordered structure, but they are capable of changing their conformation. They are served by complex enzymatic systems and are, in fact, a universal device for translating information from mRNA.

Even most primitive contemporary cells have ribosomes. Moreover, ribosomes are basically identical in all organisms living on earth. The ribosome is no less important an attribute of life than the chromosome. The origin of life cannot be understood unless we unravel the origin of ribosomes.

The off-spring of a ribosome is virtually a miracle. We ought to postulate that ribosomal synthesis came into being step-by-step, as is indicated by Crick [3]. Moreover at each successive stage, beginning from the very first, the original ribosomes should have been capable of synthesizing protein. But the ribosome itself is a complex of RNA and protein molecules. Hence, non-ribosomal protein synthesis must have been indispensable at the early stages of life.

The question arises: what were these early stages of protein synthesis, the primary ribosomes and their nucleic acids?

The old problem, whether DNA is primary and RNA is secondary, has lost its urgency. We know that RNA contains thymine and DNA uracyl. We know of double-stranded RNA's and single-stranded DNA's. The latter may serve as a template for protein synthesis. Finally, in the presence of mercury ions, it is possible to synthesize enzymatically a copolymer containing both ribose and deoxyribose [2]. Hence the conclusion: the primary nucleic acid is neither RNA nor DNA – it is an intermediate compound. It may be tentatively called XNA. What could be its most probable features?

1) XNA was normally single-stranded, incapable of forming the Watson-Crick helical structure, but capable of forming a complementary chain and replicate ion.
2) XNA could be a copolymer of ribose and deoxyribose or even contain other sugars.
3) A copolymer of a racemic mixture of sugars could not form a complementary chain. Hence XNA contained, as it does now, only D-sugars.
4) XNA could have a more varied set of bases than it has now. It is quite possible that the abundance of minor bases in the phage DNA's and RNA's is a rudiment of times immemorial.
5) The XNA chain could hardly contain more than 50–150 nucleotides, which would correspond to no more than 300 bits of information. Hence, the function of XNA could not have been other than autoreplication.

It is natural to ask the question: what was the function of a molecule which could do nothing but reproduce itself? In the primary broth it was useless. But it is quite different in the case of the protocell, regardless of whether it was a coacervate, microsphere or a molecular layer sorbed on clay.

Accumulation of substances in the coacervates has been widely discussed (see review [4]). This question must be clarified. Such accumulation is impossible without parallel binding of diffusing molecules in the protocell. No diffusion is possible against the concentration gradient; there was no active transport at that time. Accumulation occurred only in protocells in which amino acids condensed into peptides and nucleotides into nucleic acids. This is how the positive feedback was realized: synthesis intensified diffusion and vice versa.

The more rapidly synthesizing protocells intensively pumped in molecules of low molecular precursors from the primary broth. The rate of synthesis was the point of selection. Now, a paradox arises: the cell, i.e. a structure enveloped into a semipermeable membrane, came into existence prior to life itself and was an indispensable condition of initiation of life processes.

In addition to the ability to reproduce itself, XNA molecules potentially had may properties of mRNA, tRNA and rRNA. The further development is well known at the levels higher than molecular. This is the principle of differentiation of functions, well-known in biology. As a rule, newly-born biological systems are polyfunctional and it is only in the process of their differentiation that they become specialized. For example, an extremity of a primitive crustacean is organ of movement, respiration and feeding. In the course of evolution extremities become functionally specialized. We have no grounds to believe that this was not the case, provided that the differentiation of RNA cells facilitated selection of the fast growing protocells.

Could the primitive forms of non-ribosomal protein synthesis have survived as a specific rudiment (a sort of an appendix at the molecular level)? Well, this is not excluded if the relevant forms have acquired specialized functions.

The simplest way of peptide bond formation in the cell is that of an ATP-activated amino acid joining the N-terminal of the protein molecule. Such a system has been found in the cells of *Alcaligenes foecalis* [1].

The next possibility is realized through the formation of the aminoacyl-tRNA complex. Such complexes may take part in a non-ribosomal formation of the peptide bond. It is thus possible that tRNA is the first specialized derivative of XNA.

At the previous conference on the origin of life Sagan pointed out that things would have been much easier if polynucleotides had at least a weak catalytic activity. It is possible that catalytic activity was possessed by the precursors of ribosomal RNA – short-chain oligonucleotides, such as the uracil-containing nucleotide coenzyme discovered by Ito and Strominger [6].

Initially, ribosomes could have been complexes of 2–3 RNA molecules of low molecular weight and short peptides formed by non-ribosomal mechanisms. These complexes could have a primitive peptidyl-transferase moiety which is now present in 50 S ribosomal subunits.

The peptidyl-transferase center catalyzed the transfer of amino acids to the growing protein chain from aminoacyl-tRNA.

The formation of the primitive ribosome greatly enhanced the biosynthesis of protein. All the old forms gave way to ribosomal synthesis. Only some of them, strongly modified, have survived to perform specific purposes, i.e. modification of N-terminals of polypeptide chains and synthesis of short or monotonous polypeptides like those encountered in all cell wall proteins, antibiotics such as gramicidin and other specific polypeptides.

Crick [3] seems to have similar ideas on the origin of protein biosynthesis. This mechanism is probably the most acceptable at the present level of our knowledge. This scheme is undoubtedly speculative but this or any other scheme should be in agreement with the following postulates.

1) The contemporary biosynthesis machinery has gone through several gradual stages.
2) Each stage differed from the other in the higher rate of biosynthesis.
3) As a result of selection at the protocellular level this acceleration yielded a bacterial-like ribosome somewhat about 3.5–4 billion years ago.

Further evolution of ribosomes is much less speculative as the comparative biochemistry furnishes some relevant evidence. We should take into consideration some parallelism in the evolution of ribosomes and chromosomes. The DNA chain is known to elongate in the course of evolution. The common belief that there are two types of ribosomal RNA – those corresponding to the *Procaryota* ribosomes (bacteria and blue-green algae) and to the *Eucaryota* ribosomes (higher plants and animals) is not quite accurate. There are recent data on the evolution of the *Eucaryota* RNA [7].

The low molecular ribosomal RNA has the most constant length (120 nucleotides, which corresponds to the molecular weight of 40,000 daltons). The molecular weight of the small subunit varies from 0.55 million daltons in bacteria to 0.73 million daltons in *Eucaryota*. The molecular weight of RNA in a big subunit is in good agreement with the age of the taxon – in bacteria 1, in protozoa 1.30, in invertebrates 1.40, in amphibians 1.54, in birds 1.58 and in mammals (human ribosomes) it amounts to 1.75 million daltons.

In all probability, the elongation of ribosomal RNA was non-linear. By linear extrapolation of the molecular weight of RNA to zero we arrive at the conclusion that ribosome has been in existence much longer than our galaxy. The analysis of evolution of biological systems at other levels shows that the increase in their complexity may be described by the *S*-curve. This curve is asymmetric and obeys the equation of von Bertalanffi:

$$H = H_{\infty}(1 - \mathrm{e}^{-kt}) ,$$

where H is the index of the system complexity, t is time and H_{∞} and k constants. Rough graphical analysis shows that the period of intensive elongation of ribosomal RNA should have been not later than 3.5–4 billion years ago; this value in contemporary organisms approaches the H_{∞} constant.

Along with this elongation, the ribosomal complex gradually acquired protein molecules. Ribosomal proteins are in many ways similar to nuclear proteins, histones, and their content also rises, as evolution proceeds, from 40% in bacterial ribosomes to 60% in ribosomes of higher organisms.

We have paid special attention to the evolutional changes of the RNA nucleotide composition. Originally we hoped to reconstruct the base composition of primary ribosomal RNA. If its composition was largely different from equimolar, the volume of information in XNA was not great and the probability of its spontaneous formation increased.

So, we have compared the composition of the ribosomal RNA of various vertebrates and invertebrates. The G + C/A + U ratio has been found to be a good non-dimensional composition index which I am going to call the specificity coefficient.

In any taxonomic group there are older and younger species. Our conclusions are confirmed by direct paleontological evidence.

Let us take, for example, the class of insects. Here we have rather ancient orders which have been existing since Carbon, the medium age orders (originated during the Perm and Trias) and the youngest species which are undergoing evolution now.

There is a strict rule: the older groups (for example, cockroaches, ephemerids, dragonflies) have a characteristically high specificity coefficient, 1.4. In the RNA of these groups guanine–cytosine pairs prevail [8,10]. The medium age orders (beetles, orthoptera) have a lower specificity coefficient, 1.2–1.1 on average. The youngest groups (butterflies, dipterians) have RNA of equimolar composition (the specificity coefficient is equal to unity) or even RNA of the A–U type.

Similar patterns are observed within each order. For example, RNA of one of the most primitive butterflies, *Hepialus humuli*, of the same order as that of cockroaches, has a specificity coefficient of 1.40, while the medium value for the order is 1.0.

Analysis of RNA of 40 species of insects belonging to 11 orders has revealed two tendencies in the evolution of ribosomal RNA:

1) A tendency towards the accumulation of the G–C pairs resulting in the formation of the G–C type RNA.
2) A tendency towards the accumulation of the A–U pairs resulting in the prevalence of the A–U type RNA.

We have established a correlation between the nucleotide composition of ribosomal RNA and the rate of development of the insects. This dependence obeys the equation:

$$\ln N = 5.624 \frac{\mathrm{G+C}}{\mathrm{A+U}} - 1.514 \,,$$

where N is the duration of development from the egg stage till the imago stage (at 20 °C). The rapid evolution of the young groups may be accounted for by the fast changing generations [8]. For example, the period of growth of the fly at 20 °C, is two weeks, that of the cockroaches, two years. As the age of a species is measured not in years but in the number of generations, the groups with a low specificity coefficient have had more time for evolution.

For other organisms there is no direct relationship between the rate of development and the nucleotide composition of ribosomal RNA. But in some groups a similar tendency may be revealed. For example, in the class of fishes, the specificity coefficient of ribosomal RNA of the primitive form, lampreyes for example, is 1.54. The most progressive and highly organized fishes, perches, have RNA with the lower content of G–C pairs. Their specificity coefficient is 1.1–1.2. All the other fish orders investigated fall between these two extreme values [9].

At the same time the warm-blooded animals (birds and mammals) do not obey the above regularity. In mammals and birds the evolution of ribosomal RNA has taken a different route: that of acquiring ribosomal RNA of the very high G–C type (the specificity coefficient amounting to 2.0). Gazaryan and Shuppe [5] have shown that the earlier determined specificity coefficient values had been lower due to an admixture of mRNA fragment in the small ribosome subunit. It is difficult to account for these different trends. It should be taken into consideration that the ribosomes of mammals and birds function at high [37–40 °C] and stable temperatures. According to some authors the specificity coefficient of ribosomal RNA of bacteria living in hot springs is higher [11]. That may be the same mode of adaptation.

This leads to a conclusion that in the course of evolution of ribosomal RNA one and the same problem i.e. that of reliable biosynthesis of protein has taken different, sometimes opposite routes and these routes have often changed into one another or combined. Still, we are not in a position to give the nucleotide composition of ribosomal RNA in primary organisms. The majority of the contemporary ribosomal RNA's have 50–67% guanine and cytosine. Probably, the ancestory form of ribosomal RNA had a relatively high content of guanine and cytosine and each nucleotide corresponded to a lower number of bits of information than would have been the case for equimolar composition. The prevalence in RNA of some forms and even in the whole groups (for example, *Diptera* and *Protozoa*) of adenine and uracyl is likely to be a secondary phenomenon.

Further investigation along these lines will help to elucidate this question.

References

[1] Beljansky M., Bull. Soc. Chim. Biol. 43 (1961) 1017.
[2] Cannelakis E.S. and Z.W. Cannelakis, Inform. Macromol. (1963) 107.
[3] Crick F.H.C., J. Mol. Biol. 38 (1969) 367.
[4] Evreinova T.N., Accumulation of Substances and the Action of Enzymes in Coacervates (Nauka, Moscow, 1966).
[5] Gazaryan K.G. and N.G. Schuppe, Dokl. Akad. Nauk SSSR 176 (1967) 714.
[6] Ito E. and J. Strominger, J. Biol. Chem. 235 (1960) 7.
[7] Loening U.E., J. Mol. Biol. 38 (1968) 355.
[8] Mednikov B.M., Dokl. Acad. Nauk SSSR 161 (1965) 721.
[9] Mednikov B.M., A.S. Antonov and A.N. Belozersky, Dokl. Acad. Nauk SSSR 165 (1965) 227.
[10] Mednikov B.M., Problems of Evolution, Vol. 1 (Nauka, Acad. Nauk SSSR Siberian Branch, Novosibirsk, 1968) p. 47.
[11] Pace B. and L.L. Campbell, Biochemistry 57 (1967) 1110.

Chemical Evolution and the Origin of Life, eds. R. Buvet and C. Ponnamperuma

THE GENETIC CODE AND THE ORIGIN OF LIFE

P. GAVAUDAN

Station Biologique de Beau-Site, 25, Faubourg Saint-Cyprien, Poitiers, France

The nature of the present study is purely speculative and is founded on some observations on the numerical structures of the genetic code as they have been experimentally established by biochemists specializing in molecular biology. It may be asked whether the structure of the genetic code as a whole does not conform to some optimization rule so that we may ultimately draw from the observed regularities some conclusions relevant to the problem of the origin of life.

In spite of the anathema sometimes proferred against the possibility of any theoretical speculation upon the nature of the genetic code (cf. Woese [16] p. 1–2), we are not afraid of undergoing the disapprobation linked to numerological assays.

Many excellent reviews and chapters of very authoritative books (cf. Crick [3,4], Jukes [12], Woese [16], Yčas [17]) provide valuable discussions on the nature of the genetic code; it is therefore only necessary to summarize the basic theoretical features of the genetic coding. It is well known that Gamow initiated this theory with his idea of triplets of purine and pyrimidine bases in the number of $4^3 = 64$ for the specification of 20 amino acids, a fact which is necessarily followed by a degeneracy of 44 triplets. Furthermore the genetic code is of a non-overlapping type and without any ambiguous codons. The distribution of the four bases among the codons is intriguing, but beyond the scope of this discussion. The coding of the different amino acids is noteworthy because of the distribution of a redundancy which, at first sight, seems apparently very odd; we will later return to this point to discuss it in more detail.

There are also some special relationships between the constitution of the triplets and the specification of certain amino acids and their hydrophoby, an interesting observation relative to the structure of proteins.

To summarize, the genetic code is universal throughout all forms of life. On a statistical ground, and broadly speaking, the distribution of the redundancy apparently obeys rather exactly biochemical data as to the pattern of the assignment of the codons to the amino acids [7]. The origin of the assignment of particular triplets to individual amino acids is unknown. The problem of specificity is linked to that of universality. As pointed out by Crick [3,4] different theories are incurring varying degrees of censure.

The hierarchy of numerical structures in the genetic code

A model of general organization

In the past, we have briefly assumed the existence of strong physico-mathematical constraints which prescribed to the genetic code an organization linked to an optimization; its criterion is perhaps the approach of a function containing the e basis of neperian logarithms [2]. Indeed trivial, but suggestive, was the following remark on the relationship between e and the number 20 of the natural amino acids ($\sqrt[3]{20} = 2.7147176\ldots$) in association with some speculations on the yield of the coding. It was further emphasized (Besson [1]) that, considering the repartition of bases along the chain as a Poisson's process, the probability of binding any one of them on a point, and the corresponding probability at the triplet level, is not ¼ and $(1/4)^3 = 1/64$ but $1/e$ and $(1/e)^3 = 1/20$. The number 20 loses something of its magical character. Nevertheless, it is necessary to point out that the constraint is actually not exactly from 64 to 20, but rather from 64 to 23; the "crushing" is a bit less tight, degeneracy being not wholly filled by redundancy. There exists in the code, simultaneously completely and incompletely replenished, some "looseness", in a so to speak mechanical sense, introduced by the three so-called "nonsense" triplets which are not assigned to amino acids.

Thus the total number of the types of signals is raised to 23. In the absence of more accurate and ne varietur data on the exact functions of the "nonsense" triplets, more or less supposed to prescribe the end of a chain, we account for each of them as a different type of signal.

Up to the present time, the number 23 has not caught the eye, with the exception of Gurel's [10] allusion to it in his theory of the organization of the genetic code. The proximity of the number 23 to 20 is intriguing and poses a problem whose solution, as we shall see a little later, is leading to a complete definition of the structural optimization of the code. According to Gurel, all the detail of the organization of the code depends ultimately on the physical and chemical properties of the purine and pyrimidine bases. We believe that this proposition is in accordance with the idea of a unitary law prescribing the number of amino acids and types of signals, as well as the extent of degeneracy and its distribution through "nonsense" triplets, classes, and groups of amino acids.

First, let us discuss the consequences of having 23 types of signals as the price for coding the twenty amino acids. There is a kind of deficit which lowers the yield of the coding. Among different possible expressions of the yield, the following seems conspicuous; it is the common ratio: 20/23 = 0.86956. We are less interested in the absolute value of the yield being nearly 87% than in the fact that the figure (to 0.5/1000) is very near to 2 $\log_{10} e$.

This relationship between the number of amino acids and the number of types of signals in the code leads us to the definition of a model of the genetic code on the first level of structure. Hence it is quite easy to link the organization of the general structure to a series of relationships of different forms having the same significance.

For example, using 64, the total number of codons, we can write: 23/64 = 0.3593750, a value which closely approximates a slightly unorthodox and intriguing expression:

$$\frac{e-2}{2} = 0.3591409\ldots .$$

Let us also look at the $\log_e$ of 44, the number of degeneracy; taking the value of e to be 2.71828, we have:

$$\log_e 44 = 3.78419,$$

a value very close to

$$\frac{e}{e-2} = 3{,}78442.$$

It would be useless to add other such expressions; they all correspond to an empirical and very trivial approximation of the e number by a rational fraction such as:

$$\frac{2 \times 23}{64} \rightarrow e-2,$$

the approached value of e being 2.71875.

Let us point out that other numerological expressions of e do not differ greatly from the last approximation. We have, for example, $(\pi-1)+\gamma = 2.71880\ldots$, where π is given as 3.14159 and γ is Euler's constant equal to 0.57721... The first value approaches e by 0.47/1000 which is a bit better than by 3.5/1000 with the cubic root of 20, but in an unorthodox fashion relative to the finite character of the expression. Nevertheless, in spite of this restriction, the relationships presented concerning the sum of codons and amino acids as well as degeneracy and signals are quite impressive. The interdependence of all these expressions, and also of all the corresponding structures, are obligatory and almost tautological, and are almost a kind of game played with the fundamental numbers 20, 23, and 64. In particular, 23 is a "good number", for it is immediately apparent that 23/10 is very near the well known expression $\log_e 10$, since $2.3025 \ldots \times 10 = 23.0258 \ldots$

It is very likely that nature does not work with decimal numeration; the fact that we can find so easily logarithmic relations is linked to some underlying system of operation founded rather on a Principle of Continuity than upon the manipulation of strictly individualized objects, discrete values, or integers.

But returning to our initial purpose, we notice also that the three "nonsense triplets", or supernumerary signals, become quite meaningful when considered as filling elements of the code if we recall the relationship: $20 + \log_e 20 = 22.99573$, a value which is very close to 23. The code thus remains a hybrid fabric because of the existence of three nonsense signals, a number which serves at the same time as a *maximum* and a *minimum*, thus giving the exact definition of the optimization.

Let us add that however great our aversion for expressions given by the mean of exact fractions and an empirical approach to a transcendental number such as e without revealing its mathematical properties, the encountered coincidences deserve nevertheless more than pure and cold contempt.

One can recall, for example, the Titius–Bode empirical law which gives the planetary distances to the Sun by a common progression and some lucky or adjusted coefficients; this law was the object of many recent studies trying to give a more rational expression from an exponential distribution of the planets [15]. The value of such empirical expressions as those given above is to suggest a possible explanation for 23 being the "magical" number of the types of signals facing the 64 codons and to suggest a reason for the code being both filled and unfilled in an optimal fashion correlative to a certain "good" number.

The determination of the classes of codons and the system of distribution of the redundancy

Let us now study how degeneracy is exhausted in the ratio of 41/44 by the significant codons distributed into the so-called classes of triplets or coding classes. Classes [7] are the elements of distribution of all the significant codons divided into groups of redundants and singles. The former are by far more numerous than the latter, since the only two triplets which each code for one amino acid are those coding for methionine and tryptophan; 59 redundant codons are distributed into four classes, according to whether 2,3,4 or 6 triplets code for one amino acid; there are altogether five classes including that of number one.

It is important to emphasize the sum obtained with the 5 classes: $1 + 2 + 3 + 4 + 6 = 16$. Incidentally we notice that by summing $16 + 23$, i.e., the sum of classes and signals, we obtain 39, i.e., the number of all superfluous triplets added to the minimum of 20 by the redundancy system. But the by far most interesting observation is related to the ratio 16/5, which is the quotient of the sum of the classes and their number; this fraction is a reduction of 64/20, which is the quotient of the total number of triplets and the number of amino acids. Now, taking $\log_e 10 = 2.3025$ and $e = 2.71828$, then $\log_e 10/(e-2) = 3.2055$. This value is close to 3.2, i.e., our ratio 16/5; thus the organization of the coding classes is amenable to our criterion of logarithmic optimization.

Let us furthermore emphasize that obtaining the value 16 as the sum of the classes by summing five numbers is only possible in the natural case, i.e., 1,2,3,4,6, and that no arrangement with a class of number 5 is possible. Also a class of number 7 is impossible in as much as it would overvalue the sum of the five classes by at least one unity in the case of the weaker arrangement.

In his thesis devoted to the organization of the genetic code, Besson [1] has worked out a comparison between amino acids and crystallographic systems founded on the coding classes. On the one hand, the absence of a class of number 5 is striking but on the other hand, the demonstration of the cause of the eviction of the same number in crystallographic systems is a classical one. In a subsequent work (Gavaudan and Besson [7]), we emphasized again the same subject, although we did not indicate the value of 3.2 to be an index of optimization as we just have done. We only emphasized that the value 3.2 (of the ratio 16/5) was in short close to three, the value of a class of redundancy almost entirely ejected out of the code (this class contains only one amino acid: isoleucine). We stressed the fact that a code in which 20 amino acids would have been each coded by three triplets was not after all impossible; but it was also stressed that such a code should have one "extra" nonsense triplet (4 instead of 3). We want to make it clear that it is not astonishing that such a solution can't be accepted by nature in as much as an "extra" nonsense triplet would have been too "expensive"; the price is the collapsing of an optimization founded on a number of types of signals exactly equal to 23.

The present figure allows us to understand readily the organization of the 5 classes whose numbers are depicted by a system of diagonals crossing

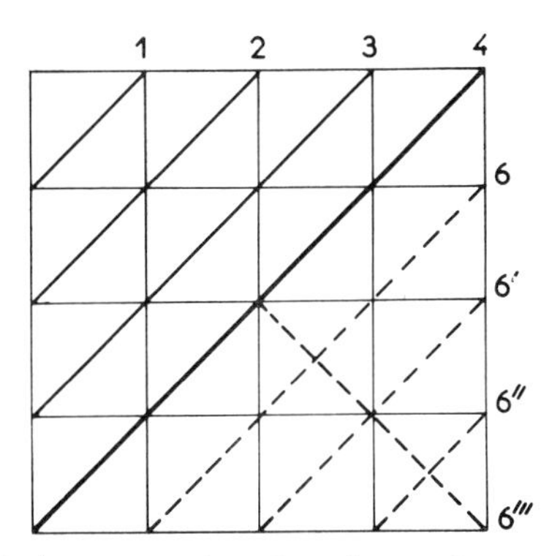

Fig. 1. Graphical representation of numbers and properties of classes.

through the 16 squares (4^2). The diagonals of numbers 1,2,3, and 4 are crossing one, two, three, four squares corresponding to the classes of the same numbers. We see also that there is no place for a class of number 5 in the board for we should obtain with this number a defect of one and an incomplete filling of the structure $4^2 = 16$. Turning now to the class of number 6, we see that there is no possibility of representing it by only one diagonal parallel to the lines of the other classes and crossing the six squares. There are two possibilities: in the first instance, the diagonal 6''' perpendicular to the four others crosses only two squares, leaving two pairs of squares, on the left and on the right; in the second instance, three diagonals 6, 6', and 6'' cross respectively three, two, and one squares. It is intriguing that this feature of heterogeneity appearing in this geometrical scheme exhibits itself also in the classical board of coding, in which three amino acids – leucine, serine and arginine – take their six codons in two different squares of four triplets, thereby adding 2 to 4. Moreover, if we rotate the diagonal 6 through an angle of 180° (around diagonal 4 as an axis), the six squares of class six occupy a position which is symmetrical relative to that formerly occupied and again replenish the board. We see that class four behaves here just as it does in the genetic code, being the center of gravity of the classes and 18,20, and 18 codons corresponding respectively to the 2, 4, and 6 classes.

Determination of the numbers of coding groups or numbers of amino acids respectively coded by each class

Nevertheless, the determination of the numbers and individual values of classes is not sufficient to give the whole definition of the coding structure.

Indeed, to each class-number should correspond a group-number. The total of the products, class by group, ought to give 20 without losing sight of the existence of only 61 significant codons. It is therefore certain that group-numbers are not meaningless, but that they also correspond to a well-defined structure according to that of the classes submitted to the underlined optimization. At first sight, indeed, the group numbers 2, 9, 1, 5, 3, do not exhibit any immediately obvious order. But, let us simplify the problem by setting apart the even classes. The following table displays the even classes and their corresponding groups:

Even classes	2	4	6
Corresponding groups defined by $2^n + 1$	$2^3 + 1 = 9$	$2^2 + 1 = 5$	$2 + 1 = 3$

The group-numbers corresponding to even classes are in accordance with a geometrical progression when the class-numbers are inversely ordered by an arithmetical progression. The sum of the products of the numbers of the two series is equal to the number of 56 triplets coding 17 amino acids by the system of even redundancy: there are now remaining five triplets (56 + 5 = 61) which ought to be attributed to three amino acids, necessarily pertaining respectively to classes one and three.

One can obtain the number 5 only with the two arrangements 1 × 3 and 2 × 1, and hence the series of groups is achieved; we see below the differences between the group-numbers:

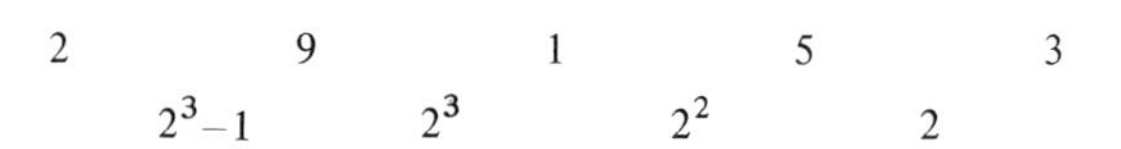

Only the natural order of the products 2 × 1, 9 × 2, 1 × 3, 5 × 4, 3 × 6, as they are existing in the code, allows 20 amino acids, 23 types of signals, and 61 significant codons corresponding to optimization. Furthermore, we recall that only the minimum group-number of one can be put in a product with class 3. It is easy to illustrate it by forming the 120 possible arrangements of products between classes and groups, the order of the classes being maintained ne varietur throughout and only the order of groups being permuted. The sums of the permuted products for the series of the different arrangements fluctuates between 42 and 89. According to the chosen criterion and to the structure of the code, only the sums equal to 61 are retained and naturally all those lower or greater than the same number are deleted. There are indeed 6 arrangements whose sums equal 61 and it is necessary to choose an ultimate criterion. We find it naturally in the observation of a minimum condition of only one amino acid coded by three triplets and this occurs only in the arrangement of the natural series of the code 2, 9, 1, 5, 3. In all the five other arrangements, there are more than one amino acid coded by three triplets. Lastly, there are also three other arrangements fulfilling the condition of only one amino acid coded by three triplets, but in these cases we find sums of 58, 64, and 69 triplets which exclude themselves ipso facto. From these three unnatural arrangements, we quote 3, 5, 1, 9, 2, which exactly gives 64 codons and which is the symmetrical and inverse disposition of the natural arrangement; but wholly filled the arrangement in question cannot correspond to our structural optimization.

Recapitulation of the remarks upon the hierarchy of structures

All the facts we have pointed out show plentifully that the genetic code is constructed in accordance with a hierarchy of structures, all associated in harmonized succession to some rule of optimization relative to a logarithmic criterion. First, the 20 amino acids, the 23 types of signals, and the 44 degenerate triplets contained in the complete set of 64 forms a primary structural level. Secondly, the optimization extends to the scale of the organization of classes and groups. We have seen how a mathematical constraint prescribes not only the number and the nature of the classes but also an eventual heterogeneity as in the case of the class six.

The concept of optimization, as presented in our first essay [2] and as herein expanded, is confirmed by the views of Mackay [13] who started from quite different considerations founded on information theory. Again the work of Gurel [10] may be quoted, whose starting point is quite different from ours. An analysis of the work of Gurel is beyond the scope of this discussion, but it may be briefly summarized as follows: starting from the weights of the atoms involved in the hydrogen bonds between the bases, Gurel has deduced the existence of a series of interdependent structures. One proceeds upward from the bases to the triplets, next to the set of 64 codons and to the "genetic table" and also to the set of the 23 signals. The main structural features of the "tableau génétique" are deduced with the aid of such mathematical tools as group transformation and group cyclic permutation. All the work of Gurel leads to the concept of an ineluctable "encasing" of structures.

Hence, to summarize, it appears that it is not possible to disregard the numerical optimization linked to the hierarchy of structures of the genetic code when speculating on its origin.

The hierarchy of structures of the genetic code and the problem of the origin of biological order

Apparently the strong interdependence of all the structural features of the genetic code has not yet been sufficiently recognized. All peculiarities are highly linked and dependent on general rules of optimization and on strong physical laws.

There are two magical numbers, 20 and 23, but we do not imply that they are the exact primitive numbers; their presence corresponds to a kind of *well or hole of order.*

Some speculations on the evolution of the number of the natural amino acids have been advanced during the last few years (Jukes [12], Crick [3,4], Schutzenberger, Gavaudan and Besson [14]).

Crick [3,4] has debated two fundamental problems relative to the primitive code: the doublet or triplet nature of the genetic code and the initial number of bases. Without hesitation Crick has rejected the assumption of a doublet primitive code [4, p. 372]. A change from a doublet code to a triplet code has been suggested by Jukes [12]; the change would have taken place in the course of evolution and certain triplets are the vestiges of doublets plus another base in the third position. We cannot do better than to quote in extenso Crick writing on the eventuality of such a transformation: "this seems highly unlikely, since it violates the Principle of Continuity. A change in the codon size necessarily makes nonsense all the previous messages and would almost certainly be lethal" [4, p. 372]. Crick has proposed a reading mechanism with three bases, among which only the first two were read; further, only a few amino acids were specifically coded.

Having adopted the principle of a primitive triplet code, Crick has nevertheless looked at the possibility of the existence of only two bases and studied the consequences of such an eventuality. Actually, in the present code, there are many amino acids coded by triplets containing only two kinds of bases; as Crick has pointed out, such a scheme "does not violate the principle of continuity" [4, p. 374], the information contained in the primaeval chain being unaltered by mutations introducing the new bases. Here also, certain triplets are the vestiges of an archaic coding system.

Let us point out some particularly important features emanating from the ideas of Crick and discuss them. It would seem that the assumption of the two bases did not appear quite satisfactory to Crick who has reintroduced the possibility of four bases at the end of his study [4, p. 377]. But as we do not immediately understand the necessity of the present number of four, a tentative justification is found in its capacity for fitting stereochemically into a double helical structure; and further, two is described as a too restrictive number. Here we are faced to a mixture of *possibility* and *necessity* from which, indeed, it is difficult to escape.

In fact, there is something strange in the choice of only two bases and of triplet codons at the origin of the genetic code. Such a hybridization of concepts leads to some perplexity and Crick has posed the question: "Why a triplet?" [4, p. 377]. It is a matter of fact that nature has not made too bad a choice with triplets chosen from among four bases. This option is not understandable without some inadmissible premonition or rather some fall in a kind of "trap" resulting from physical exigencies or, as some people believe, by natural selection. After all, it seems that our picture of the situation is not very far from Crick's opinion with regard to triplets: "however, we are inclined to suspect that the reason in this case may be a structural one" [4, p. 377]. It is also our feeling that triplet coding is an inescapable event linked to strong natural law. As a matter of fact, epistemologically speaking, the situation is not very comfortable. In all the studies, one is starting from the *postulatum* of the triplets as permutations of four bases along the initial scheme of a coding system constructed in accordance with the Gamow's *a minima hypothesis* (i.e.:4^3), especially for accounting for the magical number 20 of the natural amino acids. And then we say after that the triplet is very likely unavoidable. After the tautological assertion we are also obliged to admit that the code is universal. But we are also obliged to extend the universality into the past and we are obliged to weigh heavily the general principle of the *unity of making* of living beings with reference to the cell, even the most primitive. We have no reason to assume that the early Precambrian code in a bacterium was different from that of a contemporary microorganism. Necessarily there is a close parallel between the universality of the code in space and time and the so-called "unité de plan de composition" of the cell.

Consequently we can authoritatively state that the code is probably as old as life, the last term given in its actual meaning and excluding all more or less conjectural "protobiotic" events about which we absolutely know nothing and which we are unable to describe in realistic and adequate terms. The critic is by no means too drastic, sterile, or depressing, but is objective. The code has been edificated by ways which are quite unknown. The birth of the code appearing with life is no more astonishing than the eternal "birth of Venus". This mythology is familiar to all biologists when, studying the course of evolution, they are faced with the apparently sudden appearance of classes, groups, families, genera, and even species, without any transitional relic. For the code, the difficulty is the same as in all the other problems on the origin of fundamental structures or great functions and nowadays we can only ascertain the facts without explaining them.

It seems very unlikely that an "empodoclean" method of assembly of isolated, unrelated and heteroclit parts worked out by pure accident would have ever been the method of design of the code.

Crick [4, p. 377] further indicated that perhaps the geometry of the helix may have dictated the size of the codon and that, after all, a doublet code would not have allowed the solution of the problem of recognition with only two bases. We are reaching a conclusion which approximates that drawn by Besson [1] in his statistical approach to the structure of the genetic code. In the present study we cannot give more than a very short and certainly too elliptic sketch of an important work unfortunately not yet published.

It is assumed that the repartition of the bases along the helix obeys Poisson's law (no matter what the type of base in question is). We quote (Besson [1, p. 63–64]): "In the instance of the double helix of DNA, we take as a unitary segment a length equal to one radian, as far as the bulkiness of the helix appreciated in diameter is representative of its constitution since it indicates the space covered by the pairs of bases A–T and C–G. The maximum accuracy should now be obtained when giving to any base along the helix a linear bulkiness equal to 1.618 radian, a value equivalent to a probability of existence of 0.618 for each elementary interval of one radian, since $0.618 \times 1.618 = 1$. Such a linear bulkiness, corresponding to the side of a regular starshaped decagon, indicates a number of bases of 10 for 2π radians. As it is in fact the number of bases determined by Watson and Crick in a turn of the helix, we can apply a Poisson's process to the distribution function of the nucleotides, provided we make allowance for the fact that such a repartition is linked to the geometrical form of the helix."

Let us also emphasize another important idea relative to the repartition of the bases on a helix turn and yet quote textually the work of Besson [1, p. 79–80]. "At the ribosome level the dimension of the amino acid is relatively small compared to that of the ARN messenger molecule, and the amino acid can be depicted as a linear part. The purinic and pyrimidinic bases being separated one from another by an average value of 36°, the importance of the influence they exert on the amino acid must be a function of the cosine of the angle formed with the plane of the amino acid. Now in the instance of the helix, we postulate what follows:

1) The coding is an irreversible phenomenon always proceeding in the same direction owing to the characteristics of the two chains of whom only one codes.

2) Only the bases included in an angle of 90° can code and thus the number three is dictated."

Further, Besson has pointed out that the equipartition of bases is linked to $1/e$ as we have already seen at the beginning of the present discussion and let us recall that the same probability at the triplet level is $(1/e)^3 = 1/20$, the denominator corresponding to the number of the amino acids. Thus it is not possible to doubt that there is an imperative physico-mathematical drift towards triplet constitution and the number twenty; they are not due to "historical accidents" as said by Crick [4, p. 377]. We have discarded many questions relative to certain characteristics of the genetic code and to the origin of life. For instance, there is a rather impressive inverse correlation pointed out by us [14] between the numbers of the codons assigned to the amino acids and the molecular weights of the last; generally speaking, the lower the molecular weight, the higher the number of codons. For example, there are four codons for glycine (m.w. = 75) and only one for tryptophan (m.w. = 204). Furthermore, when one studies the statistical frequency of the amino acids of the class IV, one sees that there is a noticeable drift to the prejudice of the other classes [11]. The same class IV of the "lighter" amino acids is also characterized by its greater pool of C + G compared to class II, which itself is characterised by its higher pool of A + U; the class VI is of an intermediate nature. We don't know the cause of the relationship between molecular weights and numbers of codons. Perhaps molecular weight can be considered as a general and rather indefinite or statistical indicator of complexity of form and also of biological synthesis.

But, however great may be the interest of the preceding remarks, let us finally return to the numbers 20 and 23. The addition of one to three supernumerary amino acids to the class I is not a priori impossible since such an operation preserves the relationship of optimalization founded upon the number 23 and the structure of the code, except for the existence of one, two, or three *nonsense* codons. It is possible to imagine that some amino acids of very high molecular weight (on the order of methionine and tryptophan, or perhaps amino acids of higher molecular weights) were coded in the past or that they will be coded in the future. In the first instance, they have disappeared and in the second instance, they will appear. But it is likely that it is a mere fiction and that the existence of nonsense triplets is absolutely necessary; a perfect filling up of the code, relatively to the signification, is contradictory to the preservation of the functions of the nonsense codons. It is very likely that the functioning of such a code entirely filled by significant codons is impossible but we have no exact proof of it. Such a fiction of a code with 64 perfectly significant codons poses the critical problem of the origin and the opportune existence of nonsense codons as terminators.

We do not state that the numbers 20 and 23 were given ne varietur and in saecula saeculorum at the dawn of the life but we observe in the structure of the code a considerable urge towards them. In fact, they authorize a maximum significant utilization of the code associated with termination function.

Naturally, all our speculations are based on a perfect confidence in the results of the biochemists who have experimentally diagrammed the genetic code.

The origin of life, if linked to the origin of the genetic code, did not proceed from a kind of stroke of luck as in dicing but rather as a sequence of natural events submitted to restrictive a priori constraints generating the requisite regularities. Epigenesis is the result of conflicts between regularities prescribed both from the exterior and the interior of a system. Nature having given four bases the origin of all living regularities, among other circumstances, can be conceived of as a series of successive falls and fittings in a mathematical well of order. Thus the number 23 being optimal, the three nonsense triplets are at the bottom of the well as the final result of an ineluctable evolution. It is possible that the well is largely an energetic one. The "looseness" introduced by the three nonsense triplets, mechanically speaking as we have already written, seems as necessary and as sufficient "dans le meilleur des mondes".

As opposed to organicism and other parent conceptions of a biology with more degrees of freedom than in classical physics, reductionism is approving without conditions the thesis of neo-Darwinism and natural selection as the universal driving force. But we have many doubts about the validity of the generalized application of the biological concept of natural selection to inanimate systems. We have elsewhere stressed the same question [5,6]. But rather than perpetuate the discussion in such a direction, we are convinced that it is better to watch and accumulate more experimental data and that, until we are able to obtain the ultimate solution, the supporters of the metaphysical match must conclude a kind of gentleman's agreement.

References

[1] Besson J., Recherches sur la structure du code génétique, Thèse de Doctorat ès-Sciences (Etat), Faculté des Sciences de Poitiers, July 1967, 90 p.

[2] Besson J. and P. Gavaudan, Compt. Rend. 264 (1967) 1311–1314.

[3] Crick F.H.C., Cold Spring Harbor Symp. Quant. Biol. 3 (1966) 3–9.

[4] Crick F.H.C., J. Mol. Biol. 38 (1968) 367–379.

[5] Gavaudan P., L'évolution considérée par un Botaniste-Cytologiste, Mathematical

Challenges to Neo-Darwinian Interpretation of Evolution. The Wistar Symposium Monograph no. 5 (1967) p. 129–134.

[6] Gavaudan P., Selection naturelle, origine et evolution des etres vivant et pensant, Ann. Biol. IV part 9–10 (1967) 509–535.

[7] Gavaudan P. and J. Besson, Compt. Rend. 264 (1967) 1919–1922.

[8] Gavaudan P. and J. Besson, Compt. Rend. 268 (1969) 173–175.

[9] Gavaudan P. and J. Besson, Compt. Rend. 268 (1969) 2130–2132.

[10] Gurel O., The Structure of the Genetic Code, IBM New York Scientific Center, Report January 1969, no. 320–2963, 28p.

[11] Jukes T.H., Biophys. Biochem. Res. Commun. 19 (1965) 391–396.

[12] Jukes T.H., Molecules and Evolution (Columbia University Press, New York, London, 1966) 285p.

[13] Mackay A.L., Nature 216 (1967) 159–160.

[14] Schutzenberger M.P., P. Gavaudan and J. Besson, Compt. Rend. 268 (1969) 1342–1344.

[15] Schatzman E., Origine et evolution des Mondes (1957) 404p. (work by Weizsacker on p. 343).

[16] Woese C.R., The Genetic Code (Harper Row, New York, London, 1967) 200p.

[17] Ycas M., The Biological Code (North-Holland, Amsterdam, 1969) 360p.

Chemical Evolution and the Origin of Life, eds. R. Buvet and C. Ponnamperuma

THE ORIGINS OF BACTERIAL RESPIRATION

E. BRODA
Institute of Physical Chemistry,
University of Vienna, Austria

It is surprising that a rather acute difficulty in understanding the origin of bacterial respiration does not seem to have been noticed or discussed in the literature. To my knowledge, the difficulty was pointed out first only a short time ago [2,3]. By respiration we mean, in this context, the generation of metabolic energy through oxidative phosphorylation. In aerobic respiration, the electrons are transferred to free oxygen, in anaerobic respiration to sulfate or nitrate ion.

After Van Niel had in momentous investigations [17] elucidated the relationship between photosynthesis in bacteria and in plants, it has been generally accepted that plants, including algae – first the prokaryotic, blue-green algae – have evolved from photosynthetic bacteria. (Evolution implies, as always, mutation + selection.) The kinship is expressed in a general equation for photosynthesis:

$$2\,H_2A + CO_2 + H_2O = (CH_2O) + 2\,A + 2\,H_2O \qquad (1)$$

where A never represents oxygen in bacteria and usually represents oxygen in plants. In this way, only in the photosynthesis of plants oxygen is set free, as has also been demonstrated by Van Niel.

Therefore most authors hold that an atmosphere containing free oxygen, an "oxygenic" atmosphere, can have arisen only after the advent of the blue-greens. The geologists find that oxygen appeared in scattered places more than 2 aeons ago, and that the atmosphere as a whole became oxygenic somewhat later (see Rutten [12]; Cloud [5]). Free oxygen is aggressive, and so the organisms exposed to it had to develop mechanisms for rendering it harmless, or, better, exploiting it.

The best way of utilizing free oxygen is, of course, its application in aerobic respiration. As is well known, 1 molecule of glucose supplies 38 molecules of ATP in respiration through oxidative phosphorylation, while only

2 molecules of ATP are obtained in glycolysis. In the present context it is important that aerobic respiration could not operate before the algae had turned the atmosphere oxygenic, and therefore machinery for aerobic respiration could also not have existed before.

Some authors propose that free oxygen, produced by photolysis of water, predated algal photosynthesis. A reliable estimate of the rate of oxygen production by this process is not possible, but there are arguments against an appreciable amount of free oxygen in prealgal times. First, the most ancient rocks, including sediments, are generally in a reduced condition. For instance, Fe appears as ferrous compounds. But even if some oxygen had been set free and reacted with rocks in early times, this reaction would no doubt have been rapid, and the steady state concentration of the oxygen would therefore have been small indeed. Moreover, as has been emphasized by Wald [18], the more primitive and ancient metabolic mechanisms appear to fit life in an entirely anoxygenic environment. Machinery for reaction with oxygen, including respiration, has been grafted subsequently on the anaerobic mechanisms, essentially without interfering with them. Thus aerobic respiration was certainly a latecomer.

Presumably the original prokaryotic, blue-green algae were oxygen-tolerant rather than oxygen-exploiting. However, these algae have succumbed to respiring species in the struggle for existence. The contemporary blue-greens are aerobes. Among eukaryotic cells, facultative anaerobes are quite exceptional, and obligate anaerobes do not exist and probably never existed among them.

The blue-greens thus developed respiration long after they branched off from the bacteria. At the time of this branching, the bacteria must have been anearobes. Even more, they were probably obligate anaerobes rather than just oxygen-tolerant. The only oxygen-tolerant group of non-respiring bacteria now known are the lactic acid bacteria. They are a special case. They are often microaerophilic, and there is evidence that they have degenerated from aerobes (see Bryan-Jones and Whittenbury [4]). In general the non-respirers are obligate anaerobes, and this probably applied equally to the bacterial ancestors of the blue-green algae.

We arrive now at the difficulty mentioned at the start. It follows from the preceding considerations that the bacteria must have developed respiration and oxidative phosphorylation independently from the algae. The puzzle is even greater when it is remembered that aerobes occur among many groups of bacteria, and it cannot very well be assumed that all of them are descended from one and the same "original respirer", i.e., respiration must have been evolved many times separately by different kinds of bacteria. A truly extraordinary case of convergence that cries out for explanation.

Somebody might be tempted to escape the difficulty by postulating a descent of the respiring bacteria from respiring blue-green algae. But such a descent can reasonably be claimed only for one limited group of "bacteria" that share gliding rather than swimming movement with the blue-greens. The dividing line between the gliding bacteria and prokaryotic algae that have lost their chlorophyll is rather uncertain. Well-known members of the group are the flexibacteria. *Beggiatoa* and *Thiotrix*, organisms that oxidize sulphides, are also included. However, the gliding bacteria are remote indeed from the true bacteria from which the blue-greens are derived and whose contemporary representatives are the "eubacteria". Many aerobic species are found among the eubacteria. We cannot escape the puzzle in this way.

A solution requires a closer look at the mechanisms of respiration. The respiratory machinery of the eukaryotes, i.e., of the mitochondria, is much alike, so-to-speak standardized, in the most diverse species. The similarity with the analogous machinery of the bacteria is much more limited, as would be expected from the supposed separate origin. Moreover, there are large differences between the mechanisms for respiration in the various kinds of bacteria, namely, in respect to the composition of the electron flow chains, the pathways of electron flow, and the number of phosphorylating sites.

Nevertheless, the fact remains that we always find pyridine nucleotide (NAD), flavoproteins and cytochromes as key members of the electron flow chains. It can hardly be assumed that these substances, belonging to utterly different groups of chemical compounds, and each requiring many steps in biosynthesis, were introduced again and again de novo by the cells when respiratory chains were built up. These substances or their precursors must have preexisted in the cells.

The difficulty is not yet extreme for pyridine nucleotide or flavoprotein. These substances had been elaborated by the most ancient bacteria, probably similar, among contemporary organisms, to the clostridia. The cytochromes, never found in clostridia, are the principal stumbling block. From where did cytochromes enter respiratory chains again and again? Moreover, the electron transfer agents of the chains are not just mixed but they form an orderly assembly. Oxidative phosphorylation functions only as long as the spatial order is maintained; the particulate and periodic structure of the respiratory assemblies is clearly indispensable. So again it must be asked: Did this spatial order that includes, among other compounds, NAD, flavoprotein and the various cytochromes, arise de novo in each case? Hardly likely.

A solution may be sought through derivation of all prokaryotic respirers from photosynthetic bacteria. The machinery for photosynthesis is also always contained in ordered particles. Moreover, the photosynthetic electron flow chain always includes, like the respiratory electron flow chain, pyridine nucleotide, flavoprotein and cytochrome. Could it be that every kind of respiring bacterium is descended from one or the other species of photosynthetic bacterium?

On the basis of this hypothesis, intermediate organisms, the transitional ancestors of the aerobic organotrophs were probably similar to the facultative aerobes, e.g. *Rhodopseudomonas* spp., among the contemporary non-sulphur purple bacteria (athiorhodaceae). In analogy, the (aerobic) colourless sulphur bacteria (thiobacilli) could be derived from the coloured sulphur bacteria, probably the thiorhodaceae. It is encouraging that recently hints of oxidative phosphorylation have been found in a few of the species of the purple sulphur bacteria, notably of *Chromatium* [6].

Of course, the organic substrates must be got ready for combustion in respiration, i.e. for hydrogen atoms, equivalent to electrons, to be handed over to a respiratory chain. In bacteria, as in higher organisms, this is often done through the citric acid cycle (see Krampitz [8]). There is no particular difficulty in assuming that the reactions of this cycle were available when respiration began. These reactions are still used not only for respiration but also to provide electrons for reductions (see Stanier [13]) and to supply body constituents, and were probably so used in anoxygenic times. This aspect will not be discussed here any further.

What would, then, be the position of the anaerobic respirers, i.e., the bacteria that use sulphate or nitrate as electron acceptors and make ATP in this variant of oxidative phosphorylation? Dissimilatory reduction, as the process is also called, is probably not very important in the present context, as far as nitrate is concerned. For all nitrate respirers use alternatively ordinary oxygen respiration in aerobic conditions. In any case it is most unlikely that nitrate predated free oxygen in the biosphere. Thus nitrate respiration must have originated rather late. The same applies, whatever its relationship to nitrate respiration, to "nitrate fermentation", i.e., the bioenergetic process shown by anaerobes (*Clostridium welchii*) where the transfer of electrons to nitrate is catalyzed by a soluble enzyme system in absence of cytochromes (see Takahashi et al. [14]).

It has been suggested [7] that sulphate respiration preceded bacterial photosynthesis. Arguments for this hypothesis are the supposed similarity of the (strictly anaerobic) sulphate respirers (desulfuricants) – *Desulfovibrio* being best known – with the clostridia, and the fact that only one cytochrome has been found in them. The cytochromes of the photosynthetic chain are then thought to have arisen from this kind of cytochrome. The photosynthesizers would have added chlorophyll.

However, the geologists appear to find no sulphate in the oldest strata. In agreement therewith it is seen that on the basis of the equation:

$$SO_4^{2-} + 3\,H_2 = H_2S + 2\,H_2O + 2\,OH^-; \qquad K = 10^{14} \qquad (2)$$

the primordial waters, at least in equilibrium, can have contained practically no sulphate. (Equation (2) was written down by Urey [16], but because of a mistake in the sign Urey arrived at the opposite conclusion.) In absence of free H_2 (and with carbon present mostly as CO_2 and partly as organic compounds; see Rubey [11], Abelson [1]) sulphur was still, in equilibrium, present as sulphide in the anoxygenic biosphere.

The earliest sulphate was probably formed by coloured sulphur bacteria, namely, through the reaction:

$$H_2S + 2\ CO_2 + 2\ OH^- = SO_4^{2-} + 2(CH_2O) \qquad (3)$$

It is concluded that sulphate respiration followed rather than preceded photosynthesis, probably when the atmosphere was still anoxygenic. After free oxygen appeared, the sulphate respirers, all anaerobes, retreated into an anoxygenic environment, while in the oxygenic air the thiobacilli evolved from the coloured sulphur bacteria.

A fairly close kinship between the groups of bacteria that reduce or oxidize sulphur compounds in their bioenergetic processes is indicated by the fact that they all use forms of adenosine-5′-phosphosulphate (APS) reductase [9,10]. In the thiobacilli, the energy-rich sulphur-containing nucleotide is produced by a kind of substrate phosphorylation; it is the only chemolithotrophic process in which energy is thus conserved.

$$2\ SO_3^{2-} + 2\ AMP = 2\ APS + 4\ e^- \qquad (4)$$

The energy of the APS is used to build up ATP. However, for the disposal of the electrons cytochrome (and, terminally, oxygen) is still required, and in any case part of the ATP needed by the thiobacilli is derived from oxidative phosphorylation.

The desulfuricants, in contrast to the coloured and colourless sulphur bacteria, lack a reductive pentose phosphate (Calvin) cycle. Probably they lost it because they can in any case live only where organic compounds are abundant. For the same reason, the absence of the cycle in all non-photosynthetic organotrophic aerobic respirers is not surprising. The early photoorganotrophs may not have enjoyed such abundance, and this was probably the reason why photolithotrophs evolved from them.

The hypothesis that desulfuricants followed photosynthesizers does not require a tremendous time interval between the birth of the anaerobic and the aerobic respirers, while according to Klein and Cronquist [7] the photosynthesizers would be interposed between them, and therefore the interval would be of the order of aeons. It remains an open question whether some aerobic respirers evolved from desulfuricants.

Clearly in the end the strict aerobes have descended from the facultative (an)aerobes. Thus gram-negative strict aerobes, e.g., *Acetobacter* or *Azotobacter*, could be derived quite naturally from the gram-negative coliforms. Gram-positive facultative or strict aerobes, on the other hand may have descended from hypothetic early photosynthetic bacteria that were, like their ancestors, the clostridia, gram-positive. The gram-positive oxygen-tolerant lactic acid bacteria could then be derived from the gram-positive respirers.

The present scheme for the evolution of bacterial respiration in the following; a more general and complete scheme, somewhat differing in respect to the desulfuricants, was given in an earlier paper [3].

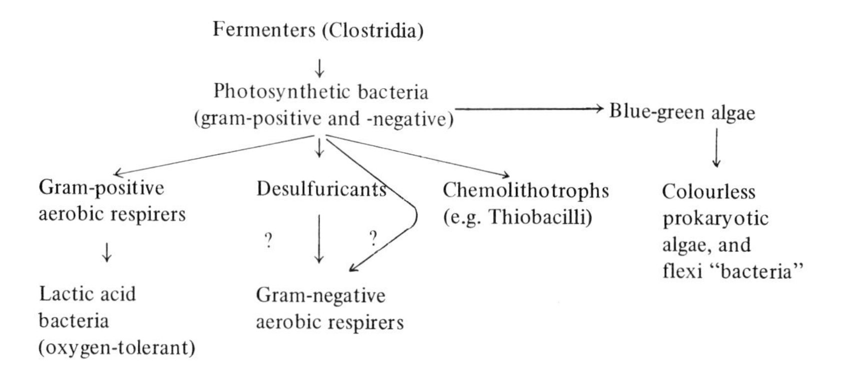

Finally, two remarks, on the basis of the symbiotic hypothesis, about consequences for the eukaryotes: First, in view of the similarity of the respiratory mechanisms in mitochondria and their relative dissimilarity in different bacteria, it seems unlikely that the mitochondria are polyphyletic, and not derived from one and the same kind of bacterium. Secondly, we note that chloroplasts, in contrast to blue-green algae, never seem to respire; probably they lost their terminal oxidase.

Explanation: 1 aeon = 10^9 years.

References

[1] Abelson P., Proc. Natl. Acad. Sci. U.S. 55 (1966) 1365.
[2] Broda E., Naturw. Rundschau 20 (1967) 14.
[3] Broda E., Progr. Biophys. Mol. Biol. 21 (1970) 146.
[4] Bryan-Jones D.G. and R. Whittenbury, J. Gen. Microbiol. 58 (1969) 247.
[5] Cloud P.E., Science 160 (1968) 729.

[6] Gibson J., Arch. Mikrobiol. 59 (1967) 104.
[7] Klein R.M. and A. Cronquist, Quart. Rev. Biol. 42 (1969) 105.
[8] Krampitz L.O., in: The Bacteria, eds. I.C. Gunsalus and R.Y. Stanier, Vol. 2 (Academic Press, New York, 1961) p. 209.
[9] Peck H.D., Ann. Rev. Microbiol. 22 (1968) 489.
[10] Peck H.D. (1966/67), quoted by Trudinger (1969).
[11] Rubey W.W., in: The Crust of the Earth, ed. A. Poldervaart, Geol. Soc. Am. Spec. Paper No. 62.
[12] Rutten M.G., The Geological Aspects of the Origin of Life on Earth (Elsevier, Amsterdam, 1962).
[13] Stanier R.Y., Bacteriol. Rev. 25 (1961) 1.
[14] Takahashi H., S. Taniguchi and F. Egami, in: Comparative Biochemistry, eds. M. Florkin and H.S. Mason, Vol. 5 (Academic Press, New York, 1963) p. 92.
[15] Trudinger P.A., Advan. Microb. Physiol. 3 (1969) 111.
[16] Urey H.C., in: Bioastronautics and the Exploration of Space, eds. N. Roadman et al. (Brooks, Texas, 1968).
[17] Van Niel C.B., Advan. Enzymol. 1 (1949) 263.
[18] Wald G., in: Recent Progress in Photobiology, ed. E.J. Bowen (Blackwell, Oxford, 1965).

Chemical Evolution and the Origin of Life, eds. R. Buvet and C. Ponnamperuma

CHANGE OF BIOCHEMICAL FUNCTIONS OF ORGANISMS IN THE EVOLUTION OF THE BIOSPHERE

E.A. BOICHENKO
V.I. Vernadsky Institute of Geochemistry and Analytical Chemistry, USSR Academy of Sciences, Moscow, USSR

Biochemical functions of organisms has been called by Vernadsky [13] the participation of processes within living cells in the migrations of elements in the biosphere. In the evolution of the biosphere, biochemical functions changed more rapidly than the other biogeochemical functions in which products of vital activity participate – organic matters and metal compounds of dead cells, gases produced by organisms. The significance of biochemical functions for the migration of elements at about the time of the origin of the biosphere was obviously not great, but gradually, in the course of organism evolution, it increased in accordance with the increase of their mass. The most important changes of biochemical functions were connected with the improvement of autotrophic organism development. They first occurred during the transition of cells from heterotrophic development and chemosynthesis to photosynthesis, as a result of which in the Proterozoic, the hydrosphere was populated by algae; and second, during the perfection of the photoautotrophic development under conditions of an increasing oxidizing environment, which allowed higher plants from the middle of the Paleozoic to populate the land and to augment many fold their general biomass in comparison with other organisms. In the modern biosphere the weight of organisms is according to Vernadsky 10^{13} tons. 99 Percent of this weight makes up green plants and only about one percent all bacteria, fungi and animals.

Vernadsky, Vinogradov and their disciples have carried out numerous analyses of the elementary composition of different organisms [13,14]. On the basis of these data, the contents of elements in all organisms of the biosphere may be calculated in tons as a percentage of their fresh weight (table 1).

A comparison of the amounts of elements in organisms with the world

Table 1
Content of elements in organisms of the biosphere.

Element	Content in organisms *	Element	Content in organisms *
O	7×10^{12}	Sr	2×10^{8}
C	1.8×10^{12}	Mn	1×10^{8}
H	1.05×10^{12}	B	1×10^{8}
Ca	5×10^{10}	I	1×10^{8}
K	3×10^{10}	Ti	8×10^{7}
N	3×10^{10}	F	5×10^{7}
Si	2×10^{10}	Zn	5×10^{7}
P	7×10^{9}	Rb	5×10^{7}
S	5×10^{9}	Cu	2×10^{7}
Mg	4×10^{9}	V, Cr, Br, Ge	$n \times 10^{7}$
Na	2×10^{9}	Ni, Pb, As, Co	$n \times 10^{6}$
Cl	2×10^{9}	Li, Mo, Y, Cs	
Fe	1×10^{9}	Se, U	$n \times 10^{5}$
Al	5×10^{8}	Hg	$n \times 10^{4}$
Ba	3×10^{8}	Ra	$n \times 10^{-1}$

* tons

reserves of raw materials shows the great geochemical signification of concentrating elements by cells [2]. Thus, the total amount of carbon in coals, oils and graphite is not greater than its content in recent organisms. The reserves of many metals in ores from various geological eras do not exceed their content in living cells. Besides, the assimilation of many elements by organisms is accompanied by a valency change in metabolic reactions, chiefly in reductions during photosynthesis and oxidations during respiration.

Types of organisms, which had arisen at different times, are characterized by a definite elementary composition. The greatest differences occurred during evolution in the content of metals, compounds of which participated especially actively in the changes of metabolism [1] (table 2).

When the mean content of iron, copper, manganese and zinc in bacteria, algae, angiosperms and mammals are compared, it is seen that their amounts markedly change. The increase of manganese content in plants and the decrease in the iron to manganese ratio is connected with participation of manganese in the production of oxygen during photosynthesis and its content in plant cells is several fold that in mammalian cells [4,5,11].

For the elucidation of biochemical functional changes of organisms in past

Table 2
Mean elementary compositions of organisms.

Organisms	Mean content (% dry weight)			
	Fe	Cu	Mn	Zn
Bacteria	0.025	0.0025	0.001	0.0002
Algae	0.180	0.003	0.004	0.007
Angiospermae	0.018	0.0019	0.018	0.0045
Mammals	0.016	0.00024	0.00002	0.016

geological eras, data on the evolution of geochemical processes of the earth are used as well as paleontological data on the time of appearance of separate types in the biosphere (table 3).

The study of old rocks of various continents gives us grounds to think that in the course of geological time on the earth's surface, the intensity of oxidation processes increased [9]. The development of organisms was

Table 3
Change of biochemical functions in the biosphere.

Era	Oxidations–reductions (E'_0 V at pH 7)	Appearance in the biosphere	Fe/Mn ratio
Archean	Oxidations of organic matters in fermentations;	*Bacteria*	
		heterotrophic	>100
	of H_2, NH_4^+, H_2S in chemosynthesis ($E'_0 < -0.10$)	autotrophic	> 10
Proterozoic	Oxidations of S, NO_2^-, Fe^{2+} in chemosynthesis;	idem	
	of various matters in photoreduction ($E'_0 < +0.55$)	*Algae*	
	of Mn^{2+} and H_2O in photosynthesis ($E'_0 > +0.82$)	Cyanophyta	>100
		Rhodophyta	> 20
Paleozoic	Localization of reactions in cell organelles	Chlorophyta	> 10
		Land plants	
		Pteridophyta	~ 3
Mesozoic Cenozoic	Localization of processes in specialized organs	Gymnospermae (needles)	~ 1
		Angiospermae (leaves)	

probably initially anaerobic as the atmosphere did not then contain oxygen. They could utilize the oxidation of organic matter of abiogenic origin and then of the inorganic matter of surrounding biosphere [7]. The gradual increase of oxidized matter in the Archean and the first half of the Proterozoic lead to the increase of the oxidation–reduction potential (E_0'V) in the environment. This enabled plant cells already in the Proterozoic to change to oxidation of water in photosynthesis with free oxygen liberation which resulted at about the middle of the Paleozoic in the creation of the earth's atmosphere similar to the present one.

In the transformation of anaerobic biosphere into the aerobic one, various types of algae and higher plants participated which appeared in subsequent geological eras [3,8]. Changes in their metabolism occurred mainly in the realization of processes with higher oxidation–reduction potential. At the same time strong differences occurred in the ratio between iron and manganese in the transition from algae to higher plants. They were conditioned by the increase of manganese in cells with a more developed photosynthetic process (table 4).

In heterotrophic metabolism, catalyzers of carboxylations could be divalent metal ions, which do not form in these reactions stable complexes either with the enzymes or with CoA [12]. During chemosynthesis the presence of iron-containing proteins – ferredoxins, flavoproteins hydrogenases and CoA – is necessary for the participation in reductions, indicating the

Table 4
Participation of metal compounds in the evolution of photosynthesis.

CO_2 assimilation	Participation of metals
Heterotrophs $CO_2 + RH \rightarrow RCOOH$	*Me*$^{2+}$ *ions* (Mg^{2+}, Mn^{2+}, Zn^{2+}) CoA
Chemosynthesis $CO_2 + 2\ RH_2 \rightarrow$ $(CH_2O) + H_2O + 2\ R$	Fe_{nh} in ferredoxins Fe_{nh} – flavin in hydrogenases CoA – its presence is necessary
Photoreduction $CO_2 + H_2O + R \xrightarrow{h\nu}$ $(CH_2O) + RO_2$	Fe_{nh} – FAD, *Mn* in chromatophores CoA – participates in reactions of CO_2 reduction
Photosynthesis $CO_2 + 2\ H_2O \xrightarrow{h\nu}$ $(CH_2O) + H_2O + O_2$	CoA, Fe_{nh}, FAD, Mn^{3+} in chloroplasts – in CO_2 reduction and H_2O oxidation

connection between acylation and electron transfer by nonheme iron compounds (Fe_{nh}) [10]. However cofactors required for these reactions did not at that time form stable complexes with each other and could be isolated separately from cells. In carbon dioxide assimilation by various algae during photoreduction with formation of peroxides and especially in photosynthesis of plant cells more complicated complexes of flavine adenine dinucleotide (FAD) participate which contained manganese as well as nonheme iron [1,6].

Comparative studies of these complexes were carried out by different methods – spectrometry, paper chromatography and electrophoresis with subsequent autoradiography, direct quantitative determination of iron, sulphur, phosphorus, thiol esters [1,11,15]. In cells of algae and angiosperms these methods have revealed that iron bound to coenzyme A and 4-phosphopantetheine, FAD, and manganese bound to galactolipids are components of such complexes. In the evolution of photosynthesis Fe_{nh} compounds were of great importance in carbon dioxide reduction while Mn^{3+} compounds were very significant in oxygen evolution. The formation of these complexes increased the influence of plant metabolism in oxidation–reduction and other processes of the biosphere.

References

[1] Boichenko E.A., Izvestia Akad. Nauk SSSR Ser. Biol. 1 (1968) 24.
[2] Boichenko E.A., G.N. Saenko and T.M. Udelnova, Geokhimiya 10 (1968) 1260.
[3] Boichenko E.A., T.M. Udelnova, S.G. Yuferova, Geokhimiya 11 (1969) 1392.
[4] Bowen H., Trace Elements in Biochemistry (Academic Press, London, New York, 1966).
[5] Kassner R. and M. Kamen, Biochim. Biophys. Acta 153 (1968) 270.
[6] Kessler E., in: Handbuch der Pflanzenphysiologie, ed. W. Ruhlands, Vol. 5 (Springer, Berlin, 1960) p. 951.
[7] Oparin A.I., The Origin of Life on the Earth (USSR Academy of Sciences, Moscow, 1957).
[8] Perelman A.I., Geochemistry of Landscape (Vysshaya Shkola, Moscow, 1966).
[9] Ronov A.B., Geokhimiya 8 (1964) 715.
[10] Singer T., ed., Biological Oxidations (Interscience, New York, London, 1968).
[11] Udelnova T.M., E.N. Kondratyeva and E.A. Boichenko., Mikrobiologia 37 (1968) 197.
[12] Vallee B. and J. Coleman, in: Comprehensive Biochemistry, eds. M. Florkin and E. Stotz, Vol. 12 (Elsevier, Amsterdam, 1964) p. 165.
[13] Vernadsky V.I., Chemical structure of the Earth's Biosphere and of its Environment (Nauka, Moscow, 1965).
[14] Vinogradov A.P., Introduction into the Geochemistry of the Ocean (Nauka, Moscow, 1967).
[15] Zarin V.E. and E.A. Boichenko, Physiol. Rastenii 16 (1969) 408.

Chemical Evolution and the Origin of Life, eds. R. Buvet and C. Ponnamperuma

INORGANIC POLYPHOSPHATES IN EVOLUTION OF PHOSPHORUS METABOLISM

I. S. KULAEV
Moscow State University, Moscow, USSR

Generally speaking, the role of inorganic polyphosphates in the evolution of life on earth can be considered from two viewpoints.

On the one hand, we have at our disposal some facts allowing us to suggest the possibility of participation of inorganic polyphosphates in abiogenic synthesis of various compounds which in the course of evolution have become the components of the living cells.

These facts, first reported in the well-known papers of Schramm [14–16], more recently by Fox [4] and Ponnamperuma [11–13] and by other authors, show that in the prebiological period inorganic polyphosphates existed on the earth which could be donors of activated phosphate in the abiogenic synthesis of peptides, polysaccharides, nucleic acids and ATP. These investigations have been discussed by Ponnamperuma and Schwartz, therefore I shall not dwell on these facts.

On the other hand, the evidence furnished by comparative biochemistry testifies to the possibility of participation of inorganic polyphosphates in the evolution of the phosphorus metabolism in living organisms. We have also succeeded in obtaining some experimental facts allowing us to outline some features of the evolution of the phosphorus metabolism and to determine the possible role of high polymer polyphosphates.

Oparin [9,10] and many other investigators believe that it was anaerobic fermentation of hexoses to lactic acid and ethyl alcohol that was the energy-supplying process in various organisms. As is well known, the main energy-producing reaction of the fermentation is that of oxidation of 3-phosphoglyceraldehyde to 3-phosphoglyceric acid. In the contemporary organisms the energy resulting from this reaction is partially accumulated in ATP whose formation is just the mechanism for coupling the majority of exergonic and endergonic reactions. As to the protobionts, we know almost nothing about their possible coupling mechanisms.

We are inclined to believe that from our experimental evidence it is possible to conclude that high polymeric polyphosphates (their structure is shown in fig. 1) could have played an important role in the coupling of exergonic and endergonic reactions in the primary living organisms.

We have established [8] that some contemporary organisms, for example, bacteria and fungi, possess an enzyme catalyzing the transport of phosphate activated via glycolytic phosphorylation from 1,3-diphosphoglyceric acid not to ADP to form ATP, as is shown in fig. 2, but directly to the high polymer polyphosphate, its chain acquiring one more phosphate residue.

In the cells of some microorganisms there is another well known enzyme, polyphosphate kinase, which catalyzes the synthesis of polyphosphate at the expense of ATP. We have demonstrated that in the most primitive of the microorganisms we have studied – micrococci and propionic bacteria [5] – the activity of polyphosphate-synthesizing system functioning without ATP is comparable to that of polyphosphate kinase. *E. coli*, a phylogenetically younger organism, possesses a polyphosphate kinase 10–20 more active than that of the enzyme responsible for the biosynthesis of polyphosphates via glycolytic phosphorylation. So, we may suggest that in the primary organisms glycolytic phosphorylation reactions yielded not ATP, but more primitive inorganic polyphosphates.

That gives grounds to believe that in protobionts formation of polyphos-

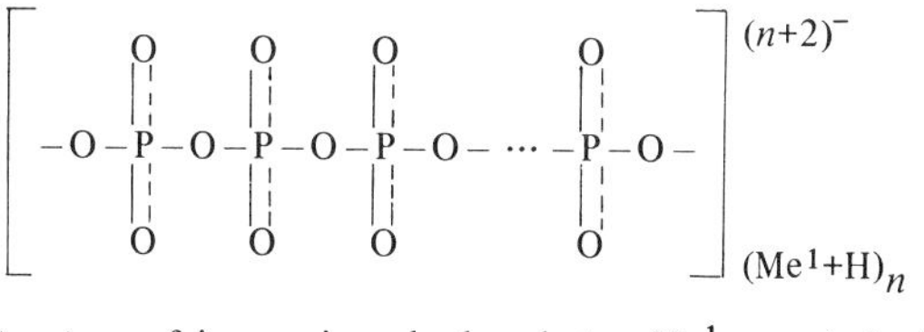

$$\left[\begin{array}{c} \quad O \quad\; O \quad\; O \qquad\quad O \\ \quad \| \quad\;\; \| \quad\;\; \| \qquad\quad\; \| \\ -O-P-O-P-O-P-O-\cdots-P-O- \\ \quad \| \quad\;\; \| \quad\;\; \| \qquad\quad\; \| \\ \quad O \quad\; O \quad\; O \qquad\quad O \end{array}\right]^{(n+2)^-}_{(Me^1+H)_n}$$

Fig. 1. The structure of inorganic polyphosphates. Me^1 = univalent cations; n = the number of condensed molecules of orthophosphate.

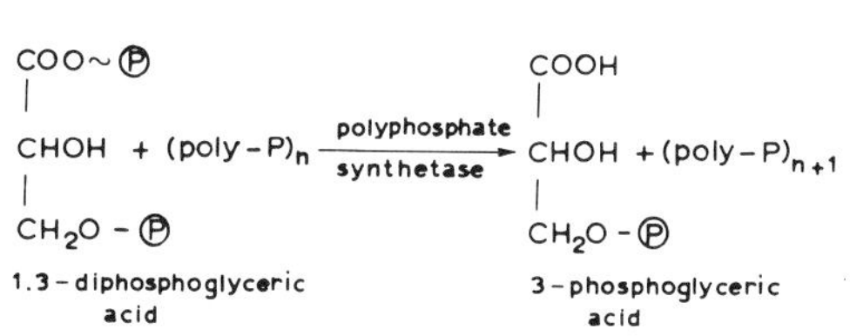

$$\begin{array}{c} COO{\sim}Ⓟ \\ | \\ CHOH \\ | \\ CH_2O-Ⓟ \end{array} + (poly\text{-}P)_n \xrightarrow[\text{synthetase}]{\text{polyphosphate}} \begin{array}{c} COOH \\ | \\ CHOH \\ | \\ CH_2O-Ⓟ \end{array} + (poly\text{-}P)_{n+1}$$

1.3-diphosphoglyceric acid → 3-phosphoglyceric acid

Fig. 2. A possible scheme of glycolysis-dependent reaction involving the participation of polyphosphate synthetase isolated from *N. crassa*.

phates must have been either the only or one of the pathways supplying energy for biological oxidation of sugars in the course of anaerobic fermentation.

Our results also suggest that in the primary organisms high polymer polyphosphates could have some functions which are known to be performed by ATP. Polyphosphates in protobionts could have been donors of energy and phosphate in the reactions in which ATP plays that role in the contemporary organisms. This suggestion is supported by the following facts: some microorganisms have been shown to possess polyphosphate glucokinase, the enzyme responsible for the transport of phosphate from polyphosphate to glucose to form glucose-6-phosphate [18–20] without participation of ATP (fig. 3).

This enzyme acts as the usual hexokinase requiring ATP to phosphorylate glucose. We have compared the activity of polyphosphate kinase and the usual ATP kinase in 38 various species of microorganisms [20,21]. Similar analysis was done by other authors for 11 other species of microorganisms [2,19]. All the 49 species studied have been found to have ATP hexokinase.

But, polyphosphate hexokinase has been found only in phylogenetically related microorganisms which in the taxonomy of Krasilnikov [6,7] are united into the special class of actinomycetes.

In other microorganisms which have been studied in this respect – true bacteria, fungi, green and blue–green algae – this enzyme has not been found.

Fig. 4 shows all the species in which we and other authors have attempted to find polyphosphate glucokinase and their position in the Krasilnikov's taxonomy of microorganisms.

And the presence of polyphosphate hexokinase only in the organisms belonging to the class of actinomycetes confirms Krasilnikov's opinion about the phylogenetic unity of this group of microorganisms.

But, the most important fact for us is that of the presence of this enzyme in the evolutionary ancient group of the microorganisms [5,17]. In micrococci, tetracocci, propionic bacteria (table 1), the polyphosphate hexokinase

CH_2OH ... + $(poly-P)_n$ $\xrightarrow[\text{hexokinase}]{\text{polyphosphate}}$ CH_2O-Ⓟ ... + $(poly-P)_{n-1}$

D-GLUCOSE D-GLUCOSE-6-PHOSPHATE

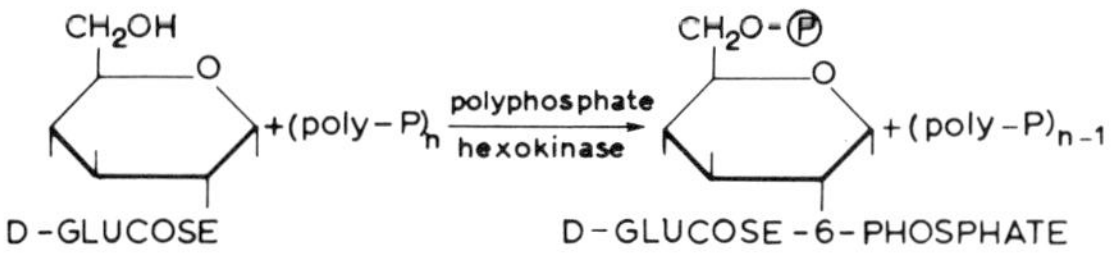

Fig. 3. A scheme of reaction involving polyphosphate hexokinase, isolated from mycobacteria and other organisms.

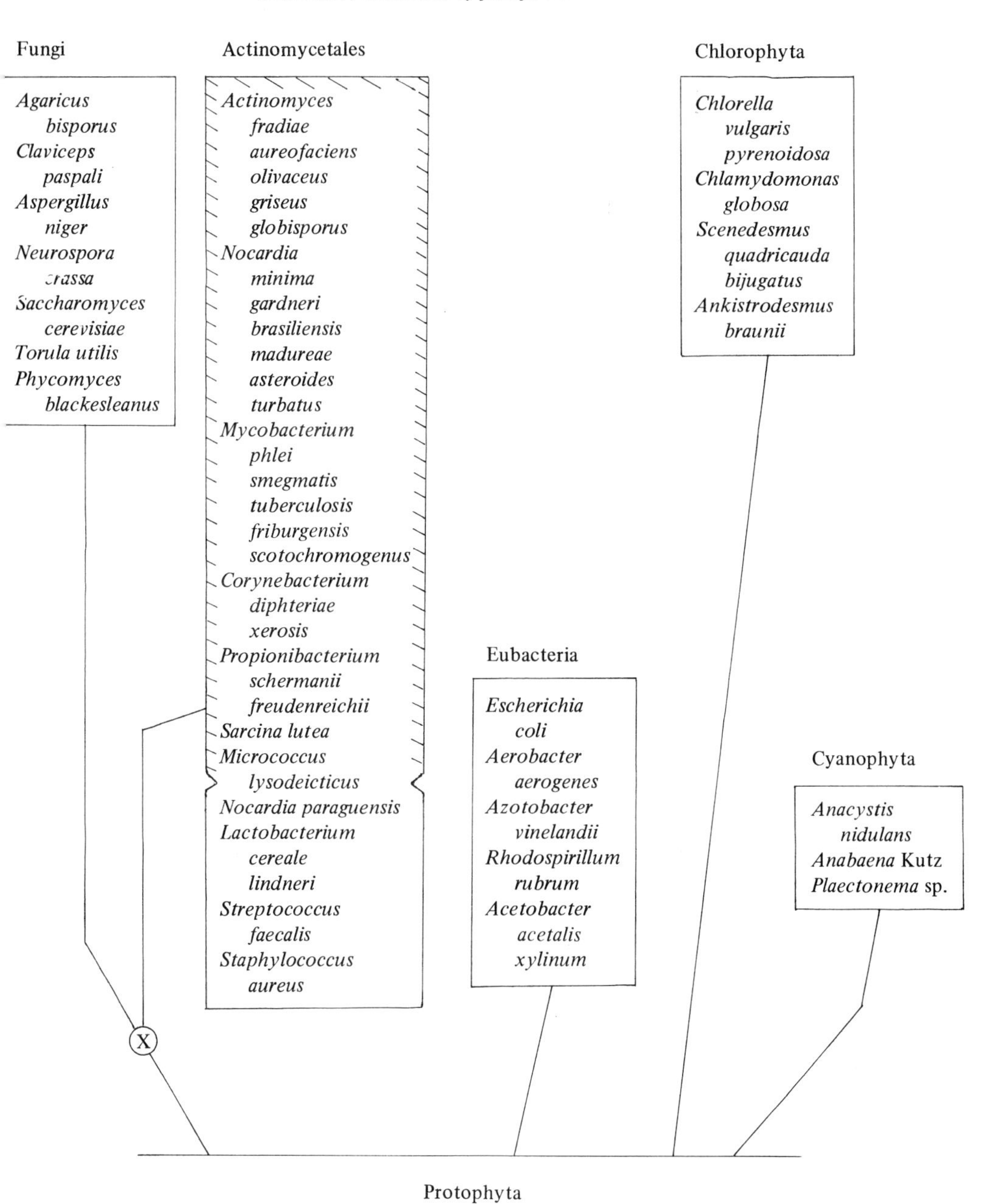

Fig. 4. Distribution of polyphosphate glucokinase in different microorganisms in accordance with the Krasilnikov's scheme of origin of microorganisms [6]. The species possessing this enzyme are shaded.

Table 1

Ratio of polyphosphate glucokinase and ATP-glucokinase specific activities in various micro-organisms (μmoles P/hr/mg protein).

Organism	Poly-P-glucokinase (A)	ATP-glucokinase (B)	A/B
Micrococcus lysodeicticus	2.0	0.4	5
Sarcina lutea	2.0	0.5	4
Propionibacterium shermanii	6.0	0.6	10
Mycobacterium tuberculosis	5.2	0.6	9
M. scotochromogenus	14.2	7.1	2
M. phlei	10.1	7.8	1.3
M. smegmatis	16.0	14.0	1.2
M. friburgensis	10.6	9.3	1.1
Corynebacterium xerosis	7.5	3.7	2
Nocardia turbatus	3.0	0.9	3.3
N.gardneri	4.0	3.4	1.2
N.brasiliensis	5.0	6.6	0.8
N.madureae	9.0	1.3	7
N.minima	10.3	9.5	1.1
N.asteroides	4.7	11.6	0.4
Actinomyces olivaceus	3.6	7.6	0.5
A.globisporus	4.8	7.3	0.6
A.fradiae	1.8	16.0	0.1
A.aureofaciens	2.3	15.1	0.1

activity exceeds several times that of ATP hexokinase. In mycobacteria [19] and proactinomycetes [20] these activities are approximately equal but in actinomycetes [20] the activity of ATP hexokinase is much higher than that of polyphosphate hexokinase.

However, Kluyver and Van Niel [5] think that in this group of micro-organisms miccococci and then tetracocci and propionic bacteria are most primitive and phylogenetically old, the true actinomycetes being the youngest phylogenetically.

If this is the case, then the results of the comparative analysis of the polyphosphate glucokinase and ATP glucokinase activities may be interpreted to mean that the utilization of high polymer polyphosphates as donors of activated phosphate for phosphorylation of glucose is evolutionally more ancient that the functioning of ATP for similar purposes.

Thus, the experimental evidence reported above confirms the idea of Belozersky [1] who in 1957 at the First International Symposium on the Origin of Life put forward the suggestion that in primary living organisms high polymer inorganic polyphosphates could play the role ATP has in the contemporary organisms.

It is possible to suggest that in the protobionts high polymer polyphosphates could function as coupling substances, for example, in the metabolic pathways presented in fig. 5.

Such or similar pathways could have existed in primary living organisms with the polyphosphates being the coupling agents between exergonic reactions of glycolysis and the endergonic reactions of sugar phosphorylation.

At the beginning of life this "duty" of polyphosphates could have been their main function but not the only one.

I would like to note that, in my opinion, polyfunctionality was one of the most important criteria for a compound to be selected in the course of the evolutionary process as a component of the living cells. For example, it is quite possible that in protobionts polyphosphates functioned not only as coupling agents but also as a relatively long-living pool of phosphorus and energy providing the organisms "privacy" in the environment. They could also detoxicate orthophosphoric acid. Being excellent ion-exchangers, they might also control cation exchange in primary organisms.

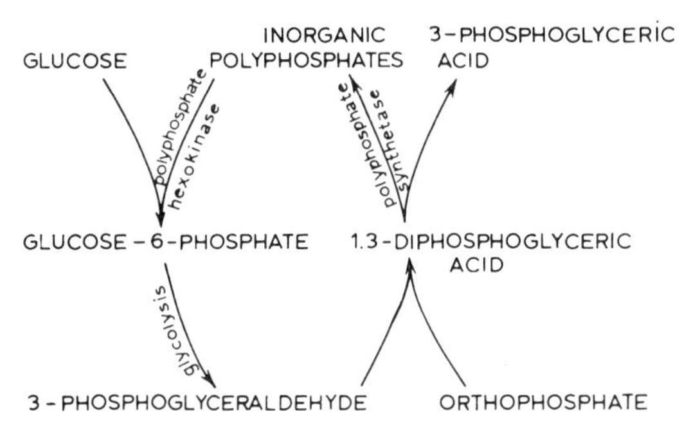

Fig. 5. The role of high-molecular polyphosphates in coupling of exergonic and endergonic reactions.

But at some stage of evolution, high polymer polyphosphates might have ceased to meet the requirements of the organism. Compounds of higher specificity were required with still more polyfunctional structure, capable of finer and more specific interactions with other components of the cells.

It seems quite reasonable that the "contactability" with other cell components is one of the most essential factors in the metabolism evolution in living organisms. It is possible that polyphosphates with their high molecular, rather monotonous and non-specific structure, could have ceased to be "convenient" for the cell when its structure and metabolism might have acquired added complexity.

The main limitation of the polyphosphates, i.e. that of their inability to make fine and specific contacts with other cellular metabolites, became incompatible with their most important function of coupling exergonic and endergonic reactions.

This function, one of the central in the entire cell metabolism, was attributed to ATP – a compound of much more specific and hence more readily recognizable structure. Besides, ATP is capable of performing other functions inaccessible to high polymer polyphosphates.

Now, quite naturally, a question arises why the high polymer polyphosphates have not been completely substituted by ATP in the course of evolution. Why do they still exist, sometimes in quite appreciable amounts, in many contemporary organisms, especially lower organisms.

In my opinion, this is first of all because the existing microorganisms in many ways depend on their environment and hence they are in great need of mechanisms of conservation of high amounts of activated phosphate. Accumulation of high polymer phosphate is just one of these mechanisms.

Cells of higher multicellular organisms do not depend so much on the environment and they could have lost in the course of evolution their ability of accumulating large amounts of inorganic polyphosphates.

But, a somewhat atavistic habit of possessing low amounts of inorganic polyphosphates, is still displayed by some higher plants and animals. In higher organisms polyphosphates serve as a source of activated phosphate for some rather specific purposes.

It is possible that in higher organisms they take part, first and foremost, in nucleic acid metabolism. This is indicated, for example, by the fact that in the cells of high animals polyphosphates are found only in cellular nuclei [3].

Thus, the experimental data obtained so far, including our results, allow us to outline some features of the phosphorus metabolism and to determine the possible role of polyphosphates in this process.

References

[1] Belozersky A.N., in: Origin of Life on Earth, eds. A.I. Oparin et al. (Pergamon Press, London, 1959) p. 322.
[2] Dirheimer G. and J.P. Ebel, Compt. Rend. Acad. Sci. 254 (1962) 2850.
[3] Griffin J.B., N.M. Davidson and R. Pennial, J. Biol. Chem. 240 (1965) 4427.
[4] Harada K. and S.W. Fox, in: Origin of Prebiological Systems, ed. S.W. Fox (Academic Press, New York, 1965).
[5] Kluyver A.J. and C.B. Van Niel, Zentr. Bakteriol. Parasitenk. Abt. II 94 (1936) 369.
[6] Krasilnikov N.A., Manual of Determinative Bacteria and Actinomycetes (Moscow, 1949).
[7] Krasilnikov N.A., in: Manual for Microbiology, Clinics and Epidemiology of Infectous Diseases (Medizina I (1962).
[8] Kulaev I.S., O. Szymona and M.A. Bobyk, Biokhimiya 33 (1968) 419.
[9] Oparin A.I., in: Origin of Life on Earth Proceedings, eds. A.I. Oparin et al. (Pergamon Press, New York, 1959).
[10] Oparin A.I., Advan. Enzymol. 27 (1965) 347.
[11] Ponnamperuma C., C. Sagan and R. Mariner, Nature 199 (1963) 222.
[12] Ponnamperuma C., in: Origin of Prebiological Systems, ed. S.W. Fox (Academic Press, New York, 1965).
[13] Rabinowitz J., S. Chang and C. Ponnamperuma, Nature 218 (1968) 442.
[14] Schramm G., in: Origin of Life on Earth Proceedings, eds. A.I. Oparin et al. (Pergamon Press, New York, 1959).
[15] Schramm G. and W. Pollmann, Angew. Chem. Intern. Ed. 2 (1962) 53.
[16] Schramm G., in: Origin of Prebiological Systems, ed. S.W. Fox (Academic Press, New York, 1965).
[17] Stanier R. and C.B. Van Niel, Arch. Microbiol. 42 (1962) 17.
[18] Szymona M., Bull. Acad. Polon. Sci. II 5 (1957) 379.
[19] Szymona M. and O. Szymona, Kulesza Acta Microbiol. Polon. 11 (1962) 287.
[20] Szymona O., S.O. Uryson and I.S. Kulaev, Biochimiya 32 (1967) 495.
[21] Uryson S.O. and I.S. Kulaev, Dokl. Akad. Nauk SSSR 183 (1968) 957.

Chemical Evolution and the Origin of Life, eds. R. Buvet and C. Ponnamperuma

INORGANIC PYROPHOSPHATE AND THE ORIGIN AND EVOLUTION OF BIOLOGICAL ENERGY TRANSFORMATION

Herrick BALTSCHEFFSKY
Bioenergetics Group, Department of Plant Physiology
and
Department of Biochemistry, University of Stockholm
Stockholm, Sweden

"The organic nature is divided into two different classes of living beings: the plants and the animals. These go from the one to the other, the plants into animals and the animals into plants, so gradually and so unobserved, that in the switch there exists beings which the scientist with equally good reason can judge to be either. One can imagine these classes of living bodies as two chains, which nature has linked together into one; but not by linking the most developed part of the one to the least perfected of the other, but by linking together the least developed parts of the plant and animal kingdoms, so that the most peculiarly organized links of both become the ends of the chain. At the junction of these chains one will find, if not exactly the transformation of inorganic nature to organic, at least the site where it is closest to this transformation. It is probable, although not quite in agreement with our theories accepted so far, that nature in the region around this point, without seed or egg, through certain colliding reasons is transformed from inorganic to organic and produces the least peculiarly built links of these chains."

J.J. Berzelius, *Föreläsninger i Djurkemien*
(Lectures on Animal Chemistry), 1806

In this presentation I would like to consider two questions concerned with prebiological and early biological energy transformation. The first question may be formulated as follows: what mechanisms of energy transformation existed at the time when chemical (prebiological) systems evolved into biochemical (biological) systems? It is closely related to the second question: what was the nature of the energy-rich compounds involved in mobilization of free energy for reproduction and other energy-requiring reactions of the earliest living systems?

Attempts to elucidate these questions may be said to include at least two basically different kinds of approach, or sources of information. One of these sources is that of the energy transformation reactions occurring in existing living organisms, especially the more primitive ones. The other is that of the more or less closely corresponding model reactions that appear possible in a chemical environment similar to the one that may have existed at the time when life originated. The area where these two approaches converge seems to be at present in a stage of rapid expansion. Thus it might be appropriate to look for common denominators which may be found within this area and which may be of particular significance in connection with the problem of the origin and evolution of biological energy transformation. I would like to focus attention upon a substance that may well be such a common denominator, namely inorganic pyrophosphate.

Inorganic pyrophosphate (PPi)

Inorganic pyrophosphate (PPi), which can be obtained for example by heating orthophosphate (Pi) to 200–300 °C, may be regarded as the simplest compound containing the pyrophosphate or P-O-P linkage that occupies such a central place in biological energy conversion, in particular in the reactions of ATP. Since the classical papers of Lipmann [23,24] in the 1940s the predominant role of ATP as primary carrier of energy between energy generating and energy requiring reactions of the living cell has been firmly established. The versatility of this compound, which may function as a "squiggling", a phosphorylating, a pyrophosphorylating, and an adenylating agent, certainly constitutes part of the explanation for its rather unique position in the biological energy conversion systems of today.

In our experimental investigations we have been concerned solely with the biochemical approach. The biological material that we have studied, in the present context, is mainly chromatophores isolated from the purple, non-sulfur photosynthetic bacterium *Rhodospirillum rubrum*. The photosynthetic apparatus of the organism is located in these chromatophores.

Our active interest in the origin and evolution of biological energy transformation came from our discovery that a light-induced formation of PPi from Pi can be shown to occur in the chromatophores [2], and from the ensuing development. Two additional findings appeared significant from the evolutionary point of view. We demonstrated, with oligomycin, that light-induced formation of PPi did not involve the adenosine phosphates [3]. Furthermore M. Baltscheffsky showed that PPi could function as a donor of biologically useful chemical energy in chromatophores even better than ATP and also in a reaction pathway not involving the adenosine phosphates [7,9]. Thus this first alternative biological system for coupling between electron transport and formation and utilization of energy-rich phosphate compounds proved to be of great potential interest in connection with energetic aspects of the origin of life.

Prebiological energy transformation

I would like to make a few comments on what appears to have been assumed and achieved within the area of prebiological existence, formation and utilization of energy-rich phosphate compounds. The comments will, no doubt, reflect the lack of first hand knowledge and understanding of a relative newcomer to the field of chemical evolution.

Jones and Lipmann [15], Lipmann [25] and Miller and Parris [25] have suggested, in the 1960s, that PPi or possibly high-molecular weight inorganic polyphosphates may be the most likely early energy-rich phosphate compounds, rather than ATP, carbamyl phosphate, and other potential candidates. Miller and Parris [25] obtained formation of PPi under conditions similar to those that might have existed on the primitive earth: a water suspension containing Pi, $CaCl_2$, KCNO and solid, finely divided apatite. Miller and Parris [25] and Lipmann [26] appeared to favor quite strongly PPi as the first biological energy-rich phosphate compound. In a rather recent paper, Ferris [12] reported that PPi was formed in aqueous solutions from cyanovinylphosphate, this compound being obtained quite easily from cyanoacetylene and orthophosphate. On the other hand, although ATP has been synthesized abiologically, its formation under completely plausible prebiological conditions has not been clearly established [27].

When we consider very generally the inorganic pyro- and polyphosphates versus the adenosine phosphates as more probable early energy carriers, two rather striking points emerge. Both appear to favor strongly the inorganic compounds. The first point is the great simplicity of the structure of PPi, as well as its higher analogs, compared to that of ATP. The second is that the abundance of inorganic orthophosphate would have been far greater than that of the adenosine phosphates.

Before turning to a presentation of the various reactions of PPi occurring in bacterial chromatophores and to a consideration on the relevance of these reactions to the problem of the origin and evolution of biological energy transformation, I would like to refer to two papers which have been of greatest interest to us in this connection. The first of these was presented at the First International Symposium on the Origin of Life on the Earth, in Moscow in 1957, and the second was presented at what is now regarded as the second meeting, in Coral Cables, Florida, in 1963. In the first, Calvin [11] envisaged a model for primitive energy conversion, assuming prebiological coupling between electron transport involving the redox system Fe^{2+}–Fe^{3+} and phosphorylation of Pi to PPi. In the second, Lipmann [25] after pointing out the probable importance of PPi as energy carrier in early biological energy transformation suggested that there may well exist what he termed "metabolic fossils" only remaining to be discovered, which might tell something about – in this case – early biological energy transformation.

Reactions of inorganic pyrophosphate in chromatophores

Our results and those of others with the reactions of PPi in chromatophores from *Rhodospirillum rubrum* have led us to assume that we have been fortunate enough to find such "metabolic fossils". In this context I would like to recall also Professor Kulaev's most stimulating presentation at this conference which brought forward his experimental evidence for the participation of inorganic polyphosphates of higher molecular weight in early biological energy transformation [21,22]. There may obviously exist still unknown close metabolic links between inorganic pyrophosphate and inorganic polyphosphate metabolism.

It appears useful to give a concentrated overall view of the eight reactions involving PPi that have thus far been discovered in chromatophores from *R. rubrum*: (1) a Mg^{2+}-stimulated inorganic pyrophosphatase (pyrophosphate phosphohydrolase, PPase) reaction [6]; (2) the above-mentioned light-induced formation of PPi [2,3]; (3) a PPi-induced reversed electron transport at the cytochrome level [7,9]; (4) a PPi-induced energy-linked transhydrogenase reaction [17,18]; (5) a PPi-induced carotenoid absorption band shift [8,10]; (6) a PPi-induced succinate linked pyridine nucleotide reduction [18]; (7) a K^{+}($+Mg^{2+}$)-stimulated PPase reaction [5] and (8) a PPi-induced energization of the chromatophore membrane as measured by the fluorescent probe techniques [1]. PPi functions as a donor of biologically useful chemical energy in five of the eight reactions (reactions 3–6 and 8).

The following scheme has earlier been shown by us to account for the metabolic connection between the reactions of PPi and other reactions of the photosynthetic energy transformation system in chromatophores [4] (minimum scheme; A and B are electron carriers and X~1 is a hypothetical energy-rich chemical intermediate):

$$
\begin{array}{ccccc}
\overset{e^-}{\rightleftharpoons} & A & \overset{e^-}{\rightleftharpoons} & B & \overset{e^-}{\rightleftharpoons} \\
 & & \upharpoonleft\downharpoonright & & \\
\text{ion movement} & \leftrightarrows & X{\sim}I & \rightleftharpoons & \text{PPi} \\
 & & \upharpoonleft\!\!\!\!+\!\!\downharpoonright & \text{oligomycin} & \\
 & & \text{ATP} & &
\end{array}
$$

With reference to what has been discussed above about prebiological formation of PPi, it is pointed out, as has been done earlier [4], that the pathway between electron transport and PPi may well have evolved first in either of the two possible directions with subsequent evolution of the reactions involved in ion movement and adenosine phosphate metabolism. Implicit in this concept of the origin and evolution of biological electron transport coupled energy transformation is a continuous, stepwise but essentially parallel evolution of electron transport and coupled energy transformation reactions in close connection.

Discussion

PPi accordingly has emerged as a very probable link which may bridge the earlier existing gap between prebiological and biological energy transformation. It seems quite possible, on the other hand, that the reactions of energy-requiring ion movement and, in particular, the energy transformation reactions involving the adenosine phosphates have evolved entirely within the biological phase of evolution.

Analysis of the relative capacities of PPi and ATP to serve as energy donors in the chromatophores shows that the two reactions, in which the initial rate of the energy-requiring reaction turns out to be as much as about ten times higher with PPi than with ATP, are the very ones that are most closely linked with the photosynthetic apparatus, namely the reversed electron transport at the cytochrome level [7,9] and the carotenoid band shift [8,10]. Indeed, it clearly emerges from table 1 that PPi plays a predominant role in the comparatively ancient photosynthetic energy transformation system in question and, on the other hand, that ATP is an about equally good or better

Table 1
Ratio for initial rates of PPi and ATP induced energy transfer reactions in chromatophores isolated from *Rhodospirillum rubrum*, measured as PPi induced rate/ATP induced rate.

Dark reaction	PPi induced rate / ATP induced rate
Oxidation of cytochrome c_2 [9]	~10
Reduction of *b*-type cytochrome [9]	~10
Carotenoid band shift [10]	~10
Energy-linked pyridine nucleotide transhydrogenase reaction [17,18]	0.8–1
Succinate-linked pyridine nucleotide reduction [19]	0.4

energy donor than PPi in the reactions more closely linked with respiratory electron transport which must be considered as more recent from the evolutionary point of view. In this connection it should be pointed out that reversed electron transport at the cytochrome level in the respiratory system of a higher organism such as yeast is driven much better by ATP than by PPi and that the effect of PPi in rat liver mitochondria is quite insignificant [7].

Finally, with respect to the questions concerned with the actual evolutionary sequences of prebiological and biological energy transformation systems, it appears that the early assumption made by Oparin [28], that anaerobic fermentation mechanisms were first on the time scale, has been rather generally accepted [20,29]. In the light of our data on the formation and utilization of PPi in chromatophores, we would like to reopen the question whether substrate-linked or electron transport-linked energy transformation came first [13,14,30] with the following tentative scheme for the origin and evolution of biological energy transformation (scheme 1). Two main assumptions are evident in the scheme as presented. The first is that *inorganic phosphates preceded adenosine phosphates* as energy carriers in both substrate level and electron transport level phosphorylation systems. The second is that *convergence* may have occurred also at the chemical stage in the evolution of energy transformation. This would diminish the importance of the question whether substrate-linked or electron transport-linked energy transformation came first. Such evolutionary convergence as depicted in the scheme could, in fact, by producing an energy conversion unit more adaptable and suitable for further development than the sum of its converging parts, have been even a necessary prerequisite for the Origin of Life.

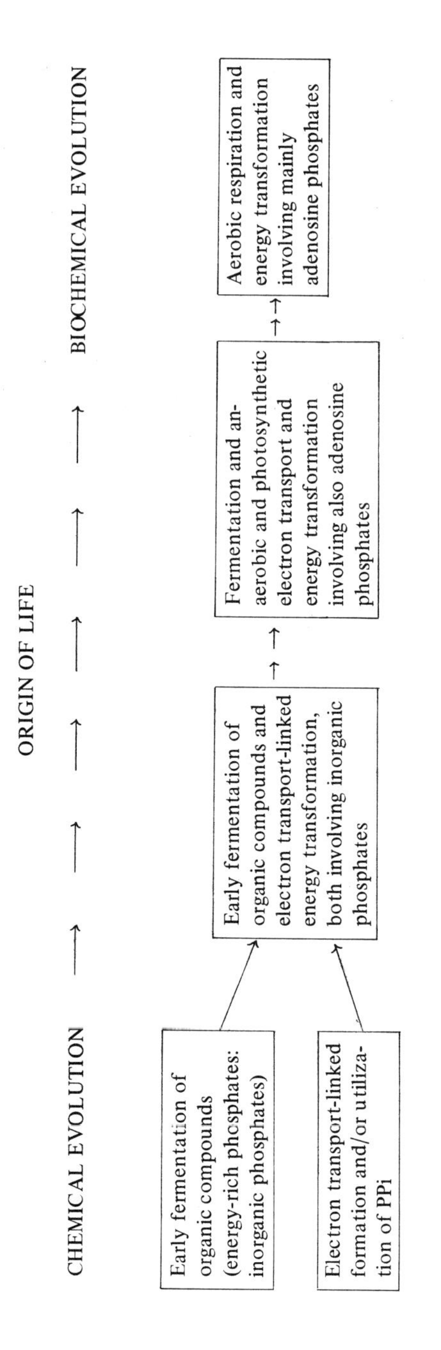

Scheme 1. Possible origin and evolution of biological energy transformation.

Acknowledgements

The light induced formation of PPi was investigated in collaboration with Dr. L.–V. von Stedingk, and, in various phases of the work, with Drs. T. Horio, Osaka, H.–W. Heldt, Munich, and M. Klingenberg, Munich. Dr. M. Baltscheffsky discovered the Mg^{2+} stimulated PPase reaction in chromatophores, the PPi induced reversed electron transport at the cytochrome level, and the carotenoid band shift. Energization of the chromatophore membrane with PPi as measured with fluorescent probes was found in collaboration with Drs. A. Azzi, Modena, and M. Baltscheffsky.

The experimental research was supported by the Swedish Natural Science Research Council, The Charles F. Kettering Foundation, Dayton, Ohio and Sigrid Juselius Stiftelse, Helsinki, Finland.

References

[1] Azzi A., M. Baltscheffsky and H. Baltscheffsky, to be published.
[2] Baltscheffsky H., L.-V. von Stedingk, H.-W. Heldt and M. Klingenberg, Science 153 (1966) 1120.
[3] Baltscheffsky H. and L.-V. von Stedingk, Biochem. Biophys. Res. Commun. 22 (1966) 722.
[4] Baltscheffsky H., Acta Chem. Scand. 21 (1967) 1973.
[5] Baltscheffsky H., M. Baltscheffsky and L.-V. von Stedingk, in: Progress in Photosynthesis Research, ed. H. Metzner, Vol. III (1969) p. 1313.
[6] Baltscheffsky M., Abstract 1st Meeting of FEBS (Academic Press, London, 1964) p. 67.
[7] Baltscheffsky M., Biochem. Biophys. Res. Commun. 28 (1967) 270.
[8] Baltscheffsky M., Nature 216 (1967) 241.
[9] Baltscheffsky M., Arch. Biochem. Biophys. 133 (1969) 46.
[10] Baltscheffsky M., Arch. Biochem. Biophys. 130 (1969) 646.
[11] Calvin M., in: Origin Life Earth Rept. Intern. Symp. Moscow (1957) p. 338.
[12] Ferris J.P., Science 161 (1968) 53.
[13] Gaffron H., in: The Origin of Prebiological Systems, ed. S.W. Fox (Academic Press, New York, London, 1965) p. 437.
[14] Granick S., Ann. N.Y. Acad. Sci. U.S. 52 (1964) 595.
[15] Jones M.E. and F. Lipmann, Proc. Natl. Acad. Sci. U.S. 46 (1960) 1194.
[16] Kalckar H.M., Chem. Rev. 28 (1941) 71.
[17] Keister D.L. and N.J. Yike, Biochem. Biophys. Res. Commun. 24 (1966) 519.
[18] Keister D.L. and N.J. Yike, Biochemistry 6 (1967) 3847.
[19] Keister D.L. and N.J. Yike, Arch. Biochem. Biophys. 121 (1967) 415.
[20] Krebs H. and H. Kornberg, Energy Transformations in Living Matter (Springer, Berlin, 1957).
[21] Kulaev I.S., in: Abiogenez Nachal'nye Stadii Evol. Zhizni, ed. A.I. Oparin, (Moscow, 1968) p. 97.

[22] Kulaev I.S., this volume, p. 458.
[23] Lipmann F., Advan. Enzymol. 1 (1941) 99.
[24] Lipmann F., in: Currents in Biochemical Research, ed. D.E. Green (Interscience, New York, 1946) p. 137.
[25] Lipmann F., in: The Origins of Prebiological Systems, ed. S.W. Fox (Academic Press, New York, London, 1965) p. 259.
[26] Miller S.L. and M. Parris, Nature 204 (1964) 1248.
[27] Miller S.L., L.E. Orgel and C. Ponnamperuma, Personal communications.
[28] Oparin A.I., The Origin of Life (transl. by S. Margulis) (Macmillan, New York, 1938).
[29] Wald G., Proc. Natl. Acad. Sci. U.S. 52 (1964) 595.
[30] Williams R.J.P., in: Current Topics in Bioenergetics, ed. D.R. Sanadi (Academic Press, New York, London, 1969) p. 79.

Chemical Evolution and the Origin of Life, eds. R. Buvet and C. Ponnamperuma

SOME INFORMATION ON THE POSSIBILITY OF PREGLYCOLYTIC WAYS IN EVOLUTION

E. PANTSKHAVA
A.N. Bakh Institute of Biochemistry,
Academy of Sciences of the USSR, Moscow, USSR

According to modern theory, glycolysis is the most ancient way of energy metabolism. Since it was widely spread among obligate anaerobes up to recent times an opinion existed that such a way of accumulating energy was the only possible and original one. However, it is difficult to agree to a number of facts regarding peculiarities of energy metabolism in some anaerobic bacteria with such an interpretation [7,8]. The first information on biochemical peculiarities of these bacteria was given by Barker [1].

Fermentation of an ethyl alcohol–acetic acid system by *Clostridium kluyveri*

Cl. kluyveri ferment neither carbon-hydrates nor amino acids and fermentation of an ethyl alcohol–acetic acid system leading to synthesis of fatty acids (butyric and capronic) is the principle mechanism supplying the microorganism with the required energy.

Oxidation of ethyl alcohol via acetyl coenzyme leads to the formation of potentially usable energy-rich thioether bonds contributing to formation of ATP [11].

Attempting to solve the problem of generation of energy in this organism, Barker supposed that energy can be released during transition of electrons in oxidation and oxido-reduction processes between ethyl alcohol and acetaldehyde on one hand and crotonyl-CoA on the other. Having discovered that oxido-reduction of crotonyl-CoA by hydrogen or $NADH_2$ is accompanied by formation of ATP, Shuster and Gunsalus proved that, during such fermentation, oxidizing phosphorylation really takes place [9].

It was also demonstrated that *Cl. kluyveri* has enzyme systems oxidizing acetaldehyde with formation of acetate and ATP [5]:

$$\text{acetaldehyde} + \text{NAD} + \text{CoA} \rightarrow \text{acetyl-CoA} + \text{NADH}_2$$

$$\text{acetyl-CoA} + \text{P}_{\text{in}} \rightarrow \text{acetylphosphate} + \text{CoA}$$

$$\text{acetylphosphate} + \text{P}_{\text{in}} + \text{ADP} \rightarrow \text{acetate} + \text{ATP}$$

Thus *Cl. kluyveri* has two points in its metabolic structure where creation of ATP is possible: 1) anaerobic oxidation of ethanol into acetate, and 2) oxido-reduction of crotonyl-CoA.

Fermentation of ethyl alcohol by *Methanobacillus omelianskii* (strain s) forming acetate and hydrogen

The above strain was isolated from the complex causing methane fermentation of ethanol and CO_2 with formation of acetate and hydrogen [4]. It was experimentally ascertained that anaerobic oxidation of ethanol into acetate passes through formation of acetaldehyde with an obligatory participation of ferredoxin and NAD [3]. Ferredoxin is required for oxidation of acetaldehyde into acetate. This strain does not display any of the well-known ways of oxidation of acetaldehyde through acetyl-CoA and acetylphosphate into acetyl.

It was believed that the energy created during anaerobic oxidation of $NADH_2$ was in the form of ATP.

We suppose that the strain produces the required energy during the oxidation of ethanol into acetate with participation of ferredoxin and NAD by the following reactions:

$$\text{ethanol} + \text{X} \rightarrow \text{acetaldehyde} + \text{H}_2\text{X}$$

$$\text{acetaldehyde} + \text{Fd}_{\text{ox}} \rightarrow \text{Fd}_{\text{red}} + \text{acetate}$$

$$\text{Fd}_{\text{red}} + \text{NAD} \rightarrow \text{Fd}_{\text{ox}} + \text{NADH}_2$$

$$\text{NADH}_2 + \text{ADP} + \text{P}_{\text{in}} \rightarrow \text{ATP} + \text{NAD} + \text{AH}_2$$

It is possible that the mentioned organism demonstrates a new variant of anaerobic production of energy without glycolysis and thioethers.

Fermentation of lactic acid, pyruvic acid and alanine by *Clostridium propionicum*

Cl. propionicum is another representative of anaerobic bacteria which does not possess glycolysis and produce the necessary energy in anaerobic oxidation of the above complexes. Final products of fermentation appear to be acetic and propionic acid. In case of alanine, ammonia is additionally discharged [2].

Based on the generally accepted data on production of energy during oxidation of pyruvic acid, Wood supposes that the only source of energy for *Cl. propionicum* lies in conversion of pyruvate into acetyl-CoA then into acetylphosphate, acetate and ATP.

Fermentation of alanine by this organism was studied more thoroughly [6].

$$3\,CH_3CHNH_2COOH + 2\,H_2O \rightarrow 3\,NH_3 + 2\,CH_3CH_2COOH + CH_3COOH + CO_2$$

It was indicated that for forming propionic acid from alanine, the presence of acetylphosphate and CoA is absolutely necessary [10]. These two combinations are known to participate in the synthesis of ATP during pyruvate oxidation. This allows us to suppose that generation of energy by *Cl. propionicum* from pyruvate proceeds as follows:

pyruvic acid $\rightarrow$ acetaldehyde + CO_2

acetaldehyde + CoA $\rightarrow$ acetyl-CoA

acetyl-CoA + P_{in} $\rightarrow$ acetylphosphate + CoA

acetylphosphate + ADP $\rightarrow$ ATP + acetate

Fermentation of purines

Fermentation of purines caused by species of Clostridium genus – *Cl. acidiurici, Cl. cylindrosporum, Cl. uracilicum* – has undergone more detailed studies [9]. These organisms do not have glycolytic exchange so anaerobic oxidation of heterocyclic nitrogen-containing compounds is in their case the only way of generating energy. The chemistry of purine fermentation was studied for xanthine oxidation by cell-free extracts of *Cl. cylindrosporum*:

xanthine $\rightarrow$ 4-ureido-5-imidazolcarbonic acid $\rightarrow$ aminoimidazol $\rightarrow$
$\rightarrow$ formimino-PH_4 $\rightarrow$ 5,10-methenyl-PH_4 $\rightarrow$ 10-formic-PH_4 $\rightarrow$
$\xrightarrow{ADP + P_{in}}$ ATP + HCOOH + PH_4

Ten successive stages were described for this process through which the organism receives the necessary energy in the form of ATP.

The above examples show that among anaerobic bacteria are organisms capable of accumulating energy not through glycolysis, but with the help of very specific reactions which can be classified along with oxidophosphorylation at the substrate level. Some of the above bacteria have very similar ways of energy metabolism, which is based on anaerobic oxidation of pyruvate into acetate.

Our attention is drawn to the fact that only four or five processes are required to form ATP, but in all the cases participation of such compounds as NAD, CoA, ferredoxin and TPP is obligatory.

Comparisons of energy exchange in a majority of anaerobic bacteria capable of glycolysis with that of the above organisms shows similarity of some stages, characteristic of one or another type of exchange:

1) Decarboxylation of pyruvic acid,
2) Formation of acetyl-CoA or acetaldehyde,
3) Oxidation of acetyl-CoA or acetaldehyde into acetate with formation of ATP.

Such comparisons raise the question of how these two stages of exchange correlate, when did they appear and whether they appear simultaneously or not?

By studying the types of energy metabolism one can suppose that, in the era of prebiological evolution, ancient protobionts could exchange energy in 2–3 stages, with obligatory participation of coenzymes:

I. a) acetaldehyde $\xrightarrow{\text{CoA}}$ acetyl-CoA

b) acetyl-CoA $\xrightarrow{P_{in}\ \text{NAD}}$ acetylphosphate

c) acetylphosphate $\xrightarrow{\text{ADP}}$ acetate + ATP

II. a) acetaldehyde $\xrightarrow{Fd_{ox}}$ acetate + Fd_{red}

b) $Fd_{red} + NAD \rightarrow Fd_{ox} + NADH_2$

c) $NADH_2 + P_{in} + ADP \xrightarrow{X} NAD + ATP + XH_2$

Aldehyde might have been present in sufficient quantities in the prebiological soup and under definite physical and chemical conditions could easily participate in various reactions. Anaerobic oxidation of acetaldehyde into acetate forming ATP could take place only with the help of such cofactors as CoA, NAD, thiamine pyrophosphate and ferredoxin. All these cofactors belong to the most ancient biochemical catalyzers. Their universality also argues in favor of the ancient origin of these processes. They are widely spread in many known metabolic cycles.

Regarding anaerobic oxidation of acetaldehyde into acetate and ATP, obligatory participation of thioether – acetyl-CoA is drawn to our attention. In this connection, it is of interest to recollect Racker, who said that in anaerobic glycolysis * which is evidently a very primitive process of forming energy, thioethers prove to be predecessors of ATP. Most probably thioethers are evolutionary ancestors of energy-rich ATP.

Based on the above we can suppose:

1) The classical form of glycolysis, having passed through the long evolution period from primitive anaerobic organisms to higher organisms, could not be primary and the only form of energy exchange.

2) Glycolysis is the result of a long biochemical evolution. It could have been preceeded by more simple types of energy metabolism not requiring preliminary phosphorylation of the substrate at the cost of ATP, for instance, anaerobic oxidation of pyruvate and acetaldehyde into acetate and ATP.

3) Evidently in the era of prebiological evolution great variety of reactions existed leading to formation of energy, which reached us as peculiarities of energy exchange in some anaerobic bacteria.

* Evidently Racker meant not the classical form of glycolysis but reactions discussed earlier in this report.

References

[1] Barker H.A., Bacterial Fermentations (New York, 1956).
[2] Barker H.A., Bacteria 2 (1961).
[3] Brill W.I. and R.S. Wolfe, Nature 212 (1966) 5059.
[4] Bryant M.P., E.A. Wolin, M.J. Wolin and R.S. Wolfe, Arch. Mikrobiol. 59 (1967) 1, 20.
[5] Burton R.M. and E.R. Stadtman, J. Biol. Chem. 202 (1953) 873.
[6] Cardon B.P. and H.A. Barker, J. Bacteriol. 52 (1946) 629.
[7] Oparin A.I., Gyzn Eio Pririda Proishogenie i Razvitie (Nauka, Moscow, 1968).
[8] Racker E., Mechanisms of Bioenergetics (Academic Press, New York, London, 1965).
[9] Shuster C.W. and I.C. Gunzalus, Federation Proc. 17 (1958) 310.
[10] Stadtman E.R., Bull Soc. Chem. Biol. 37 (1955) 931.

Chemical Evolution and the Origin of Life, eds. R. Buvet and C. Ponnamperuma

MICROBIAL EVOLUTION ON THE EARLY EARTH

Lynn MARGULIS
Department of Biology, Boston University, Boston, Massachusetts, USA

Glaessner [7] tells us "there are undisputed metazoan remains in the Vendian"....about 700–550 million years ago and that one of the main conclusions of recent intensive studies of the Precambrian record of life is:

(1) the main diversification of life occurred not earlier than in Late Proterozoic time and

(2) only primitive plants (and bacteria) are recorded definitely from many sediments of Middle and Early Proterozoic age..." (Glaessner, [7]) (The Proterozoic began 250–100 million years ago).

Translated into modern biological terms these two paleontological conclusions are:

(1) the diversification of the highest taxa (the eukaryote Kingdoms Protista, Fungi, Animalia and Plantae; Whittaker [18]) occurred in the Late Precambrian. Therefore aerobic nucleated cells containing nucleoprotein chromosomes with haploid–diploid meiosis–fertilization cycles, must have already evolved.

(2) evolution of bacteria and blue-green algae (the prokaryote Kingdom Monera; Whittaker [18]) occurred much earlier. The algae left "stromatolitic remains" and direct evidence for photosynthetic activity at least 2500 million years ago, perhaps as long ago as 3100 million years (Barghoorn and Schopf, [1]).

I would like to briefly present a model for the evolutionary relationship between Early Precambrian prokaryotic cells and Late Precambrian eukaryotic cells. Even if incorrect, the model is explicit and testable on many grounds. It was originally developed entirely independently of the fossil record to explain the behavior of "cytoplasmic genes". Implied in the model is the concept that hereditary endosymbiosis [8] has been a significant evolutionary mechanism in the origin of the eukaryotic cell.

The model theorizes that the eukaryote cell is a product of temporally ordered, specific symbioses as outlined below.

The reasons for choosing these components and this explicit order have been discussed in great detail elsewhere [9–11,15] and provides the basis for a recent book [12]. This paper will list the major conclusions of these studies and point out their relevance to problems of the origin of life.

As pointed out by Morowitz [13] all terrestrial living systems minimally contain certain organic constituents and a minimal amount of information. He estimates the minimal number of biochemical functions (enzymes) necessary for free-living replication to be about 45; in general even the smallest terrestrial cells are at least a factor of 10 larger than Morowitz's minimal cell. At least certain organelles of eukaryote cells (for example, chloroplasts and mitochondria, Cohen [4]) fulfill "minimal self-replicating criteria", that is, contain most of those items listed in table 1. The model presented here is based on the assumption that these organelles contain the set of items in table 1 precisely because they originated as free-living self-replicating cells that later became obligate symbionts in the population of complex cellular entities ancestral to all of the organisms in the four eukaryote kingdoms. These concepts are diagramed in fig. 1.

The earliest Precambrian cells were anaerobic heterotrophs. These primitive prokaryotes eventually gave rise to an enormous range of bacterial cells, among them spirochaete-like small motile heterotrophs, and bacterial photosynthesizers. Bacterial photosynthesizers and blue-green algae have a common

Table 1
Major minimal elements of the self-replicating cellular system (at least 0.1 μm diameter).

1) DNA (M.W. at least 26×10^6, Morowitz [13])
2) messenger RNA colinear with the DNA
3) polymerase enzymes:
   DNA polymerase
   DNA-dependent RNA polymerase
4) approximately 20 different activating enzymes, one each for each of the 20 amino acids
5) approximately 20 transfer RNA molecules
6) ribosomes: ribosomal RNA and protein; protein synthesis factors required for incorporation of amino acids into protein
7) lipid membrane
8) fermentable carbohydrate or other source of energy resulting in ATP production
9) particular product or activity selected by environment to perpetuate replicating system

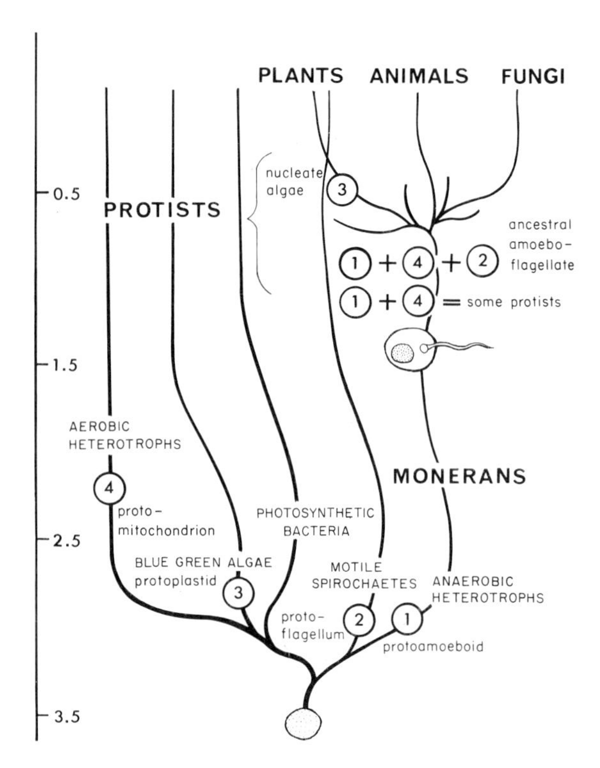

Fig. 1. Precambrian cellular evolution based on the cell symbiosis theory. (Simplified, see Margulis 1970 for details). *Ordinate*: X billions of years ago.

ancestry [6]. Mutations that permitted the use of water instead of organic hydrogen, molecular hydrogen or hydrogen sulfide as hydrogen donor in CO_2 reduction in photosynthesis led eventually to cells that eliminated molecular oxygen into the atmosphere. The blue-green algae arose from these; they intensively speciated and flourished in the Precambrian.

The production of oxygen forced the development of oxygen tolerance in many, by now highly diversified, prokaryote microbes. Oxygen toleration was superceded by oxygen utilization. Microaerophilic and aerobic bacteria evolved, including the "protomitochondrion".

The common ancestors of all eukaryotes were amoeboflagellate heterotrophs, cells that are the product of three endocellular symbioses. The ancestral forms were fundamentally aerobic; they depended on many oxygen mediated metabolic steps including the biosynthesis of steroids. It was in the ancestral eukaryote population that classical mitosis and meiosis evolved giving rise to the higher nucleated algae and plants, the fungi and the metazoans.

This model is consistent with the apparent fact that over the vast stretches of Precambrian time the earth teemed with microbial, rather than macroscopic life. It clarifies the concept that the metabolism of blue-green algae, during the middle Precambrian, was the major cause of the transition to the oxidizing atmosphere. The blue-green algae differ radically from all other algae and higher plants, except in their photosynthetic metabolism. Not until oxygen was widespread in the terrestrial atmosphere did eukaryote cells evolve. Not until the elegant classical mitotic mechanism of eukaryote chromosome segregation evolved in a large series of protists (e.g., branching of dinoflagellates, radiolarians, euglenids, cryptomonads and so forth) could the higher kingdoms (Fungi, Plant and Animal) diversify.

Along with the magnificent discoveries of ancient microbial life [3,16] this model helps push back the origin of life to a time contemporaneous with or even before the deposition of the first sedimentary rocks. Only vertebrate and angiosperm paleontologists and others concerned with morphological manifestations of evolution can hold to the idea that biological diversification is a Phanerozoic phenomenon. Those familiar with bacteria and their unbelievable range of metabolic solutions to environmental problems leading to their selection ("enrichment") in enormously varied environments realize that Phanerozoic eukaryote diversification is simply an epiphenomenon. The prokaryotes had already evolved the fundamental biological patterns: ultraviolet resistance and nucleic acid repair; genetic recombination; porphyrin syntheses and light absorption; CO_2 and N_2 fixation; sulfate reduction; carbohydrate and polymer degradations; penetration into terrestrial, aerial, fresh water, marine and hot spring environments; photosynthesis, aerobiosis and chemoautotrophy; spore formation and multicellularity, to name but a few of the virtuosities of the group as a whole. The Precambrian radiation of the prokaryotes must be perceived and understood by all those hoping to decipher the nature of the early terrestrial environment and of the organic compounds sedimented upon it.

The origin of the minimal replicating system itself with its triplet genetic code [5,14] is more remote and ancient than formerly realized. It seems that our lack of understanding of this epic-making event, which most likely occurred during the first billion years of earth history, remains the most serious barrier to our comprehension of terrestrial life and its history.

References

[1] Barghoorn E.S. and J.W. Schopf, Science 152 (1966) 758.
[2] Cloud P.E., Jr., Science 16 (1968) 729.
[3] Cloud P.E., Jr., Science 148 (1965) 27.
[4] Cohen S.S., Am. Sci. 58 (1970) 281.
[5] Crick F.H.C., J. Mol. Biol. 38 (1968) 367.
[6] Echlin P. and I. Morris, Biol. Rev. 40 (1965) 143.
[7] Glaessner M.F., Can. J. Earth Sci. 5 (1968) 585.
[8] Karakashian S.J. and R.W. Siegel, Exptl. Parasitology 17 (1965) 103.
[9] Margulis L., Science 161 (1968) 1020.
[10] Margulis L., J. Geol. 71 (1969) 606.
[11] Margulis L., Whittakers five kingdoms: Minor revisions based on considerations of the origin of mitosis evolution (1971) in press.
[12] Margulis L., Origin of Eukaryotic Cells (Yale University Press, New Haven, Conn., 1970).
[13] Morowitz H.J., Biological self-replicating systems, in: Progress in Theoretical Biology, Vol. 1 (Academic Press, New York, 1967) p. 35.
[14] Orgel L.E., J. Mol. Biol. 38 (1968) 381.
[15] Sagan L. (Margulis), J. Theoret. Biol. 14 (1967) 225.
[16] Schopf J.W. and E.S. Barghoorn, J. Paleontology 43 (1969) 111.
[17] Stanier R.Y. and D.B. van Niel, Arch. Mikrobiol. 42 (1962) 17.
[18] Whittaker R.H., Science 163 (1969) 150.

Chemical Evolution and the Origin of Life, eds. R. Buvet and C. Ponnamperuma

LIFE IN EXTREME ENVIRONMENTS

D.J. KUSHNER
Department of Biology, University of Ottawa,
Ottawa, Ontario, Canada

Many discussions on the origin of life involve model systems, based on the laws of physics and chemistry, which presumably have never changed, and on deductions that can be made about the probable state of the earth before life existed. Ingenious experiments suggest that it was gratifyingly easy for small molecules, macromolecules, and even assemblies of macromolecules superficially resembling cells to have arisen under prebiotic conditions.

The purpose of this communication is not to suggest that any of the intriguing and unlikely organisms to be discussed appeared at the beginning or very early in the development of life – almost certainly, none of them did – but rather to expand the perspectives of any model system proposed. In setting up a model one must also consider its environment. There is a natural tendency to postulate those conditions that support the life of organisms we know today (except, of course, that one would scarcely now propose a beginning system which required the presence of molecular oxygen).Studies of certain contemporary organisms show that life can exist under more extreme conditions than are normally thought possible. If such conditions should turn out to be especially favorable for the origin of living system, then, they would not oppose their continuation.

Since life on other planets is being considered here, it is also obviously important to know within just what limits life can exist in our own planet.

Usually one thinks of living beings as existing in a comfortable range of moderate temperatures, pH values near neutrality, atmospheric pressure, and low but sufficient salt concentrations. Thus, some of the conditions in which life actually does exist may be surprising. Table 1 shows the conditions under which a mixed sampling of microorganisms has been reported to live. These examples are taken from recent reviews [2,14] which should be consulted for more detailed references to earlier work and treatment of the natural distribution, physiology and origins of these and other unusual microorganisms.

Life at high temperatures

Several microorganisms can live at high temperatures, and some require high temperatures for growth. The most studied of these is *Bacillus stearothermophilus* a widely distributed organisms found in soil and food products as well as in hot springs [2,13]. It can grow at 75 °C or higher and grows poorly below 45 °C. Several of the proteins and enzymes of this and other thermophiles display unusual heat-stability. The protein-synthesizing machinery of *B. stearothermophilus* is more active at 60–70 °C than at 35 °C. Its ribosomes are more heat-stable than those of mesophilic bacteria such as *Escherichia coli*, both in structure and in amino acid-incorporating ability. Amino acid activation and formation of amino acyl-tRNA are more active at 60 °C than at 35 °C [9].

The α-amylase of *B. stearothermophilus* is quite active at 70 °C. Un-

Table 1
Some microorganisms growing in extreme environmental conditions.

Organism		Conditions in which growth is possible	Reference [a]
Name	Type [b]		
Bacillus stearothermophilus	B	Over 70°C	2, 14
Thermus aquaticus B	B	88 °C	14
Cyanidium caldarium [c]	A	80 °C, 1.0 N H_2SO_4	2, 14
Thiobacillus thiooxidans	B	0.5–5% H_2SO_4	2, 14
Ancontium velatum "Fungus D" (a *Dematiaceae* sp.)	F	2.5 N H_2SO_4 saturated with $CuSO_4$	14, 25
Bacillus circulans	B	up to pH 11.0	14
Xeromyces bisporus	F	a_w of 0.6–0.65	16
Moderately halophilic bacteria (e.g., *Vibrio costicolus, Micrococcus halodenitrificans*)	B	0.5 M to 3.5 M NaCl (a_w 0.86–0.98)	15, 16
Extremely halophilic rods and and cocci	B	3 M to 5.5 M NaCl (a_w 0.75–0.88)	15, 17
"Bacterium"	B	Saturated LiCl (a_w 0.12)	15

[a] Most references are review articles.
[b] B, bacterium; A, alga; F, fungus.
[c] However, Brock's recent work indicates that *C. caldarium* rarely grows at above a temperature of 50 °C in nature [2].

fortunately, quite recent work [23] has cast doubt on an interesting earlier theory that its stability is due to a lack of tertiary structure [20,21].

Many thermophilic blue–green algae and bacteria are found in hot springs. Brock has recently isolated *Thermus aquaticus*, a Gram-negative bacterium often filamentous in form, from hot springs at about 88 °C in Yellowstone National Park. The temperature optimum for growth of this organism is 70–75 °C [2]. Freeze and Brock [8] showed that the aldolase of *T. aquaticus* has little activity below 60 °C and optimal activity at 95 °C. It is inactivated only at 107 °C.

Membranes of thermophilic bacteria are more heat-stable than those of mesophiles [2]. Flagella of *B. stearothermophilus* are especially heat-stable and cannot be degraded by heating into subunits (flagellins) as can mesophilic flagella [13]. It is not certain why some enzymes of thermophiles need high temperatures for activity; possibly, at moderate temperatures their structural configuration is too tight.

In addition to these aerobic organisms, thermophilic clostridia have been described, as have anaerobic cellulose decomposers, hydrocarbon utilizers, sulfate reducers and others [2,14].

These examples amply demonstrate that life can exist at high temperatures, almost up to the boiling point of water. Though we think of many thermophilic bacteria as now occupying a special ecological niche, their existence and the ways they adjust to high temperatures suggest that life could quite well have originated at high temperatures if these temperatures especially favored such a process of organization.

Extremes of pH and heavy metals

Quite low or high *external* pH values are consistent with life. Bacteria and other cells, whose enzymes function best at a narrow range of pH values near neutrality, can grow over a much wider range. This is thought due to the ability of cells to control internal pH, and in some cases such control has been demonstrated [14]. Presumably, organisms that can live at extreme pH values can also control their internal pH, but there is little evidence on this point. The flagella of *Thiobacillus thiooxidans*, which oxidizes sulfide or sulfur to H_2SO_4 and which can live in 5% H_2SO_4 are unusually acid-resistant [2,6]. However, surprisingly little work has been done on the internal enzymes or on the internal pH of this organism.

The thiobacilli are often found in mines, where the acids they produce lead to leaching of ores [7]. Consequently, they also live in the presence of high concentrations of copper, lead, or uranium salts. It is suspected that they exclude the metallic cations by virtue of the positive charge on their membranes resulting from the acidity of the medium [2,7].

Extremely high resistance to acid and heavy metal ions may be found among the fungi. The original report [25] of two species which can grow in strong acid solutions (table 1) lists earlier work on fungi resistant to mercury and other ions. The mechanisms of fungal resistance to heavy metal ions have been little studied though some work has been done on copper resistance in yeast [reviewed in 14].

High solute concentration

Many organisms can live in high concentrations of salt or sugar, and such organisms have generally been of special interest for the spoilage they cause of preserved food products [15–17,24]. Two effects of solutes must be considered: their ability to tie up water and reduce the water activity (a_w); and specific effects, such as the ionic ones produced by high salt concentrations. Fungi are especially able to grow under conditions of low water activity, that is, under relatively dry conditions. A few species (the so-called "Xerophilic" fungi) seem to require lowered water activity. A number of yeasts can grow in high concentrations of salt or sugars; these have been termed "osmophilic yeasts", though the term is probably inaccurate: they tolerate but do not require high solute concentrations [reviewed in 16,24].

The relation of microorganisms, especially bacteria, to salt has been much studied [3,15,17,19]. It is necessary to distinguish between bacteria that can grow in high salt concentrations and those that require such concentrations. For example, staphylococci can grow in 10% NaCl but do not need it. Many marine bacteria need about 3% NaCl (the concentration found in seawater) for growth. A substantial proportion of them can also grow in the presence of 20% or even 30% NaCl. A small group of bacteria, not found in the open sea but in salt lakes and in salterns (ponds for making salt from sea water), needs 15–30% NaCl for growth. These "extreme halophiles" show an extreme degree of physiological adaptation to their environment. They are strict aerobes and certainly do not qualify as primitive cell types; nevertheless, they do expand our horizons on what life can be, and it is in these terms that they will be discussed.

Probably all extremely halophilic bacteria can be found within two genera, the rod shaped halobacteria and the spherical forms, which have been given different names in the past but are now frequently considered as halococci

[17]. The former group has been most studied. The halobacteria are fragile cells; their shape is easily distorted on handling; the cocci are extremely tough cells and are very difficult to break.

Extreme halophiles can survive for relatively long periods in salt solutions or in the dry state. They do not exclude the bulk of external salts: their internal salt concentration is as high as that of the external medium. Rather, they select ions, and have an exceptional ability to concentrate potassium ions. Christian and Waltho [4] suggested some years ago that *H. salinarium* growing in 4.0 M NaCl + 0.03 M KCl contained 4.7 M K^+ (that is, more than would be found in saturated KCl) and about 1.5 M Na^+. Our recent results with *H. halobium* show that this organism also has a very high K^+ content. This can be maintained for several days when non-growing cultures are suspended in salt solutions and falls only when the cells die [10]. The internal salt concentration is reflected in the function of different intracellular constituents.

Ribosomes of these organisms seem to have unique properties. Unlike other ribosomes, which are dissociated by moderately high salts, those of *H. cutirubrum* need about 4 M K^+ and 0.1 M Mg^{2+} for stability. Na^+ cannot replace K^+. If the salt concentration is drastically lowered *H. cutirubrum* ribosomes disintegrate, losing most of their proteins. These proteins, unlike those of other ribosomes are acidic instead of basic [1]. Mutual repulsion between protein and RNA is presumably responsible for the disintegration of ribosomes; a high concentration of the proper cations provides a screen enabling the parts of the ribosome to fit into an active configuration.

Membranes of extreme halophiles are unstable without salt. This also appears due to the acidic charge on the membrane proteins.

Most enzymes of extreme halophiles need salt for activity and are reversibly or irreversibly inactivated without salt. Because of the difficulty of purifying such enzymes there is little direct proof for the reasonable theory that cations act by shielding negatively-charged groups and preventing their mutual repulsion from forcing the enzyme into an inactive configuration!

Some of the effects of salt on enzyme activity and configuration are illustrated in fig. 1, which shows recent experiments with the aspartate transcarbamylase of yeast and of *H. cutirubrum*. This enzyme, which carries out one of the key steps in pyrimidine synthesis is regulated through feedback inhibition by one of the end products of this synthesis: UTP in yeast and CTP in bacteria. In yeast, moderate salt concentrations inhibit the enzyme and inhibit the feedback inhibition still more. Just the reverse happens with the enzyme from *H. cutirubrum*. High salt concentrations are needed for activity, and still more salt is needed for regulation. Enzymes

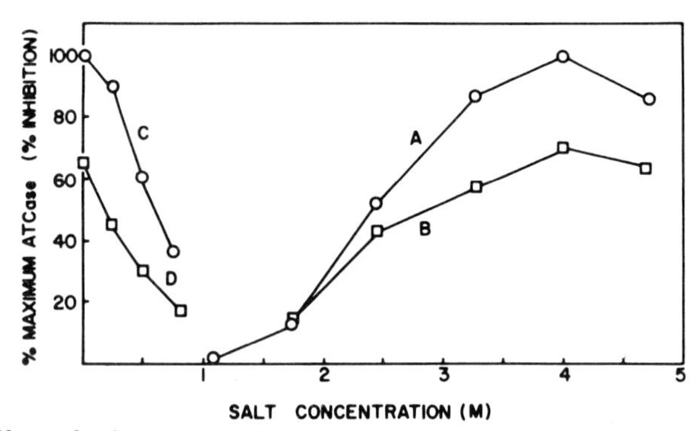

Fig. 1. Effect of salt concentration on activity and feedback inhibition of aspartate transcarbamylases (ATCases) from a yeast (*S. cerevisiae*) and an extreme halophile (*H. cutirubrum*). A, ATCase, halophile; B, percentage inhibition, CTP; C, ATCase, yeast; D, percentage inhibition, UTP [from 18].

regulated by feedback inhibition usually possess a subunit structure [5]. It is thought that those of yeast aspartate transcarbamylase are dissociated by salt, and those of the extreme halophile enzyme are dissociated when the salt concentration is lowered.

The unusual lipid composition of extreme halophiles should be mentioned, even though it is not clear how this is related to these cells' salty environment. Lipids make up 4–5% of the weight of the halobacteria, and a smaller percentage of the weight of the halococci. Of these lipids, more than 90% consists of an analog of phosphatidylglycerol phosphate together with a smaller amount of an analog of phosphatidylglycerol [11]. Instead of having ester linkages, as other phosphatides do, these lipids have ether linkages; instead of fatty acids or alcohols they have dihydrophytol groups. These compounds are confined to extremely halophilic bacteria, and so far have been found in all extreme halophiles. The extreme halophiles have only traces of fatty acids. Their high phytol content should be kept in mind since phytanes have been found in the most ancient sediments [1]. This is not to suggest that such compounds came from halophiles. Still, one can readily envisage a situation where, long ago a large pond of sea water somehow became isolated and dried in the sun to the point that it became a good breeding ground for extreme halophiles.

Only a few examples of organisms that live in extreme environments have been considered in this brief survey. Many bacteria live under immense pressure at the bottom of oceans [22]. Others are well adapted to life under cold conditions, near zero degrees centigrade, though their study might be more important in considering the end of life than its beginning.

It does not seem profitable to discuss the evolutionary position of the unusual organisms considered here. Knowing that they do exist, and something of how they exist, gives the imagination a little more freedom in conceiving ways in which life arose and in considering the kinds of life that may be found on other worlds.

Note added in proof

Quite recent work (J.K. Lanyi and J. Stevenson, J. Biol. Chem. 245 (1970) 4074) suggests that high salt concentrations may also act by preventing the unfolding of hydrophobic residues on halophilic proteins. Purified proteins, of course, would also be needed to critically assess this theory.

References

[1] Bayley S.T. and D.J. Kushner, J. Mol. Biol. 9 (1964) 654.
[2] Brock T.D., Symp. Soc. Gen. Microbiol. 17 (1969) 15.
[3] Brown A.D., Bacteriol. Rev. 28 (1964) 296.
[4] Christian J.H.B. and J. Waltho, Biochim. Biophys. Acta 65 (1962) 506.
[5] Cohen G.N., Ann. Rev. Microbiol. 19 (1965) 105.
[6] Doetsch R.N., T.M. Cook and Z. Vaituzis, Antonie van Leeuwenhoek 36 (1967) 196.
[7] Ehrlich H., in: Microbial Transformations of Minerals, eds. H. Heukelekian and N.C. Dondero (Wiley, New York, 1964) pp. 43–60.
[8] Freeze H. and T.D. Brock, J. Bacteriol. 101 (1970) 541.
[9] Friedman S.M., Bacteriol. Rev. 32 (1968) 27.
[10] Gochnauer M.B. and D.J. Kushner, Can. J. Microbiol. (1971), in press.
[11] Kates M., B. Palameta, C.N. Joo, D.J. Kushner and N.E. Gibbons, Biochemistry 5 (1966) 4092.
[12] Kenyon D.H. and G. Steinman, Biochemical Predestination (McGraw Hill, New York, 1969).
[13] Koffler H., Bacteriol. Rev. 21 (1957) 227.
[14] Kushner D.J., Exptl. Chemotherapy 2 (1964) 113.
[15] Kushner D.J., Advan. Appl. Microbiol. 10 (1968) 73.
[16] Kushner D.J., in: Inhibition and Destruction of the Microbial Cell, ed. W.B. Hugo (Academic Press, 1970) in press.
[17] Larsen H., Advan. Microbial Physiol. 1 (1967) 94.
[18] Liebl V., J.G. Kaplan and D.J. Kushner, Can. J. Biochem. 47 (1969) 1095.
[19] MacLeod R.A., Bacteriol. Rev. 29 (1965) 9.
[20] Manning G.B. and L.L. Campbell, J. Biol. Chem. 236 (1961) 2952.
[21] Manning G.B., L.L. Campbell and R.J. Foster, J. Biol. Chem. 236 (1961) 2958.
[22] Morita R.Y., Oceanography Marine Biol. Ann. Rev. 5 (1967) 187.
[23] Pfeuller S.L. and W.H. Elliott, J. Biol. Chem. 244 (1969) 48.
[24] Scott W.J., Advan. Food Res. 7 (1957) 83.
[25] Starkey R.L. and S.A. Waksman, J. Bacteriol. 45 (1943) 509.

PART VIII

EXOBIOLOGY

Chemical Evolution and the Origin of Life, eds. R. Buvet and C. Ponnamperuma

ORGANIC SUBSTANCES IN THE UNIVERSE

V.G. FESENKOV
The Meteorite Committee of the Academy of Sciences of the USSR, 3 M. Oulianova Street, Corps I, Moscow, USSR

Conditions for life in the universe are closely related to the development of our galaxy which is thought to have existed for approximately 10 billion years. The first stars born within the galaxy, consisting mainly of hydrogen, served as atomic reactors to produce heavier and heavier elements which have been gradually scattered into space. The life time of a star is dependent upon its mass. If our sun – a rather moderate star – is able to consume its hydrogen content during 10 billion years or so, a star with a mass thirty times larger will use up its hydrogen in only one million years, then it will undergo a strong contraction with a huge increase in density and temperature. During this process all the elements are formed including the heaviest ones. Then the star explodes, dispersing its substance into the surrounding space. Consequently, new-formed stars are enriched in sequence by elements with higher and higher atomic numbers.

The detailed study of radioactive elements found in the terrestrial crust, such as the various uranium isotopes, shows that the age of these elements which are necessary for the internal heating of the earth – exceeds only slightly the age of earth itself and of meteorites. The investigation of meteorites suggests that the explosion of a Supernova scattering heavy elements into space occurred immediately before the birth of our solar system. It is very well known that such explosions give rise to hypersonic waves favoring the formation of condensations within the interstellar medium and thus the birth of stars. For example, the Supernova explosion in the Cygno constellation, occurring about 70,000 years before our era, yielded supersonic waves which can still be observed nowadays as filaments radially moving off from a common center. We have discovered that these filaments are closely related to recently formed sequences of stars.

Radioactive elements which progressively accumulate in interstellar space are necessary for the thermal evolution of planets. Internal heating of a planet results in the formation of atmospheres containing water vapor, which is necessary for Life processes.

The activity of stars and the reactions which produced the atoms seem to be closely linked to the birth of life. Another role played by stars is to create in space favorable conditions for the production of organic combinations of carbon and hydrogen. All stars continuously radiate into space cosmic rays and plasma clouds containing the ionized atoms of various elements. Our sun continuously emits solar winds which reach the Jupiter orbit and even farther and represent the gaseous component of the solar crown. The crown of this solar wind also contains solid particles of different origin. Without going further on this point, we shall mention that various stars, in particular the red giants of the N species, generate in their atmospheres the fine carbon dust which is continuously expelled into space by radiative pressure and takes part as a main component in cosmic clouds. The total mass of these particles does not exceed 1% of the gaseous component of the interstellar medium. However, these particles account for the very considerable light absorption determining the general aspect of the Milky Way. These particles can generate molecular arrangements since they are irradiated by cosmic rays and emanations from stars.

Experimental proton bombing of particles consisting of silicon carbide (CSi) has been shown to produce hydroxyl groups (HO) and water. In reactions with nitrogen, it can also yield N–H containing compounds. So it is not amazing that in different areas of space, in particular in the vicinity of newly-born stars, radiometric measurements show evidence for the various organic compounds. The occurrence of H_2 and also OH within the same cosmic clouds was first discovered in 1968. NH_3 has also been found within the interstellar medium, along with H_2O and formaldehyde absorption (H_2OO) around $\lambda = 6.21$ cm with 9 overtones. These molecules are generally found in the compact interstellar clouds surrounding newly-born stars having many irregularities.

A very instructive example is the Orion Nebula containing the newly-born star cluster, Orion trapezium. Although it is encompassed within the Nebula itself it is very far away and possesses many irregularities in its cometic mass (of the order of magnitude 10^{18}–10^{20} g). Nearly all the mass of this nebula, reckoned be about 20,000 years old, is included in these irregularities. This nebula is the remnant of a cosmic cloud where the above-mentioned stars have been generated.

The cometary cloud encompassing our sun is likely to have been formed in the same way. According to reckonings this cloud, which contains many billion comets, has expanded to distances of 150,000 light years where perturbations of the galactic core and neighboring stars are fairly noticeable. These perturbations gradually scatter away the original cometary cloud which consequently must have previously been much more bulky.

Comets which represent the aforesaid conditions of the interstellar medium are known to contain many organic compounds, such as HCN, NH_3, C_2H_2, CH_4 and C_2N_2. Moving unsteadily around the common gravity center, they must collide occasionally with one another thus undergoing temporary overheatings. Within the original cloud, collisions between comets probably occur frequently. Even nowadays, after 5 billion years of comet exhaustion through continuous perturbations, they can still collide. The problem of explaining the origin of relatively recent evidences of comet hits upon the terrestrial soil should be mentioned. The last event of this kind was on June 30, 1908, the occurrence of the so-called Tunguska Meteorite which was actually a small comet. A powerful explosion, ten kilometers above sea-level, pulled down trees over a hundred kilometers in the area and the comet tail held back by the higher layers of our atmosphere produced a great brightness in the night sky, which was visible as far away as Ireland.

Mutual comet collisions in planetary nebula were a valuable source of organic compounds in the preplanetary era. Carbonaceous meteorites, allegedly originating from comets, show evidence of plentiful fairly complex organic compounds including some of the nucleic acid bases. Detailed experiments have shown that these types of organic compounds can be reproduced in reactions, approaching thermodynamic equilibrium of meteoritic material and other elements, through heating up to a few hundred degrees centrigrade followed by cooling for a few hours or at most a few days. These very conditions are similar to those which could have been achieved when the planets in our solar system were formed through comet clashes followed by temporary over-heatings.

So, prerequisite conditions for the emergence of life were existing in our planetary system even before planet formations. As can be seen, it is a general process which is determined by characteristic features in the universe, at least in our galaxy. It can be concluded that, anywhere in the Universe, life is built only on the hydrocarbon chemistry and that comets which are also emerging in the interstellar medium can serve as "parents" for this cosmic event.

Even now, collisions of comets with the earth very favorably influence the development of life as is demonstrated by the Tunguska phenomenon. The Meteorites Committee of the USSR Academy of Sciences and other institutions have carried out exhaustive investigations on this uncommon occurrence. It was established that, after such a catastrophy, growth of trees was much faster than before. Even rather old trees began growing 6 to 7 times as fast. This occurred all along the flight trajectory of the meteor where the oecogeny conditions underwent no alterations. Consequently, such phenomenon is beyond all questions related to the enrichment of this Tunguska area by the cometary material.

A more comprehensive study of comets, especially by space probes launched in their vicinity, will hopefully permit us to get more useful data to solve the problem of the occurrence of life within space and more particularly on our earth.

Chemical Evolution and the Origin of Life, eds. R. Buvet and C. Ponnamperuma

CARBONACEOUS CHONDRITES AND THE PREBIOLOGICAL ORIGIN OF FOOD

P.C. SYLVESTER-BRADLEY
University of Leicester, England

It is generally agreed that the first living organisms were anaerobic heterotrophs. Such a supposition demands the prebiological existence of food and much of the experimentation that has followed Miller's [9] first success has demonstrated how water-soluble organic molecules can be synthesized under abiotic conditions. The section on Environmental evidence of our present symposium is largely devoted to these considerations. In the early terrestrial environment these water-soluble products will eventually have reached the ocean which could have served as a reservoir of prebiological food. During prebiological time this reservoir could have gradually gained in concentration, but it is unlikely that it ever became more than an extremely dilute "prebiotic soup".

The carbonaceous chondrites afford evidence that in extraterrestrial environments another class of organic compounds has been synthesized. Some of these compounds are immiscible with water and under terrestrial conditions would be subject to geological processes of concentration. I have elsewhere suggested that life is likely to have originated first in such regions of concentration and would only secondarily have migrated to the oceanic food reservoir [21,23] – in human terms oil and bitumen form fuel rather than food. But even today there are organisms that feed on crude oil (certain yeasts and bacteria) and there is no reason to suppose that this has not always been so since life first originated. In the present paper I wish to examine more closely the conditions under which these immiscible compounds are likely to have arisen in carbonaceous chondrites, and I shall conclude with an estimate of how far such conditions could have been duplicated during the early history of the earth.

Chemical evolution

Prebiological food was a product of chemical evolution. During the present congress it has been inevitable that the particular aspect of chemical evolution studied has been that concerned with organic products. But the whole history of chemical evolution embraces far more than organic chemistry, and in its stellar, planetary and geological setting it gives rise to large-scale inorganic entities – to stars and planets, continents and oceans. A succession of environments are involved with one point in common: they are all, in the terms suggested by Morowitz [13] far from equilibrium systems standing between an energy-source and an energy-sink. They are all characterized by an increase in evolutionary organization and a decreasing entropy [22]. It is my hypothesis in this paper that the carbonaceous chondrites have frozen in them an early differentiate of planetary evolution – a stage which will have been developed on all terrestrial planets. The evolutionary stage represented by the carbonaceous chondrites will have been reached through the action of energy flowing from a source that was neither solar nor limited to the surface of the planetary body involved. This source of energy, which is stored in radioactive isotopes embodied in the planet, is ultimately derived from galactic origins, and is the main source of most geological evolution.

Environmental evidence

The carbonaceous chondrites are unmetamorphosed. They are composed of a complex, unequilibrated mixture of minerals whose origins are mutually incompatible; they include high-temperature microchondrules (olivine and pyroxene), water-soluble salts, hydrated silicates, high-molecular weight hydrocarbons and bituminous polymers. Although it is often stated that type I carbonaceous chondrites contain no chondrules, Mueller [14–16] has shown that they do contain microchondrules*, and he suggests that the aggregation of microchondrules seen in carbonaceous chondrites of types II and III characterize a method of chondrule formation correlated with a decrease in volatile constituents.

To explain the heterogeneous nature of the suites of minerals which compose carbonaceous chondrites, it is necessary to postulate their formation

* Although so far as I am aware Mueller's discovery of "microchondrules" has not yet been verified by other workers, it has been confirmed that the Orgueil meteorite does contain minute angular fragments of olivine and enstatite (K. Fredricksson, personal communication).

as the result of a succession of environmental phases, ranging from high temperature to low temperature, from dry to aqueous, and through varying intensities of reduction.

The carbonaceous chondrites are somewhat enriched, when compared with other chondrites, in mercury, boron, lead, bismuth and tellurium. On the earth some of these elements are characteristic of fumarolic and hydrothermal activity and this led Mills [10] to suggest that the history of carbonaceous chondrites is likely to have included a transient phase in which gas-flow supported particulate matter and so formed a fluidized bed. This would involve a stage in the formation of carbonaceous chondrites characterized by outgassing. Later experiments by Mills showed that the surface of a fluidized bed through which the gas flowed with varying intensity would have become scarred by craters much like those known on the surface of the moon and Mars [11], and that fluidization will have been accompanied by the build-up of very considerable electro-static charges [12]. The lunar surface is extensively scarred with craters of a very wide size range, including craters with diameters as small as a few centimeters. Some of these craters have certainly been formed by meteoritic impact, but not necessarily all of them. Analysis of Apollo samples suggests that meteoritic matter on the surface of the moon does not exceed 1 or 2%. It is possible that a large proportion of the smaller craters have been formed by fluidization during an outgassing period. The fact that Mars has a surface which (like the moon) is also cratered makes it possible that it also has craters caused by both impact and fluidization. Further, it seems legitimate to suppose that all four terrestrial planets have passed through outgassing phases.

The water-soluble salts and hydrated silicates which are included among the constituents of carbonaceous chondrites have been interpreted by Du Fresne and Anders [5] as indicating a phase in which liquid water was present.

Carbonaceous contents

Apart from a very subordinate amount of carbonate found in some C 1 carbonaceous chondrites, most of the carbon in these chondrites is in the form of organic compounds. In some samples of Orgueil these reach as much as 5% by weight. By far the largest part of these (85–95%) are not extractable by normal solvents. They are not, in consequence, easy to analyse but ESR and spectral studies suggest an aromatic skeleton [7] and the analysis of degradation products confirms the likelihood of random aromatic polymers [2,4,8]. Comparable aromatic complexes are found on the earth in the form of sporopollenin [4], kerogen, bituminous coal and melanin. These complexes are of diverse origin; what makes them comparable is their complexity, and our ignorance about the nature of the complexity. There is no difficulty in assuming that they can be synthesized both biologically and abiologically. Indeed, Pirie's joking reminder (in his discussion of Blois' paper [3]) that the unskilled chemist was always producing black "gunk" by mistake, vividly illustrates the variety of possible pathways that can lead to organic polymerization.

In addition to the insoluble polymers found amongst the carbonaceous contents of these chondrites, there is a small proportion of extractable compounds of great variety. These include paraffins, branched alkanes, cyclic alkanes, aromatics, fatty acids, benzoic acids, alcohols, purines and porphyrins, a mixture of which bears some resemblance to crude oils.

It seems most unlikely that anything beyond trace amounts among these carbonaceous contents can be ascribed to terrestrial contaminants [19]. Periods of organic synthesis must have occurred during the environmental history of the chondrites.

Organic synthesis

All the elements involved in the carbonaceous complex are volatile, and will have been present as gases or vapours before synthesis occurred. Studier, Hayatsu and Anders [20] and Hayatsu et al. [6] have conducted experiments in Chicago designed to test the possibility that synthesis took place at relatively high temperatures as a result of a mechanism akin to the Fischer-Tropsch process. They used meteoritic iron (Canyon Diablo) as a catalyst, and heated mixtures of hydrogen and carbon monoxide with and without the addition of ammonia to produce a wide range of hydrocarbons and nitrogenous organic compounds, including almost all of those recognised in carbonaceous chondrites.

Oró [17] has repeated this work, and after subjecting the products he obtained to more rigorous analytical techniques, has confirmed the efficacy of the process as one method for accounting for many of the carbonaceous components of carbonaceous chondrites. Both the Chicago group and Oró were attempting to simulate partial equilibrium conditions as they might have obtained in the solar nebula. Belsky and Kaplan [1] have concluded, from an analysis of the light hydrocarbon gases found as indigenous components of carbonaceous chondrites, and by a study of the carbon isotope ratios present, that near equilibrium processes were not responsible. It seems that Fischer-Tropsch synthesis cannot be the only process responsible. I wish to make two suggestions: (1) that Fischer-Tropsch reactions took place in only one of several environments through which the meteorites have passed during their life-history; (2) that this environment was not in the solar nebula at large but during the outgassing phase of a parent planetary body in which far-from equilibrium conditions obtained.

The synthesis of hydrocarbons and organic polymers has also been achieved from similar gas mixtures subjected to spark or corona discharge in an environment that is otherwise at a low temperature [18,25].

The Fischer-Tropsch process was used in Germany during the last war to synthesize hydrocarbon fuels on a commercial scale. The reacting gases were passed over a solid-bed catalyst. Subsequently a fluidized version of the process was developed in America which (though not commercially viable) produced a great increase of yield [24,26]. It seems certain that, if fluidization occurred as an important and recurring phenomenon during the history of the formation of the carbonaceous chondrites, then the yield of organic compounds would be much enhanced. Moreover, electro-polymerization may have aided synthesis at low temperatures [12]. There is a clear need for a series of prebiological experiments involving fluidized catalysts in a variety of environments.

Terrestrial analogues

Although the carbonaceous chondrites provide evidence for the extra-terrestrial synthesis of prebiological compounds, and although these compounds would have been available as food if life had existed in that environment, it seems most unlikely that that ever happened. Although those "organic elements" that are both organic and indigenous to the meteorites may represent a certain level of prebiological organization, there is no unequivocal evidence that life as such ever existed on the parent body from which the meteorites are derived [22].

But life certainly did develop on the earth, and it is pertinent to enquire whether processes similar to those which produced the carbonaceous contents of meteorites could have synthesized similar organic polymers. If so, could these have performed an important function as prebiological food for the first-formed organisms?

Fluidization is a not uncommon phenomenon on the earth in a volcanic environment. The main gas involved seems to be carbon dioxide, and kimberlites and carbonatites are often associated with intrusive tuffs and tuffisites that exhibit a texture suggesting fluidization. Earlier stages in the same geological environment are sometimes associated with hydrocarbons and bitumens that may represent a phase in which carbon monoxide or methane played a part. It seems not unlikely that in early Precambrian times volcanic gases were more reducing than later. In that case the abiogenic synthesis of hydrocarbons, bitumens and organic polymers may have been commonplace. A volcanic milieu is not an unlikely setting for the origin of life.

References

[1] Belsky T. and I.R. Kaplan, Geochim. Cosmochim. Acta 34 (1970) 257.
[2] Bitz M.C. and B. Nagy, Proc. Natl. Acad. Sci. U.S. 56 (1966) 1383.
[3] Blois M.S., in: The Origin of Prebiological Systems, ed. S.W. Fox (Academic Press, New York, 1965) p. 19.
[4] Brooks J. and G. Shaw, Nature 223 (1969) 754.
[5] Du Fresne E.R. and E. Anders, Geochim. Cosmochim. Acta 22 (1962) 1085.
[6] Hayatsu R., M.H. Studier, A. Oda, K. Fuse and E. Anders, Geochim. Cosmochim. Acta 32 (1968) 175.
[7] Hayes J.M., Geochim. Cosmochim. Acta 31 (1967) 1395.
[8] Hayes J.M. and K. Biemann, Geochim. Cosmochim. Acta 32 (1968) 239.
[9] Miller S.L., Science 117 (1953) 528.
[10] Mills A.A., Nature 220 (1968) 1113.
[11] Mills A.A., Nature 224 (1969) 863.
[12] Mills A.A., Nature 225 (1970) 929.
[13] Morowitz, this volume, p. 37.
[14] Mueller G., Nature 196 (1962) 929.
[15] Mueller G., Nature 218 (1968) 1239.
[16] Mueller G., Astrophys. Space Sci. 4 (1969) 3.
[17] Yang C.C. and J. Oró, this volume, p. 155.
[18] Ponnamperuma C. and N.W. Gabel, Space Life Sci. 1 (1968) 64.
[19] Smith J.W. and I.R. Kaplan, Science 167 (1970) 1367.
[20] Studier M.H., R. Hayatsu and E. Anders, Geochim. Cosmochim. Acta 32 (1968) 151.
[21] Sylvester-Bradley P.C., Discovery 25 (1964) 37.
[22] Sylvester-Bradley P.C., Proc. Geol. Assoc. 78 (1967) 137.
[23] Sylvester-Bradley P.C., Proc. Geol. Assoc. 82 (1971), in press.
[24] Turner R., Chemical Engineering Practice, Vol. 6 (Butterworths, London, 1958) p. 169.
[25] Woeller F. and C. Ponnamperuma, Icarus 10 (1969) 386.
[26] Zenz F.A. and D.F. Othmer, Fluidization and Fluid-Particle Systems (Reinhold, New York, 1960) p. 512.

Chemical Evolution and the Origin of Life, eds. R. Buvet and C. Ponnamperuma

CARBONACEOUS CHONDRITES AND THE CHEMICAL EVOLUTION OF ORGANIC COMPOUNDS

G.P. VDOVYKIN

V.I. Vernadsky Institute of Geochemistry and Analytical Chemistry, Academy of Sciences of the USSR, Moscow, USSR

The problem of the origin of living matter is connected with the formation and transformation of organic compounds under various conditions in the solar system [13,14]. Of particular interest are rare meteorites – carbonaceous chondrites which contain complicated forms of extraterrestrial organic matter.

According to results of our investigations [12], up to 0.5 percent (of the weight of meteorites) of organic matter is isolated from carbonaceous chondrites by organic solvents. This organic matter is composed of bitumen-like compounds containing up to 74 percent of "oily" components represented by hydrocarbons. The predominant form of (5–7%) carbonaceous matter in these meteorites is a polymeric and is insoluble in organic solvents. It can be isolated by treatment of meteoritic matter with HF and HCl.

The organic matter of carbonaceous chondrites is characterized by a high degree of oxidation and a heightened content of heteroatoms among which chlorine is of special interest. In the meteorite Groznaya, according to data of elementary microanalysis [8], the bitumen-like matter contains 35.19% C; 7.80% H; 49.12% N + S + Cl + O; 7.89% ash. The polymeric matter isolated from the meteorite Staroye Boriskino contains 5.47% H and 2.56% Cl [10]. According to Raya's data [2] the composition of the polymeric matter of the meteorite Orgueil is the following (in percent): C 70.39, H 4.43, Cl 1.22, F 1.25, N 1.59, S 6.91, O 9.80, ash 4.58.

Infrared spectrophotometry of the bitumen-like matter from carbonaceous chondrites shows that it is a mixture of organic compounds of a hydrocarbon and of a non-hydrocarbon character. The infrared absorption spectra of the bitumen-like matter isolated with alcohol-benzene from the carbonaceous chondrites Mighei, Staroye Boriskino and Groznaya [8,9,12] are rather similar to one another. The spectra show the presence of CH-groups (2960–2850 cm^{-1}), $-CH_3$ (1460 cm^{-1}), $=CH_2$ (1380 cm^{-1}) groups, as well as of the carbonyl group $-C=O$ (1730 cm^{-1}). In the infrared absorption spectra of the high-molecular organic matter of the carbonaceous chondrites, Mighei and Groznaya, there are also absorption bands suggestive of CH–, OH– and –C = O groups. This organic matter may be a condensed aromatic system with side groupings.

In the ultraviolet absorption spectrum of the bitumen-like matter isolated from the meteorite Mighei an absorption band with a maximum at 360 nm is noted, which is suggestive of anthracene. Anthracene, as well as 1.12–benzperylene, 3.4–benzpyrene, traces of perylene and coronene have been noted in the bitumen-like matter of the carbonaceous chondrites, Mighei, Staroye Boriskino, and Cold Bokkeveld. In the meteorites, Mighei, Staroye Boriskino and Groznaya, we have identified [17] a number of amino acids among which glycine and alanine prevail.

The organic matter of many carbonaceous chondrites was studied in detail by other investigators, especially with the use of gas chromatography and mass spectrometry. Hydrocarbons, aromatic and fatty acids, nitrogen-containing compounds, amino acids, cyclic compounds of the type of purines and pyrimidines, and porphyrines have been identified in carbonaceous chondrites. Among hydrocarbons normal and branched paraffins, olefines, and aromatic hydrocarbons have been identified [4,5,7]. Of great interest is the presence of isoprenoids, which were thought to be only characteristic of living matter. However, isoprenoids may have been synthetized from CO and D_2 by the Fischer–Tropsch reaction [7]. Cyclic nitrogen-containing compounds – adenine, dicyanodiamidine, melamine – were identified in a sample of the meteorite Orgueil [1]. These compounds have been obtained experimentally by heating CO, H_2, and NH_3 mixtures [1]. Hodgson and Baker [3] have established porphyrins in the meteorites: Orgueil (0.01 ppm), Murray (0.003 ppm), Cold Bokkeveld (0.001 ppm), and Mokoia (0.0005 ppm). By their pecularities the meteoritic porphyrins approached those synthetically obtained from simple organic compounds.

Unlike terrestrial biogenic matter, the organic compounds in meteorites appear to be optically inactive although there is insufficient material available to detect small amounts of optical activity. In some papers traces of optically active compounds and even fossilized microorganisms have been noted in the meteorite Orgueil but it was found that they are results of contamination under terrestrial conditions.

In order to more precisely determine the phase composition of the carbon matter of carbonaceous chondrites we have investigated it by methods of structural analysis [9,12,21]. This matter is represented by polymers having both an amorphous and a crystalline structure. In separate cases a very small admixture of particles with a more regulated structure is found. In the diamond-containing meteorite Novo Urei, Vdovykin [15] identified a new hexagonal carbon modification.

We have detected free organic radicals in the high molecular organic components of the carbonaceous chondrites, Orgueil, Mighei, Staroye Boriskino and Cold Bokkeveld [11,19]. On the EPR spectra they are displayed in the form of rather intensive lines of paramagnetic absorption having similar parameters.

The isotopic carbon composition in carbonaceous chondrites varies. According to our data [18] the $\delta^{13}C$ value of total C is (in percent, reduced to the PDB standard): Mighei –1.11, Staroye Boriskino –0.85, Cold Bokkeveld –1.46, Groznaya –2.05, Kainsaz –1.53. In comparison with the total C the organic matter is enriched in the light C^{12} isotope. For the bitumen-like matter of the meteorite Mighei the δC^{13} value is –2.20, of the meteorite Groznaya –2.87; for the inextractable polymeric matter of the meteorite Mighei the $\delta^{13}C$ value is –2.11, of Staroye Boriskino –2.26, of Groznaya –2.23, of Kainsaz –1.88. The enrichment of the total C in the heavy ^{13}C isotope in comparison with the C of organic matter in carbonaceous chondrites occurs at the expense of carbonates, the C of which is substantially heavier.

Thus organic compounds of carbonaceous chondrites differ from terrestrial biogenic organic matter in rocks in various ways, including the different character of the ratio of separate components and the presence of Cl [16]. Organic matters, identified in meteorites, have been obtained experimentally by synthesis from simple initial compounds, the distribution of organic matter in meteorites frequently being similar to the distribution of the synthesis products. Our experimental investigation [20] on the synthesis of amino acids during irradiation of six mixtures of simple compounds (CH_4, CO_2, NH_3, N_2, NH_4Cl, S_2 in the presence of H_2O) by protons (E = 600 meV, I = 5.4×10^{14} protons/cm^2) have shown that during radiogenic synthesis the amino acids glycine, alanine, glutamic, asparaginic acids, lysine, histidine, valine, leucine are formed. The most abundant in synthesis products and in meteorites were glycine and alanine. The greatest yield of amino acids has been noted in samples in which C initially was present in the CH_4 form and N in the NH_3 form. Their results may support the idea that the organic compounds in meteorites are radiogenic in origin.

Besides organic matter, carbonaceous chondrites contain minerals of low-temperature origin – carbonates, elementary sulfur, chlorite-serpentine minerals, etc. – which secondarily replace minerals of high-temperature origin. In various carbonaceous chondrites the amount of these minerals containing C, H, O, S, etc. increases with the increase of the organic matter content [12]. Such structure peculiarities show that the organic matter has formed simultaneously with low-temperature minerals, most likely during the meteoritic agglomeration as the result of chemical reactions from simple initial compounds.

Similar organic compounds could have also entered the earth at an early stage in its development. During differentiation and formation of the earth's crust they could again disintegrate with formation of simple compounds which in the degassing process reacted forming complex organic matter enriching the upper layers of the earth's crust. Here, on the surface, from abiogenic complex organic compounds originated living matter. Thus the results of a study of organic compounds may support the idea [6] that life appeared spontaneously on the earth.

The general similarity of organic compounds of meteorites, products of abiogenic synthesis and terrestrial biogenic material, the great abundance of C and other biophilic chemical elements in cosmic space – leads one to think [12,17] that if life exists on other planets besides the earth, at least within the ranges of the solar system it must be based on carbon and chemically similar to life on earth.

References

[1] Hayatsu R., M.H. Studier, A. Oda, K. Fuse and E. Anders, Geochim. Cosmochim. Acta 32 (1968) 175.
[2] Hayes J.M., ibid., 31 (1967) 1395.
[3] Hodgson G.W. and B.L. Baker, ibid. 33 (1969) 943.
[4] Nooner D.W. and J. Oró, ibid. 31 (1967) 1359.
[5] Olson R.J., J. Oró and A. Zlatkis, ibid. 31 (1967) 1935.
[6] Oparin A.I., Life: its Nature, Origin and Development (Nauka, Moscow, 1968).
[7] Studier M.H., R. Hayatsu and E. Anders, Geochim. Cosmochim. Acta 32 (1968) 151.
[8] Vdovykin G.P., Geochemistry 2 (1962) 152.
[9] Vdovykin G.P., Geochimiya 4 (1964) 299.
[10] Vdovykin G.P., in: Problems of Geochemistry (Nauka, Moscow, 1965).
[11] Vdovykin G.P., Meteoritika 26 (1965) 151.
[12] Vdovykin G.P., Carbon Matter of Meteorites (Organic Compounds, Diamonds, Graphite) (Nauka, Moscow, 1967).
[13] Vdovykin G.P., in: Physics of Planets (Nauka, Alma-Ata, 1967).
[14] Vdovykin G.P., in: Origin of Organic Matter in Solar System (Mir, Moscow, 1969) preface.
[15] Vdovykin G.P., Geochimiya 9 (1969) 1145.
[16] Vdovykin G.P., in: Advances in Organic Geochemistry (Pergamon Press, Oxford, 1968).

[17] Vdovykin G.P., in: Extraterrestrial Life and Methods, its Discovery (Nauka, Moscow, 1970).
[18] Vinogradov A.P., O.I. Kropotova, G.P. Vdovykin and V.A. Grinenko, Geochem. Intern. 4 (1967) 229.
[19] Vinogradov A.P. and G.P. Vdovykin, Geochem. Intern. 5 (1964) 831.
[20] Vinogradov A.P. and G.P. Vdovykin, Geochimiya 9 (1969) 1035.
[21] Vinogradov A.P., G.P. Vdovykin and N.M. Popov, Geochem. Intern. 2 (1965) 249.

Chemical Evolution and the Origin of Life, eds. R. Buvet and C. Ponnamperuma
©1971, North-Holland Publishing Company

THE PLANETS AND LIFE

Richard S. YOUNG
NASA Headquarters, Washington, D.C. 20546, USA

Exobiology is a research program, the basic objective of which is to cast light on the question of the origin and early evolution of life. There are many avenues of approach to such research. First, the area of chemical evolution; the sequence of events presumed to have taken place on the primitive earth or some other primitive planet in which macromolecules were synthesized nonbiologically and which led to the origin of life. Organic geochemistry – the study of the ancient chemical and biological fossil record of a planet – can be included under chemical evolution. Second, the study of the environmental extremes in which terrestrial forms of life are capable of surviving and growing. A third part of exobiological research is the development of life detection and organic analytical techniques for use in terrestrial and extraterrestrial environments.

A very important component of exobiological research is in relating the results of such research to contemporary terrestrial problems. We must look at the origin and early evolution of life as being inseparably interwoven with the origin and evolution of the planet, and the cause and effect relationship between a biota and its parent is one of the major problems of the contemporary earth and one to which research encompassed by exobiology can contribute a great deal of basic as well as practical information.

The subject of this paper, however, deals primarily with planetary exploration which is another part of exobiological research. Since the planet earth is overwhelmed by its biota it is almost impossible to look back into the history of earth and determine the succession of events which preceded life. The marks of life are planet wide and it is very difficult to go beyond the living record in order to study the preliving record on earth. It may be that only by going to another planet with a similar history to that of earth will we be able to get at this early record in a natural environment. Planetary exploration is not simply a program designed to detect life on another planet, in fact, one of the most important things that could be done would be to find a planet similar to the earth, such as Mars, on which life has not arisen but on which chemical evolution took place without giving rise to a living system. This planet, when studied for evidence as to why life did not arise, may turn out to be scientifically more important than a planet which has already produced a living system. Planetary exploration then, is a program in search of evidence related to planetary, prebiological as well as biological evolution.

Our solar system has nine planets. We will look briefly at these planets beginning with Mercury in an attempt to evaluate their potential for data relevant to the question of how, when and where life began. Mercury is somewhat larger than the moon and has a very high density, however, it has little if any atmosphere. The absence of an atmosphere and the very high temperature on the sun side of Mercury pretty much preclude the possibility of water and of organic molecules. Thus, Mercury does not seem a high priority objective for study, at least as far as exobiological objectives are concerned. It should be stressed at this point that we are talking about planetary exploration from the point of view of exobiology. Planetologists, physicists, geologists, etc. might put a quite different priority on planetary exploration than the exobiologists. Venus is shrouded in mystery in that the surface of the planet is not visible, being very densely cloud covered. It is approximately the same size and mass as the earth, and the atmospheric pressure at the surface is probably between 60 and 100 atm., primarily composed of carbon dioxide. The surface temperature is extremely high, approaching 900 ° to 1,000 °F. A Soviet spacecraft, Venera 4, probed the atmosphere of Venus detecting the presence of small amounts of water, carbon monoxide and oxygen in addition to carbon dioxide. Ground-based studies have also shown the presence of traces of hydrofluoric and hydrochloric acids. Venus is clearly an interesting planet, although the high surface temperatures again preclude the possibility of water or organic molecules. Some authors have speculated about the possibility of large amounts of ice in the polar caps of Venus, but the evidence for this is practically nonexistent. Other speculations have been made concerning the possibility of a bio-zone in the upper atmosphere, perhaps in the region of the cloud layer where the temperatures are in the range of 60–90 °F, and where carbon dioxide, water vapor and perhaps oxygen are present. This, however, would require a completely airborne ecology for which there is no terrestrial counterpart. Jupiter is an extremely massive planet of very low density. The atmosphere of Jupiter contains methane and ammonia, carbon dioxide, water, hydrogen and carbon monoxide, resembling in some respects, the atmosphere of a primitive planet. Therefore, the chemistry of Jupiter is of great interest since it can be postulated that here we have a primitive atmosphere on a highly energetic planet, which may well be synthesizing organic molecules today. Laboratory experiments also suggest the likelihood of high molecular weight organic molecules being synthesized on Jupiter. We would propose an orbiter or fly-by Jupiter with infrared spectroscopy aboard or an atmospheric probe with a mass spectrometer aboard, to cast some light on the chemical composition of regions in the Jovian atmosphere of great interest to exobiology. We can also consider Saturn, Uranus, Neptune and Pluto in the same category. Although little is known of these planets, it is assumed that they are at least superficially similar to Jupiter. Some of the satellites of Saturn are probably sufficiently large to have an atmosphere of their own and can therefore be considered of some interest to exobiology. Very little is known of these outer planets and it will probably be some time before a detailed exploration is made of them, so for the time being, our primary interest should probably be in the planet Jupiter. We plan a "Grand Tour" of the planets, in which a single spacecraft can be launched on a trajectory which makes use of the gravitational field of the large planets as an acceleration assisting mechanism so that all of the planets can be flown by in a single spacecraft. Such a trip is of special interest to the planetologists and physicists in that it requires an alignment of the planets that occurs once in over one hundred years.

There are two other types of bodies in the solar system that have potential for exobiology. The asteroids are of interest particularly from the point of view of chemical composition, however, it is certain to be some time before the detailed study of an asteroid is feasible. Comets have been studied spectroscopically by telescope and are found to contain carbon–hydrogen bonds indicative of organic molecules. Cometary material may have made a sizable contribution to the planets during their history and more needs to be known about the chemical composition before we can accurately assess their role in the production of planetary organic matter.

The most interesting planet in the solar system from the point of view of exobiology is Mars. Fig. 1 is an illustration of Mars as seen through the telescope and shows the gross features that have fascinated astronomers ever since the invention of the telescope. The orange–reddish color of Mars is apparent, the polar cap can be seen as can the light and dark regions. These gross features have stimulated considerable speculation about the possibilities for life on that planet. Fig. 2 shows two photographs of Mars, one during spring in the southern hemisphere and the other during summer in the southern hemisphere. It is obvious that the polar cap disappears during the course of the summer and at the same time the dark areas on the surface of the planet become progressively darker. The so called "wave of darkening" has been viewed for many years and has led many astronomers to conclude

Near earth view of Mars (late August 1956)

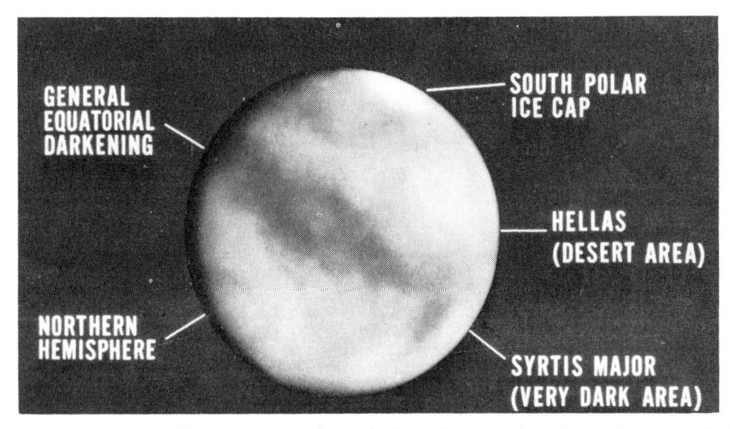

Fig. 1. An illustration of Mars as seen through the telescope that shows the gross features that have fascinated astronomers ever since the invention of the telescope.

MARS

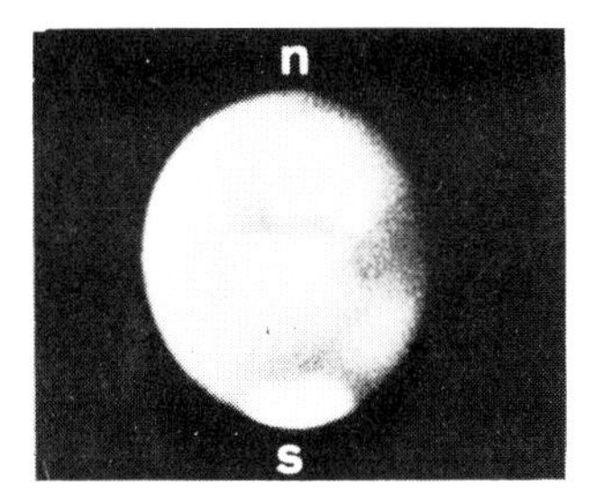

Southern Spring

Southern summer

- Day and seasons like earth
- Light atmosphere
- Atmosphere primarily CO_2, minor components H_2O, CO
- Moderate surface temperature

- Polar caps
- Dark and bright areas
- Seasonal and secular changes
- Little or no magnetic field
- 2 small satellites

Fig. 2. Two photographs of Mars, one during spring in the southern hemisphere and the other during summer in the southern hemisphere.

that this was the response of Martian vegetation to the availability of water from the pole cap. We now know from ground-based observations and from the 1969 Mariner fly-bys that the pole caps are probably not composed of water, but are primarily composed of frozen carbon dioxide, and that water comprises but a trace of the pole cap material. The day/night cycle on Mars is quite similar to that of the earth, being almost 24 hr although the seasons are almost twice as long as on the earth. Mars does have a tenuous atmosphere, around 10 mb in pressure at the mean surface, and composed primarily of carbon dioxide. The only other molecules that have been detected are water and carbon monoxide. Water appears to be present in amounts varying from zero to about 50 microns of precipitable water, which is about 1/1000th of that in the earth's atmosphere. The amount of water in the atmosphere of Mars appears to vary with the season. The temperature at the surface of the planet ranges from as high as 70 °F at the equator during the day to about −100 °F at night so that there is a tremendous diurnal freeze–thaw cycle, even at the equator. The mean temperature is probably about 40 ° below that of the earth, however, this does not preclude the possibility of biological activity. The pole caps on Mars wax and wane with the seasons and are predominantly composed of frozen carbon dioxide. There appears to be little or no magnetic field, so that the ultraviolet flux at the surface is quite high. This presents a problem for biological activity, but since ultraviolet is relatively easy to shield, it is not considered to preclude biological activity. Nitrogen was not detected in the Martian atmosphere by the Mariner spacecraft, however, atmospheric nitrogen is not a requirement for biological activity. Nitrogen may well be present in the surface material in the form of nitrate or ammonia salts. In addition, the sensitivity of the ultra-violet photometer in Mariner 6 and 7 was such that there could still be as much as 1% nitrogen in the Martian atmosphere and not be detected.

Mariner 6 and 7 flew by Mars in 1969 with a series of instruments on board, including wide and narrow angle television cameras, an infrared radiometer, an infrared spectrometer and an ultraviolet spectrometer. These flights were extremely successful and produced a great deal of photographic and spectrophotometric data from which the previously cited Martian parameters were derived. Close up photographs of Mars showed the edge of the Martian pole cap and some of the frozen CO_2 filled craters at the edge. Photographs such as this led to the conclusion that the depth of the CO_2 layer at the pole caps may be several meters. Most of Mars is heavily cratered, although some regions showed no craters at all. The total absence of craters in some areas is rather mystifying.

In 1975 the Viking mission will be launched, in which a spacecraft will be soft landed on the surface of Mars. This will be an orbiter-lander combination, the orbiter serving as the relay station for lander data, plus visual and spectrophotometric experiments to be performed in conjunction with the lander. The lander has exobiological objectives as primary to the mission, these include the direct search for biological activity, organic soil analysis, the search for water, meteorological measurements and atmospheric measurements both on the surface and during entry. All of the measurements are generally aimed at the question of life, organic molecules and those environmental parameters that are most directly relevant to the life question. There will also be a gas chromatograph–mass spectrometer designed for Viking to be used in search for organic molecules in Martian soil. This is a pyrolysis experiment in which the soil is pyrolyzed and the end products of the pyrolysis interpreted in terms of parent organic composition. In addition to these two experiments there will be a camera, seismometer and other instruments designed to monitor pressure, temperature, wind velocities, etc. on the surface.

All of these experiments must, of course, be considered preliminary. It is hoped that they will produce sufficient data to provide us with basic information about the presence of organic matter and the presence or absence of life, so that more sophisticated and detailed experiments can be designed for subsequent missions, perhaps including the eventual exploration of Mars by manned expeditions.

Chemical Evolution and the Origin of Life, eds. R. Buvet and C. Ponnamperuma

EXTRATERRESTRIAL LIFE STUDY. PROBLEM OF ITS ORIGIN AND EVOLUTION

V.A. OTROTCHENKO and L.M. MUKHIN
Institute of Space Research, Academy of Science of the USSR, Moscow, USSR

The present development of space research allows us to come close to the solution of one of the very important problems – the detection and investigation of extraterrestrial life forms.

The existence of extraterrestrial life is one of the fundamental problems of philosophy and natural sciences and related to the problem of the origin and evolution of life in the universe as well as on the earth.

The Oparin–Haldane hypothesis essentially hastened the development of investigations connected with this problem and helped the emergence of the new trend in biology [5] Due to some recent experiments, we consider that under conditions of the earth's primary atmosphere and, perhaps, partially even during the process of earth formation, synthesis of the main biologically essential compounds can occur. But despite the first success of this field many unsolved problems exist: (1) the plurality of inhabited worlds; (2) the conditions of their origin; (3) the structural peculiarities and the pathways of evolution of both organic matter and life in each specific case; (4) the interpretation of transition from organic matter to primitive life; (5) the transfer of life once formed from one planet to another.

Attempts to solve the problem of the existence of life on the near earth planets by observations directly from the earth as well as the search for traces of extraterrestrial life on meteorites falling on to our planet's surface have not led to definite results. Our information about life is based on terrestrial life. But the single origin and common conditions of all its forms strongly restrict the possible ways of evolution. The life which originated on the earth or perhaps came from outer space – the structural peculiarities of which could be dictated by random causes, affects all subsequent evolution in the sense of community of origin and the common structural features of all further forms. Their constant competition results only in systems with a definite complexity of structure remaining at each stage of evolution. The contemporary forms of life have evidently driven out intermediate forms which appeared during transition from non-living to living matter, as well as the origination of the primary forms of life and their further evolution.

Competition which is a very important factor of evolution does not permit the realization of all variants of organisms during adaptation to environmental conditions. The conditions themselves of the places of existence of terrestrial life place their own restrictions. Also it is impossible to be sure that, in conditions like the terrestrial ones, life exhausted all possible ways of evolution, not to mention the more significant changes in conditions strongly differing from the terrestrial environments.

From the properties of living systems on planets one could expect to obtain unique data on new forms of life and, perhaps, other principles of organization in conditions that are different from terrestrial ones or, at least, about life on a later or earlier stage of evolution including forms of protolife. Also the possibility of detection of life with a basis modified in comparison with terrestrial life is not ruled out. The protein–nucleic acid basis of terrestrial life is determined by the set of monomeric compounds that is single for all forms of life, which, perhaps, does not have any advantages over other ones and was separated out only in connection with earlier random formation [14].

It can be said that near our planet nature has set up some more experiments for examining both the ways and the conditions of organic matter evolution, at least, within the solar system.

It should be noted that recently by astrophysical observations the water availability and the presence of such a predecessor of complex organic compounds as formaldehyde and ammonia were found beyond the solar system [2,3,9,11].

Therefore, the investigation and the consideration of all circumstances and possibilities under which the evolution of organic matter beyond the earth occurred or, on the contrary, was stopped, may be one of the decisive factors in the correct understanding of the problems of the origin of life.

It is logically suggested that the primary synthesis of organic matter occurs first based on the gaseous atmospheric components. The atmosphere is easily mixed and thereby the presence of initial substances is continuously provided for the synthesis reaction. In addition, the atmosphere of any planet, at least in the upper layers, is affected by the constant inflow of energy at the expense of solar radiation.

Unfortunately, the chemical composition of the atmospheres and, more so, the solid matter of planets that are the nearest to the earth is undetermined

Table 1a
Venus

Gases	Spectroscopic data on gas composition up to the level of the cloud layer [4] (in bars, pressure 0.1–1 atm, temperature 240–300°K)	Space stations data	Calculated data on the surface level [4] (in bars, pressure 65 atm, temperature 650 °K)
CO_2	0.1	about 90%	
CO	9×10^{-6}		1×10^{-4}
H_2O	1×10^{-5}	$1-7\times10^{-3}$ (C_{H_2O}/C_{CO_2}) [a]	2×10^{-4}
HCl	6×10^{-8}		1.2×10^{-6}
HF	5×10^{-10}		
H_2			1.4×10^{-5}
COS			3×10^{-6}
H_2S			6×10^{-7}
NH_3			6×10^{-7}
CH_4			9×10^{-8}

[a] C is gas or vapor concentration in the atmosphere.

Table 1b
Mars

Gases	Gas composition data
CO_2	90%
H_2O	$10\mu/cm^2$ precipitated water
O_2	5×10^{-5} (C_{O_2}/C_{CO_2})
CO	?
N_2	?

Table 1c
Jupiter's atmosphere model [4a]

Gases	Pressure (log P in bars [b])	Temperature (°K)
NH_3	0.5	100
NH_3SH	1	200
H_2O, NH_4OH	2	300
NH_4Cl	3	400–500
SiO_2, Na_2SiO_3, K_2SiO_3	3–4	500–1500
Fe, Ni	4	2000

[b] P is the atmospheric pressure.

to a great extent. For many important components of the planets' atmospheres the upper limits of concentrations are only known as a consequence of limited possibilities of astronomical observation techniques. The most complete investigations of planets, at present, can be provided only by automatic space stations. In spite of the very limited information about chemical composition of the atmospheres and about the solid substances of planets, it is worthwhile to estimate the possibility of existence on the near-earth planets of the living systems or molecules which are the precursors of complex organic compounds. In table 1 data of atmospheric compositions are shown.

Venus

The experimental data about the chemical composition of the upper Venus atmosphere, its temperature, density and structure are mainly obtained by astrophysical investigations or experiments performed by spacecraft.

The availability of the noticeable amount of water vapour (0.5%) is the most significant fact. The presence of carbon dioxide and carbon monoxide, equal to 90% and 10^{-2} %, respectively, was well established. Hydrogen chloride and hydrogen fluoride have been successfully detected by astronomical experiments. The calculations [4] suggest the presence of methane and ammonia. The lack of definite data about the presence of oxygen and hydrogen does not allow a conclusion regarding the oxidative or reductive character of the planet's atmosphere. Meanwhile, these facts have a decisive value for the abiogenic synthesis problem because on the basis of all experimental experiences in this field abiogenic synthesis is assumed to be possible only under conditions of a reductive atmosphere [6].

If it is assumed that the atmosphere of Venus is of volcanic origin, then it must evidently display weak reductive properties.

This assumption becomes more evident when consideration is made of the chemical composition of the earth's magmatic and fumarole gases [10] which are given in table 2.

It should be emphasized that the availability of such components as HCl and HF in the atmosphere of Venus may serve as an additional argument for its volcanic origin. The processes of abiogenic synthesis of organic compounds may occur just in the reductive atmosphere. The possible pathways of such synthesis may be noted as applied to Venus. CO_2 is decomposed into CO and O due to the shortwave UV radiation (1600 Å) and the equilibrium ratio of CO to CO_2 which must be approximately equal to 10^{-4} at the pressure of 1.5

Table 2
Composition of the magmatic gases (in percent) at 1000–1100 °C.

Volcano	CO_2	CO	H_2	SO_2, H_2S	CH_4	N_2 and rare gases
Michara (Japan)	28.0	–	72.0	–	–	–
Oshima (Japan)	9.8	–	90.2	–	–	–
Kilauea and Mauna Loa (the Hawaian Islands)	57.0	1.6	1.7	39.7	–	–
Niiragongo (Congo)	86.7	4.6	1.5	7.2	–	–
Etna (Sicily)	28.8	0.5	16.5	34.5	1.0	18.7

atm. Due to the shortwave UV radiation from CO_2 and the atmospheric water vapors, formaldehyde is produced which is a precursor for more complicated organic compounds (sugars). This suggests the necessity of careful exploration of the main components of the atmosphere of Venus by the existing direct methods.

Mars

This planet appears to be the most promising for conducting exobiological investigations, since CO_2 and H_2O were identified in Mars and UV radiation is present on the planet's surface.Experiments of abiogenic synthesis under conditions simulating the Martian atmosphere have yielded interesting results. Even the early experiments [13] suggest the presence of acetaldehyde on Mars. The problem concerning the existence of stationary concentrations of variant organic compounds in Martian conditions is of great interest.

Mars is apparently the only planet of this solar system besides earth where the existence of living systems appears to be likely. As the temperature in the planet's equatorial regions at the warmest season can rise 20°C during the day, lack of liquid water is a main factor that limits the development of living systems. But on dry, cold planets, the water in the frozen ground also can be a solvent and a medium with a mixing coefficient with temperatures below 0°C. After freezing in the ground it is not a usual ice. At the freezing point, 0°C, according to the modern concept [7], the water structure corresponds to the regions in which its molecules having four evolving bonds, two hydrogen and 2 non-separate π-bonds of oxygen, can be linked by 4 hydrogen bonds with 4 water molecules and form comparatively stable framework. But in the region where the usual structure is broken and these four bonds cannot be formed (because of ion influence or even the water surface itself) water maintains mobility even with temperatures of the order of $-10°C$. It maintains practically the features of liquid water in the frozen ground and may transfer soluble substances [1]. Here it should be noted that some bacteria display the capability of living and reproducing at rather low temperatures in medium culture at -8–$9°C$ [8].

It is perhaps probable that the water is available from the hoar-frost which apparently falls at regular intervals. On the evaporation of the hoar-frost the temperature may be close to $0°C$. On contact with the planet's surface this layer could serve as a solvent of salts [15].

Jupiter

The data on the Jovian atmosphere is rather limited. In this case the formation of a series of organic compounds in the planet's atmosphere is apparently possible. As the atmosphere has a prominent reducting character, and methane and ammonia are present in significant amounts, the formation of such materials as hydrogen cyanide, and aldehyde is possible in the presence of water in the atmosphere of lower layers. The first experiments which have been carried out in such conditions affirm this assumption [12]. Therefore, Jupiter may also be considered as an interesting subject from the viewpoint of the exploration of the evolution of organic matter within the solar system.

In summary, it should be emphasized that only investigations of organic matter and its evolution on the other planets will completely allow answers to the questions on the origin of life on the earth.

References

[1] Ananjan A.A., Kolloidn. Zh. 14 (1952) 1.
[2] Barrett A.H., Commun. Atomic Mol. Phys. 1 (1969–1970) 5.
[3] Cheung A.C. and C.H. Tauns, Nature 221 (1969) 917.
[4] Lewis J.S., Icarus 8 (1968) 434.
[4a] Lewis J.S., Icarus 10 (1969) 393.
[5] Oparin A.I., The Origin of Prebiological Systems and of their Molecular Matrices (Academic Press, New York, London, 1965).
[6] Ponnamperuma C. and N.W. Gabel, Space Life Sci. 1 (1968) 64.

[7] Samojilov O.Y., Structures of Aqueous Solutions (Nauka, Moscow, 1957).
[8] Smart H.F., Science 82 (1935) 525.
[9] Snider L.E., Phys. Rev. Letters 22 (1969) 679.
[10] Sokolov V.A., Gases of the Earth (Nauka, Moscow, 1969).
[11] Whiteoak J.B. and F.F. Gardner, Astrophys. Letters 5 (1970) 5.
[12] Woeller F. and C. Ponnamperuma, Icarus 10 (1969) 386.
[13] Young R.S., C. Ponnamperuma and B.C. McCaw, Life Sci. Space Res. 3 (1965) 127.
[14] Aksjonov S.I., Probl. Cosmicheskoy Biol., in press.
[15] Aksjonov S.I., Biol. Nauky 7 (1968).

Chemical Evolution and the Origin of Life, eds. R. Buvet and C. Ponnamperuma

CARBON CHEMISTRY OF THE MOON

Geoffrey EGLINTON
Reader in Organic Geochemistry, University of Bristol, Bristol, England

In this talk I shall endeavour to summarize the information on the carbon chemistry of the moon up to, and including, the 'Moon issue' of *Science* published (Vol. 167) at the end of January, 1970. The full articles have been published in *Geochim. Cosmochim. Acta* (1970, Vol. XXXIV, Supplement).

The analyses of the lunar samples reported at the Houston conference in January 1970 represented the first laboratory analyses of material from another planet and are, therefore, of great significance. Never has so much laboratory effort been expended on one material in such a short time, through so many techniques and by so many people. Prior to the return of the samples, only guesses as to the organic content could be made and investigators instituted detailed preparations so that they would be able to extract, isolate and identify most types of organic matter [7]. In the event, the carbon content proved to be very small, but, even so, it still has great significance in relation to the conditions which might have supported the origin of life on earth. The evidence now available is that the lunar surface has no water, free or of crystallization, that the carbon content has always been low and that the surface is very ancient but has been modified by various erosion processes. Hence, the moon represents a particular stage of planetary evolution which is certainly not equivalent to a primitive earth but is interesting as a comparison. Thus, even if amino acids and polypeptides were to be formed, for example as a result of the levels of radiation exposure, etc. on the moon, no water would be available for the formation of proteinoid microspheres [10] and hence one possible route to living organisms.

The lunar results reported herein are also relevant to the search for life on Mars, for they would suggest that carbides and meteoritic fragments should have accumulated on the surface of Mars. The slight atmosphere, largely CO_2, will have decelerated incoming fragments but what overall planetary chemistry exists we do not yet know.

The bioscience investigators

The Bioscience group of Principal Investigators selected by NASA for the Apollo 11 mission numbered 19 scientists. Their tasks were summarized by NASA under the following headings:

1) Determine the structures and relative abundances of compounds of carbon indigenous in, and deposited on, the lunar surface.
2) Determine origin of the indigenous carbon compounds.
3) Catalog microstructures in terms of organized elements and microfossils.
4) Define presence or absence of viable lunar organisms.

The organic chemists and organic geochemists involved numbered about twelve, though there was some overlap between this classification and the other listed below. The twelve were Biemann [21], Calvin [5], Eglinton [1], Fox [10a], Halpern [12], Kaplan [14], Lipsky [16], Meinschein [19], Nagy [22,23], Oro [24], Ponnamperuma [26], Rho [27] and their collaborators. Gas chromatography, mass spectrometry and combined gas chromatography–mass spectrometry were the methods chosen by these investigators, though amino acid analysers and related equipment were also used. The quantitation of the organic matter present, either as volatile and pyrolisable material or as total carbon was studied by several investigators, notably Burlingame, Johnson, Moore and their collaborators, the techniques being combustion, pyrolysis and detection through gas chromatography and mass spectrometry [17,18,13,14,20].

Micropalaeontology, involving light and electronmicroscopy, was undertaken by three investigators – Barghoorn [4], Cloud [6], Schopf [29] and their colleagues. A search for living pathogens was made by the Lunar Receiving Laboratory as part of the quarantine operation and used a wide variety of test organisms ranging from Japanese quail to unicellular plants. Oyama [25] and his colleagues sought bacteria by providing a great variety of culture media and conditions. In all these investigations no indication of past or present life-forms was apparent.

Contamination control

The problem of contamination control has been carefully studied at Houston [9]. It is obviously of paramount importance that the returned lunar samples should be contaminated as little as possible during the Apollo missions. Further, the Bioscience investigators needed to know what unavoidable contamination might be anticipated. Thus, biolipids, micro-organisms,

Table 1
Apollo missions – contamination controls for sample handling at Houston.

Type	Contaminants	Control measures
Biological	micro-organisms (inc. viruses)	double biological barrier – quarantine
Particulate	solid particles	clean room procedures
Inorganic	solids, liquids and vapours	clean room procedures limitation of materials used
Organic	solids, liquids and vapours	measurement of contamination limitation of materials used

organic vapours and plastics are but a few of the potential hazards. Table 1 summarises the situation in regard to contamination control. The control procedures are used throughout the pre-mission preparation of the lunar sample return containers and the lunar hand tools, the mission activities on the lunar surface, the transearth flight, the recovery operations and the final handling at the Lunar Receiving Laboratory. Most of the contamination is introduced at the last stage which includes gloved handling in the cabinets, in the vacuum chamber, and in the packaging area prior to distribution. All stages were monitored for organic contamination. This was partly done by using the organic mass spectrometric analytical facility housed at the L.R.L. In this procedure, the test materials were heated to 500 °C in the mass spectrometer while the total ion current was monitored. Additional analyses were performed by extraction with very clean solvent, the residues on evaporation being studied at the Space Sciences Laboratory, University of California, Berkeley, under the direction of Dr. A.L. Burlingame. In this case standard procedures were set up involving the monitoring of all surfaces with the aim of keeping them within an arbitrary limit of 20 ng/cm^2 detectable as peaks above the base line on injection onto a suitably equipped gas chromatograph. The monitoring procedures revealed varying background levels of phthalates, hydrocarbons, silicone oils, etc., which were put on file as possible contaminants of the lunar samples.

The lunar samples

The Apollo samples are essentially titaniferous basalt with the breccias and the lunar fines as derived materials. To some extent they must represent an overall sampling of the moon resulting from long-distance transport of impact- or volcanically generated debris. The accumulated layer of dust and rocks, the 'regolith', forms the surface of the Maria. Some of the information presently available for the Apollo 11 samples is summarised in table 2.

The Apollo 12 samples were somewhat similar though the quantity of titanium and iron present was substantially less and the rocks were mostly types A and B. The cores returned from Apollo 12 showed stratification and a considerable quantity of lighter dust was returned. Some of the lighter-coloured material was thought by the astronauts to lie on the surface as lines of ejecta. There was also rather more glass found on rocks and on the surface. A summary of the samples taken on Apollo 12 and subsequently issued to Bioscience Investigators is given in fig. 1, although some of the information is still preliminary. A remarkable development in lunar exploration is apparent when one compares Apollo 11, where only about 8 min remained to the astronauts to collect documented samples, with Apollo 12, where the whole of one EVA of some three hours was devoted to a geological traverse [18]. The astronauts visited a variety of craters, including young sharp-edged and boulder-strewn craters, craters of medium age and old craters, which were classified as such because of the rounding and loss of their sharp edges,

Table 2
Lunar samples (Apollo 11) Mare Tranquilitatis (~22 kg returned).

"Fines" soil (type D)	Breccia (type C) – most samples	Holocrystalline rocks (igneous; types A and B)
Glass and mineral fragments, micrometeorites (1%–2% carbonaceous chondrite?). Age ~4.6 $\times 10^9$ years (Pb/U/Th). Solar wind gases abundant: ~0.1 cc/g, Fe and Fe, Ni globules.	Shock (impact) or melt-compacted "fines". Solar wind gases also abundant.	Basaltic, micro crystalline (Type A) and macro-crystalline (Type B). Vesicularity common. Crystallization temperature, 1000–1300°C. Oxygen pressure 10^{-13} atm. at crystallization. Water content ~0%. Low viscosity when molten. High Fe and Ti. Age ~3.7 $\times 10^9$ years ($^{87}Rb/^{87}Sr$).

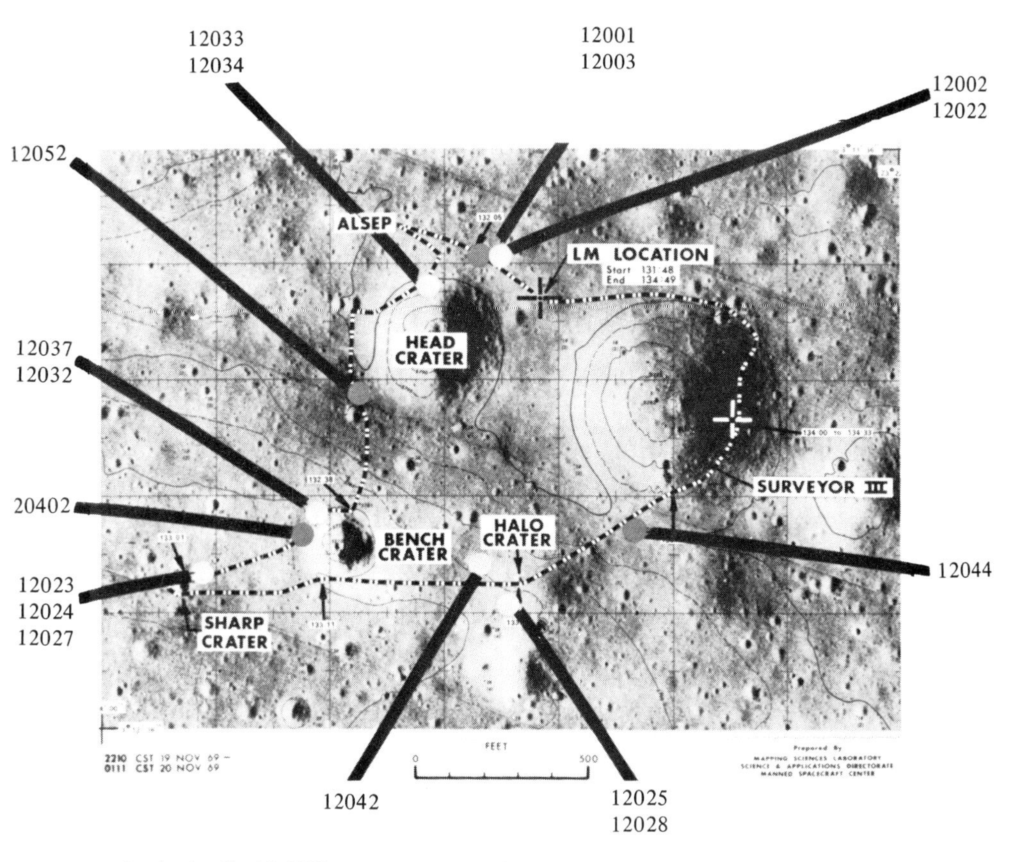

Fig. 1. Apollo 12. EVA traverse on the Oceanus Procellarum. Provisional summary of collections made at the Surveyor III site from which samples have since been allocated to the Bioscience Investigators

12001 130 ppm; selected fines (250 g). Sieved <1 cm. Aseptic, exhaust?
12002: selected sample (1529 g). Basalt.
12003: (ex 12001) 140 ppm; selected fines. Cleanest? Exhaust?
12022: selected sample (1864 g). Mound rock, crystalline
12023: 150 ppm; LESC fines. Crater rim (270 g). Bottom of trench, fine grey soil.
12024: 115 ppm; GASC fines (101 g). Trench, 8″ deep.
12025: core, top (56 g). Layers
12027: core no. 2. Bottom to trench, unopened
12028: core bottom (190 g). Layers: 115–270 ppm
12032: 60 ppm. Documented fines and fragments (310 g). Medium–dark-grey soil from side of trench? Fresh ejecta blanket?
12033: 55 ppm; documented fines (450 g). Trench bottom, 6″ deep; crater rim. Very light grey, volcanic ash?
12034: 64 ppm; documented breccia (155 g). Trench bottom
12037: 82 ppm; documented fines (145 g). Surface sample. Dark grey and granules
12040: 45 ppm; documented (319 g). Olivine dolerite
12042: 165 ppm; documented fines. Surface collection (225 g). Dark grey. Different texture, 'ray'?
12044: 44 ppm; documented fragment (92 g). Olivine basalt? Fragment?
12052: 25–65 ppm; documented (1866 g). Olivine basalt, high ilmenite.

Data for total carbon (ppm), kindly provided by C.M. Moore and collaborators.

presumably through long erosion. Conspicuous patches of light and dark 'soil' were described as having the appearance of volcanic ash. Another feature of the Apollo 12 traverse was the collection of fines from the bottom of a trench dug at the furthermost point of the traverse. This sample was taken for the Bioscience investigators as one which should be free from the exhaust products of the LM descent engine. The astronauts scooped these fines directly into a separate stainless steel container (the lunar environment special container, LESC) which they then sealed so that it remained separate from the other samples in the sample box. By comparison, the main bulk of the Apollo 12 fines was collected near the LM and will certainly have suffered some contamination from the exhaust plume of the descent engine. The LESC sample was taken some 300 meters away from the Lunar Module but it is worth noting that the Surveyor craft, which was itself some 200 meters from the LM, received a "sand blasting" from the lunar dust displaced by the Lunar Module during the final phase of its descent. We can confidently anticipate that a great deal more information concerning the carbon content of the lunar materials will be forthcoming when the variety of samples returned by Apollo 12 have been studied.

Table 3 summarises the events which probably affect, or have affected the abundance of carbon and the nature of the carbon compounds present in samples from the surface of the moon. The effects of some of these factors on the chemistry of carbon compounds can be estimated but one difficulty is

Table 3
Events on the Moon bearing on the abundance of carbon and the nature of the carbon compounds in lunar materials.

No water, free or of crystallization
High temperature (1000–1300 °C) at time of crystallization of holocrystalline rocks
High temperature at time of formation of glasses in fines
High vacuum and no free oxygen at time of formation of rocks and fines, or since
Radiation influx, cosmic rays and solar flares
Shock metamorphism, fracturing and lithification
Wide temperature range: surface $-140° \rightarrow +150$ °C
Solar wind implantation and depletion: consequential isotopic effects
Great ages of rocks and fines – approximate to age of Earth
Meteorite impact
Lunar erosion at a rate approximating to 1 mm/10^6 years
Lunar gardening
Long-distance transportation of rocks and fines
Volcanic events

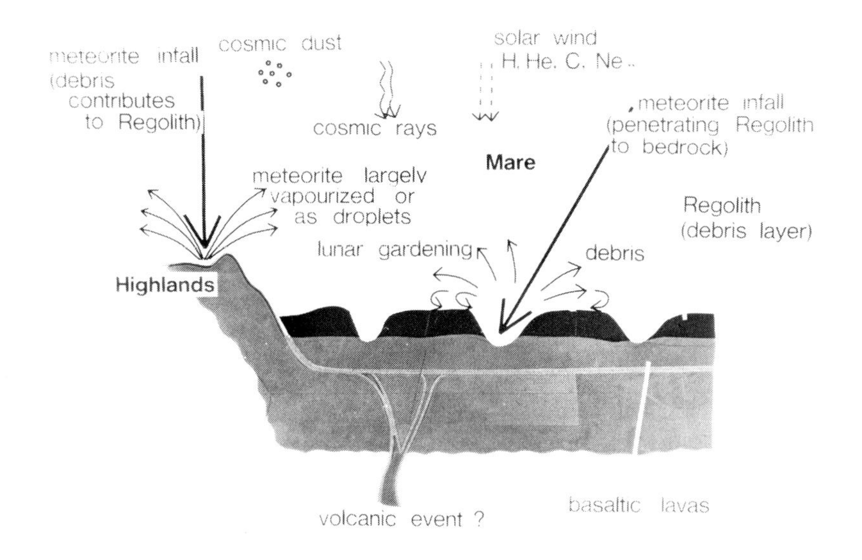

Fig. 2. Schematic representation of vertical section through the lunar surface in a typical mare region. Some of the processes and events occurring at the lunar surface are indicated.

that it is not immediately apparent what the starting materials might have been. Hence present research is confined largely to exploratory studies with the expectation of getting an overall picture of the compounds present and their rough proportions. The next stages will certainly be to survey the distribution of such compounds in relation to, firstly, the fine structure of the samples (the individual minerals, etc.), secondly the interior/exterior of samples and, thirdly, the different sites on the moon. A better appreciation of the processes by which the compounds could have been formed should then be apparent. Fig. 2 provides a diagrammatic representation of some of the processes which may affect the lunar surface.

Carbon content of the lunar samples

Although several workers measured the total carbon content of the Apollo 11 samples, the most extensive measurements were made by C.B. Moore and his collaborators [20]. They used a combustion technique at

1800 °C, analysing gas chromatographically the CO_2 produced. Sample sizes were in the region of 200 mg. The fines and the breccias contain 150–200 ppm total carbon. The crystalline rocks, on the other hand, show much lower quantities, of the order of 50 ppm. The sieved fractions of fines demonstrated that the largest fragments have lowest carbon content while the material passing 300 mesh has as high as 500 ppm of total carbon. The Apollo 12 results presently available [18] show considerable variation in the carbon content of the fines (fig. 1).

The isotopic composition of lunar carbon has been estimated by Kaplan and Smith [14]. Their results have been substantiated by other workers [8,24] but not Friedman et al. [11]. The fines, after degassing for 48 hr at 150°C under vacuum, have $\delta\ ^{13}C$ values of about +18. The values for breccias and fine-grained crystalline rocks are reportedly in the range +10 to −27, but the minus values are suggestive of terrestrial contamination. The high positive values for the fines are adjudged correct, for +18 is outside most terrestrial measurements. However, Deines [30] obtained +22 for calcite inclusions in a mica peridotite ditre if substantiated, this would indicate a fractionation process operating during the formation of igneous rocks, both lunar and terrestrial. Kaplan and Smith [14] suggested that on the moon the lighter isotope, ^{12}C, could be lost by preferential diffusion of $^{12}CH_4$, possibly formed as a result of solar wind implantation.

The twin problems of carbon content and isotope ratios will become more clearly resolved as a result of the detailed work now in progress on the Apollo 12 samples, wherein the contamination levels have been much reduced [18]. Other elements commonly associated with the study of organic compounds, i.e. hydrogen, nitrogen, oxygen, phosphorus, sulphur, have been measured in the lunar samples but work has hardly begun on interrelating element content and the presence of specific compounds.

Extraction and isolation of carbon compounds

Four main procedures have been used, singly and in combinations: ball milling in vacuo, pyrolysis, extraction with organic solvents and with water, and, finally, digestion with acids and consequent demineralisation.

The crushing of the lunar fines by ball milling in vacuo released the rare gases and certain low molecular weight hydrocarbons, notably methane and ethane. It may prove generally useful for it does not greatly alter the sample while bringing about partial release of trapped gases [1,2]. Certainly, the marked micro- and macro-vesicularity of the lunar samples is indicative of gases and fluids trapped at the time of formation.

Pyrolysis was used by several investigators and the conditions varied widely: temperatures ranged from 300 °C to 1200 °C and were achieved by stepwise, flash or continuous heating. Pyrolyses were carried out in a vacuum or under inert gas, with closed vessels and continuous gasflow. The distillates and pyrolysates were analysed by mass spectrometry directly or via a trap system. They were also analysed by GC separation and detection and by combined GC–MS. The principal findings of this approach were that most of the carbon content was readily evolved as CO and, to a lesser extent, as CO_2, especially at temperatures above 500 °C. Methane, other low molecular weight hydrocarbons and unsaturated and aromatic hydrocarbons, such as benzene, toluene and naphthalene, were also observed by several investigators. Hydrogen sulphide, sulphur dioxide, thiophenes, HCN, N_2 and ammonia were also reported in the distillates and contaminants, such as peracetic acid, ethylene oxide, phthalate and exhaust products, were recognized. Incidentally, hydrogen and the rare gases did not appear until about 350 °C showing that they were not simply adsorbed on the outside of the fines. With the exception of CO and CO_2, the pyrolysate was generally agreed to amount to only a few ppm.

Most Bioscience investigators extracted the lunar samples with organic solvents, typically a mixture of benzene and methanol which can be expected to take up most polar and non-polar compounds of an intermediate range of molecular weights. The solvent extraction procedure was also applied to demineralised samples. The consensus of opinion was that no hydrocarbons, fatty acids or other soluble medium-molecular weight compounds were individually present in excess of about 5–10 ppb. In other studies, some investigators extracted the lunar samples with water, with or without dissolution with acids. The consensus here was that amino acids, sugars, nucleic acid bases or porphyrins were not individually present in excess of about 10 ppb, although there were one or two reported which were positive, but only for very small amounts. The derivatives employed to render extracted fractions volatile and more suitable for further analysis included methyl esters, T.M.S. derivatives and trifluoro acetates.

Digestion with hydrofluoric and hydrochloric acids is a technique commonly used with terrestrial sediments, e.g. Ancient cherts. When this was applied to the lunar sample a certain amount of insoluble debris remained but no significant results were obtained using this partially demineralised material.

Analytical techniques

The principal methods employed for the Apollo 11 analyses were those current in organic geochemistry, i.e. gas chromatography, mass spectrometry, combined gas chromatography–mass spectrometry and high resolution mass spectrometry. Other sensitive techniques such as fluorescence spectrophotometry and the use of amino acid analysers and other forms of liquid chromatography were also reported. The use of such relatively insensitive techniques as infrared and NMR was attempted in one or two cases but there is no doubt that the most detailed information has been provided by high resolution mass spectrometry. The geological techniques of electron microprobe and scanning electron microscopy should afford some indication as to the whereabouts of the carbon-containing materials.

Survey of Carbon and its compounds in the lunar samples

The results summarized below and in table 4 mostly refer to 'fines', i.e. the lunar soil, returned by the Apollo 11 mission.

Carbon

There is one report of a fragment of graphite, approximately 2 mm in size [3], but no claim for any other allotropic form of carbon, such as diamond or chaoite. Finely dispersed carbon is a possibility, especially where contributed by solar wind or held in the igneous rocks since their formation: the bulk of the carbon in the lunar fines could be in this finely divided form. In the holo-crystalline rocks, elemental carbon might be present in the interstitial opaque phases.

Carbon monoxide

Carbon monoxide (identified by GC and mass spectrometry) is rapidly evolved when the fines are heated above 400 °C and nearly all the carbon is driven off as carbon monoxide (and carbon dioxide to some extent) by heating to the melting point (ca. 1100 °C). Calvin and collaborators [5] reported that approximately 66 ppm (i.e. approximately 1/3 of the total carbon) of CO are liberated on HF dissolution of the fines. However, carbon monoxide as such does not seem to be present in the rocks as only traces were liberated on crushing. Carbon (graphite), when mixed with the premelted fines, is partially converted (2%) to carbon monoxide when the mixture is remelted. Oro, in unpublished work, has recently shown that terrestrial basalts give off ca. 5 ppm of carbon monoxide and 200 ppm of carbon dioxide when heated at 750 °C.

Table 4
Summary of carbon compounds searched for or detected in the lunar fines.

Carbon compounds	Example of report [a]	Present (+) or absent (−)	Amount (maximum per individual component)	Comment
C (graphite)	Arrhenius et al. [3]	+		Only 1 particle, ~2 mm across, detected so far
CO (as bound form)	Burlingame et al. [5]	+	~60 ppm	Only report
CO_2 (as argonite?)	Agrell (1970)	+	?	Personal communication
Carbides	Ponnamperuma et al. [26]	+	~20 ppm	Cohenite? Also identified by microscopy
Alkanes				
CH_4, C_2H_6, etc.	Abell et al. [2]	+	~2 ppm	
n-C_{12}–C_{32}	Meinschein et al. [19]	−	10 ppb	Nagy's report [22] positive
Unsaturated hydrocarbons and aromatic hydrocarbons	Nagy et al. [22]	+	10 ppb	Probably only as pyrolysis product
Long-chain alcohols	Ponnamperuma et al. [26]	−	5 ppb	Attempted as volatile TMS derivative
Long-chain acids	Abell et al. [2]	−	10 ppb	Attempted as volatile TMS derivative
Sugars	Ponnamperuma et al. [26]	−	1 ppb	Attempted as *O*-TFA and *O*-TMS derivatives
Amino acids				
bound	Oro et al. [24]	−	1 ppb	Nagy et al. [22, 23] and Fox et al. [10a] claim up to 100 ppb of certain amino acids
free	Oro et al. [24]	−	0.01 ppb	Ditto
Purines and pyrimidines (bound and free)	Lipsky et al. [16]	−	10 ppb	Ion exchange and gas chromatography
Porphyrins	Rho et al. [27]	−	0.1 ppb	Ponnamperuma et al. [26] positive claim
Organo-siloxanes	Ponnamperuma et al. [26]	+	40 ppm	Only report

[a] A single example, only, given. However, where there is disagreement or only one report this is indicated in the comments. This table is a greatly simplified presentation of the reported data. Thus, CO and CO_2 are released on heating the sample.

Carbon dioxide

Carbon dioxide is liberated along with the carbon monoxide when the fines are heated above 400 °C but the evolution of the two gases are not in parallel. Carbon dioxide was identified by GC, GC–MS and mass spectrometry. The source may be carbonate.

Carbides

The first report of carbides was made at the Houston Conference by Ponnamperuma et al. [26]. They had treated the lunar fines with hydrochloric acid and found that certain gases, notably methane, ethane, ethylene and propane were evolved, corresponding to about 20 ppm by weight of carbided. The hydrocarbons were identified by gas chromatography and mass spectrometry and compared with those afforded by cohenite $(Fe,Ni)_3C$ from the Canyon Diablo meteorite.

In our own group at Bristol we have since used deuterium chloride instead of HCl and have shown that although the bulk of the hydrocarbons so released are fully labelled with deuterium, about 15% are unlabelled, indicating the presence of free alkanes in the sample. Further work is in progress [2].

Alkanes

The search [19] for alkanes in the customary geolipid range (C_{12} to C_{32}) using extracts, pyrolysates and GC–MS, and MS, was generally abortive, an upper level for individual alkanes being put at ca. 1–10 ppb. However, one report [22] gave a level of ca. 1 ppm total for *n*-alkanes C_{25}–C_{40}, but the sample may have been contaminated. The extractions were carried out on intact fines, predigested fines and on the residues after digestion with HF. Pyrolysis yields small amounts (approximately 1 ppm) of low molecular weight alkanes, mainly methane, ethane, etc. [24]. These gases are present in the fines as such (e.g. 2 ppm of methane) as has been reported by Abell et al. [2] using DCl etching. The carbides present do not interfere since they furnish deuterated hydrocarbons.

Unsaturated and aromatic hydrocarbons

Pyrolysis (e.g. at 700 °C) affords a variety of unsaturated hydrocarbons up to about C_9 and aromatic hydrocarbons (benzene, toluene and other alkyl benzenes, thiophene and alkyl thiophenes, indenes, naphthalene and methyl naphthalenes, styrenes and biphenyls [21]). The identification procedures included GC–MS [22,23], MS and HRMS [5,21]. The amounts of aromatic hydrocarbons are in the low ppm region. They were not reported in the solvent extracts and are said to be reminiscent of those recorded for heat treatment of carbonaceous chondrites. They may have been formed by pyrolysis of indigenous polymeric material.

Alcohols

None were detected above the detection limit of about 5 ppb per individual component. Methanol–benzene extracts of fines and acid-treated fines were derivatized, for example by silylation with $CF_3CON(SiMe_3)_2$ and then examined by combined GC–MS.

Fatty acids

A specific search for fatty acids in the range C_{12}–C_{32} was made by Ponnamperuma's group [26] and by Abell et al. [1], using benzene–methanol extraction followed by treatment with $CF_3CON\ (SiMe_3)_2$ or BF_3–methanol esterification. Amounts exceeding 10^{-8} g/g of individual acids were absent. Several other workers would have found these compounds had they been present.

Sugars

Extraction with water, followed by ion-exchange chromatography and derivatisation with trifluoroacetic anhydride to give the *O*-TFA derivative and with trimethylsilylating reagents to form the *O*-TMSi derivatives afforded no peaks corresponding to sugars or sugar alcohols at the limit of detection of about 6×10^{-10} g/g [26].

Amino acids and polypeptides

The search attempted by Ponnamperuma's group [26] involved:

(i) aqueous extraction, followed by ion exchange and derivatisation with trifluoracetic anhydride and *n*-BuOH–3 N HCl. Any free amino acids in excess of 10^{-11} g/g would afford distinguishable peaks on the gas chromatograph record. None were found.

(ii) acid hydrolysis (1 N HCl) of the lunar fines followed by the procedure as above. Again, no amino acids were found at a level of detection of around 10^{-9} g/g.

(iii) acid hydrolyses (6 N HCl) of the lunar fines and separation by ion exchange, derivatisation and gas chromatography, and amino acid analyser techniques. Again, no peaks appeared above the detection limit of around 10^{-9} g/g. Oro et al. [24] also used a hydrolytic step, followed by deionisation and the attempted formation of *N*-TFA isopropyl derivatives and gave a limit of no compound in excess of 10^{-7} g/g.

Fox [10] preferred to reflux the sample with water in an attempt to prepare an aqueous extract of any peptides present and followed this with acid hydrolysis (6 N HCl) of the aqueous extract, and passage through an ultrasensitive amino acid analyser. He claimed the presence of about 0.3 nmole (ca. 10^{-7} g/g) of each of the following amino acids: Gly, Ala, Glu, Ser, Asp, Thr, and $\alpha\beta$-diaminopropionic acid. These identifications were based on elution times. Nagy et al. [22,23] also used water reflux and claimed the detection of urea, Gly, Ala and ethanolamine at a level of about 3×10^{-8} g/g. Fox et al. [10] and Nagy et al. [22,23] remark that the amino acids may be the result of either (a) abiotic synthesis brought about by the rocket exhaust on the lunar surface, or (b) irradiation of indigenous carbonaceous material. Further study is required to eliminate these and other possibilities.

Nucleic acid bases etc.

Ponnamperuma et al. [26] employed hydrolysis with 1 N HCl followed by silylation with bis(trimethylsilyl) trifluoroacetamide (BSTFA) to give fractions for gas chromatographic study, but no bases could be detected in excess of 4×10^{-9} g/g. Acid hydrolysis (2.5 N HCl) followed by ion exchange was used by Lipsky et al. [16] who detected nothing in excess of 10^{-8} g/g, using a liquid chromatograph and UV absorption at 250 nm.

Porphyrins

Hodgson et al. [12] employed a benzene–methanol extraction followed by demetalation with methane sulphonic acid (MSA) fluorescence spectrophotometry of the product with an excitation wavelength of 390 nm. They reported that the spectra corresponded to about 10^{-10} g/g of porphyrin-like material similar to that detected in the dunite after exposure to the descent engine effluent during terrestrial trials.

Rho et al. [27] employed methanol, phenyl cyanide and grinding with MSA/naphthalene to obtain extracts for spectrofluorimetry. No porphyrin-like absorptions were detected at the detection limit of 10^{-10} g/g. However, pigment(s) of unknown structure fluorescing with absorption maxima near 310 and 350 nm were detected at a level of about 10^{-8} g/g.

Organosiloxanes

Certain peaks other than those of derivatives (*N*-trifluoroacetate *n*-butyl esters) of amino acids appeared in the reaction mixtures derived from the acid hydrolysates of the lunar fines. Ponnamperuma's group [26] believes these to be organosiloxanes but this startling claim is still under study.

Contaminants detected

A careful microscopic examination of the lunar fines distributed to the investigators revealed a few teflon particles and cellulose fibres amongst the mineral matter [17]. These undoubtedly originated through the handling procedures at the Lunar Receiving Laboratory. Mass spectrometry, principally high resolution mass spectrometry, revealed traces of various vacuum pumping oils (hydrocarbon and silicone), phthalate plasticizers and derivatives of sterilants (ethylene oxide and peracetic acid). Both the Berkeley group [5] and the MIT group [21] detected rocket exhaust products in the lunar fines. Certain heteroatomic species were found in the samples and in the LM exhaust products trapped on test at New Mexico.

Discussion

The amount of carbon in the lunar samples is of the order of 10s of ppm. This is many orders of magnitude less than in some meteorites and in most terrestrial rocks. Conventional biolipids such as long chain hydrocarbons, long chain fatty acids, α-amino acids, polypeptides, nucleic acid bases, porphyrins and sugars are probably not present. Certainly most investigators report them to be below the levels of detection, which usually amount to about 10 ppb of an individual compound. However, amino acids and porphyrin-like materials were detected by some investigators but not others. These disagreements should be resolved when the cleaner samples obtained from below the lunar surface on Apollo 12 are analysed and the rocket exhaust products firmly identified. Metal carbides, probably mainly cohenite $(Fe,Ni)_3C$, are present in the fines, as are small amounts of hydrocarbons such as methane. Carbon monoxide and organosiloxanes are reported as being released from the lunar fines on acid treatment.

Some of the lunar carbon, particularly the carbide fragments, should be of meteoritic origin, and continue to be so supplied to the lunar surface. Trace element abundances indicate up to 2% contribution of carbonaceous chondrite material [15]. Some of the carbon must be of solar origin – also continuously supplied: solar protons may effect in situ hydrogenation, possibly forming the methane and ethane reported as indigenous to the fines. Some of the carbon, possibly the carbon monoxide in whatever form it may be in the sample, may be of primeval origin and have been present in the rocks immediately after crystallisation. Degassing of the primitive, initially accreted, lunar material must have occurred as the planet's surface consolidated: escape of this carbon would very likely be incomplete. The low abundance of lunar carbon could in part be the result of early loss of carbon monoxide and carbon dioxide during crystallisation of the melts. Other light elements are similarly depleted. Polymeric material, of thermal and radiation-induced origin, should be present in the fines but as with most of the lunar carbon remains to be firmly characterised.

Experiments currently underway in several laboratories are designed: (a) to further explore the nature of lunar carbon whether it be elemental, matrix-bound or as carbon compounds and (b) to further determine the distribution of the materials in the fines, breccias and crystalline rocks. To distinguish between primeval, solar, meteoritic or cometary origins for the various materials it will be necessary to make detailed mineral separations and also to examine surfaces and interiors of the lunar samples. Solar wind products would be expected only in the outer 1000 Å or so of individual particles, exposed on the surface of the regolith. Churning (lunar gardening) evidently ensures that most of the lunar soil is so exposed for periods of time of the order of tens of millions of years. The contribution of carbon made to the lunar surface by the solar wind is hard to estimate (inter alia, see Moore et al., [20], Abell et al. [1,2], Nagy et al. [23]. The abundance of carbon in the sun is not known precisely and the proportion that is present in the solar wind may not be calculated with any accuracy. With the quantities of rare gases such as ^{36}Ar in the samples, as a rough guide, there remains the problem of the 'sticking factor': highly reactive carbon ions would be expected to 'stick' by reaction with matrix atoms or with the hydrogen atoms. However, methane so formed would diffuse out along crystal boundaries and shock-damaged features – as does argon. Even so, the proportion remaining after diffusion would be higher than that of unreactive elements such as argon.

Among the questions remaining to be answered are first, why should all the carbon not have been lost during the melting and crystallisation of the crystalline rocks, since melting them now releases most of the carbon as carbon monoxide? Second, in spite of the known turnover of the lunar surface, is it possible that organic materials are better preserved at depth beneath the surface layer of fine dust or at the cold polar regions where they could be condensed? One intriguing possibility is that the transient lunar phenomena which have been observed as coloured glows seen from the earth and also by the astronauts on the Apollo 11 mission could be due to emissions of gas such as methane, carbon monoxide and rare gases from the lunar surface.

References

[1] Abell P.I., G.H. Draffan, G. Eglinton, J.M. Hayes, J.R. Maxwell and C.T. Pillinger, Science 167 (1970) 757.
[2] Abell P.I., G. Eglinton, J.M. Hayes, J.R. Maxwell and C.T. Pillinger, Nature 226 (1970) 251.
[3] Arrhenius G., S. Asunmaa, J.I. Drever, J. Everson, R.W. Fitzgerald, J.Z. Frazer, H. Fujita, J.S. Hanor, D. Lal, S.S. Liang, D. Macdougall, A.M. Reid, J. Sinkankas and L. Wilkening, Science 167 (1970) 659.
[4] Barghoorn E.S., D. Philpott and C. Turnbill, Science 167 (1970) 775.
[5] Burlingame A.L., M. Calvin, J. Han, W. Henderson, W. Reed and B.R. Simoneit, Science 167 (1970) 751.
[6] Cloud P., S.V. Margolis, M. Moorman, J.M. Barker, G.R. Licari, D. Krinsley and V.E. Barnes, Science 167 (1970) 776.
[7] Draffan G.H., G. Eglinton, J.M. Hayes, J.R. Maxwell and C.T. Pillinger, Chemistry in Britain 5 (1969) 296.
[8] Epstein S. and H.P. Taylor, Jr., Science 167 (1970) 533.
[9] Flory D.A. and B.R. Simoneit, NASA Technical Report (1969).
[10] Fox S.W., this volume, p.
[10a] Fox S.W., K. Harada, P.E. Hare, G. Hinsch and G. Mueller, Science 167 (1970) 767.
[11] Friedman I., J.R. O'Neil, L.H. Adami, J.D. Gleason and K. Hardcastle, Science 167 (1970) 538.
[12] Hodgson G.W., E. Peterson, K.A. Kvenholden, E. Bunnenberg, B. Halpern and C. Ponnamperuma, Science 167 (1970) 763.
[13] Johnson R.D. and C.C. Davis, Science 167 (1970) 759.
[14] Kaplan I.R. and J.W. Smith, Science 167 (1970) 541.
[15] Keays R.R., R. Ganapathy, G.C. Laul, E. Anders, G.F. Herzog and P.M. Jeffery, Science 167 (1970) 490.
[16] Lipsky S.R., R.J. Cushley, C.G. Horvarth and W.J. McMurray, Science 167 (1970) 778.
[17] Lunar Sample Preliminary Examination Team (LSPET), Science 165 (1969) 1211.
[18] Lunar Sample Preliminary Examination Team (LSPET), Science 167 (1970) 1325.
[19] Meinschein W.G., E. Cordes and V.J. Shiner, Jr., Science 167 (1970) 753.
[20] Moore C.B., C.F. Lewis, E.K. Gibson, W. Nichiporuk, Science 167 (1970) 495.
[21] Murphy R.C., G. Preti, M.M. Nafissi-V. and K. Biemann, Science 167 (1970) 755.
[22] Nagy B., C.M. Drew, P.B. Hamilton, V.E. Modzelski, M.E. Murphy, W.M. Scott, H.C. Urey and M. Young, Science 167 (1970) 770.
[23] Nagy B., W.M. Scott, V.E. Modzelski, L.A. Nagy, C.M. Drew, M.S. McEwan, J.E. Thomas, P.B. Hamilton and H.C. Urey, Nature 225 (1970) 1028.
[24] Oro J., W.S. Updegrove, J. Gibert, J. McReynolds, E. Gil-Av, J. Ibanez, A. Zlatkis, D.A. Flory, R.L. Levy and C. Woolf, Science 167 (1970) 765.
[25] Oyama V.I., E.L. Merek and M.P. Silverman, Science 167 (1970) 773.
[26] Ponnamperuma C., K. Kvenholden, S. Chang, R. Johnson, G. Pollock, D. Philpott, I. Kaplan, J. Smith, J.W. Schopf, C. Gehrke, G. Hodgson, I.A. Breger, B. Halpern, A. Duffield, K. Krauskopf, E. Barghoorn, H. Holland and K. Keil, Science 167 (1970) 760.

[27] Rho J.H., A.J. Bauman, T.F. Yen and J. Bonner, Science 167 (1970) 754.
[28] Runcorn S.K., Science Journal 6 (1970) 27.
[29] Schopf J.W., Science 167 (1970) 779.
[30] Deines P., Geochimica Cosmochimica Acta 32 (1967) 613.

Chemical Evolution and the Origin of Life, eds. R. Buvet and C. Ponnamperuma

IN SITU ANALYSIS OF PLANETARY SURFACES FOR ORGANIC MATERIALS

K. BIEMANN
Massachusetts Institute of Technology, Cambridge, Massachusetts 02139, USA

In one's endeavour to deduce the processes that led to the evolution of life on earth one is limited to the assumption of a certain chemical model, environmental conditions and the frequency of more or less plausible, specific events. These hypotheses are highly speculative and are difficult to prove or disprove, even with elaborate, well designed experiments.

With the advent of interplanetary probes it is now going to be possible to investigate the physical, chemical and biological state of other planets in the hope of encountering one or more that is presently in a state prior to or just after the evolution of living systems. It may then be possible to observe the chemistry at and near the surface, evaluate the sources of energy available and search for the presence or absence of living systems, and their degree of similarity to terrestrial ones.

Thus, the detection and identification of organic compounds is a very important aspect of such an endeavour and the question immediately arises whether one should search for the presence or absence of a set of specific compounds or, alternatively, to perform an experiment capable of identifying a very wide range of organic compounds regardless of any terrestrial chemical or biochemical model. Here on earth we do know the chemistry of living systems and can select a set of specific compounds that must have been involved already at much earlier stages. Thus, if one were merely interested in the detection of the existence of terrestrial life elsewhere, one could perform a specific experiment, modeled along the so-called "autoanalyzers" which are presently used in terrestrial laboratories with considerable success. The amino acid analyzers are the best known example thereof. A miniaturized instrument of this type has indeed been built by Miller [8] and his associates who also were able to incorporate a technique for simultaneously detecting purine and pyrimidine bases related to those occurring in natural nucleic acids [2]. However, as already pointed out at a number of occasions during this conference, an amino acid analyzer is very reliable if one deals with a limited set of known compounds, such as the twenty or so amino acids occurring in terrestrial proteins, but is a dangerous experiment to rely upon if one deals with a system assumed to be amino acids but produced by processes other than terrestrial living systems. The retention volume alone is much too limited a parameter to uniquely identify a compound and carries only very limited structural information. Furthermore, the experimental conditions must be designed for a specific set of compounds to assure reliable separation, a precondition for successful analysis.

There is, of course, no point to conduct an experiment on another planet that can detect only terrestrial life, and even a system that is based on the same or very similar chemical processes must not necessarily employ the same compounds. For example, even if it utilizes proteins made of α-amino acids, they must not necessarily be the same 20 we know to be involved on earth but can be any one of the other hundreds or thousands of structurally possible ones. The same argument holds for the nucleic acids or their operational analogues.

For a survey analysis aimed at the detection and identification of a broad range of organic compounds not limited by the assumption of any chemical model, spectroscopic techniques, such as ultraviolet, infrared or nuclear magnetic resonance spectroscopy or mass spectrometry are much more useful. Of these, the last mentioned technique is by far the most appropriate one for a number of reasons: First, any organic compound produces a mass spectrum, the only major limitation being the requirement of volatility, at least in a good vacuum; second, it is extremely sensitive, requiring only a few nanograms of material; finally, it does not require a solvent before becoming measurable. The disadvantage is the need to achieve a good vacuum which must be maintained during the entire experiment.

One of the most important aspects of mass spectrometry is the direct interpretability of the spectrum in terms of the structure of the compound, even if one deals with an unexpected or previously unknown substance. A mass spectrum is not merely a very detailed fingerprint of a compound which has to be matched with authentic spectra, but represents a code which can be deciphered by any chemist reasonably experienced in the field. Matching with known spectra is used for automatic identification [4] or final proof of the correctness of the interpretation.

Fig. 1 represents, for example, the mass spectrum of dimethyldisulfide, a rather simple organic compound. The strong peak at m/e 94 indicates that this is the molecular weight. The difference of 15 amu to the peak at m/e 79 reveals the presence of a methyl group (CH_3 = 15 amu) which is confirmed by the peak at m/e 15. The peaks at m/e 47 and 48 correspond to cleavage of

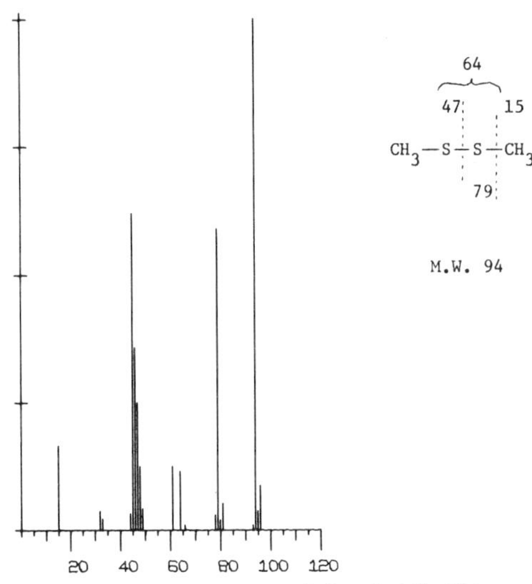

Fig. 1. Mass spectrum of dimethyldisulfide.

the S–S bond without or with abstraction of a hydrogen from the other methyl group. The peak at m/e 64 represents the two central sulfur atoms. Mass 96 is due to those molecules of dimethyldisulfide which contain one ^{34}S atom and its intensity relative to m/e 94 is a measure of the abundance of heavy sulfur. This demonstrates that one can deduce from such a spectrum at least the approximate isotopic abundance of the elements present. It is obvious from this brief discussion that the interpretation of a spectrum in terms of the structure is not very difficult and that other compounds will have very different spectra. The reader is referred to various texts for a much more detailed outline for the interpretation of mass spectra [1,3,7].

Not only can a single compound be so identified with considerable certainty, but also a simple mixture can be dealt with utilizing a single mass spectrum, because such a spectrum is an exact superposition of the spectra of the individual components. For example, a mixture of ammonia, methane, acetylene, ethane, HCN, formaldehyde and methanol can be recognized from a single mass spectrum, at least as long as the relative abundances of the components do not differ greatly. More complex mixtures of more complex compounds, however, would give a mass spectrum that may contain a peak of reasonable intensity at almost every mass and its interpretation would be difficult or impossible. It is then necessary to simplify such mixtures first and this can be done either crudely, by thermally fractionating the mixture into the mass spectrometer, or, better, by separating them first. For the latter, gas chromatography is the most powerful and most compatible technique, because it is very efficient and, like mass spectrometry, very sensitive and utilizes the material in the gas phase. The combination of gas chromatography with mass spectrometry has been developed over the past few years and is one of the most powerful techniques available (for a review, see [5]). It has the advantage for automated systems that it does not require solvents or liquid reagents. The major complication is the large amount of gas used to sweep the column, which would make the mass spectrometer inoperative, were it not removed prior to entering the spectrometer. This is even more important in the case of an instrument of limited pumping capacity, such as a miniaturized flight model.

Another problem is the separation of the organic material from the soil sample. In terrestrial laboratories this is accomplished by extraction of the inorganic matrix with organic solvents, water, acid or base. In addition to efficient extraction, one also obtains some chemical information based on the extractability of the various organic compounds under the varying conditions. Unfortunately, such an extraction is difficult to automate, particularly for repetitive experiments, and under conditions where it is important to keep the contamination of the environment with external chemicals at a minimum. The best compromise is the expulsion of the organic materials by heating the soil sample, to vaporize those components having sufficient vapor pressure and thermal stability, and to pyrolyze the more polar or polymeric substances to volatile degradation products, which still contain structural information [10].

Based on these considerations, a system (fig. 2), consisting of a series of sample ovens which can be consecutively attached to the gas stream of a small gas chromatograph which is in turn connected to a mass spectrometer via a carrier gas separator, has been developed for the exploration of the surface of Mars [11]. The spectrometer is of the Nier–Johnson type and utilizes a small ion pump. The output is fed into a digital data system which continuously records the spectra and is programmed to selectively store the significant data for telemetry to earth. Because of the very limited pumping capacity the removal of the carrier gas is particularly significant and in order to eliminate the need for additional pumps, hydrogen is chosen as the gas, which is then removed by a thin-walled palladium tube acting as the separator [6]. Although this is not quite chemically inert, because it reduces conjugated double bonds [9] this is a minor problem, since the ambiguousness of the presence or absence of a particular double bond in a compound detected on another planet is almost insignificant if one is able to deduce the entire

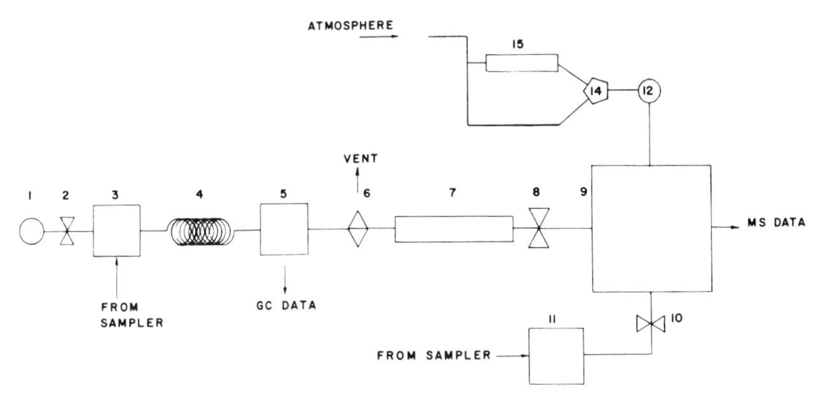

Fig. 2. Block diagram of the Viking molecular analysis experiment. (1) carrier gas (H_2) reservoir; (2,8,10,14) valves; (3,11) sample ovens; (4) gas chromatographic column; (5) cross section detector; (6) effluent divider to prevent overpressurization of the mass spectrometer; (7) carrier gas separator; (9) mass spectrometer; (12) molecular leak; (15) CO and CO_2 removal cartridge (to permit N_2 analysis).

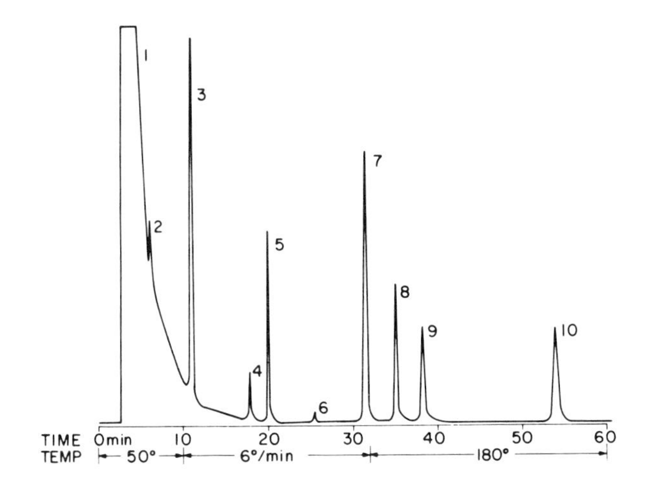

Fig. 3. Gas chromatogram of phenylalanine pyrolyzate. The peaks were identified by their mass spectra and they are: (1) carbon dioxide, water and acetonitrile; (2) benzene; (3) toluene; (4) ethyl benzene; (5) styrene; (6) cyanobenzene; (7) benzyl cyanide; (8) 4-phenyl-1-butene; (9) quinoline or isoquinoline; (10) diphenylethane. These data were obtained with the Viking molecular analysis system at the Jet Propulsion Laboratory, Pasadena, California.

structure, except the degree of saturation. Furthermore, in most cases one will be able to deduce from the behavior of the compound whether it could have contained another double bond originally.

In addition to the gas chromatographic input, there is provision to heat a soil sample and introduce the effluent directly into the mass spectrometer to detect some of the more polar substances, which cannot be passed through a gas chromatograph but are amenable to mass spectrometry without pyrolyzing them. A gas inlet system is provided for the analysis of the atmosphere for chemical and isotopic composition (fig. 2).

A gas chromatogram obtained with this system is shown in fig. 3. The mass spectrometer scans continuously, covering the range of mass 12–200 in about 5 sec, with sufficient resolution to distinguish mass 199 from 200. The entire system is packaged to weigh only about 11 kg. It represents a rather complex instrument but the structural variety of organic compounds, even on a level of relative simplicity demands a powerful system to return sufficiently specific data, warranting the effort of sending it that far. It is hoped that the results will provide an understanding of the organic chemistry on the surface of Mars and will enable one to evaluate its degree of sophistication and to design a second generation of experiments for later flights to probe the question of the existence and nature of Martian, not terrestrial, living systems with greater certainty.

Acknowledgements

The author is indebted to the members of the Viking Molecular Analysis Team, Drs. D. Anderson, L. Orgel, J. Oro, T. Owen, G. Shulman, P. Toulmin, and H. Urey, as well as Drs. R.A. Hites and P.G. Simmonds for their contributions to the subjects outlined in this paper and to the National Aeronautics and Space Administration for their support.

References

[1] Biemann K., Mass Spectrometry: Applications to Organic Chemistry (McGraw-Hill, New York, 1962).
[2] Bonnelycke B.E., K. Das and S.L. Miller, Anal. Biochem. 27 (1969) 262.
[3] Budzikiewicz H., C. Djerassi and D.H. Williams, Mass Spectrometry of Organic Compounds (Holden-Day, San Francisco, 1967).
[4] Hites R.A. and K. Biemann, in: Advances in Mass Spectrometry, Vol. 4, ed. E. Kendrick (Institute of Petroleum, London, 1968) p. 37.

[5] Leemans F.A.J.M. and J.A. McCloskey, J. Am. Oil Chem. Soc. 44 (1967) 11.
[6] Lucero D.P. and F.C. Haley, J. Gas Chromatog. 6 (1968) 477.
[7] McLafferty F.W., Interpretation of Mass Spectra (Benjamin, New York, 1966).
[8] Miller S.L. (1969) private communication.
[9] Simmonds P.G., G.R. Shoemake and J.E. Lovelock, Anal. Chem. 42 (1970) 881.
[10] Simmonds P.G., G.P. Shulman and C.H. Stembridge, J. Chem. Sci. 7 (1969) 36.
[11] Szirmay S.Z., G.R. Shoemake, C.E. Giffin and R.A. Hites, Rev. Sci. Instrum., in preparation.

LIST OF PARTICIPANTS

AMARIGLIO, A., C.N.R.S., Centre de 1er cycle, Boulevard des Aiguillettes 54, Nancy, France.

ANTONOV, A.S., Moscow State University, Corpus "A", Lenin Hills, Moscow, USSR.

BALTSCHEFFSKY, H., Bioenergetics Group, Department of Plant Physiology, University of Stockholm, Lilla Frescati, Stockholm, Sweden.

BAR-NUN, A., N.A.S.A., Ames Research Center, Moffett Field, California, USA.

BARGHOORN, E.S., The Biological Laboratories, Harvard University, 16 Divinity Ave., Cambridge, Massachusetts, USA.

BIEMANN, K., Department of Chemistry, Massachusetts Institute of Technology, Cambridge, Massachusetts, USA.

BOICHENKO, E.A., V.I. Vernadsky Institute of Geochemistry and Analytical Chemistry, USSR Academy of Sciences, Moscow, USSR.

BRODA, E., Institut für Physikalische Chemie der Universität, Waehringer Strasse 42, Vienna, Austria.

BUVET, R., Faculté des Sciences de Paris, Laboratoire d'Energétique Electrochimique, 10 rue Vauquelin, Paris, France.

CHADHA, M., N.A.S.A., Ames Research Center, Moffett Field, California, USA.

DAYHOFF, M.O., National Biomedical Research Foundation, 11200 Lockwood Drive, Silver Spring, Maryland, USA.

DOSE, K., Faculty for National Sciences of University of Frankfurt/M and Max-Planck-Institut für Biophysik, 6 Frankfurt/M, Germany.

EGLINTON, G., Organic Geochemistry Unit, School of Chemistry, University of Bristol, Cantock's close, Bristol, England.

EVREINOVA, T.N., University Biological Faculty, Leninsky Gori, Moscow, USSR.

EVSTIGNEEV, V.B., Institute of Photosynthesis, Academy of Sciences of the USSR, Leninsky Prospect 33, Moscow, USSR.

FESENKOV, V.G., Committee for Meteorites Investigation, Academy of Sciences, Moscow, USSR.

FLORKIN, M., Université de Liège, Laboratoire de Biochimie, 17 Place Delcour, Liège, Belgium.

FOX, S.W., Institute of Molecular Evolution, University of Miami, 521 Anastasia, Coral Gables, Florida, USA.

GABEL, N.W., Research Department, Illinois State Psychiatric Institute, 1601 West Taylor Street, Chicago, Illinois, USA.

GAVAUDAN, P., Station Biologique de Beau Site, 25 Faubourg Saint Cyprien, 86 Poitiers, France.

HALMANN, M.M., Isotope Department, The Weizmann Institute of Science, Rehovot, Israel.

HARADA, K., University of Miami, Institute of Molecular Evolution, Coral Gables, Florida, USA.

HOCHSTIM, A.R., Wayne State University, College of Engineering, Ries, Detroit, Michigan, USA.

KAPLAN, R.W., Institut für Mikrobiologie der Universität Frankfurt/M, Siesmayer Strasse 70, 6000 Frankfurt/M, Germany.

KARAPETYAN, N.V., Bach Institute of Biochemistry, Academy of Sciences, Leninsky Prospect 33, Moscow, USSR.

KIMBALL, A.P., University of Houston, Department of Biophysical Sciences, Houston, Texas, USA.

KORNEEVA, G.A., Bach Institute of Biochemistry, Academy of Sciences of the USSR, Leninsky Prospect 33, Moscow, USSR.

KRASNOVSKY, A.A., Bach Institute of Biochemistry, Academy of Sciences of the USSR, Leninsky Prospect 33, Moscow, USSR.

KRITSKY, M.S., Bach Institute of Biochemistry, Academy of Sciences of the USSR, Leninsky Prospect 33, Moscow, USSR.

KULAEV, I.S., Moscow University, Leninsky Gory, Moscow, USSR.

KUSHNER, D.J., Department of Biology, University of Ottawa, Ottawa, Ontario, Canada.

LIEBL, V., Institute of Microbiology, Czechoslovak Academy of Sciences, Budejovicka 1083, Prague, Czechoslovakia.

LIPMANN, F., The Rockefeller University, New York, New York, USA.

MARGULIS, L., Department of Biology, Boston University, Boston, Massachusetts, USA.

MATTHEWS, C.N., Department of Chemistry, University of Illinois at Chicago Circle, Chicago, Illinois, USA.

MEDNIKOV, B.M., Moscow State University, Corpus "A", Lenin Hills, Moscow, USSR.

MILLER, S.L., Department of Chemistry, University of California, San Diego, La Jolla, California, USA.

MITZ, M.A., N.A.S.A. Headquarters, Chief, Advanced Science Planning, Washington D.C., USA.

MOISEEVA, L.N., Bach Institute of Biochemistry, Academy of Sciences of the USSR, Leninsky Prospect 33, Moscow, USSR.

MOROWITZ, H.J., Department of Molecular Biophysics and Biochemistry, 1937 Yale Station, New Haven, Connecticut, USA.

NODA, H., Department of Biophysics and Biochemistry, Faculty of Sciences, University of Tokyo, Bunkyo-Ku, Tokyo, Japan.

OPARIN, A.I., Bach Institute of Biochemistry, Academy of Sciences of the USSR, Leninsky Prospect 33, Moscow, USSR.

ORÓ, J., Department of Biophysical Sciences, University of Houston, Houston, Texas, USA.

OSTROVSKY, D.N., Bach Institute of Biochemistry, Academy of Sciences of the USSR, Leninsky Prospect 33, Moscow, USSR.

OTROTCHENKO, V.A., The Institute of Space Research, Academy of Sciences of the USSR, Profsojusnaja, Moscow, USSR.

PANTSKHAVA, E.S., Bach Institute of Biochemistry, Academy of Sciences of the USSR, Leninsky Prospect 33, Moscow, USSR.

PAECHT-HOROWITZ, M., The Polymer Department, Weizmann Institute of Sciences, Rehovot, Israel.

PATTEE, H.H., Hansen Laboratories of Physics, Stanford University, Stanford, California, USA.

PAVLOVSKAYA, T.E., Bach Institute of Biochemistry, Academy of Sciences of the USSR, Leninsky Prospect 33, Moscow, USSR.

PONNAMPERUMA, C.A., N.A.S.A., Chief, Chemical Evolution Branch, Exobiology division, Ames Research Center, Moffett Field, California, USA.

PRIGOGINE, I., Université Libre de Bruxelles, 1 rue Héger-Bordet, Bruxelles, Belgium.

RICH, A., Massachusetts Institute of Technology, Cambridge, Massachusetts, USA.

ROSSIGNOL, M., 29 rue de Tournon, Paris, France.

SCHWARTZ, A.W., Department of Exobiology, Faculty of Sciences, University of Nijmegen, Toernooiveld, Driehuizerweg 200, Nijmegen, The Netherlands.

SEREBROVSKAYA, K.B., Bach Institute of Biochemistry, Academy of Sciences of the USSR, Leninsky Prospect 33, Moscow, USSR.

SYLVESTER-BRADLEY, P.C., Department of Geology, University of Leicester, Leicester, England.

TOUPANCE, G., Faculté des Sciences de Paris, Laboratoire Energétique Electrochimique du Professeur Buvet, 10 rue Vauquelin, Paris, France.

VASILYEVA, N.V., Bach Institute of Biochemistry, Academy of Sciences of the USSR, Leninsky Prospect 33, Moscow, USSR.

VDOVYKIN, G.P., Academy of Sciences of the USSR, V.I. Vernadsky Institute of Geochemistry and Analytical Chemistry, Moscow, USSR.

YOUNG, R.S., Chief Exobiology Program, Code S.B., N.A.S.A. Headquarters, Washington D.C., USA.

INDEX